Approximation of Elliptic
Boundary-Value Problems

Approximation of Elliptic Boundary-Value Problems

Jean-Pierre Aubin
LASTRE (Laboratoire d'Applications
des Systèmes Tychastiques Régulés)

DOVER PUBLICATIONS, INC.
Mineola, New York

This book is dedicated to

JACQUES-LOUIS LIONS

Preface

Thirty years ago the solution to a boundary-value problem could be approximated, at least in theory, by either of two competing techniques: the method of finite differences and the Rayleigh-Ritz-Galerkin method. It is fascinating to look back at the effect of the computer, and of numerical analysis, on the race between these two methods.

The first method to become practical on a large scale was that of finite differences. Young's paper in 1950 on successive overrelaxation, and the discovery of even faster iterative methods for large systems, led to the "era of the 5-point scheme."

In contrast, variational methods were still being studied (for example, in Mikhlin's early books) in a classical way. An early paper by Courant did show that the variational method, using piecewise linear trial functions, could lead to the standard 5-point scheme for Laplace's equation. However, this idea remained dormant for twelve years. In fact, the crucial step, that is, the development of an effective numerical procedure based on variational principles, was carried out by structural engineers as a natural extension of their matrix methods, and was independent not only of Courant's paper, but even of the whole Rayleigh-Ritz-Galerkin idea.

This procedure is now known as the *finite element method* and is superior in many ways to finite differences alone. Actually, it is recognized as an application of the variational method, with piecewise polynomial trial functions, which leads essentially to finite-difference equations.

The convergence and stability of the finite-element method are the subject of active research for which I hope this book will provide a foundation. More precisely, *the book is intended to imbed this combination of two important methods into the framework of functional analysis and to explain its applications to approximation of nonhomogeneous boundary-value problems for elliptic operators.*

The book is mathematically self-contained. At the beginning of Chapter 2 is a summary of the results of functional analysis to be used in the book, and Chapters 6 and 7 are devoted to an elementary study of nonhomogeneous boundary-value problems for elliptic operators.*

The Introduction summarizes the main results established in the book.

Chapter 1 introduces the variational method and the finite difference method in the simple case of second-order differential equations.

Chapters 2 and 3 deal with abstract approximations of Hilbert spaces and linear operators, and Chapters 4 and 5 study finite element approximations of Sobolev spaces.†

The remaining four chapters consider the several methods for approximating nonhomogeneous boundary-value problems for elliptic operators.

This book grew out of a set of lecture notes of a one-year graduate course given at Purdue University in 1968–1969 and was completed during two summers spent at the Mathematical Research Center of the University of Wisconsin. I am very grateful to these schools for the ideal environment they provided.

I express my deep thanks to Professors S. Parter and G. Strang, whose support and encouragement have been necessary to the completion of this work.

This book is dedicated to Professor J. L. Lions, who has profoundly influenced my thinking, as evidenced, for instance, by the way I study nonhomogeneous boundary-value problems.

JEAN-PIERRE AUBIN

Paris, France
August 1971

---

* These chapters can be used as an introduction to the three volumes of *Nonhomogeneous boundary-value problems and applications* by J. L. Lions and E. Magenes, Springer-Verlag, 1972; French edition, Dunod, 1968.

† This material can serve as an introduction to the forthcoming book *An analysis of the finite element method* by G. Strang and G. Fix, Prentice-Hall, and to a forthcoming book of P. Ciarlet and P. A. Raviart.

Contents

Approximation of Elliptic Boundary-Value Problems

Introduction

1. Aim and Scope

In this book we look at the problem of approximating solutions of linear differential problems in a stringent way. Instead of studying convergence properties of a given class of approximate schemes we construct approximate problems fulfilling given requirements, such as:

- Stability and optimal stability of approximate problems.
- Convergence of the solutions of approximate problems.
- Achievement of the minimum lack of consistency.
- Achievement of the optimal rate of convergence.
- Sparsity of the matrices of the approximate problems.
- Optimal behavior of the condition number of these matrices.
- Regularity of the convergence.

Later we give a precise meaning to these items. But one permanent constraint of "simplicity," not listed above, should be kept in mind: the construction of approximate problems must be achieved, that is, the finite dimensional problem must be computable.

This approach enables us to use "abstract methods" and to embed a part of "numerical analysis" in the framework of "functional analysis." It is for this reason that we begin by studying problems of approximation of solutions of differential problems in an abstract way and point out several well defined notions and useful principles.

The payoff of this effort of abstraction will become obvious when we apply the abstract results to nonhomogeneous boundary-value problems: the technical difficulties of numerical analysis are concentrated in a few areas of our study.

To keep this book to a reasonable length we restrict the scope of applications to the study of nonhomogeneous boundary-value problems for elliptic linear differential operators.

1

Since a good theory of nonhomogeneous boundary-value operators can be made in the framework of Hilbert spaces (by using Sobolev spaces), we study only approximation of solutions of operational equations in Hilbert spaces. In particular, this enables us to obtain more results in the simplest possible way.

Naturally, many of the results can be extended to the case of Banach spaces. However, we do not think that obtaining the convergence results in the L^p spaces for $p \neq 2$ is worth the additional difficulties so long as we restrict our study to linear problems.

2. Neumann Problems

We now describe a nonhomogeneous Neumann problem and show how it can be viewed as an operational equation in Hilbert spaces.

Let $\Omega \subset R^n$ be a smooth bounded open subset and Γ its boundary. Let $a_{pq}(x)$ be functions in $L^\infty(\Omega)$ and let us consider the differential operator Λ defined by

$$(1) \qquad \Lambda u = \sum_{|p|,|q| \leq k} (-1)^{|q|} D^q(a_{pq}(x) D^p u)$$

where $p = (p_1, \ldots, p_n), q = (q_1, \ldots, q_n)$ are multi-integers, $|p| = p_1 + \cdots + p_n$, and $D^p u = \partial u^p / \partial x_1^{p_1} \cdots \partial x_n^{p_n}$ is the derivative of order p of u (in the sense of distributions).

We denote by $\gamma_j u = \partial^j u / \partial n^j$ the jth-order normal derivative of u (for $j = 0$, $\gamma_0 u$ is the restriction of u to Γ). These operators γ_j map spaces of functions defined on Ω into spaces of functions defined on Γ. We shall give examples of such spaces of functions.

We denote by $H^s(R^n)$ the Sobolev space of order s defined by

$$(2) \quad H^s(R^n) = \{u \in L^2(R^n) \qquad \text{such that} \qquad (1 + |y|^2)^{s/2} \hat{u}(y) \in L^2(R^n)\}$$

where $s \geq 0$ and $\hat{u}(y)$ denotes the Fourier transform of $u(x)$. [If $s < 0$, we denote by $H^s(R^n)$ the dual of $H^{-s}(R^n)$.]

If $s = k$ is a positive integer, $H^k(R^n)$ can be defined by

$$(3) \quad H^k(R^n) = \{u \in L^2(R^n) \qquad \text{such that} \qquad D^p u \in L^2(R^n) \qquad \text{for} \quad |p| \leq k\}.$$

If $\Omega \subset R^n$ is a smooth bounded open subset, we define $H^s(\Omega)$ to be the space of restrictions to Ω of functions of $H^s(R^n)$.

On the other hand, we can identify the boundary Γ of Ω to R^{n-1} by means of local coordinates and, roughly speaking, define $H^s(\Gamma)$ to be isomorphic to the Sobolev space $H^s(R^{n-1})$.

The theory of boundary-value problems is based on the following "trace theorem":

THEOREM 1 The operator $\gamma = (\gamma_1, \ldots, \gamma_{k-1})$ is a continuous linear operator from $H^k(\Omega)$ *onto* $\prod_{0 \le j \le k-1} H^{k-j-1/2}(\Gamma)$. Furthermore, the kernel $H_0^k(\Omega)$ of γ is the closure in $H^k(\Omega)$ of the space of infinitely differentiable functions with compact support in Ω. ▲

The second tool used in the theory of boundary-value problems is the Green formula.

We associate with the operator Λ the bilinear form $a(u, v)$ defined by

$$(4) \qquad a(u, v) = \sum_{|p|,|q| \le k} \int_\Omega a_{pq}(x) D^p u D^q v \, dx$$

and the space $H^k(\Omega, \Lambda)$ defined by

$$(5) \qquad H^k(\Omega, \Lambda) = \{u \in H^k(\Omega) \quad \text{such that} \quad \Lambda u \in L^2(\Omega)\}.$$

THEOREM 2 There exist unique operators $\delta_j \in L(H^k(\Omega), H^{k-j-1/2}(\Gamma))$ (for $k \le j \le 2k - 1$) associated with the operators Λ and γ_j such that the Green formula

$$a(u, v) = \int_\Omega \Lambda u \cdot v \, dx + \sum_{0 \le j \le k-1} \int_\Gamma \delta_{2k-j-1} u \gamma_j v \, d\sigma(x)$$

holds when u ranges over $H^k(\Omega, \Lambda)$ and v ranges over $H^k(\Omega)$. ▲

Now we are able to define a Neumann problem associated with Λ. Let $f \in L^2(\Omega)$ and $t_j \in H^{k-j-1/2}(\Gamma)$ be given for $k \le j \le 2k - 1$. We say that u is a solution of the Neumann problem for $\Lambda + \lambda$ if

$$(6) \qquad \begin{cases} \text{(i)} \ u \in H^k(\Omega, \Lambda), \\ \text{(ii)} \ \Lambda u + \lambda u = f, \\ \text{(iii)} \ \delta_j u = t_j \quad \text{for} \quad k \le j \le 2k - 1. \end{cases}$$

Theorems 1 and 2 enable us to give an equivalent formulation of this Neumann problem. Let us set

$$(7) \qquad \begin{cases} \text{(i)} \quad (u, v) = \int_\Omega u(x) v(x) \, dx, \ \langle f, g \rangle = \int_\Gamma f(x) g(x) \, d\sigma(x), \\ \text{(ii)} \quad l(v) = (f, v) + \sum_{0 \le j \le k-1} \langle t_{2k-j-1}, \gamma_j v \rangle. \end{cases}$$

THEOREM 3 Any solution u of the Neumann problem (6) is a solution of the variational problem

$$(8) \qquad \begin{cases} \text{(i)} \ u \in H^k(\Omega), \\ \text{(ii)} \ a(u, v) + \lambda(u, v) = l(v) \quad \text{for any} \quad v \in H^k(\Omega), \end{cases}$$

and conversely. ▲

Variational equation (8) has a meaning, since $a(u, v) + \lambda(u, v)$ is a continuous bilinear form on the space $V = H^k(\Omega)$ and $l(v)$ is a continuous linear form on V. Let us set $H = L^2(\Omega)$. Then we are led to study the following type of problems. Let V and H be two Hilbert spaces such that

(9)
$$\begin{cases} \text{(i)} & \text{the injection from } V \text{ into } H \text{ is compact,} \\ \text{(ii)} & V \text{ is dense in } H. \end{cases}$$

Let $a(u, v)$ be a continuous bilinear form of V and $l(v)$ be a continuous linear form of V. Find u satisfying

(10)
$$\begin{cases} \text{(i)} & u \in V, \\ \text{(ii)} & a(u, v) + \lambda(u, v) = l(v) \qquad \text{for any } v \in V. \end{cases}$$

The Lax-Milgram and Riesz-Fredholm theorems imply the following abstract result:

THEOREM 4   Let us assume that $a(u, v)$ is V-elliptic, that is, that there exists a positive constant c such that

(11)
$$a(v, v) \geq c \, \|v\|_V^2 \qquad \text{for any } v \in V.$$

Assume also (9). If λ does not belong to a countable set of isolated eigenvalues, then there exists a unique solution of (10). ▲

To apply Theorem 4 to prove existence and uniqueness of the Neumann problem (6) we must give sufficient conditions to ensure the $H^k(\Omega)$-ellipticity of the form $a(u, v)$ defined by (4). Inequality (12) is such a condition:

(12)
$$\sum_{|p|,|q| \leq k} a_{pq}(x) z^p z^q \geq c \sum_{|p| \leq k} |z^p|^2 \qquad \text{for almost every } x \in \Omega.$$

(There are also other weaker conditions.)

Therefore, by using the variational formulation (8) of the Neumann problem (6) and Theorem 4 we prove in an easy way the existence and uniqueness of the solution of a Neumann problem. We employ the same approach for approximation purposes, but before we proceed, let us take one further step.

Let (f, v) be the duality pairing on $V' \times V$, which is the extension of the inner product of the Hilbert space H. Then we can make the following identifications:

(13)
$$V \subset H = H' \subset V'.$$

The data of the continuous bilinear form $a(u, v)$ amount to the data of the continuous linear operator $A \in L(V, V')$ defined by

(14)
$$a(u, v) = (Au, v) \qquad \text{for any } u, v \in V.$$

Therefore, the variational problem (10) is equivalent to the operational equation

$$(15) \qquad Au + \lambda u = l$$

where l is given in V'.

By summing up these results we have transformed the Neumann problem (6) into an operational equation (15) where $V = H^k(\Omega)$. We cannot, however, give any analytic expression of the operator A (which "involves" the differential operator Λ and the Neumann operators δ_j). This motivates the study of approximation of solutions of operational equations in Hilbert spaces, that is, of solutions u of equations

$$(16) \qquad Au = f$$

where A is a continuous operator from a Hilbert space V into a Hilbert space F and f is given in F.

3. Introduction of Internal Approximations

The general problem of approximation amounts to associating with problem (16) a family of approximate problems

$$(17) \qquad A_h u_h = f_h$$

where h is a parameter going to 0, A_h is a linear operator from a Hilbert space V_h into a Hilbert space F_h, and f_h is a given element of F_h such that the solution u_h "converges" to the solution u of (16).

Naturally, for practical purposes, V_h and F_h will be finite-dimensional spaces $R^{n(h)}$, so that (17) amounts to the construction of the entries of a matrix A_h and the components of the vector f_h.

The first task is to define the "convergence" of $u_h \in V_h$ to $u \in V$. This can be done in two ways:

(1) by introducing a "discrete" norm $\|u_h\|_{V_h}$ of V_h and a linear operator r_h (called restriction) from V onto V_h, calling $\|u_h - r_h u\|_{V_h}$ the "discrete error," saying that u_h converges discretely to u if the discrete error converges to 0;

(2) by introducing an isomorphism p_h (called "prolongation") from V_h onto a closed space P_h (called "space of approximants") of V and saying that u_h converges to u if $p_h u_h$ converges to u in V.

Hence we say that the data (V_h, p_h, r_h) are an "approximation" of the space V.

Now let us consider approximations (V_h, p_h, r_h) and (F_h, q_h, s_h) of the spaces V and F, respectively.

The first question that arises is: *under what conditions on A_h does the discrete convergence of f_h to f imply the discrete convergence of u_h to u?*

These conditions are:

- *The stability of the operators A_h*—there exists a constant S independent of h such that $\|u_h\|_{V_h} \leq S \|A_h u_h\|_{F_h}$ for any $u_h \in V_h$.
- *The consistency of A_h to A*—the lack of consistency $\|A_h r_h u - s_h A u\|_{F_h}$ converges to 0 for any $u \in V$.

Actually, we can state the following principle of equivalence:

THEOREM 5 The stability of the operators A_h and the consistency of A_h to A are sufficient conditions for the discrete convergence of f_h to f to imply the discrete convergence of u_h to u. These conditions are necessary if we assume that there exists a constant M independent of h such that $\|A_h u_h\|_{F_h} \leq M \|u_h\|_{V_h}$ for any u_h in V_h. ▲

We now estimate the discrete error. For this purpose we assume that the solution u of (16) belongs to a space U contained in V with a stronger topology (i.e., *the solution u is smoother than required*). We define the global lack of consistency $\Phi_U(A_h)$ by

$$(18) \qquad \phi_U(A_h) = \sup_{\|u\|_U \leq 1} \|A_h r_h u - s_h A u\|_{F_h}.$$

Then the discrete error obeys the inequality

$$(19) \qquad \|u_h - r_h u\|_{V_h} \leq M(\|f_h - s_h f\|_{F_h} + \Phi_U(A_h) \|u\|_U).$$

By choosing $f_h = s_h f$ we cancel the contribution due to the discretization of the data f in estimate (19).

Is it possible to minimize the global lack of consistency $\Phi_U(A_h)$?

THEOREM 6 There exists a prolongation p_h (dependent on r_h and the space U but independent of the operator A) such that

$$(20) \qquad \Phi_U(s_h A p_h) \leq \Phi_U(A_h) \qquad \text{for any operator} \quad A_h \in L(V_h, F_h). \qquad ▲$$

This theorem motivates the introduction of approximate problems of the form

$$(21) \qquad (s_h A p_h) u_h = s_h f$$

obtained from the data of a restriction s_h from F onto F_h and a prolongation p_h mapping V_h into V. We say that such approximate problems are *internal approximations* of the problem (16). If $V_h = F_h = R^{n(h)}$, the operators p_h and s_h are of the following form:

$$(22) \qquad \begin{cases} \text{(i)} \quad p_h u_h = \sum_j u_h{}^j \mu_h{}^j & \text{where} \quad u_h = ((u_h{}^j))_j \in R^{n(h)}, \\ \text{(ii)} \quad s_h f = ((\lambda_h{}^j, f))_j \end{cases}$$

where $(\mu_h{}^j)_j$ and $(\lambda_h{}^j)_j$ are linearly independent sequences of elements of V and the dual F' of F, respectively. In this case, the entries of the matrix $s_h A p_h$ are equal to $(\lambda_h{}^i, A\mu_h{}^j)$.

4. Properties of Internal Approximations

By choosing approximate operators A_h which minimize the global lack of consistency, we introduced the method of construction that associates with equation (16) its internal approximation (21). This method depends on the data of a prolongation p_h mapping V_h into V and a restriction s_h mapping F onto F_h.

We associate with p_h and s_h approximations $(V_h, p_h, \hat{r}_h)$ and $(F_h, \hat{q}_h, s_h)$ of V and F in the following way. First, the "optimal restriction" $\hat{r}_h$ (associated with p_h and V) is defined by

$$(23) \qquad \|u - p_h \hat{r}_h u\|_V = \inf_{v_h \in V_h} \|u - p_h v_h\|_V \qquad \text{for any} \quad u \in V.$$

Then the "optimal prolongation" $\hat{q}_h$ (associated with s_h and F) is defined by

$$(24) \qquad s_h \hat{q}_h f_h = f_h; \qquad \|\hat{q}_h f_h\|_F = \inf_{s_h f = f_h} \|f\|_F \qquad \text{for any} \quad f_h \in F_h.$$

We supply the discrete spaces V_h and F_h with the discrete norms

$$\|u_h\|_{V_h} = \|p_h u_h\|_V \qquad \text{and} \qquad \|f_h\|_{F_h} = \|\hat{q}_h f_h\|_F.$$

We say that the approximations $(V_h, p_h, \hat{r}_h)$ of V are "convergent" if

$$\lim \|u - p_h \hat{r}_h u\|_V = 0 \qquad \text{for any} \quad u \in V.$$

On the other hand, let us assume that the operators $s_h A p_h$ are invertible and let

$$(25) \qquad S(s_h, A) = \sup_{v_h \in V_h} \frac{\|p_h v_h\|_V}{\|\hat{q}_h s_h A p_h v_h\|_F}$$

be the norm of $(s_h A p_h)^{-1}$. We say that the operators $s_h A p_h$ are "stable" if the norms $S(s_h, A)$ of $(s_h A p_h)^{-1}$ are bounded. Here again a principle of equivalence holds:

THEOREM 7 Let us assume that A is an isomorphism from V onto F. Then $p_h u_h$ converges to u if and only if the approximations $(V_h, p_h, \hat{r}_h)$ of V are convergent and if the operators $s_h A p_h$ are stable. Furthermore, the following error estimate holds:

$$(26) \qquad \|u - p_h u_h\|_V \leq (1 + S(s_h, A) \|A\|) \|u - p_h \hat{r}_h u\|_V$$

where $\|A\|$ denotes the norm of A in $L(V, F)$. ▲

Theorem 7 implies that if we assume the stability, then $p_h u_h$ converges to u with the same speed as the best approximant $p_h \hat{r}_h u$ of u by elements of $P_h = p_h V_h$.

Therefore, we must choose prolongations p_h satisfying optimal or quasi-optimal properties of convergence. Once p_h has been chosen, it remains to select restrictions s_h that satisfy optimal or quasi-optimal stability properties.

Let us begin by choosing p_h. Let U be a Hilbert space contained in V with a stronger topology. We associate with p_h the error function $e_U{}^V(p_h)$ defined by

$$(27) \qquad e_U{}^V(p_h) = \sup_{\|u\|_U \leq 1} \|u - p_h \hat{r}_h u\|_V.$$

If V_h is a finite-dimensional space and if p_h maps V_h into U, we define the stability function $s_U{}^V(p_h)$ by

$$(28) \qquad s_U{}^V(p_h) = \sup_{v_h \in V_h} \frac{\|p_h v_h\|_U}{\|p_h v_h\|_V}$$

Usually, the stability functions converge to ∞ and the error functions converge to 0 when the dimension $n(h)$ of V_h increases; for example, when the approximations $(V_h, p_h, \hat{r}_h)$ are convergent and when the injection from U into V is compact.

Then, if the solution u of (16) belongs to U, we deduce from (26) the estimate

$$(29) \qquad \|u - p_h u_h\|_V \leq M e_U{}^V(p_h) \|u\|_U.$$

We are led to the study of the two following problems:

1. Does there exist an optimal prolongation $\tilde{p}_h$ such that

$$(30) \qquad e_U{}^V(\tilde{p}_h) = \inf_{p_h \in L(V_h, V)} e_U{}^V(p_h)$$

2. Give sufficient conditions to make the prolongations quasi-optimal, that is, there must be a constant M independent of h satisfying

$$(31) \qquad e_U{}^V(p_h) \leq M e_U{}^V(\tilde{p}_h).$$

In other words, by choosing quasi-optimal prolongations we are certain to obtain an optimal rate of convergence for a given "degree of regularity" (i.e., for a given choice of a space U contained in V).

Let us recall that if $V_h = R^{n(h)}$, the expression $E_U{}^V(n(h)) = \inf_{p_h} e_U{}^V(p_h)$ is called *the $n(h)$-width of the injection from U in V* (or, equivalently, of the unit ball of U). We later construct such optimal prolongations achieving the $n(h)$-width, but first let us state a solution of problem 2:

THEOREM 8 Let us assume that the dimension $n(h)$ of V_h is finite and that p_h maps V_h into U. If there exists a constant M independent of h such that

(32) $$\sup_h e_U^V(p_h) s_U^V(p_h) = M,$$

then the error functions obey the estimates

$$e_U^V(p_h) \leq M E_U^V(n(h) - 1). \qquad \blacktriangle$$

We say that prolongations p_h satisfying (32) are "quasi-optimal" (for the injection from U into V). In this case, Theorem 8 implies that rough estimates of the error functions and of the stability functions leading to (32) are sufficient to know that we have obtained approximations behaving as the $n(h)$-width. This is quite useful, since the computation of the n-widths for concrete spaces is often difficult. On the other hand, we can choose among these quasi-optimal prolongations the ones that satisfy other properties (simplicity, sparsity of the matrix $s_h A p_h$, etc.).

5. Stability, Optimal Stability, and Regularity of the Convergence

Let us assume that we have chosen prolongations p_h (which are quasi-optimal, for instance). We are looking for restrictions s_h such that the operators $s_h A p_h$ are stable and, if possible, for restrictions s_h minimizing the norm $S(s_h, A)$ of the operators $(s_h A p_h)^{-1}$ for a class of operators A mapping V into F.

Let K be a given isometry from V onto F' and let us consider the class of operators $A \in L(V, F)$ satisfying

(33) $$\begin{cases} \text{(i)} \ (Au, Kv) \leq M \, \|u\|_V \, \|v\|_V & \text{for any } u, v \text{ in } V, \\ \text{(ii)} \ (Av, Kv) \geq c \, \|v\|_V^2 & \text{for any } v \text{ in } V. \end{cases}$$

THEOREM 9 Let K be an isometry from V onto F', p_h a given prolongation, and L_h any isomorphism from V_h' onto F_h. Then the restrictions $\tilde{s}_h = L_h p_h' K'$ mapping F onto F_h satisfy

(34) $$S(\tilde{s}_h, A) \leq M c^{-1} S(s_h, A)$$

for any s_h and any operator A satisfying (33), and the operators $\tilde{s}_h A p_h$ are stable. $\qquad \blacktriangle$

Let us state other results regarding stability.

THEOREM 10   Let us assume that the isomorphism $A = B + C$ from V onto F is the sum of an isomorphism B and of a compact operator C.

Let us assume that V_h and F_h are finite-dimensional spaces and that the approximations $(V_h, p_h, \hat{r}_h)$ and $(F_h, \hat{q}_h, s_h)$ of V and F are convergent. Then, if the operators $B_h = s_h B p_h$ are stable, so are the operators $A_h = s_h A p_h$ for h small enough.   ▲

We prove also the results regarding the regularity of the convergence. For instance, let us state the following result:

THEOREM 11   Let us assume that the isomorphism $A \in L(V, F)$ is also an isomorphism from U onto E, where $U \subset V$ and $E \subset F$. Also, let us assume that p_h maps V_h into U and that

$$(35) \qquad \begin{cases} \text{(i)} & \sup_h e_U{}^V(p_h) s_U{}^V(p_h) \leq M, \\[2mm] \text{(ii)} & \lim \| u - p_h \hat{r}_h u \|_U = 0 \qquad \text{for any } u \in U. \end{cases}$$

Then if the operators $s_h A p_h$ are stable for the norms of V and F, they remain stable for the norms of U and E, and $p_h u_h$ converges to u in U whenever the data f belong to E.   ▲

We prove an analogous theorem in the case where we can extend $A \in L(V, F)$ to an isomorphism from W onto G, where $V \subset W$ and $F \subset G$.

6.   The Case of Operators Mapping a Hilbert Space onto Its Dual

We have seen that the solution of a Neumann problem is actually the solution of an operational equation $Au = f$ where A is an operator mapping a Hilbert space V into its dual $F = V'$.

First, we have to chose prolongations p_h from V_h into V satisfying given requirements (e.g., we can choose quasi-optimal prolongations for the injection of a space U into V).

Second, we must select a restriction s_h satisfying the quasi-optimal property of stability (34). If the operator A is V-elliptic, property (33) is satisfied for $K = I$ (since $F = V'$). Therefore Theorem 9 implies that we must choose $\tilde{s}_h = p_h'$ (since we can take $L_h = I$). In this case, then the internal approximation of the equation

$$(36) \qquad Au + \lambda u = l$$

is the discrete operational equation

$$(37) \qquad p_h' A p_h u_h + \lambda p_h' p_h u_h = p_h' l$$

which is equivalent to the discrete variational equation

(38) $a(p_h u_h, p_h v_h) + \lambda(p_h u_h, p_h v_h) = l(p_h v_h)$ for any $v_h \in V_h$

where $a(u, v) = (Au, v)$. Now, let us assume that there exists a Hilbert space H identified with its dual such that

(39) $V \subset H = H' \subset V'$,   the injections being dense and compact.

THEOREM 12   Let $(V_h, p_h, \hat{r}_h)$ be convergent approximations of V, where the dimension $n(h)$ of V_h is finite. Assume that A is V-elliptic, that $A + \lambda$ is an isomorphism from V onto V', and that (39) holds. Then for h small enough there exists a unique solution u_h of (37) and a constant M independent of h such that

(40) $\|u - p_h u_h\|_V \leq M \|u - p_h \hat{r}_h u\|_V.$

Furthermore, if the solution u belongs to $U \subset V$ and if $X \subset V$, we obtain the estimates

(41) $\sup_{v \in X} \dfrac{|a(u - p_h u_h, v)|}{\|v\|_X} \leq M e_X{}^V(p_h) e_U{}^V(p_h) \|u\|_U.$

Finally, if we assume that p_h maps V_h into U and if the prolongations p_h are quasi-optimal for the injection from U into V, then

(42) $\|u - p_h u_h\|_V \leq M E_U{}^V(n(h) - 1) \|u\|_U,$

and $p_h u_h$ converges to u in U whenever the solution u belongs to U.   ▲

We also estimate the condition number, which plays a role in the study of methods for solving discrete problem (37). We suppose that the finite-dimensional space V_h is identified with its dual by a canonical inner product $(u_h, v_h)_h$. Let $|u_h|_h = \sqrt{(u_h, u_h)_h}$ be the canonical norm and $|A_h|$ (resp. $|A_h^{-1}|$) the norm of A_h (resp. A_h^{-1}) when V_h and V_h' are supplied with the canonical norm $|u_h|_h$. The product $|A_h| \, |A_h^{-1}|$ is called the "condition number" of the matrix A_h.

THEOREM 13   Suppose that there exists a constant k such that

$$k^{-1} |v_h|_h \leq |p_h v_h|_H \leq k |v_h|_h \quad \text{for any} \quad v_h \in V_h.$$

Under the assumptions of Theorem 12 there exists a constant M such that

(43) $|A_h| \, |A_h^{-1}| \leq M(s_V{}^H(p_h))^2.$ ▲

7. Finite-Element Approximations of Sobolev Spaces

We must construct approximations of Sobolev spaces $H^m(R^n)$ and $H^m(\Omega)$, since these spaces are the Hilbert spaces involved in the study of boundary-value problems.

We construct the class of so-called finite-element approximations of Sobolev spaces and characterize the finite-element approximations that are quasi-optimal.

Let us associate with the parameter $h = (h_1, \ldots, h_n)$ the knots $jh = (j_1 h_1, \ldots, j_n h_n)$ where the multi-integers $j = (j_1, \ldots, j_n)$ range over Z^n.

The discrete space V_h associated with the Sobolev spaces $V = H^m(R^n)$ is the space $l^2(Z^n)$ of sequences $u_h = (u_h{}^j)_{j \in Z^n}$ satisfying

$$(44) \qquad |u_h|_h = (h_1 \cdots h_n)^{1/2} \Big(\sum_{j \in Z^n} |u_h{}^j|^2 \Big)^{1/2}.$$

Let $\mu \in H^m(R^n)$ be a function with compact support. We associate with it the prolongation p_h defined by

$$(45) \qquad p_h u_h = \sum_{j \in Z^n} u_h{}^j \mu \left(\frac{x_1}{h_1} - j_1, \ldots, \frac{x_n}{h_n} - j_n \right).$$

Finally, if $\lambda \in L^\infty(R^n)$ is a function with compact support, we define a restriction r_h by

$$(46) \; r_h u = \left(\int \frac{1}{h_1 \cdots h_n} \lambda \left(\frac{x_1}{h_1} - j_1, \ldots, \frac{x_n}{h_n} - j_n \right) u(x_1, \ldots, x_n) \, dx_1 \cdots dx_n \right)_{j \in Z^n}.$$

We say that such approximations (V_h, p_h, r_h) are finite-element approximations of $H^m(R^n)$ associated with the functions μ and λ.

Theorem 14 characterizes the convergent finite-element approximations.

THEOREM 14   The following two statements are equivalent:

1. The finite-element approximations associated with μ are convergent approximations of $H^m(R^n)$.

2. The function μ satisfies the criterion of m-convergence

$$(47) \qquad \sum_{k \in Z^n} \frac{k^j}{j!} \mu(x - k) = \sum_{0 \le k \le j} b^k \frac{x^{j-k}}{(j-k)!} \qquad \text{for any} \quad |j| \le m$$

where $b^0 = 1$.

Furthermore, if λ is related to μ by

$$(48) \qquad \int \mu(x)\lambda(y)(x - y)^k \, dx \, dy = \begin{cases} 1 & \text{if} \quad k = 0, \\ 0 & \text{if} \quad |k| \le m, \end{cases}$$

then there exists a constant M such that

(49) $\|u - p_h r_h u\|_{H^k(R^n)} \leq M |h|^{s-k} \|u\|_{H^s(R^n)}$

$$\text{for}\quad 0 \leq h \leq s \leq m + 1, k \leq m.$$

Finally, if μ satisfies the stability property that there is no $y \in (-\pi, +\pi)$ such that

(50) $\int e^{ix(y+2\pi j)} \mu(x) \, dx = 0 \qquad \text{for any}\quad j \in Z^n,$

then there exists a constant M such that

(51) $\|p_h u_h\|_{H^s(R^n)} \leq M |h|^{-(s-k)} \|p_h u_h\|_{H^k(R^n)} \qquad \text{for}\quad 0 \leq k \leq s \leq m.$ ▲

In other words, if the function μ satisfies the criterion of m-convergence and the stability property, the finite-element approximations associated with μ *are quasi-optimal approximations* of Sobolev spaces $H^k(R^n)$ for the injection from $H^s(R^n)$ into $H^k(R^n)$ (for $0 \leq k \leq s \leq m$).

The criterion of m-convergence shows how polynomials play a crucial role in approximations of Sobolev spaces.

Before giving examples of convergent finite-element approximations, let us consider their "number of levels," which is equal to the number of multi-integers k such that the measure of the intersection of the supports of $\mu(x)$ and $\mu(x - k)$ is positive. This number of levels is closely related to the sparsity of the matrices of the internal approximations of differential problems involving these prolongations.

EXAMPLE 1 $[(2m + 1)^n]$-*level   piecewise-polynomial   approximations*
These approximations are the finite-element approximations associated with the function $\mu = \pi_{(m+1)}$, where π denotes the characteristic function of the cube $(0, 1)^n$ and $\pi_{(m+1)}$ the $(m + 1)$-fold convolution of π.

These prolongations satisfy the criterion of m-convergence and the stability property, and their number of levels is equal to $(2m + 1)^n$. Finally, the restriction of $p_h u_h$ to each cube $(jh, (j + (1)h)$ is a polynomial of multi-degree m.

EXAMPLE 2 $[2(2m)^n - (2m - 1)^n]$-*level piecewise-polynomial approxi-mations*   These approximations are the finite-element approximations associated with the function $\mu = \psi_* \pi_{(m)}$, where ψ is the measure defined by

$$(\psi, u) = \int_0^1 u(x, \ldots, x) \, dx.$$

These prolongations satisfy the criterion of m-convergence and the stability property, and their number of levels is equal to $2(2m)^n - (2m - 1)^n$. The restriction of $p_h u_h$ to each simplex is a polynomial of multidegree m.

Clearly the difference between these two numbers of levels is $(2m + 1)^n -$ $2(2m)^n + (2m - 1)^n = \sum_{1 \le k \le n/2} \binom{n}{2k} (2m)^{n-2k}$, that is, it is 0 if $n = 1$, 2 if $n = 2$, $12m$ if $n = 3$.

Knowing the approximations of Sobolev spaces $H^m(R^n)$, we can construct approximations of the Sobolev spaces $H^m(\Omega)$ when Ω is a smooth bounded open subset of R^n.

We introduce the grid $\mathscr{R}_h{}^\mu(\Omega)$ which is the subset of multi-integers j such that the intersection of Ω and the support of $\mu(x/h - j)$ are not empty.

We choose the discrete space $V_h = H_h{}^\mu(\Omega)$ to be the space of sequences $u_h = (u_h{}^j)_{j \in \mathscr{R}_h \mu(\Omega)}$ defined on $\mathscr{R}_h{}^\mu(\Omega)$.

The prolongation p_h is defined by

$$p_h u_h = \sum_{j \in \mathscr{R}_h \mu(\Omega)} u_h{}^j \mu\left(\frac{x}{h} - j\right) \qquad \text{(restricted to } \Omega\text{)},$$

and we can choose $\hat{r}_h$ to be the optimal restriction associated with p_h in $H^m(\Omega)$.

We deduce from Theorem 14 the following result:

THEOREM 15 Let us assume that μ satisfies the criterion of m-convergence. Then the finite-element approximations $(H_h{}^\mu(\Omega), p_h, \hat{r}_h)$ are convergent approximations of the Sobolev spaces $H^k(\Omega)$ for $k \le m$.

Furthermore, there exists a constant M such that

$$(52) \quad \|u - p_h \hat{r}_h u\|_{H^k(\Omega)} \le M |h|^{s-k} \|u\|_{H^s(\Omega)} \text{ for } 0 \le k \le s \le m + 1, k \le m$$

If Ω satisfies a property of μ-stability, which we do not define here, and if μ satisfies the stability property, then there exists a constant M and a subsequence of h such that

$$(53) \quad \|p_h u_h\|_{H^s(\Omega)} \le M |h|^{-(s-k)} \|p_h u_h\|_{H^k(\Omega)} \qquad \text{for } 0 \le k \le s \le m. \quad \blacktriangle$$

In other words, if Ω satisfies the property of μ-stability and if μ satisfies the criterion of m-convergence and the stability property, the finite-element approximations are quasi-optimal approximations of the Sobolev spaces $H^k(\Omega)$ for the injection from $H^s(\Omega)$ into $H^k(\Omega)$ when $0 \le k \le s \le m$.

8.   Approximation of Nonhomogeneous Neumann Problems

Let us consider the solution of the Neumann problem

$$(54) \quad \begin{cases} \text{(i)} \quad u \in H^k(\Omega, \Lambda), \\ \text{(ii)} \quad \Lambda u + \lambda u = f, \qquad \text{where } \Lambda u = \sum_{|p|, |q| \le k} (-1)^{|q|} D^q(a_{pq}(x) D^p u), \\ \text{(iii)} \quad \delta_j u = t_j \qquad \text{for } k \le j \le 2k - 1, \end{cases}$$

where f is given in $L^2(\Omega)$, t_j in $H^{k-j-\frac{1}{2}}(\Gamma)$, the bilinear form $a(u, v) = \sum_{|p|, |q| \leq k} \int_\Omega a_{pq}(x) D^p u D^q v \, dx$ is $H^k(\Omega)$-elliptic, and λ is not an eigenvalue of the Neumann problem. By using Theorems 12 and 14 we can approximate this solution by the solution $u_h \in H_h{}^\mu(\Omega)$ of the discrete variational equation

$$(55) \quad a(p_h u_h, p_h v_h) + \lambda(p_h u_h, p_h v_h) = l(p_h v_h) \qquad \text{for any} \quad v_h \in H_h{}^\mu(\Omega),$$

where

$$l(v) = (f, v) + \sum_{0 \leq j \leq k-1} \int_\Gamma t_{2k-j-1}(x) \gamma_j v \, d\sigma(x).$$

We can write (55) as

$$(56) \qquad\qquad A_h u_h = l_h$$

where the entries $d_h(i, j)$ of A_h are equal to

$$(57) \qquad a\left(\mu\left(\frac{x}{h} - j\right), \mu\left(\frac{x}{h} - i\right) \right) + \lambda\left(\mu\left(\frac{x}{h} - j\right), \mu\left(\frac{x}{h} - i\right) \right)$$

and the components $l_h{}^j$ of l_h are equal to

$$(58) \qquad \int_\Omega f(x) \mu\left(\frac{x}{h} - j\right) dx + \sum_{0 \leq i \leq k-1} \int_\Gamma t_{2k-i-1}(x) \gamma_i \mu\left(\frac{x}{h} - j\right) d\sigma(x).$$

The form of $a(u, v)$ reveals that the largest number of nonzero entries in each row and each column of A_h is at most equal to the number of levels of the prolongation p_h.

THEOREM 16   Let us assume that μ satisfies the criterion of m-convergence for $m \geq k$. If the solution u of the Neumann problem belongs to $H^s(\Omega)$ with $k \leq s \leq m + 1$, then there exists a constant M such that

$$(59) \quad \begin{cases} \text{(i)} \ \|u - p_h u_h\|_{H^k(\Omega)} \leq M \, |h|^{s-k} \|u\|_{H^s(\Omega)}, \\ \text{(ii)} \ \|\Lambda(u - p_h u_h)\|_{H^{-m-1}(\Omega)} \leq M \, |h|^{m+1+s-2k} \|u\|_{H^s(\Omega)}. \end{cases}$$

Furthermore, let us assume that the solution u of the Neumann problem belongs to $H^{2k}(\Omega)$ whenever $t_j \in H^{2k-j-\frac{1}{2}}(\Gamma)$ for $k \leq j \leq 2k - 1$. Then there exists a constant M such that

$$(60) \quad \begin{cases} \text{(i)} \ \|u - p_h u_h\|_{L^2(\Omega)} \leq M \, |h|^{s+\min(0, m+1-2k)} \|u\|_{H^s(\Omega)} \\ \qquad\qquad\qquad\qquad\qquad\qquad \text{for} \ k \leq s \leq m + 1, \\ \text{(ii)} \ \|\gamma_j(u - p_h u_h)\|_{H^{-j-1/2}(\Gamma)} \leq M \, |h|^{s+\min(0, m+1-2k)} \|u\|_{H^s(\Omega)} \\ \qquad\qquad\qquad\qquad\qquad\qquad \text{for} \ k \leq s \leq m + 1. \end{cases}$$

▲

Now let us study the regularity of the convergence. Under convenient regularity assumptions, we can prove that

(61) $\begin{cases} \text{(i) } u \text{ belongs to } H^{2k}(\Omega) & \text{whenever } t_j \text{ belongs to } H^{2k-j-\frac{1}{2}}(\Gamma), \\ \text{(ii) } u \text{ belongs to } L^2(\Omega) & \text{whenever } t_j \text{ belongs to } H^{-j-\frac{1}{2}}(\Gamma). \end{cases}$

THEOREM 17 Let us assume that μ satisfies the criterion of m-convergence for $m \geq 2k$, the stability property, and that Ω satisfies the property of μ-stability. Also, let us assume that the regularity properties (61) hold. Then we can prove the regularity of the convergence:

(62) $\begin{cases} \text{(i) } \lim \|u - p_h u_h\|_{H^{2k}(\Omega)} = 0 & \text{if } t_j \in H^{2k-j-\frac{1}{2}}(\Gamma), \\ \text{(ii) } \lim \|u - p_h u_h\|_{H^k(\Omega)} = 0 & \text{if } t_j \in H^{k-j-\frac{1}{2}}(\Gamma), \\ \text{(iii) } \lim \|u - p_h u_h\|_{L^2(\Omega)} = 0 & \text{if } t_j \in H^{-j-\frac{1}{2}}(\Gamma). \end{cases}$

Furthermore, if $n(h)$ is the dimension of $H_h^\mu(\Omega)$ and if $E_{2k}^k(n)$ denotes the n-width of the injection from $H^{2k}(\Omega)$ into $H^k(\Omega)$, then there exists a constant M such that

(63) $$\|u - p_h u_h\|_{H^k(\Omega)} \leq M E_{2k}^k(n(h) - 1) \|u\|_{H^{2k}(\Omega)}. \qquad \blacktriangle$$

9. Approximations of Nonhomogeneous Dirichlet Problems

For simplicity, we consider the Dirichlet problem

(64) $\begin{cases} \text{(i) } \Lambda u = -\Delta u + \lambda u = f, & \text{where } f \text{ is given in } L^2(\Omega), \\ \text{(ii) } \gamma_0 u = u|_\Gamma = t, & \text{where } t \text{ is given in } H^{\frac{1}{2}}(\Gamma). \end{cases}$

The "usual" variational formulation of this Dirichlet problem is

(65) $\begin{cases} \text{(i) } u \in H^1(\Omega), \\ \text{(ii) } a(u, v) + \lambda(u, v) = (f, v) & \text{for any } v \in H_0^1(\Omega), \\ \text{(iii) } \gamma_0 u = t. \end{cases}$

Since γ_0 maps $H^1(\Omega)$ onto $H^{\frac{1}{2}}(\Gamma)$, there exists (at least) a function $u_1 \in H^1(\Omega)$ such that $\gamma_0 u_1 = t$. If we set $w = u - u_1 \in H_0^1(\Omega)$, then w is the solution of the variational equation

(66) $\begin{cases} \text{(i) } w \in H_0^1(\Omega), \\ \text{(ii) } a(w, v) + \lambda(w, v) = l(v) & \text{for any } v \in H_0^1(\Omega), \end{cases}$

where $l(v) = (f, v) - a(u_1, v) - \lambda(u_1, v)$. Conversely, we can prove that if w is a solution of (66), $u = w + u_1$ is the solution of the Dirichlet problem

(64). Therefore, we can approximate the solution u of (64) by approximating the solution w of (66) if we solve the two following problems:

1. Find $u_1 \in H^1(\Omega)$ such that $\gamma_0 u_1 = t$.
2. Construct approximations of $H_0^1(\Omega)$.

Except for the case where $t = 0$, it is hopeless to solve problem 1. We construct finite-element approximations of the Sobolev space $H_0^1(\Omega)$ which are convergent and quasi-optimal for the injection from $H_0^s(\Omega)$ into $H_0^k(\Omega)$ when μ satisfies the criterion of m-convergence but not quasi-optimal for the injection from $H^s(\Omega) \cap H_0^k(\Omega)$ into $H_0^k(\Omega)$.

Since we need the latter property for proving meaningful error estimates and the regularity of the convergence, we are led to find other variational formulations which bypass problems 1 and 2.

There is a difference between the Neumann problem and the Dirichlet problem; namely, in the usual variational formulations of these boundary-value problems, the Dirichlet conditions are "forced," whereas the Neumann conditions are "natural." In other words, the Dirichlet problem is equivalent to a variational equation on a closed subspace of $H^m(\Omega)$ defined by homogeneous boundary-value conditions, and the Neumann problem is equivalent to a variational equation on the whole Sobolev space $H^m(\Omega)$.

Therefore, we give examples of variational equations on spaces of functions (*which do not satisfy homogeneous boundary conditions*) that are equivalent or approximatively equivalent to the Dirichlet problem.

A. Perturbation method   We approximate the solution u of the Dirichlet problem by the solution u_ε of the variational equation

$$(67) \begin{cases} \text{(i)} \ u_\varepsilon \in H^1(\Omega), \\ \text{(ii)} \ a(u_\varepsilon, v) + \lambda(u_\varepsilon, v) + \varepsilon^{-1}\langle \gamma_0 u, \gamma_0 v \rangle = (f, v) + \varepsilon^{-1}\langle t, \gamma_0 v \rangle \\ \hspace{5cm} \text{for any} \ v \in H^1(\Omega), \end{cases}$$

and we discretize this variational equation on $H^1(\Omega)$ by using finite-element approximations of $H^1(\Omega)$: find $u_h \in H_h^\mu(\Omega)$ satisfying

$$(68) \begin{aligned} a(p_h u_h, p_h v_h) + \lambda(p_h u_h, p_h v_h) + \varepsilon(h)^{-1}\langle \gamma_0 p_h u_h, \gamma_0 p_h v_h \rangle \\ = (f, p_h v_h) + \varepsilon(h)^{-1}\langle t, \gamma_0 p_h v_h \rangle \quad \text{for any} \ v_h \in H_h^\mu(\Omega). \end{aligned}$$

THEOREM 18   Let us assume that μ satisfies the m-criterion of convergence for $m \geq 1$. Let us choose $\varepsilon(h) = M |h|^{m+\frac{1}{2}-\alpha}$, where $\alpha > 0$. Then there exists a constant M_α such that

$$(69) \begin{cases} \text{(i)} \ \|u - p_h u_h\|_{H^1(\Omega)} \leq M |h|^{[2(m+1)/4]-\alpha/2} \|u\|_{H^{m+1}(\Omega)}, \\ \text{(ii)} \ \|\Lambda(u - p_h u_h)\|_{H^{-m-1}(\Omega)} \leq M |h|^m \|u\|_{H^1(\Omega)}, \\ \text{(iii)} \ \|\gamma_0(u - p_h u_h)\|_{L^2(\Gamma)} \leq M |h|^{m+\frac{1}{2}-\alpha} \|u\|_{H^{m+1}(\Omega)}. \end{cases}$$

Furthermore,

(70) $M^{-1}|h|^{\alpha-m-\frac{1}{2}}\gamma_0(u - p_h u_h)$ converges to $\dfrac{\partial u}{\partial n}$ in $L^2(\Gamma)$.   ▲

B. Least-squares method   We assume that

(71) $\begin{cases} \text{(i) } (\Lambda, \gamma_0) \text{ is an isomorphism from } H^2(\Omega) \text{ onto } L^2(\Omega) \times H^{3/2}(\Gamma), \\ \text{(ii) } (\Lambda, \gamma_0) \text{ is an isomorphism from } H^{m+1}(\Omega) \text{ onto} \\ \qquad\qquad\qquad\qquad\qquad\qquad H^{m-1}(\Omega) \times H^{m+1/2}(\Gamma). \end{cases}$

On the other hand, the solution u of the Dirichlet problem (64) satisfies the variational equation

(72) $(\Lambda u, \Lambda v) + \beta(h)^{-2}\langle\gamma_0 u, \gamma_0 v\rangle = (f, \Lambda v) + \beta(h)^{-2}\langle t, \gamma_0 v\rangle$
$$\text{for any} \quad v \in H^2(\Omega).$$

Therefore, we approximate u by the solution u_h of the discrete variational equation

(73) $\begin{aligned} (\Lambda p_h u_h, \Lambda p_h v_h) &+ \beta(h)^{-2}\langle\gamma_0 p_h u_h, \gamma_0 p_h v_h\rangle \\ &= (f, \Lambda p_h v_h) + \beta(h)^{-2}\langle t, \gamma_0 p_h v_h\rangle \quad \text{for any} \quad v_h \in H_h^\mu(\Omega). \end{aligned}$

THEOREM 19   Let us assume that μ satisfies the criterion of m-convergence for $m \geq 2$. Let us choose $\beta(h) = \sqrt{M}\,|h|^{3/2}$. Then the following error estimates hold

(74) $\begin{cases} \text{(i) } \|\Lambda(u - p_h u_h)\|_{L^2(\Omega)} \leq M\,|h|^{m-1}\,\|u\|_{H^{m+1}(\Omega)}, \\ \text{(ii) } \|u - p_h u_h\|_{L^2(\Omega)} \leq M\,|h|^{m-1+\min(2,\,m-1)}\,\|u\|_{H^{m+1}(\Omega)}, \\ \text{(iii) } \|\gamma_0(u - p_h u_h)\|_{L^2(\Gamma)} \leq M\,|h|^{m+1/2}\,\|u\|_{H^{m+1}(\Omega)}. \end{cases}$ ▲

C. Conjugate problem (I)   We prove that the Dirichlet problem (64) is equivalent to the following variational equation on $H^1(\Omega, \Delta)$:

(75) $\begin{cases} \text{(i) } u \in H^1(\Omega, \Delta), \\ \text{(ii) } (\Delta u, \Delta v) + \lambda a(u, v) = (f, \Delta v) + \lambda\left\langle t, \dfrac{\partial v}{\partial n}\right\rangle \\ \qquad\qquad\qquad\qquad\qquad\qquad \text{for any} \quad v \in H^1(\Omega, \Delta). \end{cases}$

Since we also prove that the finite-element approximations $(H_h^\mu(\Omega), p_h, \hat{r}_h)$ associated with a function μ satisfying the criterion of m-convergence for $m \geq 2$ are convergent approximations of the space $H^1(\Omega, \Delta)$, we can approximate the solution u of the Dirichlet problem by the solution $u_h \in H_h^\mu(\Omega)$ of the discrete variational equation

$$(76) \quad (\Delta p_h u_h, \Delta p_h v_h) + \lambda a(p_h u_h, p_h v_h) = (f, \Delta p_h v_h) + \lambda \left\langle t, \frac{\partial p_h v_h}{\partial n} \right\rangle$$

$$\text{for any} \quad v_h \in H_h{}^\mu(\Omega).$$

THEOREM 20 Let us assume that μ satisfies the criterion of m-convergence for $m \geq 2$. Then there exists a constant M such that

$$(77) \quad \begin{cases} \text{(i)} \ \|\Delta(u - p_h u_h)\|_{L^2(\Omega)} \leq M \, |h|^{m-1} \, \|u\|_{H^{m+1}(\Omega)}, \\ \text{(ii)} \ \|u - p_h u_h\|_{L^2(\Omega)} \leq M \, |h|^{m-1+\min(2, m-1)} \, \|u\|_{H^{m+1}(\Omega)}. \end{cases} \quad \blacktriangle$$

D. Conjugate problem (II) We can construct another variational formulation of the Dirichlet problem (64) by using the splitting $-\Delta = -\text{div. grad}$ and the fact that the operator $D^* = -\text{div}$ is the formal adjoint of the operator $D = \text{grad}$.

Let us introduce the space

$$(78) \quad \mathbf{H}^1(\Omega, D^*) = \{\mathbf{u} \in (L^2(\Omega))^n \quad \text{such that} \quad D^*\mathbf{u} \in L^2(\Omega)\}.$$

We prove the following Green formula:

$$(79) \quad [\mathbf{u}, Dv] = (D^*\mathbf{u}, v) + \langle \beta \mathbf{u}, \gamma_0 v \rangle \quad \text{when} \quad \mathbf{u} \in \mathbf{H}^1(\Omega, D^*), v \in H^1(\Omega),$$

where $[\mathbf{u}, \mathbf{v}] = \sum_{1 \leq i \leq n} (u^i, v^i)$, $\beta \mathbf{u} = \mathbf{u} \cdot \mathbf{n}$, $\mathbf{n}$ being the normal to the boundary Γ of Ω.

Let us consider the following variational equation on $\mathbf{H}^1(\Omega, D^*)$:

$$(80) \quad \begin{cases} \text{(i)} \ \mathbf{u} \in \mathbf{H}^1(\Omega, D^*), \\ \text{(ii)} \ (D^*\mathbf{v}, D^*\mathbf{v}) + \lambda[\mathbf{u}, \mathbf{v}] = (f, D^*\mathbf{v}) + \langle t, \beta \mathbf{v} \rangle \\ \qquad\qquad\qquad\qquad\qquad \text{for any} \quad \mathbf{v} \in \mathbf{H}^1(\Omega, D^*). \end{cases}$$

We prove: if u is the solution of the Dirichlet problem (64), then $\mathbf{u} = Du$ is the solution of the variational equation (80). Conversely, if $\mathbf{u}$ is the solution of the variational equation (80), then $u = \lambda^{-1}(f - D^*\mathbf{u})$ is the solution of the Dirichlet problem (64).

In order to approximate the solution $\mathbf{u}$ of the variational equation (80), we must construct approximations of the domain $\mathbf{H}^1(\Omega, D^*)$ of the divergence operator D^*. If μ is a function of $H^m(R^n)$ with compact support, we set

$$\mu_i(x) = \int_0^1 \mu(x_1, \ldots, x_i - y_i, \ldots, x_n) \, dy_i$$

and

$$
(81) \quad
\begin{cases}
\text{(i)} \quad p_h{}^i u_h = \sum_{j \in \mathcal{R}_h{}^{\mu_i}(\Omega)} u_h{}^j \mu_i\left(\frac{x}{h} - j\right), \\[2mm]
\text{(ii)} \quad \mathbf{H}_h{}^\mu(\Omega, D^*) = \prod_{1 \le i \le n} H_h{}^{\mu_i}(\Omega); \ \mathbf{u}_h = (u_h{}^i)_{1 \le i \le n}, \\[2mm]
\text{(iii)} \quad \mathbf{p}_h \mathbf{u}_h = (p_h{}^i u_h{}^i)_{1 \le i \le n},
\end{cases}
$$

where $\mathbf{p}_h$ is a prolongation mapping $\mathbf{H}_h{}^\mu(\Omega, D^*)$ into $\mathbf{H}^1(\Omega, D^*)$ and satisfying

$$
(82) \quad D^* \mathbf{p}_h \mathbf{u}_h = p_h D_h^* \mathbf{u}_h \quad \text{where} \quad D_h^* \mathbf{u}_h = -\sum_{1 \le i \le n} \nabla_{h_i} u_h{}^i.
$$

In other words, the approximations $[\mathbf{H}_h{}^\mu(\Omega, D^*), \mathbf{p}_h, \hat{r}_h]$, (where $\hat{r}_h$ is the optimal restriction associated with $\mathbf{p}_h$) are convergent approximations of the space $\mathbf{H}^1(\Omega, D^*)$.

Therefore we can approximate the solution $\mathbf{u}$ of (80) by the solution $\mathbf{u}_h \in \mathbf{H}_h{}^\mu(\Omega, D^*)$ of the discrete variational equation

$$
(83) \quad
\begin{cases}
(p_h D_h^* \mathbf{u}_h, p_h D_h^* \mathbf{v}_h) + \lambda \sum_{1 \le i \le n} (p_h{}^i u_h{}^i, p_h{}^i v_h{}^i) \\[2mm]
= (f, p_h D_h^* \mathbf{v}_h) + \int_\Gamma t(x) \mathbf{n} \cdot \mathbf{p}_h \mathbf{v}_h \, d\sigma(x) \quad \text{for any} \ \mathbf{v}_h \in \mathbf{H}_h{}^\mu(\Omega, D^*).
\end{cases}
$$

THEOREM 21   Let us assume that μ satisfies the criterion of $m - 1$ convergence, that the solution u of the Dirichlet problem (64) belongs to $H^{m+1}(\Omega)$, and that Δu belongs to $H^m(\Omega)$. Then there exists a constant M such that

$$
(84) \quad
\begin{cases}
\text{(i)} \quad \| \Delta u - p_h D_h^* \mathbf{u}_h \|_{L^2(\Omega)} \le M \, |h|^m \, \|u\|_{H^{m+1}(\Omega)}, \\[2mm]
\text{(ii)} \quad \| D_i u - p_h{}^i u_h{}^i \|_{L^2(\Omega)} \le M \, |h|^{m+1} \, \|u\|_{H^{m+1}(\Omega)}
\end{cases}
\quad \blacktriangle
$$

Theorem 21 shows that each derivative $D_i u$ of the solution u of the Dirichlet problem is separately approximated by $p_h{}^i u_h{}^i$. Naturally, we can prove that $\mathbf{p}_h \mathbf{u}_h$ converges to the solution $\mathbf{u}$ of (80).

10. A Posteriori Error Estimates

The error estimates that appeared previously only point out the rate of convergence when the parameter h goes to 0; no estimation of the actual error between the solution u of the boundary-value problem and the approximate solution is provided. We can estimate such actual error by using two different approximations of the solution u. For instance, let us consider the solution u of the Dirichlet problem (64).

THEOREM 22 Let u be the solution of the Dirichlet problem (64) and $v \in H^1(\Omega)$ any approximant of u satisfying the boundary condition $\gamma_0 v = t$. If $w \in H^1(\Omega, \Delta)$ is any approximant of u, then

$$(85) \quad \begin{cases} \|u - v\|^2_{H^1(\Omega)} + \lambda \|u - v\|^2_{L^2(\Omega)} \\ \qquad \leq \lambda^{-1} \|f - \lambda v - \Delta w\|^2_{L^2(\Omega)} + \|v - w\|^2_{H^1(\Omega)}. \end{cases}$$

If $w \in H^1(\Omega, D^*)$ is any approximant of Du, then

$$(86) \quad \begin{cases} \|u - v\|^2_{H^1(\Omega)} + \lambda \|u - v\|^2_{L^2(\Omega)} \\ \qquad \leq \lambda^{-1} \|f - \lambda v - D^* w\|^2_{L^2(\Omega)} + \|Dv - w\|^2_{(L^2(\Omega))^n} \quad \blacktriangle \end{cases}$$

The right-hand sides of the a posteriori estimates (85) and (86) *do not involve the unknown solution u* and, each right-hand side converges to 0 when v and w converge to the solution u of (64) and when w converges to the solution u of the variational equation (80).

11. External and Partial Approximations

We have seen that the homogeneous Neumann problem for $\Lambda = -\Delta + 1$ is equivalent to the variational equation

$$(87) \quad \begin{cases} a(u, v) = \sum_{1 \leq i \leq n} \int_\Omega D_i u D_i v \, dx + \int_\Omega uv \, dx \\ \qquad = \int_\Omega fv \, dx \qquad \text{for any} \quad v \in H^1(\Omega). \end{cases}$$

Then $a(u, v)$ is the sum of the bilinear forms $a_i(u, v)$ continuous on the spaces $V_i = H(\Omega, D_i) = \{u \in L^2(\Omega)$ such that $D_i u \in L^2(\Omega)\}$. On the other hand, the space $V = H^1(\Omega)$ is the intersection of the spaces $H(\Omega, D_i)$ for $i = 1, \ldots, n$.

If we consider prolongations p_h^i mapping a discrete space V_h in the spaces $V_i = H(\Omega, D_i)$ and a prolongation p_h^0 mapping V_h into $V_0 = L^2(\Omega)$, then we can introduce the discrete variational equation

$$(88) \quad \sum_{1 \leq i \leq n} a_i(p_h^i u_h, p_h^i v_h) + a_0(p_h^0 u_h, p_h^0 v_h) = (f, p_h^0 v_h) \qquad \text{for any} \quad v_h \in V_h.$$

We wish to study the convergence properties of such discrete variational equations, which is actually a particular case of the problem of approximating an operator A mapping a closed subspace of a space $\bar{V}$ onto a factor space of a space $\bar{F}$ by using approximations of the spaces $\bar{V}$ and $\bar{F}$ instead of approximations of the closed space and the factor space. Namely, let $\bar{A} \in L(\bar{V}, \bar{F})$

be a continuous linear operator from $\bar{V}$ into $\bar{F}$ and

(89) $\begin{cases} \text{(i) } V \text{ be a closed subspace of } \bar{V}, \\ \text{(ii) } F = \bar{F}/F_0 \text{ be a factor space of } \bar{F}, \\ \text{(iii) } A = \mu\bar{A}\pi \in L(V, F) \end{cases}$

where π is the canonical injection from V into $\bar{V}$ and μ is the canonical surjection from $\bar{F}$ onto $\bar{F}/F_0$.

Letting $g \in \bar{F}$ and $f = \mu g \in F$ be given data, we approximate the solution u of the equation

(90) $Au = f$ (or, equivalently, $\mu\bar{A}\pi u = \mu g$)

by the solution u_h of the discrete equation

(91) $s_h \bar{A} p_h u_h = s_h g,$

where

$\begin{cases} \text{(i) } p_h \text{ is a prolongation mapping } V_h \text{ into } \bar{V} \text{ (instead of } V), \\ \text{(ii) } s_h \text{ is a restriction mapping } \bar{F} \text{ (instead of } F = \bar{F}/F_0) \text{ onto } F_h. \end{cases}$

Let $\hat{r}_h$ be the optimal restriction associated with p_h in $\bar{V}$ and $\hat{q}_h$ the optimal prolongation associated with s_h in $\bar{F}$, and we introduce the following definitions:

• approximations $(V_h, p_h, \hat{r}_h)$ of $\bar{V}$ are said to be "convergent in V" if

$\lim \|u - p_h \hat{r}_h u\|_{\bar{V}} = 0$ for any u belonging to the closed subspace
$$V \text{ of } \bar{V},$$

• approximations $(V_h, p_h, \hat{r}_h)$ of $\bar{V}$ are said to be "external approximations of the closed subspace V of $\bar{V}$" if

$$p_h u_h \text{ converges weakly to } u \text{ in } \bar{V} \text{ implies that } u \text{ belongs to } V$$

• approximations $(F_h, \hat{q}_h, s_h)$ of $\bar{F}$ are said to be external approximations of the factor space $F = \bar{F}/F_0$ if

$$\lim \|\hat{q}_h s_h f\|_{\bar{F}} = 0 \quad \text{for any } f \text{ in } F_0.$$

We can furnish results analogous to the results regarding internal approximations. For instance, a theorem of equivalence holds:

THEOREM 23   If A is an isomorphism from V onto F, the two following properties are equivalent:

(92) $\begin{cases} \text{(i) there exists a unique solution } u_h \text{ of (91),} \\ \text{(ii) } \lim \|u - p_h u_h\|_{\bar{V}} = 0, \end{cases}$

and

$$(93) \quad \begin{cases} \text{(i) the operators } s_h \bar{A} p_h \text{ are stable,} \\ \text{(ii) the approximations } (V_h, p_h, \hat{r}_h) \text{ are convergent in } V. \\ \text{(iii) the approximations } (F_h, \hat{q}_h, s_h) \text{ are external approximations of } F. \end{cases}$$

▲

We can also estimate the errors in the following way. First introduce spaces U and $\bar{E}$ such that

$$(94) \qquad U \subset V; \bar{E} \subset \bar{F},$$

the injections being continuous, and set

$$E = \bar{E}/E_0 \quad \text{where} \quad E_0 = \bar{E} \cap F_0.$$

We define the error function and the external error function by

$$(95) \quad \begin{cases} \text{(i)} \quad e_U{}^{\bar{V}}(p_h) = \sup_{u \in U} \dfrac{\| u - p_h \hat{r}_h u \|_{\bar{V}}}{\| u \|_U}, \\ \text{(ii)} \quad t_{E_0}{}^{\bar{F}}(s_h) = \sup_{f \in E_0} \dfrac{\| \widehat{\hat{q}_h s_h f} \|_{\bar{F}}}{\| f \|_{\bar{E}}}. \end{cases}$$

THEOREM 24 Suppose that the solution u of (90) belongs to U and that both g and $\bar{A}u$ belong to $\bar{E}$. Then, if the operators $s_h \bar{A} p_h$ are stable, there exists a constant M such that

$$(96) \qquad \| u - p_h u_h \|_V \leq M(e_U{}^V(p_h) \| u \|_U + t_{E_0}{}^{\bar{F}}(s_h) \| g - \bar{A}u \|_{\bar{E}}). \qquad ▲$$

EXAMPLE: **Partial approximations.** The approximate problem (88) is a partial case of the approximate problem (91), where we take

$$(97) \quad \begin{cases} \text{(i)} \quad \bar{V} = \Pi \, V_i, \; \bar{F} = \bar{V}' = \Pi \, V_i', \\ \text{(ii)} \quad V = \cap \, V_i = \{u = (u, \ldots, u)\}, \\ \qquad\qquad\qquad\qquad F = V' = \bar{V}'/V^\perp \text{ (i.e., } F_0 = V^\perp), \\ \text{(iii)} \quad \bar{A}\bar{u} = (A_0 u_0, \ldots, A_n u_n) \\ \qquad\qquad\qquad \text{where} \quad (A_i u, v) = a_i(u, v), \; i = 0, \ldots, n \\ \text{(iv)} \quad p_h \text{ is defined by } p_h u_h = (p_h{}^0 u_h, \ldots, p_h{}^n u_h), \\ \text{(v)} \quad s_h \text{ is defined by } s_h f = \sum_{0 \leq i \leq n} (p_h{}^i)' f_i \quad \text{where} \quad f = (f_0, \ldots, f_n). \end{cases}$$

In other words, when we identify the intersection of spaces V_i with the diagonal of the product space $\prod V_i$, we say that approximations of $\prod V_i$ which are external approximations of V are partial approximations of V.

12. General Outline

In Chapter 1 we study in an elementary way approximations of Neumann problems for ordinary second-order differential equations in order to motivate the general theory.

Chapter 2 is devoted to the approximations of Hilbert spaces, and, in Section 1, to a summary of the results of functional analysis, that will be needed later.

Chapter 3 deals with approximations of operators and operational and variational equations.

Finite-element approximations and piecewise-polynomial approximations of Sobolev spaces are constructed in Chapter 4 (in the case of functions of one variable) and in Chapter 5 (in the case of functions of several variables).

Chapter 6 considers an elementary theory of variational boundary-value problems, and pertinent examples are gathered in Chapter 7.

We approximate the Neumann type problems in Chapter 8, and other boundary-value problems are covered in Chapter 9 (by using perturbed approximations and least-squares approximations).

Chapter 10 is devoted to conjugate problems of boundary-value problems and their applications to obtaining a posteriori error estimates and to approximation of Dirichlet problems.

Chapter 11 extends some of the previous results to the case of external and partial approximations, and the book ends with a set of comments.

CHAPTER 1

Approximation of Solutions of Neumann Problems for Second-Order Linear Differential Equations

In this chapter we illustrate the basic techniques that are developed later in the case of the simplest kind of boundary-value problem, namely, the Neumann problem for a second-order linear differential equation.

Section 1 of this chapter is devoted to the construction of the variational formulation of the Neumann problem. In Section 2 we introduce the approximations of a Hilbert space and we prove a first theorem of convergence. Then we construct the piecewise-linear approximations of the Sobolev spaces.

Finally, in Section 4, we use these approximations for constructing approximate schemes of the Neumann problem.

1. WEAK SOLUTIONS OF NEUMANN PROBLEMS FOR SECOND-ORDER LINEAR DIFFERENTIAL OPERATORS

1-1. The Neumann Boundary-Value Problem

Let us state the classical Neumann boundary-value problem for a second-order linear differential equation by calling $a(x)$ and $b(x)$ two real-valued functions and denoting the derivative by $D = d/dx$.

If $f(x)$ is a given function and α_0 and α_1 are given numbers, we can look for the solution (if any) of

$$(1\text{-}1) \quad \begin{cases} \text{(i)} & -D(a(x)Du(x)) + b(x)u(x) = f(x); \quad 0 \le x \le 1, \\ \text{(ii)} & Du(0) = \alpha_0, \ Du(1) = \alpha_1. \end{cases}$$

A priori we have to assume that the functions a and b are continuously differentiable. This is a serious restriction, since in several concrete examples the coefficients a and b of the equation are piecewise-constant functions. We thus are obliged to extend the notion of solution of (1-1), prove the existence and the uniqueness of this solution, and approximate it in a convenient sense. In doing so, we shall actually simplify those questions and use methods general enough to be applied to more sophisticated problems.

1-2. Definition of Distributions

Let $u(x)$ be a locally integrable function defined on an open interval I. This means, first, that u is measurable on I and, second, that for any closed bounded (i.e., compact) subinterval $[\alpha, \delta]$, the following integral is convergent:

$$(1\text{-}2) \qquad \int_\alpha^\delta u(x)\, dx < +\infty.$$

As usual, we identify two functions that are almost everywhere equal. Let us notice also that such a function u is well defined if and only if we know its value on every test function ϕ, namely:

$$(1\text{-}3) \qquad \int_I u(x)\phi(x)\, dx \qquad \text{for any} \quad \phi \in \Phi_0$$

where Φ denotes the space of indefinitely differentiable functions and Φ_0 the subspace of these functions with compact support (i.e., vanishing outside a compact interval).

This suggests to extend the notion of function in the following way.

Definition 1-1 We say that u is a *distribution* (or a *generalized function*) if u is a linear functional on the space Φ_0 (defined by its value (u, ϕ) on each $\phi \in \Phi_0$) continuous in the sense where (u, ϕ_n) converges to 0 whenever

(1-4)   the supports of the ϕ_n are contained in a compact $J \subset I$,

and

(1-5)   the derivatives $D^k\phi_n$ converge to 0 uniformly over J.   ▲

Then (1-3) shows that a locally integrable function defines a distribution. From now on, we view any "usual" function as a distribution.

1-3. Weak Derivatives of a Distribution

We can extend the notion of derivative thus.

Definition 1-2   The distribution Du defined by

(1-6) $(Du, \phi) = -(u, D\phi)$ for any $\phi \in \Phi_0$.

is said to be the weak derivative (or, in short, the derivative of u). Obviously, the weak derivative of a differentiable function coincides with the usual derivative (by the integration by parts formula).   ▲

We are now able to give a meaning to the problem (1-1).

1-4.   Variational Formulation of the Problem

Let us assume that the coefficients a and b, the function f, and the solution u of (1-1) are "smooth" and that $a(0)$ and $a(1)$ are different from 0. Multiplying [1-1(i)] by a "test function" $\phi \in \Phi$, integrating by parts and using [1-1(ii)], we obtain the new relation

(1-7) $\begin{cases} \displaystyle\int_I a(x)Du(x)D\phi(x)\, dx + \int_I b(x)u(x)\phi(x)\, dx \\[2mm] \displaystyle = \int_I f(x)\phi(x)\, dx + a(1)\alpha_1\phi(1) - a(0)\alpha_0\phi(0) \qquad \text{for any} \quad \phi \in \Phi. \end{cases}$

Conversely, we can check that if a function u satisfies (1-7) for any $\phi \in \Phi$, u satisfies (1-1), and we say that (1-7) is the *variational formulation* of the problem and that (1-1) is the *operational formulation*. In so doing, we have halved the order of the derivatives operating on u and replaced the system (1-1) by a single equation put in a variational form.

1-5.   Weak Solutions of the Neumann Boundary-Value Problem

The variational formulation can be used to share the regularity requirements between the solution u and the test function ϕ. This is necessary in order for the integrals involved in (1-7) to acquire a meaning. In other words, the smoother are the test functions, the "weaker"* the solution u can be. We return to this problem, providing more details.

Nevertheless, let us give the choice where u and ϕ have the same order of regularity. It is clear that the integrals involved in (1-7) have a meaning when we assume:

(1-8) $\begin{cases} \text{(i) } a(x), b(x) \in L^\infty(I), \\ \text{(ii) } u(x), \phi(x) \in L^2(I), \\ \text{(iii) } Du(x), D\phi(x) \in L^2(I), \\ \text{(iv) } f(x) \in L^2(I). \end{cases}$

___
* The smaller the number of smoothness requirements that need to be imposed on u.

In this case, we have to define the numbers $a(0)$ and $a(1)$ when $a(x) \in L^\infty(I)$: We shall deduce from Theorem 6.2-1 of Chapter 6 the following proposition:

PROPOSITION 1-1 Let us assume (1-8) and

$$(1\text{-}9) \qquad\qquad D[a(x)Du(x)] \in L^2(I).$$

Then there exist two numbers $a(0)$ and $a(1)$ such that

$$(1\text{-}10) \quad \left\{ \int_I a(x)DuD\phi \, dx + \int_I D(a(x)Du)\phi \, dx \right.$$
$$= a(1)Du(1)\phi(1) - a(0)Du(0)\phi(0). \quad \blacktriangle$$

Therefore, we shall assume that $a(0)$ and $a(1)$ are defined by this proposition.

Hence the assumption (1-8) gives a meaning to the variational equation (1-7). If a solution exists, we say that it is a "weak solution," as opposed to a "smooth" (or a "strong" or "usual" one). We also say that u is a "generalized" solution. The use of the word "smooth" (or "weak") is a convenient way to say that a function has (has not) a certain order of regularity made precise by the context.

Conversely, a weak solution is also a solution of an operational equation. Indeed, taking $\phi \in \Phi_0$, we deduce from (1-7) that:

$$(1\text{-}11) \qquad -D(a(x)Du(x)) + b(x)u(x) = f(x) \in L^2(I).$$

Therefore (1-9) holds and, by Proposition 1-1, we can integrate by parts. Hence

$$(1\text{-}12) \quad (Du(1) - \alpha_1)a(1)\phi(1) = (Du(0) - \alpha_0)a(0)\phi(0) \qquad \text{for any} \quad \phi \in \Phi.$$

Let us define

$$(1\text{-}13) \qquad H^1(I) = \{u(x) \in L^2(I) \quad \text{such that} \quad Du(x) \in L^2(I)\}.$$

We summarize these results in:

THEOREM 1-1 Let us assume

$$(1\text{-}14) \quad \begin{cases} \text{(i)} \ \ a(x), b(x) \in L^\infty(I); f(x) \in L^2(I), \\ \text{(ii)} \ \ a(1) \quad \text{and} \quad a(0) \neq 0. \end{cases}$$

Then the problems represented by

$$(1\text{-}15) \quad \begin{cases} \text{(i)} \ \ u(x) \in H^1(I), \\ \text{(ii)} \ \ \int_I a(x)Du \, Dv \, dx + \int_I b(x)uv \, dx = \int_I f(x)v(x) \, dx \\ \qquad\qquad + a(1)\alpha_1 v(1) - a(0)\alpha_0 v(0) \qquad \text{for any} \quad v(x) \in H^1(I) \end{cases}$$

and

$$(1-16) \quad \begin{cases} \text{(i)} \ u(x) \in H^1(I); \ D(a(x)Du(x)) \in L^2(I), \\ \text{(ii)} \ -D(a(x)Du) + b(x)u(x) = f(x), \\ \text{(iii)} \ Du(1) = \alpha_1; \ Du(0) = \alpha_0, \end{cases}$$

and equivalent. ▲

1-6. Sobolev Spaces

Definition 1-3 We say that the space $H^m(I)$

$$(1-17) \quad H^m(I) = \{u \in L^2(I) \quad \text{such that} \quad D^j u \in L^2(I) \quad \text{for} \quad 0 \leq j \leq m\}$$

is the *Sobolev space* of order m (on the interval I), and supply it with the following inner product and norm:

$$(1-18) \quad ((u, v))_m = \sum_{j=0}^{m} (D^j u, D^j v); \ \| u \|_m = \left(\sum_{j=0}^{m} |D^j u|^2 \right)^{\frac{1}{2}} = \sqrt{((u, u))_m} \quad ▲$$

We set

$$(1-19) \quad \begin{cases} (u, v) = ((u, v))_0 = \int_I u(x)v(x) \, dx; \\ |u| = \|u\|_0 = \left(\int_I |u(x)|^2 \, dx \right)^{\frac{1}{2}}. \end{cases}$$

THEOREM 1-2 The Sobolev space $H^m(I)$ is a Hilbert space. ▲

Proof We have to prove that it is complete (i.e., that any Cauchy sequence u_μ of $H^m(I)$ is convergent). By definition of the norm of $H^m(I)$, the sequences $D^j u_\mu$ are Cauchy sequences of the space $L^2(I)$ and thus converge to an element u^j of $L^2(I)$. We have to prove that $u^j = D^j u^0$ for $0 \leq j \leq m$. But for any $\phi \in \Phi_0$, the equalities

$$(D^j u_\mu, \phi) = (-1)^j (u_\mu, D^j \phi)$$

converge to

$$(u^j, \phi) = (-1)^j (u^0, D^j \phi) = (D^j u^0, \phi) \quad \text{for} \quad \phi \in \Phi_0 \quad \text{and} \quad 0 \leq j \leq m.$$

Therefore, the Cauchy sequence u_μ converges to u^0 because the convergence in $H^m(I)$ is the convergence in $L^2(I)$ of all the derivatives $D^j u_\mu$ to the derivative $D^j u^0$. ■*

---

* The symbol ■ denotes the end of a Proof or a Remark.

Thus we replace in our study the space of continuously differentiable functions (which is a nonreflexive Banach space) by Sobolev spaces. In so doing, we are able not only to use the properties of the Hilbert spaces but also to simplify the proofs of the forthcoming results.

The following result gives some information about the properties of functions of Sobolev spaces.

THEOREM 1-3 If $u \in H^m(I)$, then for any $j \leq m - 1$, the derivatives $D^j u$ are Hölderian functions:

$$|D^j u(x) - D^j u(y)| \leq |D^{j+1} u| \sqrt{|x - y|}. \qquad \blacktriangle$$

Proof By the Cauchy-Schwarz inequality, we obtain

$$|D^j u(x) - D^j u(y)| = \left| \int_y^x D^{j+1} u(t) \, dt \right| \leq \sqrt{|x - y|} \left(\int_I |D^{j+1} u(t)|^2 \, dt \right)^{\!1/2}. \qquad \blacksquare$$

We continue our consideration of Sobolev spacs $H^m(I)$, and, more generally, of spaces $H^m(R^n)$, $H^m(\Omega)$, $H_0^m(\Omega)$ in Sections 2.1-11, 2.1-14, 5.2-1, 5.3-1, and 6.3.

1-7. The Lax-Milgram Theorem

The problem (1-15) is a particular case of an abstract variational problem. Let us set

$$(1-20) \quad \begin{cases} \text{(i)} & V = H^1(I), \\[2mm] \text{(ii)} & a(u, v) = \displaystyle\int_I a(x) Du \, Dv \, dx + \int_I b(x) uv \, dx, \\[2mm] \text{(iii)} & l(v) = \displaystyle\int_I f(x) \, v(x) \, dx + \alpha_1 a(1) v(1) - \alpha_0 a(0) v(0). \end{cases}$$

We can check that $a(u, v)$ is a continuous bilinear form on $V \times V$ and that $l(v)$ is a continuous linear form on V.

Let us consider an abstract variational problem in which V is a Hilbert space, $a(u, v)$ is a continuous bilinear form on $V \times V$, and $l(v)$ is a continuous linear form on V. When we look for u satisfying

$$(1-21) \quad \begin{cases} \text{(i)} & u \in V, \\[2mm] \text{(ii)} & a(u, v) = l(v) \quad \text{for any} \quad v \in V, \end{cases}$$

the Lax-Milgram theorem gives a sufficient condition for existence and uniqueness:

THEOREM 1-4   Let us assume that the form $a(u, v)$ is *V-elliptic;* that is, that

(1-22) There exists a constant $c > 0$ such that $a(v, v) \geq c \|v\|^2$ for all $v \in V$.

Then there exists a unique solution u of the variational problem (1-21). ▲

We prove Theorem 1-4 in Section 2-4 by a constructive method, when V is a separable Hilbert space, and in Section 2.1-16, in the general case. ■

We deduce from this theorem

COROLLARY 1-1   Let us assume that

$$(1\text{-}23) \quad \begin{cases} \text{(i)} \ a(x) \quad \text{and} \quad b(x) \in L^\infty(I); f(x) \in L^2(I), \\ \text{(ii) there exists } c > 0 \quad \text{such that} \\ \qquad a(x) \geq c, \, b(x) \geq c \text{ almost everywhere in } I. \end{cases}$$

Then there exists a unique solution of the problem (1-15) [or (1-16)]. ▲

Proof   The assumption [(1-23)(ii)] implies the $H^1(I)$-ellipticity of the form $a(u, v)$ defined by [(1-20)(ii)] since

$$a(v, v) = \int_I a(x) |Dv(x)|^2 \, dx + \int_I b(x) |v(x)|^2 \, dx \geq c \|v\|_1^2.$$

2.   APPROXIMATION OF AN ABSTRACT VARIATIONAL PROBLEM

We begin the study of a first process of approximation of variational equations. But before we do this, we must use an example to define the term "approximation of a Hilbert space V."

2-1.   The Galerkin Approximation of a Separable Hilbert Space

Let us assume that V is a *separable* Hilbert space (i.e., that there exists a denumerable sequence spanning a dense subspace of V). Therefore, there exists a complete orthonormal basis $(\mu_j)_j$ $(1 \leq j)$ of the Hilbert space V. Setting $h = 1/n$, we associate with h and this basis (μ_j) the triple (V_h, p_h, r_h), where

$$(2\text{-}1) \quad \begin{cases} \text{(i)} \ V_h = R^n \text{ whose elements are denoted } u_h = (u_h^j); \, 1 \leq j \leq n, \\ \text{(ii)} \ p_h u_h = \sum_{j=1}^n u_h^j \mu_j; \quad p_h \text{ maps } V_h \text{ into } V, \\ \text{(iii)} \ r_h u = (((\mu_j; \, u)))_j; \, 1 \leq j \leq n; \quad r_h \text{ maps } V \text{ onto } V_h. \end{cases}$$

We denote by $P_h = p_h V_h$ the subspace spanned by $\mu_1, \ldots, \mu_n$. The operator $p_h r_h$ is the orthogonal projector onto P_h

$$(2\text{-}2) \quad \|u - p_h r_h u\| = \inf_{u_h \in P_h} \|u - u_h\| = \inf_{v_h \in V_h} \|u - p_h v_h\| \; ; p_h r_h u = \sum_{j=1}^{n} ((\mu_j, u))\mu_j.$$

Definition 2-1 We say that (V_h, p_h, r_h) are the *Galerkin approximations of V associated with the basis* (μ_j), which are *convergent* because

$$(2\text{-}3) \qquad \lim_{h \to 0} \|u - p_h r_h u\| = 0 \qquad \text{for any} \quad u \in V. \qquad \blacktriangle$$

We supply the space V_h with the norm $\|u_h\|_{V_h} = \|p_h u_h\|$.

2-2. Approximation of a Hilbert Space

Definition 2-2 More generally, we define an *approximation* (V_h, p_h, r_h) associated with a parameter h converging to 0 by the following items:

$$(2\text{-}4) \quad \begin{cases} \text{(i)} \ V_h \text{ is a Hibert space,} \\ \text{(ii)} \ p_h \text{ is an isomorphism from } V_h \text{ onto its closed range } P_h \text{ in } V, \\ \text{(iii)} \ r_h \text{ is a linear operator from } V \text{ onto } V_h. \end{cases} \qquad \blacktriangle$$

In particular, we can choose the optimal $\hat{r}_h$ defined by

$$(2\text{-}5) \qquad \|u - p_h \hat{r}_h u\| = \inf_{u_h \in P_h} \|u - u_h\| = \inf_{v_h \in V_h} \|u - p_h v_h\|.$$

If p_h^{-1} denotes the inverse of p_h mapping P_h onto V_h and if t_h is the orthogonal projector onto P_h, then $\hat{r}_h = p_h^{-1} \cdot t_h$.

Definition 2-3 It is useful to define V_h as the *discrete space*, p_h as the *prolongation*, r_h as the *restriction*, and P_h as the *space of approximants*.
$$\blacktriangle$$

Since it may be difficult to characterize the optimal restriction $\hat{r}_h$, we cannot count on being able to choose this restriction in the concrete examples.

As before, we say that the approximations (V_h, p_h, r_h) are *convergent* if and only if

$$(2\text{-}6) \qquad \lim_{h \to 0} \|u - p_h r_h u\| = 0 \qquad \text{for any} \quad u \in V.$$

Usually we supply the discrete space V_h with the *discrete norm* $\|u_h\|_h = \|p_h u_h\|$, but there are many other possible choices.

In Chapter 11, we define the more general notion of external approximation.

2-3. Internal Approximation of a Variational Equation

If V is a Hilbert space, $a(u, v)$ a continuous bilinear form on $V \times V$, and $l(v)$ a continuous linear form on V, we can give a procedure for approximating the solution of the following variational equation: look for u in V satisfying

$$(2\text{-}7) \qquad a(u, v) = l(v) \qquad \text{for any} \quad v \in V.$$

Definition 2-4 If V_h is a discrete space and p_h a prolongation from V_h into V associated with a parameter h, we define *internal approximation of the variational equation* (2-7) by the following discrete variational equation on the space V_h:

$$(2\text{-}8) \qquad a(p_h u_h, p_h v_h) = l(p_h v_h) \qquad \text{for any} \quad v_h \in V_h. \qquad \blacktriangle$$

This has a meaning since $a(p_h u_h, p_h v_h)$ is a continuous bilinear form on $V_h \times V_h$ and $l(p_h v_h)$ is a continuous linear form on V_h. $\qquad\blacksquare$

Let us give the operational formulation of (2-8) when

$$(2\text{-}9) \qquad V_h = R^{n(h)} \text{ whose elements are denoted by } v_h = (v_h^j)_j.$$

Then any isomorphism p_h from $R^{n(h)}$ into V is associated with a basis $(\lambda_h^j)_j$ of $n(h)$ elements λ_h^j of the space V by

$$(2\text{-}10) \qquad p_h u_h = \sum_j u_h^j \lambda_h^j.$$

PROPOSITION 2-1 Let us assume (2-9) and (2-10), calling A_h the matrix of elements

$$(a(\lambda_h^j, \lambda_h^i))_{i,j}$$

and l_h the vector of components $l(\lambda_h^j)$. Then the variational equation (2-8) is equivalent to the discrete operational equation

$$(2\text{-}11) \qquad A_h u_h = l_h. \qquad \blacktriangle$$

Proof Indeed, (2-8) and (2-10) imply

$$\sum_{i,j} a(\lambda_h^i, \lambda_h^j) u_h^i v_h^j = \sum_j l(\lambda_h^j) v_h^j = \sum_j (A_h u_h)^j v_h^j.$$

(This process of construction of approximate equation is generalized in Chapter 11.)

2-4. Existence, Uniqueness, and Convergence Properties

Here we prove the theorem of Lax-Milgram and the convergence of the solutions of (2-8) to the one of (2-7).

THEOREM 2-1   Let us assume V separable and $a(u, v)$ continuous and V-elliptic:

(2-12) $|a(u, v)| \le M \|u\| \|v\|$; $a(v, v) \ge c \|v\|^2$ for any $v \in V$.

Then the variational equations (2-7) and (2-8) have unique solutions u and u_h belonging to V and V_h, respectively. The error is estimated by

(2-13) $$\|u - p_h u_h\| \le Mc^{-1} \|u - p_h \hat{r}_h u\|$$

and converges to 0 like the best approximation of u by the elements of P_h when the approximations $(V_h, p_h, \hat{r}_h)$ are convergent.   ▲

Proof   Let us first notice that the V-ellipticity of $a(u, v)$ implies the V_h-ellipticity of $a(p_h u_h, p_h v_h)$.

1. *Proof of uniqueness*
If v and w are two solutions of (2-7), then

$$c \|v - w\|^2 \le a(v - w, v - w) = 0$$

and thus $v = w$. Therefore any V-elliptic variational equation has at most one solution.

2. *Proof of existence*
We begin by proving that (2-8) has a (unique) solution when V_h is a finite-dimensional space. Indeed, the V_h-ellipticity of $a(p_h u_h, p_h v_h)$ implies that the operator A_h defined in Proposition 2-1 is positive definite and thus invertible.

We deduce from the existence of a solution in the finite-dimensional case the existence of a solution in the case where V is a separable Hilbert space using the convergent Galerkin approximations (V_h, p_h, r_h) introduced in Section 2-1.

By the V-ellipticity, $p_h u_h$ lies in a ball of V because

$$c \|p_h u_h\|^2 \le a(p_h u_h, p_h u_h) = l(p_h u_h) \le \|l\|_* \|p_h u_h\|$$

where $\|l\|_* = \sup |l(v)| / \|v\|$.

Since any ball of a Hilbert space is weakly compact (see Theorem 2.1-10) we can extract from the sequence h a subsequence k such that

$p_k u_k$ converges weakly to an element u of V as k converges to 0.

Let us fix v in V and take $v_h = r_h v$ in (2-8). Since $p_k r_k v$ converges strongly to v and $p_k u_k$ converges weakly to u, the equalities

$$a(p_k u_k, p_k r_k v) = l(p_k r_k v)$$

converge to

$$a(u, v) = l(v) \qquad \text{for any} \quad v \in V.$$

Therefore this weak limit u is the (unique) solution of (2-7).

3. *Convergence and estimate of the error*

Now let $(V_h, p_h, \hat{r}_h)$ be any approximation of V, and u and u_h the solutions of (2-7) and (2-8), respectively. Then, since

$$a(u - p_h u_h, p_h v_h) = l(p_h v_h) - l(p_h v_h) = 0 \qquad \text{for any} \quad v_h \in V_h,$$

we obtain

$$a(u - p_h u_h, u - p_h u_h) = a(u - p_h u_h, u - p_h \hat{r}_h u),$$

which implies by the continuity and the V-ellipticity of $a(u, v)$ that

$$c \, \| u - p_h u_h \|^2 \leq M \, \| u - p_h \hat{r}_h u \| \, \| u - p_h u_h \|.$$

We thus have proved our theorem. ∎

Let us emphasize the problem of estimates of errors.

2-5. Estimates of Global Error

The inequality (2-13) gives a first estimate of the error between the solutions u and u_h of (2-7) and (2-8), which depends on the unknown solution u. Thus the problem arises of finding estimates depending only on a subset which we know a priori to contain the solution u.

Definition 2-5 Let us introduce a Hilbert space U contained in V with a stronger topology. Then we associate with U the *global error* defined by

$$\sup_{\| u \|_U \leq 1} \| u - p_h u_h \|_V = \sup_{u \in U} \frac{\| u - p_h u_h \|_V}{\| u \|_U}. \qquad \blacktriangle$$

Definition 2-6 The *error function of the subspace* P_h or of the *prolongation* p_h is denoted and defined by

$$(2\text{-}14) \qquad e_U^V(P_h) = \sup_{\| u \|_U \leq 1} \, \inf_{u_h \in P_h} \| u - u_h \|_V = \sup_{\| u \|_U \leq 1} \| u - t_h u \|_V$$

where t_h denotes the orthogonal projector onto P_h. When $P_h = p_h V_h$ and $t_h = p_h \hat{r}_h$, we can also write

$$(2\text{-}15) \qquad e_U^V(P_h) = e_U^V(p_h) = \sup_{\| u \|_U \leq 1} \| u - p_h \hat{r}_h u \|_V. \qquad \blacktriangle$$

We deduce from the Theorem 2-1 the following:

THEOREM 2-2 Let us assume V separable and $a(u, v)$ V-elliptic. If the solution u of (2-7) actually belongs to a subspace U of V, the following estimate of the global error $u - p_h u_h$ holds:

$$(2\text{-}16) \qquad \|u - p_h u_h\| \leq Mc^{-1} \|u\|_U \, e_U^V(p_h). \qquad \blacktriangle$$

Does this error converge to 0? To answer this question, we can either estimate the error function for given examples or use Theorem 2-3. Let us assume that the injection from U into V is *compact*—that is, that the unit ball of U is compact in V. (If the interval I is bounded, the injection from $H^m(I)$ into $H^n(I)$ is compact for $m > n$; see Theorem 6.3-10.)

THEOREM 2-3 Assume the injection from U into V compact and the approximations $(V_h, p_h, \hat{r}_h)$ convergent. Then the error functions $e_U^V(p_h)$ converge to 0. $\qquad \blacktriangle$

Proof If not, there would exist $\varepsilon > 0$ such that $e_U^V(p_h) \geq \varepsilon > 0$ for h small enough. On the other hand, since the unit ball of U is compact in V, the supremum involved in the definition of the error function is achieved on a point $u(h)$:

$$e_U^V(p_h) = \|u(h) - p_h \hat{r}_h u(h)\|; \quad \|u(h)\|_U \leq 1.$$

Then, since the sequence $u(h)$ is compact in V, we can extract a subsequence k from h such that

$u(k)$ converges to an element u, strongly in V, and weakly in U.

Thus we can write

$$
\begin{aligned}
0 < \varepsilon &\leq \|u(k) - p_k \hat{r}_k u(k)\| \\
&\leq \|u(k) - u\| + \|u - p_k \hat{r}_k u\| + \|p_k \hat{r}_k(u - u(k))\| \\
&\leq \|u - p_k \hat{r}_k u\| + 2 \|u - u(k)\|, \text{ which converges to 0.}
\end{aligned}
$$

We thus have obtained a contradiction. $\qquad \blacksquare$

2-6. What Kind of Approximations Should Be Chosen?

The introduction of error functions furnishes a way of comparing different approximations of a space V. Assuming that the spaces U and V are fixed, we look for the prolongation for which the error function is smallest. Let us set

$$(2\text{-}17) \qquad E_U^V(V_h) = \inf_{p_h \in L(V_h, V)} e_U^V(p_h).$$

If the dimension of V_h is equal to $n(h) < \infty$, we can write equivalently

$$(2\text{-}18) \qquad E_U^V(n(h)) = e_U^V(V_h) = \inf_{\dim(P_h) = n(h)} e_U^V(P_h).$$

Definition 2-7   We say that $E_U^V(V_h)$ is the V_h-*width of the injection from U into V* and that $E_U^V(n(h))$ is the $n(h)$-*width of the injection from U into V.*

▲

One of our next problems is to characterize the approximation achieving the $n(h)$-width and, more generally, to recognize among all approximations the ones behaving like the $n(h)$-width: There exists a constant k such that

$$(2\text{-}19) \qquad e_U^V(p_h) \leq kE_U^V(n(h) - 1); \; \dim(V_h) = n(h).$$

(See Theorems 2.2-3 and 2.3-2.)

Using such approximations in the construction of internal approximation of variational equations, we deduce the estimate

$$(2\text{-}20) \quad \|u - p_h u_h\| \leq Mc^{-1}k \, \|u\|_U \, E_U^V(n(h) - 1); \; \dim(V_h) = n(h),$$

where only the constant k (and not the order of convergence $E_U^V(n(h) - 1)$) will depend on the choice of the approximation. (The study of internal approximations of variational equations continues in Section 3.1-7.)

3.   EXAMPLES OF APPROXIMATIONS OF SOBOLEV SPACES

3-1.   Piecewise-Linear Approximations of the Sobolev Space $H^1(I)$

Simple approximations of $H^1(I)$ can be constructed which will, in turn, play an important role in the construction of finite-difference schemes approximating a differential equation. Let us set $h = 1/n$, associating with it the $n + 1$ knots jh $(0 \leq j \leq n)$ of the interval $I = (0, 1)$, and introducing the following functions:

$$(3\text{-}1) \quad \begin{cases} \text{(i)} \quad \theta_h^j = \text{characteristic function of the interval } (jh, (j + 1)h), \\[2mm] \text{(ii)} \; \lambda_h^0(x) = \left(-\dfrac{x}{h} + 1\right)\theta_h^0(x), \\[2mm] \text{(iii)} \; \lambda_h^j(x) = \left(\dfrac{x}{h} - j + 1\right)\theta_h^{j-1}(x) + \left(-\dfrac{x}{h} + j + 1\right)\theta_h^j(x), \\[4mm] \hspace{6cm} 1 \leq j \leq n - 1, \\[2mm] \text{(iv)} \; \lambda_h^n(x) = \left(\dfrac{x}{h} - n + 1\right)\theta_h^{n-1}(x). \end{cases}$$

The piecewise-linear approximations of $H^1(I)$ are defined by

$$(3\text{-}2) \quad \begin{cases} \text{(i)} \qquad V_h^1 = R^{n+1}; \; u_h = (u_h^j)_{0 \leq j \leq n} \in V_h^1, \\[2mm] \text{(ii)} \; p_h^1 u_h(x) = \displaystyle\sum_{j=0}^n u_h^j \lambda_h^j(x) \quad \text{for} \quad u_h \in V_h^1, \\[2mm] \text{(iii)} \qquad r_h^0 u = (u(jh))_j; \, 0 \leq j \leq n \quad \text{for } u \in H^1(I). \end{cases}$$

LEMMA 3-1 The function $p_h{}^1 u_h$ is a continuous piecewise-linear function equal to

$$(3\text{-}3) \qquad p_h{}^1 u_h = \sum_{j=0}^{n-1} \left[\left(u_h{}^j + (u_h{}^{j+1} - u_h{}^j) \left(\frac{x}{h} - j \right) \right) \right] \theta_h{}^j(x). \qquad \blacktriangle$$

Proof The proof of this lemma is left as an exercise. ■

In addition, let us define the piecewise-constant approximations of the space $L^2(I)$

$$(3\text{-}4) \qquad \begin{cases} \text{(i)} \quad V_h{}^0 = R^n, \\[2mm] \text{(ii)} \quad p_h{}^0 u_h(x) = \displaystyle\sum_{j=0}^{n-1} u_h{}^j \theta_h{}^j(x), \\[2mm] \text{(iii)} \quad r_h{}^1 u = \left(\dfrac{1}{h} \displaystyle\int_{jh}^{(j+1)h} u(x)\, dx \right)_{0 \le j \le n-1} \end{cases}$$

Let ∇_h be the finite-difference operator defined by

$$(3\text{-}5) \qquad (\nabla_h u_h)^j = \frac{1}{h}(u_h{}^{j+1} - u_h{}^j); \; 0 \le j \le n-1.$$

It maps $V_h{}^1$ into $V_h{}^0$.

LEMMA 3-2 The operators $D, \nabla_h, p_h{}^1, p_h{}^0, r_h{}^1,$ and $r_h{}^0$ are related by the following commutation formulas:

$$(3\text{-}6) \qquad \begin{cases} \text{(i)} \quad D p_h{}^1 u_h = p_h{}^0 \nabla_h u_h \\[2mm] \text{(ii)} \quad \nabla_h r_h{}^0 u = r_h{}^1 Du \end{cases} \qquad \blacktriangle$$

Proof The proof of this lemma is left as an exercise. (See Remark 3-2.) ■

3-2. Estimates of Error Functions of Piecewise-Linear Approximations

In this section, any positive constant independent of h is denoted by c.

THEOREM 3-1 The following estimates hold:

$$(3\text{-}7) \qquad \begin{cases} \text{(i)} \quad |u - p_h{}^1 r_h{}^0 u| \le ch\,|Du| & \text{if } \; u \in H^1(I), \\[2mm] \text{(ii)} \quad |u - p_h{}^1 r_h{}^0 u| \le ch^2\,|D^2 u| & \text{if } \; u \in H^2(I), \end{cases}$$

and therefore, if we set $U = H^2(I)$, $V = H^1(I)$, and $H = L^2(I)$,

$$(3\text{-}8) \qquad e_V{}^H(p_h{}^1) = e_1{}^0(p_h{}^1) \le ch; \; e_U{}^H(p_h{}^1) = e_2{}^0(p_h{}^1) \le ch^2. \qquad \blacktriangle$$

Proof We can write $u - p_h^1 r_h^0 u$ in the following form:

$$\sum_{j=0}^{n-1} \theta_h^j \cdot \left\{ u(x) - u(jh) - [u((j+1)h) - u(jh)]\left(\frac{x}{h} - j\right) \right\}.$$

If $u \in H^1(I)$, we can replace $u(x)$ by $u(jh) + \int_{jh}^x Du(t)\, dt$ for any j. Thus

$$u - p_h^1 r_h^0 u = \sum_{j=0}^{n-1} \theta_h^j \left[\int_{jh}^x Du(t)\, dt - \left(\frac{x}{h} - j\right) \left(\int_{jh}^{(j+1)h} Du(t)\, dt \right) \right].$$

On each interval $(jh, (j+1)h)$ the Cauchy-Schwarz inequality implies

$$(3\text{-}9) \quad \begin{cases} \text{(i)} \ \left| \int_{jh}^x Du(t)\, dt \right|^2 \leq h\left(\frac{x}{h} - j\right) \int_{jh}^x |Du(t)|^2\, dt \\[2mm] \qquad\qquad\qquad\qquad \leq h \int_{jh}^{(j+1)h} |Du(t)|^2\, dt, \\[4mm] \text{(ii)} \ \left(\frac{x}{h} - j\right)^2 \left| \int_{jh}^{(j+1)h} Du(t)\, dt \right|^2 \leq h \int_{jh}^{(j+1)h} |Du(t)|^2\, dt. \end{cases}$$

Hence

$$(3\text{-}10) \qquad |(u - p_h^1 r_h^0 u)(x)|^2 \leq 4h \sum_{j=0}^{n-1} \theta_h^j \int_{jh}^{(j+1)h} |Du(t)|^2\, dt.$$

If we now integrate (3-10) on $(0, 1)$, we obtain [(3-7)(i)].

If we assume that $u \in H^2(I)$, we obtain in an analogous way [(3-7)(ii)], replacing $u(x)$ by its Taylor expansion on each interval $(jh, (j+1)h)$:

$$u(x) = u(jh) + (x - jh)\, Du(jh) + \int_{jh}^x (x - t)\, D^2 u(t)\, dt.$$

Using the Cauchy-Schwarz inequality, we obtain the following estimates:

$$(3\text{-}11) \quad \begin{cases} \text{(i)} \ \left| \int_{jh}^x (x - t)\, D^2 u(t)\, dt \right|^2 \leq \frac{h^3}{3} \int_{jh}^{(j+1)h} |D^2 u(t)|^2\, dt, \\[4mm] \text{(ii)} \ \left(\frac{x}{h} - j\right)^2 \left| \int_{jh}^{(j+1)h} ((j+1)h - t)\, Du(t)\, dt \right|^2 \\[4mm] \qquad\qquad\qquad\qquad \leq \frac{h^3}{3} \int_{jh}^{(j+1)h} |D^2 u(t)|^2\, dt, \end{cases}$$

and then

$$(3\text{-}12) \qquad |(u - p_h{}^1 r_h{}^0 u)(x)|^2 \leq \tfrac{4}{3} h^3 \sum_{j=0}^{n-1} \theta_h{}^j \int_{jh}^{(j+1)h} |D^2 u(t)|^2 \, dt.$$

Integrating (3-12) on (0, 1), we obtain [(3-7)(ii)]. ∎

THEOREM 3-2   The following estimate holds:

$$(3\text{-}13) \qquad |D(u - p_h{}^1 r_h{}^0 u| \leq ch \, |D^2 u| \qquad \text{if} \quad u \in H^2(I),$$

and thus, if $U = H^2(I)$ and $V = H^1(I)$,

$$(3\text{-}14) \qquad e_U{}^V(p_h{}^1) = e_2{}^1(p_h{}^1) \leq ch. \qquad \blacktriangle$$

Proof The proof is left as an exercise. Use Lemma 3-2 and estimates analogous to those used previously. ∎

Finally, it is an other exercise to prove:

THEOREM 3-3

$$(3\text{-}15) \quad \begin{cases} \text{(i)} & \lim_{h \to 0} |u - p_h{}^1 r_h{}^0 u| = 0 & \text{if} \quad u \in L^2(I), \\[2mm] \text{(ii)} & \lim_{h \to 0} |u - p_h{}^0 r_h{}^1 u| = 0 & \text{if} \quad u \in L^2(I), \\[2mm] \text{(iii)} & \lim_{h \to 0} |D(u - p_h{}^1 r_h{}^0 u)| = 0 & \text{if} \quad u \in H^1(I). \end{cases} \qquad \blacktriangle$$

Remark 3-1 When we construct piecewise-polynomial (of degree m) approximations of the Sobolev spaces $H^m(I)$, the foregoing results will appear as particular cases. (See Chapters 4 and 5.) ∎

4. EXAMPLES OF APPROXIMATE EQUATIONS

4-1. Construction of a Finite-Difference Scheme

In this section we apply Theorem 2-1 to problem (1-15)—(or 1-16)—using piecewise-linear approximations of the space $V = H^1(I)$. (See Sections 3-1 and 3-2.)

Let u_h be the solution of

$$(4\text{-}1) \quad \begin{cases} \text{(i)} \ u_h = (u_h{}^j)_j \in R^{n+1}; \ h = \dfrac{1}{n+1}, \\[2mm] \text{(ii)} \ \displaystyle\int_I a(x)\, Dp_h{}^1 u_h \, Dp_h{}^1 v_h \, dx + \int_I b(x) p_h{}^1 u_h \cdot p_h{}^1 v_h \, dx \\[3mm] \qquad = \displaystyle\int_I f(x) p_h{}^1 v_h \, dx + \alpha_1 a(1) p_h{}^1 v_h(1) - \alpha_0 a(0) p_h{}^1 v_h(0) \\[3mm] \qquad\qquad\qquad\qquad\qquad\qquad\qquad\qquad \text{for any} \ \ v_h \in V_h \end{cases}$$

We thus deduce (by Theorems 3-1–3-2) the following corollary:

COROLLARY 4-1 If we assume (1-23), then there exists a unique solution u of (1-15) and a unique solution u_h of (4-1) such that

$$(4\text{-}2) \qquad \lim_{h \to 0}(|u - p_h{}^1 u_h|^2 + |Du - p_h{}^0 \nabla_h u_h|^2) = 0.$$

If, moreover, the coefficient $a(x)$ satisfies

$$(4\text{-}3) \qquad Da(x) \in L^\infty(I),$$

then the solution u belongs to the space $H^2(I)$ and we can estimate the error in the following way:

$$(4\text{-}4) \qquad (|u - p_h{}^1 u_h|^2 + |Du - p_h{}^0 \nabla_h u_h|^2)^{1/2} \le c \, \|u\|_2 \, h. \qquad \blacktriangle$$

Let us now compute the coefficients of the matrix A_h and the vector f_h of the operational formulation of (4-1), setting

$$(4\text{-}5) \quad \begin{cases} \text{(i)} \ b_j = -\dfrac{1}{h^2}\displaystyle\int_{jh}^{(j+1)h} a(x)\, dx + \int_{jh}^{(j+1)h} b(x)\left(\dfrac{x}{h} - j\right)\left(\dfrac{x}{h} - j - 1\right) dx; \\[3mm] \qquad\qquad\qquad\qquad\qquad\qquad\qquad\qquad\qquad 0 \le j \le n-1, \\[3mm] \text{(ii)} \ a_0 = \dfrac{1}{h^2}\displaystyle\int_0^h a(x)\, dx + \int_0^h b(x)\left(\dfrac{x}{h} - 1\right)^2 dx, \\[3mm] \text{(iii)} \ a_j = \dfrac{1}{h^2}\displaystyle\int_{(j-1)h}^{(j+1)h} a(x)\, dx + \int_{(j-1)h}^{(j+1)h} b(x)\left(\dfrac{x}{h} - j - 1\right)^2 dx; \\[3mm] \qquad\qquad\qquad\qquad\qquad\qquad\qquad\qquad\qquad 1 \le j \le n-1, \\[3mm] \text{(iv)} \ a_n = \dfrac{1}{h^2}\displaystyle\int_{(n-1)h}^1 a(x)\, dx + \int_{(n-1)h}^1 b(x)\left(\dfrac{x}{h} - n - 1\right)^2 dx, \end{cases}$$

and

$$\text{(4-6)} \begin{cases} \text{(i)} \ l_h^0 = \int_0^h f(x)\left(-\frac{x}{h}+1\right) dx - \alpha_0 a(0), \\[2ex] \text{(ii)} \ l_h^j = \int_{(j-1)h}^{jh} f(x)\left(\frac{x}{h}-j+1\right) dx \\[2ex] \qquad\qquad + \int_{jh}^{(j+1)h} f(x)\left(-\frac{x}{h}+j+1\right) dx; \ 1 \le j \le n-1, \\[2ex] \text{(iii)} \ l_h^n = \int_{(n-1)h}^1 f(x)\left(\frac{x}{h}-n+1\right) dx + \alpha_1 a(1). \end{cases}$$

COROLLARY 4-2 The solution u_h of (4-1) is also the solution of $A_h u_h = l_h$, where A_h is the matrix

$$\text{(4-7)} \quad \begin{Vmatrix} a_0 & b_0 & 0 & 0 & 0 & \cdots & \cdots & 0 & 0 \\ 0 & b_0 & a_1 & b_1 & 0 & \cdots & \cdots & 0 & 0 \\ 0 & 0 & 0 & 0 & b_{j-1} & a_j & b_j & 0 & 0 \\ 0 & 0 & 0 & 0 & 0 & 0 & 0 & b_{n-1} & a_n \end{Vmatrix}$$

and l_h the vector of components l_h^j defined by (4-6). ▲

Proof By Lemma 3-1 and 3-2, we can write

$$a(p_h u_h, p_h v_h) = \sum_{j=0}^{n-1} \int_{jh}^{(j+1)h} g_j(x) \, dx$$

where

$$\begin{cases} g_j(x) = (u_h^{j+1} - u_h^j)(v_h^{j+1} - v_h^j)\left[\frac{a(x)}{h^2} + b(x)\left(\frac{x}{h}-j\right)^2\right] + u_h^j v_h^j b(x) \\[2ex] \qquad + (u_h^j v_h^{j+1} - 2v_h^j u_h^j + u_h^{j+1} v_h^j) b(x)\left(\frac{x}{h}-j\right) \end{cases}$$

Therefore, by (4-5), we obtain

$$a(p_h u_h, p_h v_h) = (b_0 u_h^1 + a_0 u_h^0) v_h^0 + \sum_{j=1}^{n-1} (b_j u_h^{j+1} + a_j u_h^j + b_{j-1} u_h^{j-1}) v_h^j$$

$$+ (a_n u_h^n + b_{n-1} u_h^{n-1}) v_h^n$$

$$= \sum_{j=0}^{n} (A_h u_h)_j v_h^j.$$

4-2. A Simpler Finite-Difference Scheme

Not all finite-difference schemes are internal approximations of (1-15). For example, the bilinear form $a(u, v)$ is the sum of the form $\int a(x)Du\, Dv\, dx$, which is defined on $H^1(I) \times H^1(I)$ and of the form $\int b(x)uv\, dx$, which is defined on $L^2(I) \times L^2(I)$. On the other hand, $\int f(x)v(x)\, dx$ is defined on $L^2(I)$.

To obtain an approximation of (1-15), we can replace the functions u and v in $\int a(x)\, Du\, Dv\, dx$ by approximants in $H^1(I)$ and replace the functions u and v in $\int b(x)uv\, dx$ and $\int f(x)v\, dx$ by *approximants belonging only in the larger space* $L^2(I)$. If we perceive "simpler" functions in a "larger space," we can hope to obtain in this way "simpler" approximations of (1-15). Naturally, we shall obtain a "weaker" convergence. (Roughly, the convergence holds in the space containing both the solution and the approximants.)

Let us consider for example the following approximate problem: look for $u_h = (u_h{}^j)_j \in R^{n+1}$ satisfying the following variational equation:

$$(4\text{-}8) \quad \int a(x)\, Dp_h{}^1 u_h\, Dp_h{}^1 v_h\, dx + \int b(x) p_h{}^0 u_h p_h{}^0 v_h\, dx$$

$$= \int f(x) p_h{}^0 v_h\, dx$$

$$\text{for any} \quad v_h \in V_h.$$

(We choose $\alpha_0 = \alpha_1 = 0$ for simplicity.)

Check as an exercise that this is a well defined elliptic variational equation which thus has a unique solution u_h. We learn in Chapter 11 that u_h converges to the solution u of (1-15) in some sense. This approximate solution is also the solution of the operational equation $A_h u_h = l_h$ where A_h is the matrix (4-7) with the following coefficients:

$$(4\text{-}9) \quad \begin{cases} \text{(i)} \;\; b_j = -\dfrac{1}{h^2} \displaystyle\int_{jh}^{(j+1)h} a(x)\, dx;\; 0 \leq j \leq n-1, \\[3mm] \text{(ii)} \;\; a_0 = \dfrac{1}{h^2} \displaystyle\int_0^h a(x)\, dx + \int_0^h b(x)\, dx, \\[3mm] \text{(iii)} \;\; a_j = \dfrac{1}{h^2} \displaystyle\int_{(j-1)h}^{(j+1)h} a(x)\, dx + \int_{jh}^{(j+1)h} b(x)\, dx;\; 1 \leq j \leq n-1, \\[3mm] \text{(iv)} \;\; a_n = \dfrac{1}{h^2} \displaystyle\int_{(n-1)h}^1 a(x)\, dx, \end{cases}$$

and where l_h is the vector of coefficients

$$(4\text{-}10) \qquad l_h{}^j = \int_{jh}^{(j+1)h} f(x)\, dx; \ 0 \le j \le n \le 1; \ l_h{}^n = 0.$$

The convergence properties of the approximate solutions u_h of (4-8) are described in Theorem 11.4-1; see also Remark 11.4-4.

CHAPTER 2

Approximations of Hilbert Spaces

In this chapter, which is devoted to the study of approximations of Hilbert spaces (introduced in Sections 1.2-1 and 1.2-2), we associate with the error function of a prolongation its stability function, defining as well the quasi-optimal optimal approximation (Section 2).

We study in Section 3 the approximations achieving the n-width: these optimal approximations are Galerkin approximations. Section 4 introduces the optimal prolongation (restriction) associated with a given prolongation (restriction) and examines their properties.

Since we shall now be using several results of functional analysis, we have devoted Section 1-1 of this chapter to summary of the theorems to be used later.

1. HILBERT SPACES AND THEIR DUALS

Since we employ extensively the duality theory in the framework of Hilbert spaces, we begin by defining the dual of a Hilbert space, the duality pairing, and the canonical isometry from this space onto its dual.

We do not always identify a Hilbert space with its dual, and therefore we refer to the spaces for which this identification holds as "pivot spaces." Indeed, we deal with several Hilbert spaces ordered by inclusion and only one is chosen to be a pivot space.

After deducing the Hahn–Banach theorem and a frequently employed criterion of density from the theorem of projections we characterize the duals of dense subspace, closed subspaces, factor spaces, product spaces, and the domains of operators.

Also, we prove several useful consequences of the uniform boundedness theorem and of the Banach theorem. Finally, recalling the Riesz–Fredholm alternative, we devote Section 1-16 to the study of V-elliptic and coercive operators.

It should be pointed out, however, that we do not prove here the theorem of projections, the Riesz theorem, the uniform-boundedness theorem, the

Banach theorem, or and the Riesz–Fredholm theorem (see, e.g., Dunford–Schwartz [1958–1971], Treves [1967], Yosida [1968], etc).

1-1. Dual of a Hilbert Space and Canonical Isometry

If V is a Hilbert space, we shall denote its dual by V' and set

$$(1\text{-}1) \qquad (f, v) = (f, v)_V = f(v) \; ; f \in V', v \in V,$$

the value of the continuous linear functional f at v. The bilinear form (f, v) is called the *duality pairing on* $V' \times V$. If $\|u\|$ is the norm on V, the space V' is supplied with the dual norm

$$(1\text{-}2) \qquad \|f\|_* = \sup_{v \in V} \frac{|(f, v)|}{\|v\|} \; ; |(f, v)| \le \|f\|_* \|v\|.$$

By the Riesz theorem, we know that there exists an isometry J from V onto its dual V'. If $((u, v))$ denotes the inner product on V, this isometry is defined by:

$$(1\text{-}3) \qquad ((u, v)) = (Ju, v) \qquad \text{for any} \quad u, v \in V.$$

The dual space V' is a Hilbert space for the inner product

$$(1\text{-}4) \qquad ((f, g))_* = ((J^{-1}f, J^{-1}g)) = (f, J^{-1}g); f, g \in V',$$

and we can check that the norms $\|f\|_* = \sqrt{((f,f))_*}$ and $\|v\| = \sqrt{((v, v))}$ are related by (1-2).

If we want to identify V with its dual V', we have to identify J with the identity mapping or, equivalently, the inner product $((u, v))$ with the duality pairing $(f, v) = ((f, v))$ on $V \times V$. In this case, we say that V is a *pivot space*.

We do not necessarily consider a Hilbert space to be a pivot space. Indeed, when the duality pairing is "simpler" than the inner product, it is preferable to use the duality pairing rather than the inner product (e.g., in such problems as the transposition of operators). Indeed, let V and W be two Hilbert spaces, V' and W' be their duals for the duality pairings $(f, v)_V$ and $(g, u)_W$, respectively. If A is a continuous linear operator from V into W, its transpose A' mapping W' into V' is defined by

$$(1\text{-}5) \quad (A'g, v)_V = (g, Av)_W \qquad \text{for any} \quad g \in W' \quad \text{and} \quad v \in V.$$

If we consider now the spaces V and W as pivot spaces, we say that the transpose of A (for the duality pairings identified with the inner products) is the adjoint A^* of A: the operator A^* mapping W into V defined by

$$(1\text{-}6) \quad ((A^*g, v))_V = ((g, Av))_W \qquad \text{for any} \quad g \in W \quad \text{and} \quad v \in V.$$

If J and K are the canonical isometries from V onto V' and from W onto W', the adjoint A^* is related to the transpose A' by:

$$(1\text{-}7) \qquad\qquad A' = JA^*K^{-1}.$$

1-2. Example: Finite-Dimensional Hilbert Spaces

Let us examine the Hilbert space $V = R^n$ (with the canonical basis) supplied with the inner product

$$(1\text{-}8) \qquad\qquad ((u, v)) = \sum k_{ij} u^i v^j$$

where the matrix $K = (k_{ij})$ is symmetric and positive definite. In addition, consider the inner product associated with the identity matrix

$$(1\text{-}9) \qquad\qquad (u, v) = \sum u^i v^i.$$

Because of its simplicity, this inner product is usually used as duality pairing on $R^n \times R^n$. Those two inner products are related by

$$((u, v)) = (Ku, v).$$

If $A = (a_{ij})$ is a matrix, its transpose A' [with respect to the duality pairing (1-9)] is the matrix $A' = (a_{ji})$ and its adjoint A^* is the matrix $A^* = KA'K^{-1}$. The dual V' of V is the space R^n supplied with the inner product

$$(1\text{-}10) \qquad\qquad ((f, g))_* = \sum k_{ij}^* f^i g^j$$

where the k_{ij}^* are the entries of the matrix K^{-1}.

1-3. Hahn–Banach Theorem

If V is a Hilbert space, the Hahn–Banach theorem is a simple consequence of the existence of orthogonal projectors on closed subspaces. Recall, for example, that the norm of an orthogonal projector is equal to 1 and that its transpose is also an orthogonal projector.

THEOREM 1-1 (*Hahn–Banach*) Let P be a closed subspace of a Hilbert space V. If f is a continuous linear functional defined on P, there exists a continuous linear functional g on V extending f with the same norm as f:

$$(1\text{-}11) \qquad g(u) = f(u) \quad \text{for all} \quad u \in P; \; \|g\|_{V'} = \|f\|_{P'}. \qquad\qquad \blacktriangle$$

Proof Let t be the orthogonal projector onto P and $t' \in L(P', V')$ its transpose. Then the functional $g = t'f$ satisfies (1-11). First

$$(1\text{-}12) \quad g(u) = (t'f, u) = (f, tu) = (f, u) = f(u), \quad \text{for all} \quad u \in P.$$

Second

(1-13) $$\|f\|_{P'} = \sup_{v \in P} \frac{|(g, v)|}{\|v\|} \leq \sup_{v \in V} \frac{|(g, v)|}{\|v\|} = \|g\|_{V'}.$$

Finally, since $\|tv\| \leq \|v\|$ for any $v \in V$, we deduce

(1-14) $$\|g\|_{V'} = \sup_{v \in V} \frac{|(g, v)|}{\|v\|} \leq \sup_{v \in V} \frac{|(f, tv)|}{\|tv\|} = \|f\|_{P'}.$$

(For the Hahn–Banach theorem in general vector spaces, see Treves [1967], p. 181.) ∎

Another consequence of the existence of orthogonal projectors is provided in the following important characterization of a dense subspace.

THEOREM 1-2 Let V be a Hilbert space and D a subspace of V. Then D is dense in V if and only if any continuous linear functional on V vanishing identically on D vanishes identically on V also. ▲

Proof The necessity is obvious; let us prove the sufficiency. If $f \in V'$ vanishes on D, it vanishes also on its closure $\bar{D}$. Let t be the orthogonal projector on $\bar{D}$. Then $t'f = 0$ implies that $f = 0$ [since $(t'f, v) = (f, tv) = 0$]. Then the orthogonal projector $t' = 1$ and thus $t = 1$, which implies that $\bar{D} = V$. ∎

1-4. Dual of a Dense Subspace

Let U and V be two Hilbert spaces such that

(1-15) $\qquad U \subset V$, the injection is continuous and dense.

That is to say, the canonical injection π from U into V is continuous (i.e., $\|u\|_V \leq k \|u\|_U$ for any $u \in U$) and that U is dense in V. Let U' and V' be the duals of U and V.

THEOREM 1-3 If we assume (1-15), then there exists an operator π' from V' into U' such that

(1-16)
$$\begin{cases} \text{(i) } \pi' \text{ is a continuous linear operator,} \\ \text{(ii) } \pi' \text{ is injective} \\ \text{(iii) } \pi'V' \text{ is dense in } U'. \end{cases}$$
▲

Proof Let us define π' by

(1-17) $(\pi'f, u)_U = (f, \pi u)_V$ for any $u \in U.$

Since π is the canonical injection, $\pi u = u$ for any $u \in U$, and π' is the transpose of π. It is clear that π' is continuous and linear. It is injective because, if $\pi'f = 0$, then $(\pi'f, u)_U = (f, u)_V = 0$ for any $u \in U$. But since U is dense in V, $(f, u)_V = 0$ for any $u \in V$; therefore, $f = 0$.

We use Theorem 1-2 to prove the density of $\pi'V'$ in U' by noting that, since U is a Hilbert space, a linear form u on U' is an element of U. Thus let $u \in U'' = U$ be a linear form on U' vanishing on the subspace $\pi'V'$. Then $(\pi'f, u)_U = (f, u)_V = 0$ for any $f \in V'$. Thus $u = 0$ and $\pi'V'$ is dense in U'. ∎

Since π' is a one-to-one map which is the transpose of the canonical injection from U into V, π' can be identified with the canonical injection from V' into U'. In so doing, we identify the duality pairing $(f, v)_U$ on $U' \times U$ with the (unique) extension by continuity of the duality pairing $(f, v)_V$ defined on $V' \times U \subset V' \times V$:

THEOREM 1-4 Let U and V be two Hilbert spaces satisfying (1-15). Then we can identify the dual V' with a dense subspace of U' and the duality pairing $(f, v)_U$ on $U' \times U$ with the extension of $(f, v)_V$; in short:

(1-18) $V' \subset U'$, the injection is dense and continuous. ▲

1-5. Imbedding of a Space into Its Dual

If we choose for the Hilbert space V a pivot space $V = V' = H$ supplied with an inner product (u, v)—identified with duality pairing (f, v) on $H \times H$— we obtain the following theorem:

THEOREM 1-5 Let U be a Hilbert space and H be a pivot space such that

(1-19) $U \subset H$, the injection is continuous and dense.

Then we can identify U and H with dense subspaces of the dual U'

(1-20) $U \subset H \subset U',$

the injections are continuous and dense, and the duality pairing on $U' \times U$ is identified with the unique extension of the inner product (u, v) of H. ▲

Proof We apply Theorem 1-4 with $V = V' = H$. We have to prove the density of U into U' again using Theorem 1-2. Let us consider a continuous linear functional $u \in U'' = U$ on U', vanishing on U. Thus $(u, v) = 0$ for any $v \in U$ and, since U is dense in H, $(u, v) = 0$ for any $v \in H$. Therefore, $u = 0$ and U is dense in U'. ■

Remark 1-1 Theorems 1-3, 1-4, and 1-5 are actually consequences of Theorem 1-2 and the fact that U is equal to its bidual U''. Such spaces are called *reflexive spaces*. Since Theorem 1-2 holds for reflexive spaces (see Treves [1967], p. 86), Theorems 1-3, 1-4, and 1-5 hold when U is reflexive. ■

1-6. Example: Imbedding of Spaces of Functions into Spaces of Distributions

If I is an interval, we shall generally choose for the pivot space of functions defined on I the space $L^2(I)$ supplied with the inner product

$$(u, v) = \int_I u(x)v(x) \, dx.$$

Let us consider now the space Φ_0 of indefinitely differentiable functions with compact support. The topology of Φ_0 is the one for which ϕ_n converges to 0 if

(1-21) $\begin{cases} \text{(i) support } (\phi_n) \subset J \subset I & \text{where } J \text{ is a compact interval,} \\ \text{(ii) the derivatives } D^j\phi_n \text{ converge uniformly to 0 on } J \text{ for any } j \end{cases}$

(see Trêves [1967], p. 131).

Supplied with such a topology, the space Φ_0 is a reflexive space. Theorem 1-5 permits us to make more precise what we said in Section 1.1-2 (see also Trêves [1967], pp. 243 and 302). Indeed, the space Φ_0 is a dense subspace of the space $L^2(I)$ and thus

COROLLARY 1-1 We can identify the spaces Φ_0 and $L^2(I)$ with dense subspaces of the space of distributions Φ_0', dual of Φ_0, and the duality pairing (f, v) on $\Phi_0' \times \Phi_0$ with the extension of the inner product $(u, v) = \int_I u(x)v(x) \, dx$:

(1-22) $\Phi_0 \subset L^2(I) \subset \Phi_0'$, the injections are continuous and dense. ▲

But there is yet another consequence of Theorem 1-5.

Definition 1-1 We say that a Hilbert space V of distributions is "normal" if:

(1-23) $\Phi_0 \subset V$; the injection is continuous and dense. ▲

COROLLARY 1-2   If V is a normal space of distributions, its dual V' satisfies

(1-24) $V' \subset \Phi'_0$, the injection is continuous and dense.

If, moreover, a normal space V of distributions is dense in $L^2(I)$, we have

(1-25) $\Phi_0 \subset V \subset L^2(I) \subset V' \subset \Phi'_0$,

the injections are continuous and dense.   ▲

In order to characterize the duals of spaces of distributions—for instance, the duals of Sobolev spaces—we need the characterization of closed subspaces, factor spaces, and domains of operators.

1-7.   Dual of Closed Subspaces and Factor Spaces

Let P be a closed subspace of a Hilbert space V, and denote by $P^\perp$ the closed subspace of the functionals $f \in V'$ such that

(1-26) $P^\perp = \{f \in V' \text{ such that } (f, v) = 0 \quad \text{for any } v \in P \}$.

We say that $P^\perp$ is the *orthogonal* of P in V', and this orthogonal is related to the *Hilbertian orthogonal* $P^\oplus$ of P defined by

(1-27) $P^\oplus = \{u \in V \quad \text{such that} \quad ((u, v)) = 0 \quad \text{for any } v \in P\}$.

LEMMA 1-1   If J is the canonical isometry from V onto V', then the space $P^\oplus$ is related to $P^\perp$ by

(1-28) $P^\perp = JP^\oplus$.   ▲

Proof   Indeed, $((u, v)) = (Ju, v)$; therefore $u \in P^\oplus$ if and only if $Ju \in P^\perp$.
■

THEOREM 1-6   The dual of a closed subspace P of V is isometric to the factor space $V'/P^\perp$ and the dual of the factor space V/P is isometric to $P^\perp$.
▲

Proof   We deduce the second statement from the first by transposition, since $V'' = V$ and $P^{\perp\perp} = P$.

Let π be the canonical injection from P into V and π' mapping V' into P' its transpose.

1. π' is onto.

Indeed, if $p' \in P'$, we know by the Hahn-Banach theorem that a functional $f \in V'$ exists such that

$$(1\text{-}29) \qquad (f, v)_V = (p', v)_P \qquad \text{for any} \quad v \in P; \ \|f\|_* = \|p'\|_{P'}.$$

Therefore $p' = \pi'f$, since when $v \in P$, $\pi v = v$, and

$$(\pi'f, v)_P = (f, \pi v)_V = (f, v)_V = (p', v)_P \qquad \text{for any} \quad v \in P.$$

2. The kernel of π' is equal to $P^\perp$.

Indeed, $\pi'f = 0$ if and only if $(\pi'f, v)_P = (f, \pi v)_V = (f, v)_V = 0$ for any $v \in P$.

3. The isometry θ.

If φ denotes the canonical surjection from V' onto $V'/P^\perp$, we can write

$$\pi'f = \theta\varphi f \qquad \text{for any} \quad f \in V',$$

where θ is an isomorphism from $V'/P^\perp$ onto P'. Therefore, any $g \in V'/P^\perp$ can be written in the form $g = \varphi f$ and, by definition of a norm of a factor space, we have

$$\|g\|_{V'/P^\perp} = \inf_{\varphi f = g} \|f\|_* = \inf_{h \in P^\perp} \|f - h\|_* \qquad \text{when} \quad g = \varphi f.$$

Therefore, $\|\theta g\|_{P'} \leq \|g\|_{V'/P^\perp}$, since if $g = \varphi f$ and $h \in P^\perp$, we get

$$\|\theta g\|_{P'} = \|\pi'f\|_{P'} = \sup_{v \in P} \frac{|(f - h, v)|}{\|v\|} \leq \|f - h\|_*.$$

4. Let p' belong to P' and f be the functional satisfying (1-29). Then $p' = \pi'f = \theta\varphi f = \theta g$ if $g = \varphi f$. Thus

$$\|g\|_{V'/P^\perp} \leq \|f\|_* = \|p'\|_{P'} = \|\theta g\|_{P'}$$

and thus

$$\|\theta g\|_{P'} = \|g\|_{V'/P^\perp} \qquad \text{for any} \quad g \in V'/P^\perp. \qquad \blacksquare$$

1-8. Applications to Error Estimates

THEOREM 1-7   Let P be a closed subspace of V. Then, if t is the orthogonal projector onto P,

$$\|u - tu\|_V = \inf_{v \in P} \|u - v\|_V = \sup_{f \in P} \frac{|(f, u)|}{\|f\|_{V'}}.$$

If $U \subset V$ and the injection is continuous and dense, and if P is closed in both U and V, then inequality

(1-30)
$$\inf_{v \in P} \|u - v\|_V \le M \|u\|_U \qquad \text{for any} \quad u \in U$$

implies

(1-31)
$$\inf_{v \in P} \|u - v\|_V \le M \inf_{v \in P} \|u - v\|_U \qquad \text{for any} \quad u \in U. \qquad \blacktriangle$$

Proof If φ denotes the canonical surjection from V onto V/P, then $\|\varphi u\|_{V/P} = \inf_{v \in P} \|u - v\|_V$. On the other hand, since V/P is isometric to the dual of $P^\perp$, we deduce that

$$\|\varphi u\|_{V/P} = \|\varphi u\|_{(P^\perp)'} = \sup_{f \in P^\perp} \frac{|(f, u)|}{\|f\|_{V'}}.$$

To prove the second part of the theorem, let i be on the injection from U into V and φ the surjection from V onto V/P. Then (1-30) implies that $\|\varphi i\|_{L(U, V/P)} \le M$. Let us denote by $P_X^\perp$, the orthogonal of P in $X'(X = U, V)$. Then

$$\|\phi i\|_{L(U, V/P)} = \|i' \varphi'\|_{L(P_V^\perp, U')} = \sup_{f \in P_V^\perp} \frac{\|f\|_{U'}}{\|f\|_{V'}} \le M,$$

since φ' is the canonical injection from $P_V^\perp$ into V' and i' is the canonical injection from V' into U'.

On the other hand, if j is the canonical injection from U/P into V/P, its transpose j' is the canonical injection from $P_V^\perp$ into $P_U^\perp$. Therefore,

$$\|j\|_{L(U/P, V/P)} = \|j'\|_{L(P_V^\perp, P_U^\perp)} = \sup_{f \in P_V^\perp} \frac{\|f\|_{U'}}{\|f\|_{V'}} \le M$$

by assumption (1-30). We end the proof by noticing that

$$\|j\|_{L(U/P, V/P)} = \sup_{u \in U} \left[\frac{\inf\limits_{v \in P} \|u - v\|_V}{\inf\limits_{v \in P} \|u - v\|_U} \right].$$

1-9. Dual of a Product

In considering a finite number of Hilbert spaces V_j supplied with the inner products $((u_j, v_j))_j$, the space $V = \prod V_j$ is a Hilbert space for the inner product $((u, v)) = \sum ((u_j, v_j))_j$ if $u = (u_j)_j$ denotes an element of V.

If V_j' denotes the dual of V_j for the duality pairing $(f_j, v_j)_j$, it is easy to verify that the space $V' = \prod V_j'$ can be identified with the dual of V for the duality pairing

$$(1\text{-}32) \qquad (f, v) = \sum (f_j, v_j)_j$$

and is a Hilbert space for the inner product

$$(1\text{-}33) \qquad ((f, g))_* = \sum ((J_j^{-1} f_j, J_j^{-1} g_j))_j.$$

The canonical isometry from V onto V' is the operator J defined by

$$Ju = (J_j u_j)_j.$$

1-10. Dual of Domains of Operators

Let V be a vector space and consider a finite number of operators A_j mapping V into Hilbert spaces V_j supplied with inner products $((u_j, v_j))_j$. We say that V is the *domain of the operators* A_j when it is supplied with the "graph-inner product"

$$(1\text{-}34) \qquad ((u, v)) = \sum ((A_j u, A_j v))_j.$$

We often identify the space V with its *graph* $G(V)$ defined by

$$(1\text{-}35) \qquad G(V) = \theta V \subset \prod V_j \qquad \text{where} \quad \theta u = (A_j u)_j \in \prod V_j.$$

THEOREM 1-8   Let us assume that $((u, v))$ defined by (1-34) is positive definite. Then the domain V of the A_j is a Hilbert space if and only if its graph $G(V)$ is closed, that is to say,

$$(1\text{-}36) \qquad \begin{array}{l} \text{if } u_n \in V \text{ satisfies } A_j u_n \text{ converges to } u_j \text{ in } V_j \text{ for all } j, \\ \text{then, there exists } u \in V \text{ such that } u_j = A_j u \text{ for all } j. \end{array}$$

The dual V' of V can be identified with $\prod V'_j / G(V)^\perp$. Therefore we can associate with any functional f of V' functionals $f_j \in V'_j$ such that

$$(1\text{-}37) \qquad (f, v) = \sum (f_j, A_j v)_j; \; f = \sum A'_j f_j. \qquad \blacktriangle$$

Proof   It is clear that the map θ defined by (1-35) is an isometry when V is supplied with (1-34) and $\prod V_j$ with the inner product of a product of spaces. Therefore V is complete if and only if $G(V)$ is complete (i.e., if $G(V)$ is closed in $\prod V_j$).

Therefore, by Theorem 1-6, we can identify V' [or $G(V)'$] with the factor space $\prod V'_j / G(V)^\perp$. Thus if φ is the canonical surjection from $\prod V'_j$ onto this factor space and if $g = \varphi f$, we can write

$$(g, v) = (\varphi f, v) = \sum_j (f_j, A_j v)_j \qquad \text{for any} \quad v \in V.$$

The decomposition $f = \sum A_j' f_j$ is not unique: two decompositions are equivalent if $\varphi f = \varphi h$; that is, if and only if

$$\sum (f_j - h_j, A_j v)_j = 0 \qquad \text{for any} \quad v \in V. \qquad \blacksquare$$

1-11. Examples: Dual of Sobolev Spaces $H_0^m(I)$

We denote by $H^{-m}(R)$ the dual of the Sobolev space $H^m(R)$ on the real line R. We can consider $H^m(R)$ as the domain of the operators $A_j = D^j$ $(0 \leq j \leq m)$ mapping $H^m(R)$ into $V_j = L^2(R)$. Therefore, by Theorem 1-8, a functional $f \in H^{-m}(R)$ is represented by functions f_j of $L^2(R)$ in this way:

$$(1\text{-}38) \qquad \begin{cases} (f, v) = \displaystyle\sum_{j=0}^{m} (f_j, D^j v) \\[2mm] \qquad = \displaystyle\sum_{j=0}^{m} \int_R f_j(x) D^j v(x)\, dx \\[2mm] \qquad = \displaystyle\int_R \left(\sum_{j=0}^{m} (-1)^j D^j f_j \right) v(x)\, dx. \end{cases}$$

We prove later (Theorem 6.3-3) that $\Phi_0(R)$ is dense in $H^m(R)$. Therefore (1-38) has meaning by the definition of weak derivatives.

COROLLARY 1-3   A distribution f belongs to $H^{-m}(R)$ if and only if there exist $m + 1$ functions f_j of $L^2(R)$ such that $f = \sum_{j=0}^{m} D^j f_j$. The canonical isometry from $H^m(R)$ onto $H^{-m}(R)$ is the differential operator J defined by

$$(1\text{-}39) \qquad\qquad Ju = \sum_{j=0}^{m} (-1)^j D^{2j} u. \qquad\qquad \blacktriangle$$

Proof   Indeed,

$$((u, v)) = \int_R \sum_{j=0}^{m} D^j u D^j v\, dx = \int_R \left(\sum_{j=0}^{m} (-1)^j D^{2j} u \right) v(x)\, dx. \qquad \blacksquare$$

Let us consider the space $H^m(I)$ where $I = (a, b)$ is a bounded open interval and the closed subspace $H_0^m(I)$ of functions u of $H^m(I)$ satisfying

$$(1\text{-}40) \quad H_0^m(I) = \{u \in H^m(I) \text{ such that } D^j u(a) = D^j u(b) = 0 \\ \text{for} \quad 0 \leq j \leq m - 1\}.$$

We shall see later that $\Phi_0(I)$ is dense in $H_0^m(I)$ but not in $H^m(I)$ (Theorem 6.3-1). We deduce in the same way the following Corollary:

COROLLARY 1-4   A distribution f belongs to the dual $H^{-m}(I) = (H_0^m(I))'$ of $H_0^m(I)$ if and only if there exist $m + 1$ functions f_j of $L^2(I)$ such that $f = \sum_{j=0}^{m} D^j f_j$. The canonical isometry from $H_0^m(I)$ onto $H^{-m}(I)$ is the differential operator $Ju = \sum_{j=0}^{m} (-1)^j D^{2j} u$. ▲

1-12. Properties of Bounded Sets of Operators; Uniform Boundedness

In this section we furnish some necessary properties of bounded sets of operators. We say that a family of operators t_h mapping a Hilbert space U into a Hilbert space V is *bounded* if and only if

$$(1\text{-}41) \qquad \sup_h \sup_{\|u\|_U \leqslant 1} \|t_h u\|_V = \sup_h \|t_h\|_{L(U,V)} \leq M < +\infty$$

THEOREM 1-9   Let $\{t_h\}$ be a family of bounded operators mapping U into V.

1. If D is a dense subset of U and if

$$(1\text{-}42) \qquad t_h u \text{ converges to } tu \text{ in } V \quad \text{for any} \quad u \in D \subset U,$$

then t_h converges to t uniformly over every compact $K \subset U$:

$$(1\text{-}43) \qquad \sup_{u \in K} \|t_h u - tu\|_V \qquad \text{converges to 0 with } h$$

and in particular, t_h converges pointwise to t.

2. If t_h converges pointwise to t and if a sequence u_n converges to u in U when n goes to ∞, then

$$(1\text{-}44) \quad t_h u_n \text{ converges to } tu \text{ when } h \text{ converges to 0 and } n \text{ goes to infinity.} ▲$$

Proof   Let M be the upper bound of the norms of the t_h, K be a compact subset of U and $\varepsilon > 0$ be fixed. Since K is compact, there exist n points u_i such that for any $u \in K$, some u_i satisfies $\|u - u_i\| \leq \varepsilon/(4M + 1)$. Since D is dense in U, there exist n points $v_i \in D$ such that $\|u_i - v_i\| \leq \varepsilon/(4M + 1)$. Finally (1-42) implies that there exists h_0 such that $\|t_h v_i - tv_i\| \leq \varepsilon/(4M + 1)$ for $h < h_0$ and $i = 1, \ldots, n$. Writing

$$t_h u - tu = t_h(u - u_i) + t_h(u_i - v_i) + t_h v_i - tv_i + t(v_i - u_i) + t(u_i - u),$$

we deduce that $\|t_h u - tu\| \leq \varepsilon$ whenever $h \leq h_0$.

The second part of the theorem is simpler: For any $\varepsilon > 0$, by assumption there exist h_0 and N such that $\|t_h u - tu\| \leq \varepsilon/(M + 1)$ for $h \leq h_0$ and $\|u_n - u\| \leq \varepsilon/(M + 1)$ for $n \geq N$. Then $\|t_h(u_n - u) + t_h u - tu\| \leq \varepsilon$ for $h \leq h_0$ and $n \geq N$. ■

THEOREM 1-10 (*uniform boundedness*)   A family of operators t_h is bounded if and only if we can associate with any $u \in U$ a constant $M(u)$ such that

(1-45) $\sup_h \|t_h u\|_V \leq M(u) < +\infty$ for any $u \in U$.   ▲

Proof See Treves [1967], p. 347.   ■

Let us translate these results for continuous linear functionals:

THEOREM 1-11

1. A subset $B \subset V'$ is bounded if and only if

(1-46) $\sup_{f \in B} |(f, v)| \leq M(v)$ for any $v \in V$.

2. A bounded family $\{f_h\}$ of continuous linear forms converges weakly to f in V' if

(1-47) (f_h, v) converges to (f, v) for any v belonging to a dense subset D.

3. If a bounded sequence f_h converges weakly to f in V' and if a sequence u_n converges to u in V, then

(1-48) (f_h, u_n) converges to (f, u) when h converges to 0 and n to ∞.

4. A weakly closed bounded set is weakly compact.   ▲

Proof The weak topology on $V' = L(V, R)$ is by definition the pointwise converges topology on $L(V, R)$. Thus the first three statements are translations of Theorems 1-9 and 1-10. In order to prove the last one, note that we can consider V' as a subspace of the product space R^V. (We identify R^V with the space of functionals on V supplied with the pointwise convergence topology.) Then a weakly closed subset B of V' (i.e., of R^V) is compact if and only if their projections are compact. But the projections of B are the sets $\{(f, u)\}_{f \in B}$. Since B is bounded, these projections are compact intervals. Thus B is compact in R^V (i.e., in V').   ■

1-13. Banach Theorem

THEOREM 1-12 (*the Banach theorem*) Let U and V be two Hilbert spaces and A a one-to-one continuous linear operator from U onto V. Then its inverse A^{-1}, a linear operator from V onto U, is also continuous.   ▲

Proof See Trêves [1967], p. 172.   ■

Let us deduce the following consequence:

THEOREM 1-13 Let A be a continuous linear operator mapping a Hilbert space U onto a Hilbert space V. Then here exists a continuous right-inverse B of A; that is, an operator $B \in L(V, U)$ such that

$$(1\text{-}49) \qquad\qquad ABv = v \qquad \text{for any} \quad v \in V. \qquad\qquad \blacktriangle$$

Proof Let P be the kernel of A, φ be the canonical map from U onto U/P. The restriction of φ to the Hilbertian orthogonal subspace $P^\oplus$ of P is an isomorphism; Let ψ denote its inverse.

Then we can write $A = \tilde{A}\varphi$, where $\tilde{A}$ is one-to-one, continuous; its inverse exists by Theorem 1-12, and $B = \psi(\tilde{A})^{-1}$ is a continuous right inverse of A. $\qquad\qquad\blacksquare$

Finally, let us prove the following properties of the transpose:

THEOREM 1-14 Let A be a continuous linear operator from a Hilbert space U into a Hilbert space V and $A' \in L(V', U')$ its transpose. Then

$$(1\text{-}50) \qquad\qquad \ker (A') = A(U)^{\perp}.$$

The operator A' is one-to-one if and only if $A(U)$ is dense in V, and the range $A(U)$ of A is closed if and only if the range $A'(V')$ of A' is closed. $\quad\blacktriangle$

Proof The relation (1-50) is quite obvious: indeed, $A'f = 0$ if and only if $(A'f, u) = (f, Au) = 0$ for any $u \in U$.

Then $A(U)$ is dense in V (by Theorem 1.1-2) if and only if $A(U)^{\perp} = 0$; that is, by (1-50), if and only if $\ker (A') = 0$.

Let us assume that A maps U onto V, and that B is a right inverse of A. Then $A'(V')$ is closed. Indeed, if $A'f_n$ converges to g in U', then $f_n = B'A'f_n$ converges to $f = B'g$ in V'. Thus $A'f_n$ converges both to g and to $A'f$; this implies that $g = A'f$; in other words $A'(V')$ is closed. Now, if the range of A is closed, A maps U onto the Hilbert space $W = A(U)$. The previous statement implies that the range of A' is closed. $\qquad\qquad\blacksquare$

1-14. Dual of Sobolev Spaces $H^m(I)$

Let $I = (a, b)$ be bounded open interval. The space $H^m(I)$ is no longer a normal space of distributions, but nevertheless, we can characterize the elements of $H^m(I)'$.

COROLLARY 1-5 The dual $H^m(I)'$ of $H^m(I)$ is the space of functionals f that can be written in the form

$$(1\text{-}51) \qquad f = \sum_{0 \le j \le m} D^j f_j + \sum_{0 \le j \le m-1} (\alpha_j \delta_{(a)}^{(j)} + \beta_j \delta_{(b)}^{(j)}),$$

where $f_j \in L^2(I)$ and $\delta_{(c)}^{(j)}$ is the jth derivative of the Dirac measure at c. $\quad\blacktriangle$

Proof If we construct an isomorphism ω from $H^m(I)$ onto its closed range in H^m, then by transposition, the dual $H^m(I)'$ will be the range of the transpose ω' of ω. Formula (1-51) will follow from the characterization of ω'.

There exist m functions $\alpha_j(x)(0 \leq j \leq m - 1)$ of H^m with compact support contained in $(a - \varepsilon, a + \varepsilon)$ and satisfying

$$(1\text{-}52) \qquad D^k \alpha_j(a) = \begin{cases} 0 & \text{if } j \neq k \\ 1 & \text{if } j = k \end{cases} \qquad \text{for } 0 \leq j, k \leq m - 1.$$

For instance, we can take fundamental Hermite polynomials of interpolation at the knots $a - \varepsilon$, a, and $a + \varepsilon$ and extend them by 0 outside of $(a - \varepsilon, a + \varepsilon)$. In the same way, there exist m functions $\beta_j(x)(0 \leq j \leq m - 1)$ of H^m with compact support in $(b - \varepsilon, b + \varepsilon)$ satisfying

$$(1\text{-}53) \qquad D^k \beta_j(b) = \begin{cases} 0 & \text{if } j \neq k \\ 1 & \text{if } j = k \end{cases} \qquad \text{for } 0 \leq j, k \leq m - 1.$$

We choose ε small enough so that $a + \varepsilon < b$ and $b - \varepsilon > a$, whereupon the following function $\omega(u)$ belongs to H^m if $u \in H^m(I)$.

$$(1\text{-}54) \qquad \omega(u)(x) = \begin{cases} u(x) & \text{if } a < x < b \\ \displaystyle\sum_{0 \leq j \leq m-1} D^j u(a) \alpha_j(x) & \text{if } x \leq a \\ \displaystyle\sum_{0 \leq j \leq m-1} D^j u(b) \beta_j(x) & \text{if } b \leq x. \end{cases}$$

Then ω is a continuous linear operator from $H^m(I)$ into H^m. On the other hand, ω is a right inverse of the operator ρ which associates with $u \in H^m$ its restriction ρu to I, which belongs to $H^m(I)$.

Therefore, ω is an isomorphism from $H^m(I)$ onto its closed range in H^m, and, by Theorem 1-14, its transpose ω' maps H^{-m} onto $H^m(I)'$. Let us characterize ω': if $f \in H^{-m}$,

$$(\omega'(f), u) = (f, \omega(u)) = \int_I f(x)u(x)\, dx + \sum_{0 \leq j \leq m-1} [\alpha_j D^j u(a) + \beta_j D^j u(b)]$$

$$= \left(\rho f + \sum_{0 \leq j \leq m-1} (\alpha_j \delta_{(a)}^{(j)} + \beta_j \delta_{(b)}^{(j)}), u \right),$$

where

$$\alpha_j = \int_{-\infty}^a f(x)\alpha_j(x)\, dx \quad \text{and} \quad \beta_j = \int_b^{+\infty} f(x)\beta_j(x)\, dx.$$

On the other hand, Corollary 1-3 implies that $f = \sum_{0 \leq j \leq m} D^j f_j$ where $f_j \in L^2$. Then (1-51) is proved. ∎

1-15. The Riesz–Fredholm Alternative

Let V and F be two Hilbert spaces, $B \in L(V, F)$ an isomorphism from V onto F, and $C \in L(V, F)$ a compact operator—that is, an operator satisfying

(1-55) C maps bounded subsets of V into compact subsets of F.

The Riesz–Fredholm theory answers the question. For what (real) values of λ is the operator $A_\lambda = B + \lambda C$ an isomorphism from V onto F?

THEOREM 1-15 (*Riesz–Fredholm*) Let us assume that B is an isomorphism from V onto F and that C is a compact operator from V into F. Then $A_\lambda = B + \lambda C$ is an isomorphism from V onto F except when λ belongs to a countable subset $S(B, C)$ with no finite accumulation point.

If $\lambda \in S(B, C)$, the kernel of A_λ is a finite-dimensional subspace and the range of A_λ is closed. The transpose $A'_\lambda = B' + \lambda C'$ is an isomorphism from F' onto V' if and only if λ does not belong to $S(B, C)$. If λ belongs to $S(B, C)$, then the dimension of the kernel of A'_λ is equal to the dimension of the kernel of A_λ and the range of A'_λ is closed.   ▲

Definition 1-2   We say that $S(B, C)$ is the "spectrum" of C relative to B and that $\lambda \in S(B, C)$ is an "eigenvalue."   ▲

We deduce from this theorem the Riesz–Fredholm alternative:

COROLLARY 1-6   Making the assumptions of Theorem 1-15, we consider the equation

(1-56) $A_\lambda u = Bu + \lambda Cu = f.$

Either $\lambda \notin S(B, C)$, whereupon exists a unique solution of (1-56) for any $f \in F$, or $\lambda \in S(B, C)$, whereupon solutions of (1-56) do exist if f belongs to the orthogonal of the kernel of $A'_\lambda = B' + \lambda C'$.   ▲

1-16. V-Elliptic and Coercive Operators

In considering the case of operators mapping a Hilbert space V into its dual $F = V'$, we can associate with such an operator $A \in L(V, V')$ a continuous bilinear form $a(u, v)$ on V by setting

(1-57) $a(u, v) = (Au, v)$ where (f, v) is the duality pairing on $V' \times V$.

Conversely, the data of a continuous bilinear form $a(u, v)$ on V define an operator $A \in L(V, V')$ by

(1-58) Au is the continuous linear form $v \to a(u, v)$ for any $u \in V$.

We shall denote the transpose of $A \in L(V, V')$ by A^* instead of A'. It is clear that $A^* \in L(V, V')$ and that it is associated with the continuous bilinear form $a_*(u, v) = a(v, u)$.

Definition 1-3 We say that $A \in L(V, V')$ is *V-elliptic* if there exists a positive constant c such that

$$(1\text{-}59) \qquad a(v, v) \geq c \, \|v\|^2 \qquad \text{for any} \quad v \in V \qquad \blacktriangle$$

Now, we assume that V is a dense subspace of a pivot space H (i.e., a space identified with its dual—see Sections 1-1 and 1-5).

Then Theorem 1-5 implies that

$$(1\text{-}60) \qquad V \subset H \subset V', \text{ the injections being dense and continuous.}$$

Definition 1-4 In this case, we say that $A \in L(V, V')$ is (V, H)-*coercive* if there exist a positive constant c and a constant λ such that

$$(1\text{-}61) \qquad a(v, v) + \lambda \, |v|^2 \geq c \, \|v\|^2 \qquad \text{for any} \quad v \in V$$

where $|v| = \sqrt{(v, v)}$ is the norm of H. $\qquad \blacktriangle$

In other words, A is (V, H)-*coercive* if $A + \lambda$ is V-elliptic for λ large enough. $\blacksquare$

When V is separable, the Lax-Milgram theorem implies that a V-elliptic operator is an isomorphism (see Theorem 1-1-4 and 1.2-1). The Lax-Milgram theorem holds even when V is not separable.

THEOREM 1-16 Any V-elliptic operator is an isomorphism from V onto V'. If the injection from V into a pivot space H is dense and compact and if A is a (V, H)-coercive operator, then $A + \lambda$ and $A^* + \lambda$ are isomorphisms from V onto V' when λ does not belong to a countable subset $S(A)$ of isolated points. $\qquad \blacktriangle$

Proof Let us prove the first statement, calling A a V-elliptic operator. Then A is one to one since, if $Av = 0$, $\|v\|^2 \leq c^{-1}(Av, v) = 0$ and thus $v = 0$. The image of A is closed. Indeed, if Au_n is a Cauchy sequence of V', then u_n is a Cauchy sequence of V because

$$\|u_n - u_m\|^2 \leq c^{-1}[A(u_n - u_m), u_n - u_m] \leq c^{-1} \, \|Au_n - Au_m\|_{V'} \, \|u_n - u_m\|_V.$$

Then u_n converges to an element u of V and Au_n converges to Au. Therefore $A(V)$ is a complete subspace of V' and thus is closed.

Finally, the range of A is dense. Indeed, if $u \in V = (V')'$ is a continuous linear functional on V' vanishing on the range of A, then $\|u\|^2 \leq c^{-1}(Au, u) = 0$, and thus $u = 0$. Then the first statement of Theorem 1-16 follows from Theorem 1-2 of Section 1-3.

Then A is a one-to-one continuous operator mapping V onto V' and thus is an isomorphism from V onto V' by Theorem 1-12 of Section 1-13. Actually, the norm of A^{-1} is less than or equal to c^{-1}, since

$$\|v\|^2 \leq c^{-1} \|Av\|_{V'} \|v\|_V.$$

Now we consider a (V, H)-coercive operator A, where $A + \lambda$ is an isomorphism from V to V'. Since the injection from V into H is compact, the injection from V into V' is a compact operator. Therefore, by Theorem 1-15, $A + \lambda + \mu$ is an isomorphism from V onto V' when μ does not belong to a countable subset of isolated points.

Hence the proof of Theorem 1-16 is completed, and we say that $S(A)$ is the "spectrum" of the operator A defined by $a(u, v)$. ∎

2. QUASI-OPTIMAL APPROXIMATIONS

In this continuation of the study of approximations begun previously, we introduce the stability function

$$s_U^V(p_h) = \sup_{v_h \in V_h} \frac{\|p_h v_h\|_U}{\|p_h v_h\|_V}$$

of a prolongation that compares the behavior of the discrete norms $\|p_h u_h\|_U$ and $\|p_h v_h\|_V$, where $U \subset V$. These stability functions are related to the error functions by duality relations.

A simple but important theorem shows that

$$1 \leq s_U^V(p_h) e_U^V(p_k)$$

whenever $\dim V_h < \dim V_k$ and $U \subset V$. This implies in particular that

$$s_U^V(p_h) \geq \frac{1}{E_U^V[n(h) - 1]} \quad \text{when} \quad \dim V_h = n(h),$$

where $E_U^V(n)$ is the n-width of the injection from U into V.

This inequality implies also that if the approximations are quasi-optimal for the injection from U into V—that is, if the following inequality holds—

$$\sup_h e_U^V(p_h) s_U^V(p_h) \leq M,$$

then the error function $e_U^V(p_h)$ behaves like the n-width. In other words, the error function satisfies

$$e_U^V(p_h) \leq M E_U^V[n(h) - 1] \quad \text{where} \quad n(h) \text{ is the dimension of } V_h.$$

We introduce also the notion of truncation error $e_U{}^V(p_h r_h)$ and study its relation to the error function $e_U{}^V(p_h) = e_U{}^V(p_h \hat{r}_h)$.

2-1. Stability Functions

Let p_h be a prolongation mapping a discrete space V_h into a Hilbert space V and q_h a prolongation mapping the same discrete space V_h into another Hilbert space U.

Definition 2-1 It is convenient to relate these two prolongations by the stability function

$$(2\text{-}1) \qquad\qquad s_U{}^V(q_h, p_h) = \sup_{v_h \in V_h} \frac{\|q_h v_h\|_U}{\|p_h v_h\|_V}$$

(allowed to take infinite values).

In the case of $p_h = q_h$, we set

$$(2\text{-}2) \qquad s_U{}^V(p_h) = s_U{}^V(P_h) = \sup_{v_h \in V_h} \frac{\|p_h v_h\|_U}{\|p_h v_h\|_V} = \sup_{v \in P_h} \frac{\|v\|_U}{\|v\|_V},$$

where $P_h = p_h V_h$. ▲

These stability functions are finite if $V \subset U$ with a stronger topology. Usually we assume that

$$(2\text{-}3) \qquad\qquad U \subset V, \text{ the injection is continuous and dense.}$$

In this case, the stability functions are finite whenever the dimension of the discrete space V_h is finite, for all the norms of a finite-dimensional space are equivalent.

The stability functions play an important role. First, we show that there are duality relations between stability functions and error functions

$$(2\text{-}4) \quad e_U{}^V(p_h) = e_U{}^V(P_h) = \sup_{u \in U} \inf_{v_h \in V_h} \frac{\|u - p_h v_h\|_V}{\|u\|_U} = \sup_{u \in U} \inf_{v \in P_h} \frac{\|u - v\|_V}{\|u\|_U}$$

we introduced in Section 1.2-5.

2-2. Duality Relations Between Error and Stability Functions

THEOREM 2-1   Let U and V be two Hilbert spaces satisfying (2-3), P a closed subspace of $U \subset V$, and $P^\perp$ its orthogonal. Then

$$(2\text{-}5) \qquad (i)\ e_U{}^V(P) = s_{U'}^{V'}(P^\perp), \qquad (ii)\ s_U{}^V(P) = e_{U'}^{V'}(P^\perp). \qquad ▲$$

Proof It is enough to prove the first relation, because the second is obtained by replacing U and V by their duals and P by its orthogonal. Let π be the injection from U into V and φ the canonical surjection from V onto the factor space V/P. Since the norm

$$\|\varphi u\|_{V/P} = \inf_{v \in P} \|u - v\|_V,$$

we deduce that

$$e_U^V(P) = \|\varphi \pi\|_{L(U,V/P)} = \|\pi' \varphi'\|_{L((V/P)',U')}$$

because the norm of an operator is equal to the norm of its transpose. But φ' is the canonical isometry from $(V/P)'$ onto $P^\perp$. Thus if we identify the dual of V/P with $P^\perp$, we identify φ' with the identity. On the other hand, by Theorem 1-5, π' is identified with the canonical injection from V' into U' because the injection π from U into V is dense. Therefore $\pi'\varphi'$ is the canonical injection from $P^\perp$ supplied with the norm of V' into U' and we deduce

$$e_U^V(P) = \sup_{f \in P^\perp} \frac{\|f\|_{U'}}{\|f\|_{V'}} = s_{U'}^{V'}(P^\perp). \quad \blacksquare$$

2-3. Estimates of the Stability Functions

We now prove the main result of this section.

THEOREM 2-2 If V is a Hilbert space, P a subspace of dimension n and Q a subspace of dimension $n - 1$ of a Hilbert space $U \subset V$, then

$$(2\text{-}6) \qquad\qquad 1 \leq s_U^V(P) e_U^V(Q). \qquad\qquad \blacktriangle$$

Proof First of all, since the dimension of Q is smaller than that of P, we know that

$$(2\text{-}7) \qquad\qquad P \cap Q^\oplus \neq \{0\}.$$

Indeed, if $(\pi_j)_{1 \leq j \leq n-1}$ is a basis of Q, the functionals $((v, \pi_j))$ are independent on the n-dimensional vector space P and thus, there exists an element $u \in P$ different from 0 such that

$$((u, \pi_j)) = 0 \qquad \text{for} \quad 1 \leq j \leq n - 1.$$

In other words, there exists u such that

$$u \in P, u \neq 0; \quad \|u\|_V = \inf_{v \in Q} \|u - v\|_V$$

For such a u we have $\|u\|_U / \|u\|_V \leq s_U^V(P)$ and thus

$$1 \leq s_U^V(P) \frac{\|u\|_V}{\|u\|_U} = s_U^V(P) \inf_{v \in Q} \frac{\|u - v\|_V}{\|u\|_U} \leq s_U^V(P) e_U^V(Q). \quad \blacksquare$$

COROLLARY 2-1 If the n-width of the injection from U into V converges to 0 when n goes to ∞, the stability functions $s_U^V(P_h)$ converge to ∞ whenever the dimension of P_h goes to ∞. ▲

Proof Taking the infimum of (2-6) on the family of spaces Q of dimension $n(h) - 1$, we obtain

$$(2\text{-}8) \qquad 1 \leq s_U^V(P_h) E_U^V[n(h) - 1]; \; \dim (P_h) = n(h). \qquad ■$$

2-4. Quasi-Optimal Approximations; Estimate of the Error Function

Definition 2-2 Let (V_h, p_h, r_h) be a family of approximations of the Hilbert space V. We say that these approximations are *quasi-optimal* (for the injection from U into V) if $P_h = p_h V_h \subset U$ and if

$$(2\text{-}9) \qquad \sup_h s_U^V(p_h) e_U^V(p_h) \leq M < + \infty. \qquad ▲$$

Before considering the next theorem, we must recall the definition of the n-width introduced in Section 1.2-6:

$$(2\text{-}10) \qquad E_U^V(n) = \inf_{\dim (P)=n} e_U^V(P).$$

THEOREM 2-3 If (V_h, p_h, r_h) is a family of quasi-optimal approximations for the injection from U into V and if the dimension of V_h is equal to $n(h)$, then the error function of the approximations is estimated in the following way:

$$(2\text{-}11) \qquad e_U^V(p_h) \leq M E_U^V[n(h) - 1],$$

where

$$(2\text{-}12) \qquad M = \sup_h e_U^V(p_h) s_U^V(p_h). \qquad ▲$$

Remark 2-1 Even if we do not know accurate estimates of the n-width of the injection from U into V, the property of quasi-optimality of a family of approximations informs us that the behavior of the error function is optimal.

Proof It follows from (2-12) and (2-8) that

$$(2\text{-}13) \qquad e_U^V(p_h) \leq \frac{M}{s_U^V(p_h)} \leq M E_U^V[n(h) - 1].$$

2-5. Truncation Errors and Error Functions

In the definition of approximations (see Section 1.2-2) we introduced restrictions r_h, which are not necessarily optimal. One of the reasons involves spaces; namely, if $U \subset V$, and if $P_h = p_h V_h \subset U$, we can consider the approximations (V_h, p_h, r_h) as approximations of both the spaces U and V. Then the optimal restriction r_h^U for U is no longer the one for V and we cannot consider the error function to be the norm of $1 - p_h \hat{r}_h$ in $L(U, V)$ (see Section 1.2-14). Thus we must introduce the notion of *truncation error*, defined as follows:

Definition 2-3 The function

$$(2\text{-}14) \qquad e_U{}^V(p_h r_h) = \sup_{u \in U} \frac{\|u - p_h r_h u\|_V}{\|u\|_U} = \|1 - p_h r_h\|_{L(U,V)}.$$

is called the truncation error of the approximations (V_h, p_h, r_h). ▲

To compare the behavior of the truncation error with that of the error function, we need Theorem 2-4.

THEOREM 2-4 If we assume that $r_h p_h = 1$ and $U \subset V$, then

$$(2\text{-}15) \qquad e_U{}^V(p_h) \leq e_U{}^V(p_h r_h) \leq e_V{}^V(p_h r_h) e_U{}^V(p_h).$$ ▲

Remark 2-2 If $r_h p_h = 1$, the operators $p_h r_h$ and $1 - p_h r_h$ are projectors. If those projectors are bounded in $L(V, V)$, then $e_V{}^V(p_h r_h)$ is bounded, and the truncation error and the error function behave the same.

Proof The first inequality of (2-15) follows from the definition of error functions. To prove the second one, we deduce from $r_h p_h = 1$ that

$$u - p_h r_h u = (1 - p_h r_h)(u - p_h v_h) \qquad \text{for any} \quad v_h \in V_h.$$

Therefore $\|u - p_h r_h u\|_V \leq e_V{}^V(p_h r_h) \|u - p_h v_h\|_V$. Taking the infimum on V_h and the supremum on the unit ball of U, we obtain

$$\sup_{\|u\|_U \leq 1} \|u - p_h r_h u\|_V \leq \sup_{\|u\|_U \leq 1} \inf_{v_h \in V_h} \|v - p_h v_h\|_V,$$

which is nothing but the second inequality of (2-15). ■

3. OPTIMAL APPROXIMATIONS

In our examination of approximations achieving the n-width of the injection from a Hilbert space U into V, we characterize these approximations when the injection from U into V is dense and compact.

These optimal approximations are the Galerkin approximations associated with the eigenvectors μ_j of the operator $A = K^{-1}J$, where K is the canonical isometry from U onto U' and J the canonical isometry from V onto V'. (This operator A is a symmetric compact operator from V into V, and thus there exists an orthonormal basis of eigenvectors μ_j and a decreasing sequence of eigenvalues α_j converging to 0.)

In other words, these optimal approximations are defined by

- $V_h = R^{n+1}$ where $h = \dfrac{1}{n+1}$,

- $p_h u_h = \displaystyle\sum_{0 \le j \le n} u_h^{\,j} \mu_j$,

- $r_h u = \{((\mu_j, u))\}_{0 \le j \le n}$

and we obtain

$$e_U^V(p_h) = E_U^V(n+1) = \alpha_{n+1}^{1/2}, \quad s_U^V(p_h) = \alpha_n^{-1/2}.$$

We begin by recalling the properties of symmetric compact operators. Then we construct the optimal Galerkin approximations, and finally we associate with U and V a family of spaces H_θ ($\theta \ge 0$) which are the domains of the operators $A^{-\theta/2}$. It will become clear that the optimal Galerkin approximations are convergent approximations of these spaces H_θ.

Unfortunately, the use of these optimal Galerkin approximations is limited for two reasons: first, the actual computation of the eigenvectors μ_j of $A = K^{-1}J$ is often impossible, and second, the matrices of the approximate problems obtained by using these approximations are not sparse.

3-1. Eigenvalues and Eigenvectors of Symmetric Compact Operators

Let U and V be two Hilbert spaces such that

(3-1) $U \subset V$; the injection is dense and compact.

Let $K \in L(U, U')$ and $J \in L(V, V')$ be the canonical isometries associated with the inner products of U and V:

(3-2) $((u, v))_U = (Ku, v)$ and $((u, v))_V = (Ju, v)$.

Since $V' \subset U'$ (see Theorem 1-5), we can view $J \in L(V, V')$ as a continuous linear operator from U into U'. Furthermore, $J \in L(U, U')$ is a compact operator because the injection from U into V is compact. On the other hand, $K \in L(U, U')$ is an isomorphism.

The fact that both K and J are symmetric positive operators permits to improve the conclusions of Theorem 1-16.

THEOREM 3-1 Assuming (3-1), let $K \in L(U, U')$ and $J \in L(V, V')$ be the canonical isometries. Then the spectrum $S(K, J)$ of K relative to J is a countable sequence $(\lambda_j)_{j \geq 0}$ such that

$$(3\text{-}3) \qquad 0 \leq \lambda_0 \leq \cdots \leq \lambda_j \leq \cdots ; \lim_{j \to \infty} \lambda_j = \infty.$$

Furthermore, there exists a countable sequence of eigenvectors $(\mu_j)_{j \geq 0}$ which is an orthonormal basis of the space V. ▲

Proof See Trêves [1967], p. 490. ■

If we set $\alpha_j = \lambda_j^{-1}$, the sequences $(\alpha_j)_{j \geq 0}$ and $(\mu_j)_{j \geq 0}$ satisfy the following properties:

$$(3\text{-}4) \qquad \begin{cases} \text{(i)} & J\mu_j = \alpha_j K\mu_j \quad \text{for any } j \geq 0, \\[2mm] \text{(ii)} & u = \sum_{j \geq 0} ((\mu_j, u))_V \mu_j \quad \text{for any } u \in V, \\[2mm] \text{(iii)} & \alpha_0 \geq \cdots \geq a_j \geq \cdots ; \lim_{j \to \infty} \alpha_j = 0, \\[2mm] \text{(iv)} & ((\mu_j, \mu_k))_V = \delta_{jk} \quad \text{and} \quad ((\mu_j, \mu_k))_U = \alpha_j^{-1} \delta_{jk}, \end{cases}$$

where δ_{jk} is the Kronecker symbol.

3-2. Optimal Galerkin Approximations

The Galerkin approximation associated with the orthonormal basis $(\mu_j)_j$ of eigenvectors of the operator $A = K^{-1}J$ is the approximation that achieves the n-width of the injection from U into V. We must first realize that they are convergent approximations of both the spaces U and V. Let us set $h = 1/(n + 1)$ and

$$(3\text{-}5) \qquad \begin{cases} \text{(i)} & V_h = R^{n+1}, \\[2mm] \text{(ii)} & p_h u_h = \sum_{j=0}^{n} u_h^{\,j} \mu_j, \\[2mm] \text{(iii)} & r_h u = \{((\mu_j, u))_V\}_j = \{\alpha_j((\mu_j, u))_U\}_j; \ 0 \leq j \leq n, \end{cases}$$

since $((\mu_j, u))_V = (J\mu_j, u) = \alpha_j(K\mu_j, u) = \alpha_j((\mu_j, u))_U$.

Then we can write

$$(3\text{-}6) \qquad p_h r_h u = \sum_{j=0}^{n} ((\mu_j, u))_V \mu_j = \sum_{j=0}^{n} ((\sqrt{\alpha_j}\mu_j, u))_U \sqrt{\alpha_j}\, \mu_j.$$

Because the basis $(\sqrt{\alpha_j}\mu_j)_j$ is orthonormal in U, we deduce the following:

LEMMA 3-1 Let us assume (3-1). The Galerkin approximations associated with the orthonormal basis of eigenvectors of $K^{-1}J$ are convergent approximations of the spaces U and V:

(3-7)
$$\begin{cases} \lim_{h \to 0} \|u - p_h r_h u\|_V = 0 & \text{for any } u \in V; \\ \lim_{h \to 0} \|u - p_h r_h u\|_U = 0 & \text{for any } \mu \in U. \end{cases} \qquad \blacktriangle$$

3-3. Convergence and Optimality Properties

THEOREM 3-2 Again assuming (3-1), the Galerkin approximations associated with the orthonormal basis of eigenvectors of $K^{-1}J$ achieve the n-width of the injection from U into V

(3-8)
$$\begin{cases} \text{(i)} \quad e_U^V(p_h) = E_U^V(n+1) = \sqrt{\alpha_{n+1}} \, ; \quad h = \dfrac{1}{n+1}, \\[2mm] \text{(ii)} \quad s_U^V(p_h) = \dfrac{1}{\sqrt{\alpha_n}}, \\[2mm] \text{(iii)} \quad \lim_{h \to 0} s_U^V(p_h) \|u - p_h r_h u\|_V = 0 \quad \text{for any } u \in U. \end{cases} \qquad \blacktriangle$$

Proof

1. $e_U^V(p_h) = \sqrt{\alpha_{n+1}}$.

This property follows from (3-6). Indeed, by [(3-3)(i)]

$$\|u - p_h r_h u\|_V^2 = \sum_{j=n+1}^{\infty} |((\mu_j, u))_V|^2 = \alpha_{n+1} \sum_{j=n+1}^{\infty} \frac{\alpha_j}{\alpha_{n+1}} |((\mu_j, u))_U|^2$$

$$\leq \alpha_{n+1} \sum_{j=0}^{\infty} |((\mu_j, u))_U|^2 = \alpha_{n+1} \|u\|_U^2.$$

On the other hand, by (3-4)

$$\|\mu_{n+1}\|_V = \|\mu_{n+1} - p_h r_h \mu_{n+1}\|_V = \sqrt{\alpha_{n+1}} \|\mu_{n+1}\|_U.$$

Therefore

$$\frac{\|u - p_h r_h u\|_V}{\|u\|_U} \leq \sqrt{\alpha_{n+1}} = \frac{\|\mu_{n+1} - p_h r_h \mu_{n+1}\|_V}{\|\mu_{n+1}\|_U} = e_U^V(p_h).$$

2. $s_U^V(p_h) = \dfrac{1}{\sqrt{\alpha_n}}$.

If $u \in P_{n+1} = p_h R^{n+1}$, its norm in U can be estimated in this way:

$$\|u\|_U{}^2 = \sum_{j=0}^{n} |\sqrt{\alpha_j}((\mu_j, u))_U|^2 = \frac{1}{\alpha_n} \sum_{j=0}^{n} \frac{\alpha_n}{\alpha_j} |((\mu_j, u))_V|^2$$

$$\leq \alpha_n^{-1} \sum_{j=0}^{n} |((\mu_j, u))_V|^2 = \alpha_n^{-1} \|u\|_V{}^2.$$

On the other hand, by (3-4), $\mu_n \in P_{n+1}$ and $\|\mu_n\|_U = \alpha_n^{-1/4} \|\mu_n\|_V$; thus

$$\frac{\|u\|_U}{\|u\|_V} \leq \alpha_n^{-1/2} = \frac{\|\mu_n\|_U}{\|\mu_n\|_V} \qquad \text{for any} \quad u \in P_{n+1}.$$

3. $E_U{}^V(n+1) = e_U{}^V(p_h)$.

Indeed, we deduce from 1 and 2 that

(3-9) $$1 = s_U{}^V(P_{n+1}) e_U{}^V(P_n).$$

However, we deduce from Theorem 2-2 that $1 \leq s_U{}^V(P_{n+1}) e_U{}^V(Q_n)$ for any space Q_n of dimension n. Those two relations imply that

(3-10) $$e_U{}^V(P_n) = \frac{1}{s_U{}^V(P_{n+1})} \leq e_U{}^V(Q_n) \qquad \text{for any} \quad Q_n \text{ of dimension } n,$$

that is to say, P_n achieves the n-width.

Finally, the proof of [(3-8)(iii)] is analogous to that of 1 and is left as an exercise. ∎

3-4. Spaces H_θ

The optimal Galerkin approximations remain optimal Galerkin approximations of a family of spaces H_θ, which we define.

Let U and V be two spaces satisfying (3-1). If θ is a positive real number, we shall denote by H_θ the subspace of elements $u \in V$ such that the following series converge:

(3-11) $$\|u\|_\theta = \|A^{-\theta/2}u\|_V = \left(\sum_{j=0}^{\infty} \alpha_j^{-\theta} |((\mu_j, u))_V|^2 \right)^{1/2} < +\infty.$$

It is left as an exercise to verify that H_θ is a Hilbert space.

In other words, the spaces H_θ are the domains of the operators $A^{-\theta/2}$ defined formally by $A^{-\theta/2}u = \sum_{j=0}^{\infty} \alpha_j^{-\theta/2}((\mu_j, u))_V \mu_j$. Notice that $(\alpha_j^{\theta/2}\mu_j)_j$ is an orthonormal basis of H_θ when it is supplied with the inner product $((u, v))_\theta = ((A^{-\theta/2}u, A^{-\theta/2}v))_V$. Therefore we can write

(3-12) $$p_h r_h u = \sum_{j=0}^{n} ((\alpha_j^{\theta/2}\mu_j, u))_\theta \alpha_j^{\theta/2}\mu_j,$$

and thus the Galerkin approximations associated with the basis $(\mu_j)_j$ are convergent approximations of the spaces H_θ.

The proof of the following theorem is left as an exercise.

THEOREM 3-3 The Galerkin approximations associated with the orthonormal basis of eigenvectors of $K^{-1}J$ achieve the n-width of the injection from H_{θ_1} into H_{θ_0} when $\theta_1 > \theta_0$. The error and stability functions are equal to

$$
(3\text{-}13) \quad
\begin{cases}
\text{(i)} \quad e_{\theta_1}^{\theta_0}(p_h) = E_{\theta_1}^{\theta_0}(n+1) = (\sqrt{\alpha_{n+1}})^{\theta_1-\theta_0}, \\[2mm]
\text{(ii)} \quad s_{\theta_1}^{\theta_0}(p_h) = \left(\dfrac{1}{\sqrt{\alpha_n}}\right)^{\theta_1-\theta_0}, \\[2mm]
\text{(iii)} \quad \lim_{h\to 0} s_{\theta_1}^{\theta_0}(p_h) \, \|u - p_h r_h u\|_{\theta_0} = 0, \quad \text{for any} \quad u \in H_{\theta_1}.
\end{cases}
\quad \blacktriangle
$$

Proof The proof is analogous to the proof of Theorem 3-2. ■

Remark 3-1 (Interpolation spaces) When $\theta = 0$, the space H_0 coincides with V and for $\theta = 1$, the space H_1 coincides with U.

When $0 \le \theta \le 1$, the spaces H_θ are usually denoted by

$$
(3\text{-}14) \qquad\qquad H_\theta = U^\theta V^{1-\theta} = [U, V]_{1-\theta}.
$$

The norm of $U^\theta V^{1-\theta}$ satisfies the inequality

$$
(3\text{-}15) \qquad\qquad \|u\|_\theta \le \|u\|_U^\theta \, \|u\|_V^{1-\theta} \qquad \text{for any} \quad u \in H_\theta.
$$

Indeed, we can write

$$
\|u\|_\theta^2 = \sum_{j=0}^\infty \alpha_j^{-\theta} |((\mu_j, u))_V|^{2\theta} |((\mu_j, u))_V|^{2(1-\theta)}.
$$

We then apply the Hölder inequality with $p = 1/\theta$ and $p' = 1/(1-\theta)$ $(1/p + 1/p' = 1)$ and we obtain (3-15).

Then the spaces $U^\theta V^{1-\theta}$ are *interpolation spaces* between U and V and play a fundamental role in the theory of partial differential equations. (See J. L. Lions and E. Magenes [1969], chapter 1.)

For $\theta \in (0, 1)$, the spaces H_θ are examples of spaces of order $1 - \theta$ between U and V (see Section 4-6). ■

Remark 3-2 (V is a pivot space) If $V = H_0 = H = H'$ is a pivot space (i.e., a Hilbert space identified with its dual), we obtain the inclusions

$$
(3\text{-}16) \qquad\qquad H_\theta \subset H \subset (H_\theta)'
$$

by Theorem 1-5 of Section 1-4. Indeed, the injections of one space H_θ into another are dense, since the basis (μ_j) spans dense subspaces in each space

H_θ. Therefore, if we set $H_{-\theta} = (H_\theta)'$, we can write

(3-17) $H_\theta \subset H_\delta \subset H_0 \subset H_{-\delta} \subset H_{-\theta}$ $(\theta \geq \delta)$.

Next we check that the canonical isometry from H_θ onto $H_{-\theta}$ is equal to $A^{-\theta} = K^\theta$ (since $J = 1$ and V is identified with its dual) when $H_{-\theta}$ is supplied with the inner product $((A^{\theta/2}u, A^{\theta/2}v))_V = ((u, v))_{-\theta}$. Then

(3-18) $\|u\|_{-\theta} = \left(\sum_{j=0}^\infty \alpha_j^{\,\theta} |((\mu_j, u))_V|^2 \right)^{\!\frac{1}{2}}$

The operators $K^{(\sigma-\delta)/2}$ are isometries from H_α onto H_δ whenever $\alpha > \delta$ (α, δ are real numbers).

It is left as an exercise to verify that when V is a pivot space, Theorem 3-3 can be extended for spaces H_θ with negative indices. These Galerkin approximations have the following property of commutation:

(3-19) $K^\theta p_h r_h = p_h r_h K^\theta$ for any θ.   ■

Remark 3-3   The example of Galerkin approximation of Section 1.3-1 is an example of optimal Galerkin approximations when $U = H^1(I)$ and $V = H = L^2(I)$.   ■

4.   OPTIMAL RESTRICTIONS AND PROLONGATIONS; APPLICATIONS

We now require concrete knowledge either of the prolongation p_h or the restriction r_h of a given approximation.

Therefore, if p_h (respectively r_h) is given, we only have to know the existence of a restriction r_h (respectively, a prolongation p_h) associated with p_h (respectively r_h). We shall choose r_h to be the optimal restriction $\hat{r}_h$, defined by

$$\|u - p_h \hat{r}_h u\| = \inf_{v_h \in V_h} \|u - p_h v_h\|$$

and p_h to be the optimal prolongation $\hat{p}_h$, defined by

$$r_h \hat{p}_h v_h = v_h, \|\hat{p}_h v_h\| = \inf_{r_h v = v_h} \|v\| \quad \text{for any } v_h \in V_h.$$

These definitions are consistent: if $\hat{r}_h$ is the optimal restriction associated with p_h, then p_h is the optimal prolongation associated with $\hat{r}_h$, and conversely. If $(V_h, p_h, \hat{r}_h)$ are approximations of V where $\hat{r}_h$ is the optimal restriction associated with p_h, then $(V'_h, \hat{r}'_h, p'_h)$ are approximations of V' and $\hat{r}'_h$ is the optimal prolongation associated with p'_h.

Furthermore, if J is the canonical isometry from V onto V', we obtain the following formulas:

$$\hat{r}_h = (p'_h J p_h)^{-1} p'_h J, \qquad \hat{p}_h = J^{-1} r'_h (r_h J^{-1} r'_h)^{-1}, \qquad J p_h \hat{r}_h = (p_h \hat{r}_h)' J.$$

Also, we can characterize the error functions and the stability functions of a given prolongation p_h as eigenvalues of operators.

Finally, in Section 4-6, we introduce spaces W of order θ between V and H; that is, spaces W satisfying

$$\|v\|_W \leq c \, \|v\|_V^{1-\theta} \, \|v\|_H^{\theta} \qquad \text{for any} \quad v \in V,$$

and we deduce analogous inequalities for the norms of operators mapping a space X into the spaces V, W, and H or mapping these spaces V, W, H into a space Y.

4-1. Optimal Restrictions and Prolongations

Let V and V_h be, respectively, a Hilbert space and a discrete Hilbert space and V' and V_h' their duals for the duality pairings (f, v) and $(f_h, v_h)_h$. We denote by $\|f\|_*$ and $\|f_h\|_{h*}$ the dual norms of V' and V_h'.

If a prolongation p_h mapping V_h into V is given, we associate with it a discrete norm and a restriction in the following way.

Definition 4-1 We call the *discrete norm* (associated with p_h in V) the norm

(4-1) $$\|v_h\|_h = \|p_h v_h\|.$$ ▲

For such a discrete norm, the norm of p_h is equal to 1.

Definition 4-2 We say that $\hat{r}_h$ is the *optimal restriction* (associated with p_h in V) if

(4-2) $$\|u - p_h \hat{r}_h u\| = \inf_{v_h \in V_h} \|u - p_h v_h\|.$$ ▲

Such a restriction $\hat{r}_h$ satisfies the following properties:

(4-3) $$\hat{r}_h p_h = 1; \quad \|u_h\|_h = \|p_h u_h\| = \|p_h \hat{r}_h u\| \leq \|u\|$$

for any $u \in V$ such that $\hat{r}_h u = u_h$, since $p_h \hat{r}_h$, being the orthogonal projector, has a norm equal to 1.

In several cases (problem of interpolation, construction of approximate operators, etc.) a restriction r_h mapping V onto V_h is given and we have the problem of associating with it a discrete norm and an optimal prolongation.

Since r_h maps V onto V_h, its transpose r_h' is a one-to-one map and the following norm has a meaning:

(4-4) $$\|v_h\|_h = \sup_{f_h \in V_h'} \frac{|(f_h, v_h)_h|}{\|r_h' f_h\|_*}$$

Definition 4-3 The norm $\|v_h\|_h$ defined by (4-4) is the discrete norm associated with r_h in V.   ▲

THEOREM **4-1**   When V_h is supplied with the discrete norm (4-4) associated with r_h, the norm of r_h is equal to 1.   ▲

Proof   Indeed

$$\|r_h u\|_h = \sup_{f_h} \frac{|(f_h, r_h u)_h|}{\|r'_h f_h\|_*} = \sup_{f_h} \frac{|(r'_h f_h, u)|}{\|r'_h f_h\|_*} \leq \|u\|.$$ ■

Definition 4-4   We say that $\hat{p}_h$ is the *optimal prolongation* (associated with r_h in V) if

(4-5) $\quad r_h \hat{p}_h = 1; \; \|\hat{p}_h v_h\| = \|v_h\|_h \leq \|v\|$ $\quad$ for any $\quad v$ such that $r_h v = v_h$.   ▲

We shall see that such a prolongation exists and satisfies the property

(4-6) $$\|u - \hat{p}_h r_h u\| = \inf_{v_h \in V_h} \|u - \hat{p}_h v_h\|.$$

The aim of this section is to construct these operators in terms of the canonical isometry J. But first, we must understand the duality relations between the problems of finding the optimal restriction and the optimal prolongation.

4-2.   Dual Approximations

Let us consider an approximation (V_h, p_h, r_h) of the Hilbert space V. By transposition, we can associate with it an approximation of the dual V' of V. Indeed, since p_h is an isomorphism from V_h into its range $P_h = p_h V_h \subset V$, its transpose $s_h = p'_h$ maps V' onto V'_h and, since r_h maps V onto V_h, its transpose $q_h = r'_h$ maps V'_h onto its range $Q_h = q_h V'_h = (\ker r_h)^\perp \subset V'$. Therefore $q_h = r'_h$ can play the role of a prolongation and $s_h = p'_h$ can play the role of a restriction.

Definition 4-5   The approximation (V'_h, q_h, s_h) of V' is the *dual approximation* of (V_h, p_h, r_h) when

(4-7) $\quad \begin{cases} \text{(i)} \quad q_h = r'_h \text{ is defined by } (q_h f_h, v) = (f_h, r_h v)_h \\ \qquad\qquad\qquad \text{for any } f_h \in V'_h \quad \text{and} \quad v \in V, \\ \text{(ii)} \quad s_h = p'_h \text{ is defined by } (s_h f, v_h)_h = (f, p_h v_h) \\ \qquad\qquad\qquad \text{for any } f \in V' \quad \text{and} \quad v_h \in V_h. \end{cases}$ ▲

Returning to the problem of optimal restriction and optimal prolongation, we notice that if a restriction r_h is given from V onto V_h, a prolongation $q_h = r_h'$ is given from V_h' into V'. Then we must ask, *Is the optimal restriction associated with q_h in V' the transpose of the optimal prolongation associated with r_h in V?* This question is answered in the following section.

4-3. Construction of Optimal Prolongations and Restrictions

THEOREM 4-2   Let r_h be a given restriction from V onto V_h. The optimal prolongation associated with r_h is equal to

$$(4\text{-}8) \qquad \hat{p}_h = J^{-1}r_h'(r_hJ^{-1}r_h')^{-1},$$

where J is the canonical isometry from V onto V'. Then $\hat{p}_h r_h$ is the orthogonal projector onto $\hat{P}_h = \hat{p}_h V_h = (\ker r_h)^{\oplus} = J^{-1}Q_h$, where $Q_h = r_h'V_h' = (\ker r_h)^{\perp}$.

The transpose $\hat{s}_h = (\hat{p}_h)'$ is the optimal restriction associated with $q_h = r_h'$ in V':

$$(4\text{-}9) \qquad \hat{s}_h = (r_hJ^{-1}r_h')^{-1}r_hJ^{-1} = (q_h'J^{-1}q_h)^{-1}q_h'J^{-1},$$

and the following commutation formula holds:

$$(4\text{-}10) \qquad J\hat{p}_h r_h = q_h \hat{s}_h J. \qquad\qquad \blacktriangle$$

Proof   The first point to verify is that the operator $r_hJ^{-1}r_h'$ is invertible. Actually, it is the canonical isometry from V_h' onto V_h when V_h' is supplied with the inner product

$$(4\text{-}11) \qquad ((f_h, g_h))_{h*} = ((r_h'f_h, r_h'g_h))_*.$$

Indeed,

$$(4\text{-}12) \qquad ((f_h, g_h))_{h*} = (f_h, r_hJ^{-1}r_h'g_h)_h,$$

since

$$((r_h'f_h, r_h'g_h))_* = (r_h'f_h, J^{-1}r_h'g_h) = (f_h, r_hJ^{-1}r_h'g_h)_h.$$

Therefore $(r_hJ^{-1}r_h')^{-1}$ is the canonical isometry from V_h onto V_h', and it is possible to check that the discrete norm (4-4) associated with r_h is nothing other than the dual norm of $\|f_h\|_{h*} = \|r_h'f_h\|_*$

$$(4\text{-}13) \qquad \|v_h\|_h = \|(r_hJ^{-1}r_h')^{-1}v_h\|_{h*}.$$

Thus we can define $\hat{p}_h$ by (4-8) and it is clear that $r_h\hat{p}_h = 1$. On the other hand, $\|\hat{p}_h v_h\| = \|v_h\|_h$ because by (4-13),

$$\|\hat{p}_h v_h\| = \|J\hat{p}_h v_h\|_* = \|r_h'(r_hJ^{-1}r_h')^{-1}v_h\|_* = \|(r_hJ^{-1}r_h')^{-1}v_h\|_{h*}.$$

The projector $\hat{p}_h r_h$ is the orthogonal projector onto $\hat{P}_h = \hat{p}_h V_h$, since

$$((\hat{p}_h r_h v, \hat{p}_h v_h)) = (r'_h (r_h J^{-1} r'_h)^{-1} r_h v, \hat{p}_h v_h) = ((r_h J^{-1} r'_h)^{-1} r_h v, v_h)_h$$
$$= (v, JJ^{-1} r'_h (r_h J^{-1} r'_h)^{-1} v_h) = ((v, \hat{p}_h v_h)).$$

We deduce from this that

$$\|\hat{p}_h v_h\| = \|\hat{p}_h r_h v\| \le \|v\| \qquad \text{for any} \quad v \quad \text{such that} \qquad r_h v = v_h$$

and that:

$$\|u - \hat{p}_h r_h u\| = \inf_{v_h \in V_h} \|u - \hat{p}_h v_h\|.$$

We have proved that the operator $\hat{p}_h$ defined by (4-8) satisfies the properties (4-5) and (4-6). In particular, $\hat{p}_h$ is the optimal prolongation associated with r_h.

Let us consider now the transpose $\hat{s}_h = (\hat{p}_h)'$ of $\hat{p}_h$. We have to verify that $q_h \hat{s}_h$ is the orthogonal projector onto $Q_h = q_h V'_h$. We first check (4-9) and deduce from it that

$$(4\text{-}14) \qquad\qquad \hat{p}_h r_h J^{-1} = J^{-1} q_h \hat{s}_h.$$

Therefore,

$$((q_h \hat{s}_h g, q_h f_h))_* = (J^{-1} q_h \hat{s}_h g, q_h f_h) = (\hat{p}_h r_h J^{-1} g, q_h f_h)$$
$$= (g, J^{-1} q_h f_h) = ((g, q_h f_h))_*.$$

That is to say, $q_h \hat{s}_h$ is the orthogonal projector onto Q_h. ∎

4-4. Miscellaneous Remarks

Remarks 4-1 Convergence properties

THEOREM 4-3   Let (V_h, p_h, r_h) be approximations of the Hilbert space V and $(V'_h, \hat{q}_h, s_h)$ their dual approximations of the space V'. If $\hat{r}_h$ is the optimal restriction associated with p_h (and thus $\hat{q}_h$ the optimal prolongation associated with s_h), then the approximations $(V_h, p_h, \hat{r}_h)$ are convergent if and only if the dual approximations $(V'_h, \hat{q}_h, s_h)$ are.   ▲

Proof This a consequence of (4-14), since

$$\|f - \hat{q}_h s_h f\|_* = \|J^{-1}(1 - \hat{q}_h s_h) f\| = \|(1 - p_h \hat{r}_h) J^{-1} f\|.$$ ∎

The convergence of approximations (V_h, p_h, r_h) does not imply the convergence of the dual approximations when r_h is no longer the optimal restriction associated with p_h (or p_h the optimal prolongation associated with r_h).   ∎

THEOREM 4-4 If the approximations (V_h, p_h, r_h) are convergent, the dual approximations converge weakly:

(4-15) $q_h s_h f$ converges weakly to f in V', uniformly over bounded sets of V'. ▲

Proof Indeed

$$\sup_{\|f\|_* \leq 1} |(f - q_h s_h f, v)| = \sup_{\|f\|_* \leq 1} |(f, v - p_h r_h v)| = \|v - p_h r_h v\|. \quad \blacksquare$$

Remarks 4-2 (*V_h is a finite-dimensional space*) Let V_h be the space $R^{n(h)}$ supplied with the duality pairing $(f_h, v_h)_h = \sum_j f_h^j v_h^j$. Then any restriction r_h from V onto V_h is associated with a sequence $(\varphi_h^j)_j$ of $n(h)$ continuous linear forms φ_h^j belonging to V'.

(4-16) $$r_h u = ((\varphi_h^j, u))_j \in R^{n(h)}$$

and the transpose $q_h = r_h'$ of r_h is equal to

(4-17) $\qquad q_h f_h = \sum f_h^j \varphi_h^j \qquad$ where $\quad f_h = (f_h^j)_j \in V_h' = R^{n(h)}$.

Therefore Q_h is the finite-dimensional subspace of V' spanned by the forms φ_h^j and, since $\hat{P}_h = J^{-1} Q_h$, it is spanned by the elements $\mu_h^j = J^{-1} \varphi_h^j \in V$. In order to construct $\hat{p}_h$, we take into account that the elements of the matrix of $r_h' J^{-1} r_h$ are equal to $(J^{-1} \varphi_h^i, \varphi_h^j)$. Then, if (α_i^j) denotes the coefficients of its inverse matrix, (4-8) implies that

(4-18) $\quad \hat{p}_h u_h = \sum_{i,j} \alpha_i^j u_h^j J^{-1} \varphi_h^i = \sum_j u_h^j \lambda_h^j \quad$ where $\quad \lambda_h^j = \sum_i \alpha_i^j J^{-1} \varphi_h^i$.

Therefore, we can write the orthogonal projectors $\hat{p}_h r_h$ and $q_h \hat{s}_h$ in the following form:

(4-19) $$\hat{p}_h r_h u = \sum_j (\varphi_h^j, u) \lambda_h^j \; ; \quad q_h \hat{s}_h f = \sum_j (f, \lambda_h^j) \varphi_h^j. \quad \blacksquare$$

Remark 4-3 (*Optimal right inverse of an operator*) We can give a more general interpretation of Theorem 4-2. ■

THEOREM 4-5 Let L be an operator mapping a Hilbert space V onto a space W. Among all the right inverses of L mapping W into V, the operator M defined by

(4-20)
$$M = J^{-1} L' (L J^{-1} L')^{-1}$$

where J is the canonical isometry from V onto V'

is the right inverse of L of the minimal norm when W is supplied with the norm

$$\|w\|_W = \sup_{\|L'g\|_* \leq 1} |(g, w)_W|. \quad \blacktriangle$$

Proof   Theorem 4-5 is a restatement of Theorem 4-2, when we set $W = V_h$, $L = r_h$, and $M = \hat{p}_h$.   ■

Remark 4-4 (*Characterization of orthogonal projector onto $P^{\oplus}$*)   If P is a closed subspace of a Hilbert space V, we can consider it either as the range of an operator p, which is an isomorphism of a space E onto P, or as the kernel of an operator L mapping V onto a space W. For instance, we can always choose $E = P$, where p is the canonical injection, $W = V/P$, and L the canonical surjection. Actually, in concrete cases E, p, W, and L are given concrete spaces and operators. Then Theorem 4-2 gives a way to construct the orthogonal projectors onto P and $P^{\oplus}$ in terms of p and L.   ■

THEOREM 4-6   Let us assume that $P = pE$ where p is an isomorphism from E onto $P \subset V$. Then the orthogonal projector t onto P is equal to

$$(4\text{-}21) \qquad\qquad t = p(p'Jp)^{-1}p'J.$$

If P is the kernel of an operator L mapping V onto W, then the orthogonal projector $1 - t$ onto $P^{\oplus}$ is equal to

$(4\text{-}22)$ $1 - t = J^{-1}L'(LJ^{-1}L')^{-1}L = ML$ where M is defined by (4-20).   ▲

Proof   The first statement is (4-9) with V' replaced by V, V_h' by E, q_h by p, and J^{-1} by J.

In order to prove that $u - tu = MLu$, we first check to make certain that $tu \in P$ (since $Ltu = Lu - LMLu = 0$). Second, assuming that $r_h = L$ and $\hat{p}_h = M$, we use (4-3) to verify that

$$\|u - tu\| = \|MLu\| = \|ML(u - v)\| \le \|u - v\| \qquad \text{for any} \quad v \in P$$

since $v \in P$ if and only if $Lv = 0$.   ■

4-5.   Characterization of Error and Stability Functions

Here we characterize the error and stability functions as eigenvalues of operators involving the canonical isometries of the Hilbert spaces U and V and the optimal restrictions and prolongations. Let U and V be two Hilbert spaces satisfying

$(4\text{-}23) \qquad U \subset V$; the injection is continuous and dense,

and let K and J denote the cononical isometries of these spaces.

Let $(V_h, p_h, \hat{r}_h)$ be approximations of V where $\hat{r}_h$ is the optimal restriction associated with p_h in V, and denote by e and s, respectively, the error and stability functions of $P_h = p_h V_h$:

$$(4\text{-}24) \quad \begin{cases} \text{(i)} \quad e = e_U{}^V(p_h) = \sup_{\|u\|_U \leq 1} \|u - p_h \hat{r}_h u\|_V, \\[2mm] \text{(ii)} \quad s = s_U{}^V(p_h) = \sup_{v_h \in V_h} \dfrac{\|p_h v_h\|_U}{\|p_h v_h\|_V}. \end{cases}$$

THEOREM 4-7 If the injection from U into V is compact and dense, the error function $e_U{}^V(p_h)$ is the square root of the largest eigenvalue of the operator $(1 - p_h \hat{r}_h)K^{-1}J(1 - p_h \hat{r}_h)$ mapping $P_h^{\oplus}$ into itself.

If V_h is a finite-dimensional space such that

$$(4\text{-}25) \qquad P_h = p_h V_h \subset U \qquad \text{and} \qquad KP_h \subset V',$$

the stability function $s_U{}^V(p_h)$ is the square root of the largest eigenvalue of the operator $p_h \hat{r}_h J^{-1} K p_h \hat{r}_h$ mapping P_h into itself. ▲

Proof We begin by proving the second part of the theorem. Since V_h is a finite-dimensional space, the supremum

$$(4\text{-}26) \quad s^2 = s_U{}^V(p_h)^2 = \sup \frac{\|p_h v_h\|_U{}^2}{\|p_h v_h\|_V{}^2} = \|p_h u_h\|_U{}^2; \ \|p_h u_h\|_V = 1$$

is achieved at a point u_h of V_h. Therefore, for any $w_h \in V_h$ and for any $\theta > 0$, the following inequality holds:

$$s^2 \|p_h(u_h + \theta w_h)\|_V{}^2 - \|p_h(u_h + \theta w_h)\|_U{}^2 \geq 0$$

or equivalently:

$$2\theta[s^2((p_h u_h, p_h w_h))_V - ((p_h u_h, p_h w_h))_U] \geq -\theta^2[s^2\|p_h w_h\|_V{}^2 - \|p_h w_h\|_U{}^2].$$

Dividing by θ and taking the limit as θ goes to zero, we get

$$s^2((p_h u_h, p_h w_h))_V - ((p_h u_h, p_h w_h))_U$$
$$= (s^2 p_h' J p_h u_h - p_h' K p_h u_h, w_h)_h \geq 0 \qquad \text{for any} \quad w_h \in V_h.$$

Therefore, since w_h and $-w_h$ belong to V_h, we deduce that s^2 is an eigenvalue of the operator

$$(4\text{-}27) \qquad (p_h' J p_h)^{-1}(p_h' K p_h)u_h = s^2 u_h \ ; \ \|p_h u_h\|_V = 1.$$

It is the largest one because, if λ^2 is another eigenvalue, we get

$$p_h' K p_h v_h = \lambda^2 p_h' J p_h v_h \qquad \text{and thus} \qquad \lambda^2 = \frac{\|p_h v_h\|_U{}^2}{\|p_h v_h\|_V{}^2} \leq s^2.$$

Since $KP_h \subset V'$ (and thus, $JJ^{-1}Kp_h = Kp_h$) and since p_h and J are isomorphisms, we deduce that s^2 is also the eigenvalue of the following operator:

$$p_h(p_h'Jp_h)^{-1}p_h'JJ^{-1}Kp_h\hat{r}_hp_hu_h = s^2p_hu_h.$$

Therefore, since $\hat{r}_h = (p_h'Jp_h)^{-1}p_h'J$ by Theorem 4-2, and setting $u = p_hu_h \in P_h$ then s^2 is the largest eigenvalue:

$$p_h\hat{r}_hJ^{-1}Kp_h\hat{r}_hu = s^2u; \quad \|u\|_V = 1.$$

To prove the first part of Theorem 4-7, we consider $P_h = \ker L$ as the kernel of an operator L mapping V onto W.

Therefore L' is an isomorphism from W' onto $P_h^{\perp}$ and, by Theorem 2-1, we get

$$(4\text{-}28) \qquad e^2 = e_U^V(P_h)^2 = s_{U'}^{V'}(P_h^{\perp})^2 = \sup_{g \in W'} \frac{\|L'g\|_{U'}^2}{\|L'g\|_{V'}^2} = \|L'f\|_{U'}^2$$

with $\|L'f\|_{V'} = 1$. Indeed, the supremum is achieved at a point f of W' because the injection from V' into U' is compact. (An operator is compact if and only if its transpose is also compact; thus the injection from V' into U' is compact, since it is the transpose of the injection from U into V, which is compact by assumption.)

Therefore, in the same way that we deduced (4-27) from (4-26), we deduce from (4-28) that e^2 is the largest eigenvalue of

$$(4\text{-}29) \qquad (LJ^{-1}L')^{-1}LK^{-1}L'f = e^2f; \quad \|L'f\|_{V'} = 1.$$

But since $J^{-1}L'$ is an isomorphism from W' onto $P_h^{\oplus}$, e^2 is also the largest eigenvalue of

$$(4\text{-}30) \qquad J^{-1}L'(LJ^{-1}L')^{-1}LK^{-1}JJ^{-1}L'f = e^2J^{-1}L'f.$$

Thus, since the orthogonal projector $1 - p_h\hat{r}_h$ onto $P_h^{\oplus}$ is equal to $J^{-1}L'(LJ^{-1}L')^{-1}L$ by theorem 4-6; we deduce from (4-30)—setting $u = J^{-1}L'f = (1 - p_h\hat{r}_h)u \in P_h^{\oplus}$—that e^2 is the largest eigenvalue of

$$(4\text{-}31) \qquad (1 - p_h\hat{r}_h)K^{-1}J(1 - p_h\hat{r}_h)u = e^2u; \quad \|u\|_V = 1; \ u \in P_h^{\oplus}. \qquad \blacksquare$$

4-6. Spaces of Order θ

Let V, W, and H be Hilbert spaces satisfying

$$(4\text{-}32) \qquad V \subset W \subset H, \text{ the injections being dense and continuous.}$$

Definition 4-6 We say that W is a space of order θ between V and H ($0 \leq \theta \leq 1$) if there exists a constant c such that

$$(4\text{-}33) \qquad \|v\|_W \leq c \, \|v\|_V^{1-\theta} \, \|v\|_H^{\theta} \qquad \text{for any} \quad v \in V. \qquad \blacktriangle$$

For instance, the spaces H_θ we introduced in Section 3-3 are spaces of order $1 - \theta$ between $H_1 = U$ and $H_0 = V$ when $0 \le \theta \le 1$. (See Remark 3-1 of Section 3-4.)

We shall see that Sobolev spaces H^s are spaces of order θ between Sobolev spaces H^m and H^k with $k \le s \le m$ (see Theorem 6.3-7 of Section 6.3-9).

THEOREM 4-8   Let $A \in L(X, V) \cap L(X, H)$. If W is a space of order θ between V and H, then $A \in L(X, W)$ and

$$(4\text{-}34) \qquad \|A\|_{L(X,W)} \le c \, \|A\|_{L(X,V)}^{1-\theta} \, \|A\|_{L(X,H)}^{\theta}.$$

Let $B \in L(V, Y) \cap L(H, Y)$. If W' is a space of order $1 - \theta$ between H' and V', then $B \in L(W, Y)$ and

$$(4\text{-}35) \qquad \|B\|_{L(W,Y)} \le c \, \|B\|_{L(V,Y)}^{1-\theta} \, \|B\|_{L(H,Y)}^{\theta}. \qquad \blacktriangle$$

Proof   Inequality (4-34) is quite obvious: indeed, (4-33) implies that

$$\|Av\|_W \le c \, \|Av\|_V^{1-\theta} \, \|Av\|_H^{\theta} \le c \, \|A\|_{L(X,V)}^{1-\theta} \, \|A\|_{L(X,H)}^{\theta} \, \|v\|_X$$

and thus, (4-34).

On the other hand, if $B \in L(V, Y) \cap L(H, Y)$, its transpose $A = B'$ belongs to $L(Y', V') \cap L(Y', H')$. Since W' is a space of order $1 - \theta$ between H' and V', we deduce from (4-34) that

$$\begin{aligned}
\|B\|_{L(W,Y)} = \|B'\|_{L(Y',W')} &\le c \, \|B'\|_{L(Y',H')}^{\theta} \, \|B'\|_{L(Y',V')}^{1-\theta} \\
&= c \, \|B\|_{L(V,Y)}^{1-\theta} \, \|B\|_{L(H,Y)}^{\theta}. \qquad \blacksquare
\end{aligned}$$

Later, we make use of the following consequences of Theorem 4-8.

PROPOSITION 4-1   Let W be a space of order θ between V and H. If U is a subspace of V, then

$$(4\text{-}36) \qquad e_U^{\,W}(p_h r_h) \le c e_U^{\,V}(p_h r_h)^{1-\theta} e_U^{\,H}(p_h r_h)^{\theta}. \qquad \blacktriangle$$

Let q_h be a prolongation mapping V_h into U and p_h be a prolongation mapping V_h into V, related by the stability function

$$s_U^{\,V}(q_h, p_h) = \sup_{v_h} \frac{\|q_h v_h\|_U}{\|p_h v_h\|_V}.$$

PROPOSITION 4-2   Let K be a space of order θ between U and V. Then

$$(4\text{-}37) \qquad s_K^{\,V}(q_h, p_h) \le c s_U^{\,V}(q_h, p_h)^{1-\theta} s_V^{\,V}(q_h, p_h)^{\theta}. \qquad \blacktriangle$$

CHAPTER 3

Approximation of Operators

This chapter is devoted to the general properties of approximation of a solution $u \in V$ of the equation $Au = f$ (f being given in F) by solutions $u_h \in V_h$ of equations $A_h u_h = f_h$ (f_h being given in F_h).

Section 1 covers internal approximations and their properties, and Section 2 deals with the problem of the regularity of the convergence which is solved by using quasi-optimal approximations.

Finally, in Section 3, we consider the general case, defining the discrete convergence and the lack of consistency, stating a theorem of convergence, and showing that the discrete operators which minimize the lack of consistency are internal approximations.

1. INTERNAL APPROXIMATIONS

Let $A \in L(V, F)$ and $f \in F$ be given, and $u \in V$ the solution of the operational equation $Au = f$. Then letting p_h be a prolongation from V_h into V and s_h a restriction from F onto F_h, we say that the discrete equation

$$s_h A p_h u_h = s_h f$$

is the internal approximation of the equation $Au = f$.

We define the stability of the operators $s_h A p_h$ as follows: there exists a constant S independent of h such that

$$\|p_h u_h\|_V \leq S \|\hat{q}_h s_h A p_h u_h\|_F \qquad \text{for any} \quad u_h \text{ in } V_h,$$

where $\hat{q}_h$ is the optimal prolongation associated with s_h.

We shall prove that if the solution u of $Au = f$ belongs to $U \subset V$, there exists a constant M such that

$$\|u - p_h u_h\|_V \leq M \|u - p_h \hat{r}_h u\|_V \leq M e_U^V(p_h) \|u\|_U$$

and that, if $F \subset G$, then

$$\|A(u - p_h u_h)\|_G \leq M e_F^G(\hat{q}_h s_h) \|u - p_h u_h\|_V.$$

82

The assumption of stability is not only sufficient but also a necessary condition: If A is an isomorphism, the following two conditions are equivalent:

1. The $s_h A p_h$ is an isomorphism and, for any $f \in F$, $p_h u_h$ converges to u.

2. The approximations $(V_h, p_h, \hat{r}_h)$ are convergent and the operators $s_h A p_h$ are stable.

Therefore, we devote the rest of the section to obtaining sufficient conditions of stability. For instance, if an isomorphism $A = B + C$ is the sum of an isomorphism B and a compact operator C, then, under convenient assumptions, the operators $s_h A p_h$ are stable whenever the operators $s_h B p_h$ are stable.

We prove that internal approximations of V-elliptic operators and invertible coercive operators are stable.

When p_h is given (and thus, when the rate of convergence $e_U^V(p_h)$ is fixed), we shall solve the problem of optimal and quasi-optimal stability by constructing restrictions $\bar{s}_h$ that minimizes the norms of $(s_h A p_h)^{-1}$ either for a given operator A or for a class of operators A.

1-1.   Construction of an Internal Approximate Equation

Let V and F be two Hilbert spaces and let A be an operator from V into F. We shall approximate a solution $u \in V$ (if any) of the equation

$$(1\text{-}1) \qquad\qquad\qquad Au = f.$$

In order to construct an approximation of (1-1), we shall associate the following items with a parameter h converging to 0:

$$(1\text{-}2) \qquad \begin{cases} \text{(i) a discrete space } V_h, \\ \text{(ii) a prolongation } p_h \text{ mapping } V_h \text{ into } V, \end{cases}$$

and

$$(1\text{-}3) \qquad \begin{cases} \text{(i) a discrete space } F_h, \\ \text{(ii) a restriction } s_h \text{ mapping } F \text{ onto } F_h. \end{cases}$$

Among operators A_h mapping V_h into F_h and elements $f_h \in F_h$, we shall choose the following ones associated with A and f:

$$(1\text{-}4) \quad \begin{cases} \text{(i) } A_h = s_h A p_h \text{ mapping } V_h \text{ into } F_h \text{ (called } internal\ approximation\ of\ A), \\ \text{(ii) } f_h = s_h f \in F_h. \end{cases}$$

Definition 1-1 We say that the approximate equation

(1-5) $$u_h \in V_h \; ; \; A_h u_h = s_h A p_h u_h = s_h f = f_h$$

is the *internal approximation of* (1-1). ▲

By Theorem 2.4-2, we can associate with p_h an optimal restriction $\hat{r}_h$ and with s_h an optimal prolongation $\hat{q}_h$. Thus we assume from now on that the internal approximation of (1-1) is defined by approximations $(V_h, p_h, \hat{r}_h)$ and $(F_h, \hat{q}_h, s_h)$ of spaces V and F.

1-2.   The Case of Finite-Dimensional Discrete Spaces

Let us consider the case where

(1-6) $$V_h = F_h = R^{n(h)}$$

We have seen that any prolongation p_h is associated with a sequence of $n(h)$ linearly independent elements $\lambda_h{}^j$ of V and that any restriction s_h is defined by a sequence of $n(h)$ linearly independent elements $\phi_h{}^j$ of F', dual of F

(1-7) $$p_h u_h = \sum_j u_h{}^j \lambda_h{}^j \; ; \; s_h f = \{(\phi_h{}^j, f)\}_j.$$

Then the entries of the matrix of A_h are equal to $(\phi_h{}^i, A\lambda_h{}^j)$ and the components of f_h are equal to $(\phi_h{}^j, f)$ since we can write (1-5) in the form

(1-8) $$\sum_j (\phi_h{}^i, A\lambda_h{}^j) u_h{}^j = (\phi_h{}^i, f) \qquad \text{for any index } i.$$

Remark 1-1 The sequences $\{\lambda_h{}^j\}_j$ and $\{\phi_h{}^j\}_j$ have to be "simple" enough to permit computation of the entries of A_h and the components of f_h. ■

Remark 1-2 Besides the problem of constructing approximate equations minimizing the error (or having an optimal behavior of error), we shall have to deal with the problem of constructing approximate operators A_h whose matrices are as sparse as possible. Indeed, statements of convergence or estimates of error are not the only points of our study. Since the actual size of the matrix A_h is large, we must find constructive methods for solving (1-5), and having matrices with a small number of nonzero entries is important for this purpose. ■

1-3.   The Case of Operators from V onto V′

In Section 1.2 we studied internal approximations of variational equations, which are related to internal approximations of operational equations in the following way: if $a(u, v)$ is a continuous bilinear form on $V \times V$, then $A \in L(V, V')$ is the operator associated with $a(u, v)$ by

$$(1\text{-}9) \qquad\qquad Au : v \to (Au, v) = a(u, v).$$

Therefore, if $f \in F = V'$, any solution of $Au = f$ is also solution of the variational equation

$$(1\text{-}10) \qquad\qquad a(u, v) = (f, v) \qquad \text{for any} \quad v \in V.$$

Consider now an approximation $(V_h, p_h, \hat{r}_h)$ of V. By duality, we use the dual approximations $(V'_h, \hat{q}_h, s_h)$ of $F = V'$ where $s_h = p'_h$ and $\hat{q}_h = \hat{r}'_h$ (see Section 2.4-2). Then the bilinear form $a_h(u_h, v_h)$ associated with $A_h = p'_h A p_h$ is equal to $a(p_h u_h, p_h v_h)$. In other words, (1-5) is equivalent to

$$(1\text{-}11) \quad a_h(u_h, v_h) = a(p_h u_h, p_h v_h) = (f, p_h v_h) = (p'_h f, v_h)_h \quad \text{for any} \quad v_h \in V_h,$$

which is the internal approximation of the variational equation (1-10).

Let us recall that if V_h is supplied with the norm $\|v_h\|_h = \|p_h v_h\|$ associated with p_h, the dual norm on $F_h = V'_h$ is the norm associated with the restriction $s_h = p'_h$. The choice of the dual approximations of V' is motivated by Theorem 1-9 (see Remark 1-5).

1-4. Stability of Internal Approximations of Operators

Let $A_h = s_h A p_h$ be internal approximations of an isomorphism A from V onto F defined by approximations $(V_h, p_h, \hat{r}_h)$ and $(F_h, \hat{q}_h, s_h)$ of V and F. Since p_h and s_h are given, we supply discrete spaces V_h and F_h with the discrete norms associated with p_h and s_h (see Section 2.4-1):

$$(1\text{-}12) \qquad \|u_h\|_{V_h} = \|p_h u_h\|_V ; \ \|f_h\|_h = \sup_{g_h \in F_h'} \frac{|(g_h, f_h)_h|}{\|s'_h g_h\|_{F'}} = \|\hat{q}_h f_h\|_F$$

Definition 1-2 We say that the *internal approximations* $A_h = s_h A p_h$ are *stable* (for the norms of V and F) if there exists a constant $S > 0$ independent of h such that

$$(1\text{-}13) \qquad \|p_h u_h\|_V \le S \|s_h A p_h u_h\|_{F_h} = S \|\hat{q}_h s_h A p_h u_h\|_F$$

for any $u_h \in V_h$ and for any h. ▲

Then, since the norm of s_h is equal to 1, stable internal approximations satisfy the inequalities

$$(1\text{-}14) \quad S^{-1} \|p_h u_h\|_V \le \|\hat{q}_h s_h A p_h u_h\|_F \le M \|p_h u_h\|_V \qquad \text{for any} \quad u_h \in V_h,$$

where M denotes the norm of operator A.

In particular, the inequality (1-13) implies that $A_h = s_h A p_h$ is an isomorphism from V_h onto F_h:

LEMMA 1-1   If the operators $A_h = s_h A p_h$ are stable internal approximations of A, the approximate equation (1-5) has a unique solution.   ▲

Actually, the discrete norm $\|\hat{q}_h f_h\|_F$ associated with the restriction s_h is often difficult to characterize in concrete examples. The following theorem gives sufficient conditions for stability.

THEOREM 1-1   If there exist norms $\|f_h\|_{F_h}$ on F_h and constants c and S independent of h such that

$$(1\text{-}15) \quad \begin{cases} \text{(i) } \|s_h f\|_{F_h} \le c \, \|f\|_F & \text{for any } f \in F, \\ \text{(ii) } \|p_h u_h\|_V \le S \, \|s_h A p_h u_h\|_{F_h} & \text{for any } u_h \in V_h, \end{cases}$$

then the operators $A_h = s_h A p_h$ are stable.   ▲

Proof   Indeed, taking $f = \hat{q}_h f_h$ in [(1-15)(i)], we obtain

$$(1\text{-}16) \qquad \|f_h\|_{F_h} = \|s_h \hat{q}_h f_h\|_{F_h} \le c \, \|\hat{q}_h f_h\|_F \qquad \text{for any } f_h \in F_h.$$

Therefore, setting $f_h = s_h A p_h u_h$ in [(1-15)(ii)], we get

$$(1\text{-}17) \qquad \|p_h u_h\|_V \le Sc \, \|\hat{q}_h s_h A p_h u_h\|_F \qquad \text{for any } u_h \in V_h.$$

1-5.   Convergence and Error Estimates

Our first problem is to give sufficient (and, if possible, necessary) conditions in order that the following statements may hold:

$$(1\text{-}18) \quad \begin{cases} \text{(i) there exists a unique solution } u_h \text{ of the internal approximate} \\ \quad \text{equation } s_h A p_h u_h = s_h f \quad \text{for any } f \in F, \\ \text{(ii) } p_h u_h \text{ converges strongly to a solution } u \text{ of the initial equation} \\ \quad Au = f \quad \text{for any } f \in F. \end{cases}$$

On the other hand, it is quite "natural" to assume the following conditions:

$$(1\text{-}19) \quad \begin{cases} \text{(i) the approximations } (V_h, p_h, \hat{r}_h) \text{ of } V \text{ are convergent,} \\ \text{(ii) the operators } A_h = s_h A p_h \text{ are stable (for the norms of } V \text{ and } F\text{).} \end{cases}$$

First we prove that the conditions (1-19) are sufficient.

THEOREM 1-2 Let $u \in V$ be a solution of the equation $Au = f$, where f is given in F, and let u_h be the solution of the internal approximate equation $s_h A p_h u_h = s_h f$. Then the conditions (1-19) imply the statements (1-18). Furthermore, we obtain the inequality

$$(1-20) \qquad \|u - p_h u_h\|_V \leq M \|u - p_h \hat{r}_h u\|_V,$$

which shows that the behavior of the error is the same as the behavior of the best approximation of the solution u by approximants of $P_h = p_h V_h$.

Therefore, if the solution u belongs to a subspace $U \subset V$, the following estimate holds:

$$(1-21) \qquad \|u - p_h u_h\|_V \leq M e_U^V(p_h) \|u\|_U \qquad \blacktriangle$$

Proof The stability of the A_h implies [(1-18)(i)] by Lemma 1-1, and [(1-18)(ii)] follows from inequality (1-20) and from [(1-19)(i)]. It remains to be shown that estimate (1-20) follows from the stability assumption. Indeed, since $s_h A p_h u_h = s_h f = s_h A u$, we deduce from the stability

$$(1-22) \quad \begin{cases} \|p_h(u_h - \hat{r}_h u)\|_V \leq S \|\hat{q}_h s_h A p_h(u_h - \hat{r}_h u)\|_F \\ \qquad = S \|\hat{q}_h s_h A(u - p_h \hat{r}_h u)\|_F \leq S M \|u - p_h \hat{r}_h u\|_V \end{cases}$$

where M is the norm of A in $L(V, F)$.

Finally, estimate (1-21) follows from (1-20) and the very definition of the error function. ∎

Theorem 1-2 shows that neither the convergence of $p_h u_h$ to u nor the estimate of the error $\|u - p_h u_h\|$ involves the assumption of convergence of the approximations $(F_h, \hat{q}_h, s_h)$ of F. We have only to choose s_h for the stability assumption to hold (see Section 1-8). Nevertheless, convergence properties of the approximations $(F_h, \hat{q}_h, s_h)$ play a role in estimates of the error $u - p_h u_h$ in larger spaces.

THEOREM 1-3 Let G be a Hilbert space containing F with a weaker topology. If u and u_h are solutions of $Au = f$ and $s_h A p_h u_h = s_h f$, then we obtain the following error estimate:

$$(1-23) \quad \begin{cases} \|A(u - p_h u_h)\|_G \leq M e_F^G(q_h s_h) \|u - p_h u_h\|_V \\ \qquad \qquad \qquad \text{for any prolongation} \quad q_h. \end{cases}$$

Furthermore, if we assume that (1-19) holds and that the solution u belongs to a Hilbert space U contained in V, we obtain

$$(1-24) \qquad \|A(u - p_h u_h)\|_G \leq M e_F^G(q_h s_h) e_U^V(p_h) \|u\|_U. \qquad \blacktriangle$$

Proof Since $q_h s_h A(u - p_h u_h) = 0$, we can write

(1-25) $A(u - p_h u_h) = (1 - q_h s_h) A(u - p_h u_h)$.

Then inequality (1-23) follows from (1-25) and the definition of the error functions. If we assume the stability, then Theorem 1-2 and (1-23) imply (1-24). ∎

Finally, we prove the converse of Theorem 1-2, known as the theorem of equivalence.

THEOREM 1-4 Let us assume that A is an isomorphism from V onto F. Then the conditions (1-19) (i.e., convergence of the approximations of V and stability of the operators $s_h A p_h$) are equivalent to the conditions (1-18). ▲

Proof Assume (1-18). Since $\|u - p_h \hat{r}_h u\|_V \leq \|u - p_h u_h\|_V$ by definition of an optimal restriction, we deduce that the approximations are convergent.

Now, let us prove the stability of the operators $s_h A p_h = A_h$. The statement [(1-18)(ii)] amounts to saying that

(1-26) $p_h u_h = p_h A_h^{-1} s_h f$ converges to $u = A^{-1} f$ for any $f \in F$.

Then by the uniform boundedness theorem (see Theorem 2.1-10), there exists a constant S independent of h such that

(1-27) $\|p_h u_h\|_V = \|p_h A_h^{-1} s_h f\|_V \leq S \|f\|_F$,   for any $f \in F$.

Now, if we take $f = \hat{q}_h f_h$ for some $f_h \in F_h$, we obtain

(1-28) $\begin{cases} \|p_h u_h\|_V = \|p_h A_h^{-1} s_h \hat{q}_h f_h\|_V = \|p_h A_h^{-1} f_h\| \\ \qquad\qquad\qquad\qquad \leq S \|\hat{q}_h f_h\|_F = \|\hat{q}_h s_h A p_h u_h\|_F, \end{cases}$

where u_h is the solution of $s_h A p_h u_h = A_h u_h = f_h$. ∎

1-6. Approximation of a Sum of an Isomorphism and a Compact Operator

Let $B \in L(V, F)$ be an isomorphism and $C \in L(V, F)$ be a compact operator. The Riesz-Fredholm alternative (see Section 2.1-15) implies that $A_\lambda = B + \lambda C$ is an isomorphism from V onto F when λ does not belong to a countable subset $S(B, C)$ of isolated points.

We shall show that if A_λ is an isomorphism from V onto F, the stability of the internal approximations $B_h = s_h B p_h$ of B implies the stability of the internal approximations $s_h A_\lambda p_h$ of A_λ.

THEOREM 1-5 Let us assume that $A \in L(V, F)$ is an isomorphism and that $A = B + C$, where B is an isomorphism and C a compact operator.

We assume also that V_h and F_h are finite-dimensional spaces and that the approximations $(V_h, p_h, \hat{r}_h)$ of V and $(F_h, \hat{q}_h, s_h)$ of F are convergent.

If the internal approximations $B_h = s_h B p_h$ of B are stable, then the internal approximations $A_h = s_h A p_h$ of A are stable for h small enough, and the conclusions of Theorem 1-2 hold. ▲

Proof First we prove that the operators $A_h = s_h A p_h$ are invertible for h small enough. Since the spaces V_h and F_h are finite dimensional, this amounts to proving that the A_h's are one to one, that is, that the subsets N_h defined by

$$(1\text{-}29) \qquad N_h = \{u_h \in V_h \quad \text{such that} \quad A_h u_h = 0 \quad \text{and} \quad \|p_h u_h\|_V = 1\}$$

are empty.

Since the union of the subsets $p_h N_h$ is contained in the unit ball, and since the unit ball is weakly compact (see Theorem 2.1-11), there exists a subsequence $p_k u_k$ (where $u_k \in N_k$) which converges weakly to u in V.

Therefore, $A p_k u_k$ converges weakly to Au in F and $\hat{q}_k s_k A p_k u_k$ converges weakly to Au in F because

$$(1\text{-}30) \qquad \begin{cases} (\hat{q}_k s_k A p_k u_k, g) = (A p_k u_k, (\hat{q}_k s_k)'g) \\ \qquad\qquad\qquad\qquad \text{converges to } (Au, g) \text{ for any } g \in F'. \end{cases}$$

(See Theorems 2.4-3 and 2.1-11.) Then $Au = 0$ and, since A is an isomorphism, $u = 0$.

Since any subsequence $p_k u_k$ of the union of $p_h N_h$ converges weakly to 0 in V, the family $p_h u_h$ (where u_h ranges over N_h) converges weakly to 0 in V and $C p_h u_h$ converges strongly to 0 in F. Therefore

$$(1\text{-}31) \qquad \varepsilon(h) = \sup_{u_h \in N_h} \|C p_h u_h\|_F$$

converges to 0.

On the other hand, the stability of the operators B_h implies

$$(1\text{-}32) \quad 1 = \|p_h u_h\|_V \le M \|\hat{q}_h s_h B p_h u_h\|_F = M \|\hat{q}_h s_h C p_h u_h\|_F \le M\varepsilon(h)$$

for any $u_h \in N_h$. Therefore, if $\varepsilon(h) < 1/M$, the inequality implies that N_h is empty.

In order to prove that the isomorphisms A_h are stable (i.e., that the norms $c(h)$ of A_h^{-1} are bounded), let us assume that $c(h)$ converges to ∞. Since the dimension of V_h is finite, there exists $f_h \in F_h$ such that

$$(1\text{-}33) \qquad c(h) = \|p_h A_h^{-1} f_h\|_V = \|p_h u_h\|_V \quad \text{and} \quad \|\hat{q}_h f_h\|_F = 1.$$

Then if we set $v_h = u_h/c(h)$ and $g_h = f_h/c(h) = s_h A p_h\, v_h$, we obtain a contradiction by proving that

$$(1\text{-}34) \qquad\qquad \hat{q}_h s_h B p_h v_h \text{ converges strongly to 0 in } F.$$

Indeed, in this case the stability of the operators B_h would imply that

$$(1\text{-}35) \qquad 1 = \|p_h v_h\| \leq M \|\hat{q}_h s_h B p_h v_h\|_F \text{ converges to 0.}$$

Let us prove (1-34). Because the subsequence $p_h v_h$ belongs to the unit ball of V, we can extract a subsequence $p_k v_k$ converging weakly to v in V. As previously, this implies that $\hat{q}_k s_k A p_k v_k$ converges weakly to Av. Since $\|\hat{q}_h g_h\|_F \leq 1/c(h)$, $\hat{q}_h g_h$ converges strongly to 0, and thus $\hat{q}_k s_k A p_k v_k = \hat{q}_k g_k$ converges to 0. Then $Au = 0$ and $u = 0$. Moreover, the family $p_h v_h$ itself converges to 0. Therefore, since C is compact, we deduce that $\hat{q}_h s_h B p_h u_h = \hat{q}_h g_h - \hat{q}_h s_h C p_h u_h$ converges strongly to 0. ∎

1-7. Approximation of Coercive and V-Elliptic Operators

Let V be a Hilbert space, $F = V'$, $(V_h, p_h, \hat{r}_h)$ approximations of V, and $(V_h', \hat{r}_h', p_h')$ the dual approximations of V'. We saw in Section 1-3 that the internal approximation of the equation $Au = f$ is the discrete variational equation

$$(1\text{-}36) \qquad a(p_h u_h, p_h v_h) = (f, p_h v_h) \qquad \text{for any} \quad v_h \in V_h,$$

where $a(u, v) = (Au, v)$; and now consider the cases of

$$(1\text{-}37) \qquad\qquad A \text{ is } V\text{-elliptic}$$

and

$$(1\text{-}38) \quad \begin{cases} \text{(i) } V \subset H = H' \subset V', \text{ the injections being dense and compact,} \\[4pt] \text{(ii) } A \text{ is a } (V, H)\text{-coercive operator,} \\[4pt] \text{(iii) } A \text{ is an isomorphism from } V \text{ onto } V', \\[4pt] \text{(iv) } V_h \text{ is a finite-dimensional space.} \end{cases}$$

(See Section 2.1-16.)

THEOREM 1-6 Let $(V_h, p_h, \hat{r}_h)$ be convergent approximations of a Hilbert space V, A an operator mapping V into V', u a solution of the equation $Au = f$ where f is given in V', and u_h a solution of (1-36).

We assume either (1-37) or (1-38). Then there exists a unique solution u_h of (1-36) for h sufficiently small, and $p_h u_h$ converges to u in V. Furthermore, the following inequality holds:

$$(1\text{-}39) \qquad \|u - p_h u_h\|_V \leq M \|u - p_h \hat{r}_h u\|_V.$$

If the solution u belongs to $U \subset V$ and if $X \subset V$, we deduce the following estimate of the error:

$$(1\text{-}40) \qquad \sup_{v \in X} \frac{|a(u - p_h u_h, v)|}{\|v\|_X} \leq M e_X^V(p_h) e_U^V(p_h) \|u\|_U \qquad \blacktriangle$$

Proof We deduce Theorem 1-6 from Theorems 1-2 and 1-3. For that purpose, we must prove that the internal approximations $A_h = p'_h A p_h$ of A are stable. Beginning with A being V-elliptic, we obtain the inequalities

$$(1\text{-}41) \quad \begin{cases} c \, \|p_h u_h\|_V^2 \leq a(p_h u_h, p_h u_h) = (p'_h A p_h u_h, u_h)_h \\ \qquad \leq \|p'_h A p_h u_h\|_{V_h'} \|p_h u_h\|_V = \|f'_h p'_h A p_h u_h\|_{V'} \, \|p_h u_h\|_V, \end{cases}$$

since f'_h is the optimal prolongation associated with p'_h in V'. (See Theorem 2.4-2.)

Now assume (1-38), which amounts to saying that $A + \lambda$ is V-elliptic. Then $A = (A + \lambda) - \lambda$ is the sum of the isomorphism $A + \lambda$ and of the compact operator $-\lambda$. On the other hand, we know that the internal approximations $p'_h(A + \lambda)p_h$ of the V-elliptic operator $A + \lambda$ are stable. Then Theorem 1-5 implies that the internal approximations $p'_h A p_h$ of A are stable for h small enough. Thus in both cases Theorem 1-2 implies inequality (1-39).

We can deduce (1-40) from Theorem 1-3. Actually, since $a(u - p_h u_h, p_h v_h) = 0$ for any $v_h \in V_h$, inequality (1-40) follows from

$$(1\text{-}42) \quad \begin{cases} |a(u - p_h u_h, v)| = |a(u - p_h u_h, v - p_h f_h v)| \\ \qquad \leq M \, \|u - p_h u_h\|_V \, \|v - p_h f_h v\|_V \\ \qquad \leq M \, \|u - p_h f_h u\|_V \, \|v - p_h f_h v\|_V \\ \qquad \leq e_U^{\,V}(p_h) e_X^{\,V}(p_h) \, \|u\|_U \, \|v\|_X. \end{cases} \quad \blacksquare$$

When the dimension of the discrete space V_h is finite, the space V_h is identified with its dual by a canonical duality pairing $(u_h, v_h)_h$ which coincides with an inner product. Let $|u_h|_h = (u_h, u_h)_h^{1/2}$ be the associated pivot norm.

Definition 1-3 The condition number $\chi(A_h)$ of a matrix A_h is the product

$$(1\text{-}43) \qquad \chi(A_h) = |A_h|_h \, |A_h^{-1}|_h$$

where

$$|A_h|_h = \sup_{v_h} \frac{|A_h v_h|_h}{|v_h|_h}. \qquad \blacktriangle$$

Let us assume that there exists a constant k such that

$$(1\text{-}44) \qquad k^{-1} \|p_h v_h\|_H \leq |v_h|_h \leq k \, \|p_h v_h\|_H.$$

THEOREM 1-7 If we assume (1-38) and (1-44), then there exists a constant M such that

$$(1\text{-}45) \qquad \chi(A_h) \leq M s_V^{\,H}(p_h)^2. \qquad \blacktriangle$$

Proof   Since $\|v_h\|_{V_h} = \|p_h v_h\|_V \leq k s_V{}^H(p_h) |v_h|_h$, we deduce by transposition that

(1-46)
$$|v_h|_h \leq k s_V{}^H(p_h) \|v_h\|_{V_{h'}}.$$

Furthermore,

(1-47)
$$\begin{cases} \|A_h u_h\|_{V_{h'}} = \sup_{v_h} \dfrac{|(A_h u_h, v_h)|}{\|p_h v_h\|_V} = \sup_{v_h} \dfrac{|a(p_h u_h, p_h v_h)|}{\|p_h v_h\|_V} \\ \qquad\qquad \leq M \|p_h u_h\|_V \leq M k s_V{}^H(p_h) |u_h|_h, \end{cases}$$

where M is the norm of A. Therefore (1-46) and (1-47) imply that

(1-48)
$$|A_h|_h \leq M k^2 s_V{}^H(p_h)^2.$$

On the other hand, since the operators A_h are stable, there exists a constant c such that

(1-49)
$$\|A_h^{-1} f_h\|_{V_h} \leq c^{-1} \|f_h\|_{V_{h'}}.$$

Since $|v_h|_h \leq k \|p_h v_h\|_H \leq k\rho \|p_h v_h\|_V = k\rho \|v_h\|_{V_h}$, where ρ is the norm of the injection from V into H, we deduce by transposition that

(1-50)
$$\|f_h\|_{V_{h'}} \leq k\rho |f_h|_h.$$

Then we deduce from (1-49) and (1-50) that

(1-51)
$$|A_h^{-1} f_h|_h \leq c^{-1} k^2 \rho^2 |f_h|_h.$$

This implies that $\chi(A_h) \leq M c^{-1} k^4 \rho^2 s_V{}^H(p_h)^2$.   ∎

1-8.   Optimal and Quasi-Optimal Stability

Let A be an isomorphism from V onto F and $(V_h, p_h, \hat{r}_h)$ and $(F_h, \hat{q}_h, s_h)$ approximations of V and F, respectively. We can approximate the solution $u \in V$ of

(1-52)
$$Au = f \qquad \text{where } f \text{ is given in } F$$

by the solution $u_h \in V_h$ of its internal approximation

(1-53)
$$s_h A p_h u_h = s_h f.$$

Denote by

(1-54)
$$S(s_h, A) = \sup_{v_h \in V_h} \frac{\|p_h v_h\|_V}{\|\hat{q}_h s_h A p_h v_h\|_F}$$

the norm of $(s_h A p_h)^{-1}$, which is bounded whenever the operators A_h are stable. In this case, we have proved the error estimate

(1-55)
$$\|u - p_h u_h\|_V \leq (1 + S(s_h, A) \|A\|_{L(V, F)}) \|u - p_h \hat{r}_h u\|_V.$$

This inequality implies that the convergence properties depend only on the choice of the prolongations p_h when the stability holds.

Then the problem arises of finding a restriction $\tilde{s}_h$ which minimizes $S(s_h, A)$ either for a given operator A or for a class of operators A. In the first case, we shall prove the following result:

THEOREM 1-8   Let J be the canonical isometry from F onto F' and L_h any isomorphism from V'_h onto F_h.

If p_h is a given prolongation from V_h into V, the restrictions $\tilde{s}_h = L_h p'_h A'J$ satisfy the following "optimal stability" property:

(1-56) $S(\tilde{s}_h, A) \leq S(s_h, A)$ for   any restriction s_h mapping F onto F_h,

and the operators $\tilde{s}_h A p_h$ are stable.   ▲

Remark 1-3   If we choose the restriction $\tilde{s}_h = L_h p'_h A'J$, then the internal approximation (1-53) of (1-52) is equivalent to the discrete variational equation

(1-57) $(JAp_h u_h, Ap_h v_h) = (Jf, Ap_h v_h)$ for any $v_h \in V_h$.   ■

In the second case, consider the following subset $K(c, M)$ of operators $A \in L(V, F)$ satisfying the inequalities

(1-58) $\begin{cases} \text{(i)} \ |(Au, Kv)| \leq M \|u\|_V \|v\|_V & \text{for any} \ \ u, v \in V, \\ \text{(ii)} \ (Av, Kv) \geq c \|v\|_V{}^2 & \text{for any} \ \ v \in V, \end{cases}$

where K is a given isometry from V onto F'.

THEOREM 1-9   Let K be a given isometry from V onto F', $K(c, M)$ the subset of operators $A \in L(V, F)$ satisfying (1-58), and L_h any isomorphism from V'_h onto F_h. If p_h is a given prolongation from V_h into V, the restrictions $\tilde{s}_h = L_h p'_h K'$ satisfy the following "quasi-optimal stability" property:

(1-59) $$S(\tilde{s}_h, A) \leq Mc^{-1}S(s_h, A)$$

for any operator $A \in K(c, M)$ and any restriction s_h from F onto F_h; furthermore the operators $\tilde{s}_h A p_h$ are stable.   ▲

Remark 1-4   If we choose the restriction $\tilde{s}_h = L_h p'_h K'$, then the internal approximation (1-53) of (1-52) is equivalent to the discrete variational equation

(1-60) $(Ap_h u_h, Kp_h v_h) = (f, Kp_h v_h)$ for any $v_h \in V_h$.   ■

Remark 1-5 This theorem motivates the choice of the restriction $\tilde{s}_h = p'_h$ in the case of V-elliptic operators from V onto $F = V'$. Indeed, in this case, we can take $K = 1 \in L(V, V)$ and $L_h = 1 \in L(V'_h, V'_h)$. ∎

Proof of Theorems 1-8 and 1-9 If $s_h A p_h$ is invertible, then

$$(1\text{-}61) \qquad S(s_h, A) = \|(s_h A p_h)^{-1}\|_{L(F_h, V_h)} = \|(p'_h A' s'_h)^{-1}\|_{L(V_{h'}, F_h)}$$

is finite.

However, the dual norm of $\|\hat{q}_h f_h\|_F$ is equal to the norm $\|s'_h f_h\|_{F'}$ (see Theorem 2.4-2). Therefore,

$$(1\text{-}62) \qquad S(s_h, A) = \sup_{f_h} \frac{\|s'_h f_h\|_{F'}}{\|\hat{r}'_h p'_h A' s'_h f_h\|_{V'}} = S(M_h s_h, A)$$

for any isomorphism M_h from F_h onto itself.

Let J be a F-elliptic operator from F onto F' and consider the restriction

$$(1\text{-}63) \qquad \tilde{s}_h = L_h p'_h A' J \qquad \text{where} \quad L_h = s_h A p_h (p'_h A' J A p_h)^{-1}.$$

Therefore

$$(1\text{-}64) \qquad \tilde{s}_h A p_h = s_h A p_h.$$

On the other hand, since J is F-elliptic and since $(J')^{-1} \tilde{s}'_h f_h = A p_h L'_h$, we deduce from (1-64) that

$$(1\text{-}65) \quad \begin{cases} c\|\tilde{s}'_h f_h\|_{F'}{}^2 \leq ((J')^{-1} \tilde{s}'_h f_h, \tilde{s}'_h f_h) = (A p_h L'_h f_h, \tilde{s}'_h f_h) = (L'_h f_h, p'_h A' \tilde{s}'_h f_h) \\ \qquad = (L'_h f_h, p'_h A' s'_h f_h) = ((J')^{-1} \tilde{s}'_h f_h, s'_h f_h) \leq M\|\tilde{s}'_h f_h\|_{F'} \|s'_h f_h\|_{F'}, \end{cases}$$

where M is the norm of $(J')^{-1}$ and c the constant of ellipticity of $(J')^{-1}$.

Therefore, we deduce from (1-62) and (1-65) that

$$(1\text{-}66) \qquad S(\tilde{s}_h, A) \leq Mc^{-1} S(s_h, A).$$

This inequality does not depend on the particular choice of L_h made in (1-63). Indeed, if K_h is another isomorphism from V'_h onto F_h, then $K_h = M_h L_h$ where $M_h = K_h L_h^{-1}$, and, we deduce from (1-62) that $S(K_h p'_h A' J, A) = S(L_h p'_h A' J, A)$.

If we choose the restriction $\tilde{s}_h = L_h p'_h A' J$, the approximate equation (1-53) is equivalent to the discrete equation $p'_h A' J A p_h u_h = p'_h A' J f$; that is, (1-53) is equivalent to the discrete variational equation

$$(1\text{-}67) \qquad (J A p_h u_h, A p_h v_h) = (J f, A p_h v_h) \qquad \text{for any} \quad v_h \in V_h.$$

Therefore, we deduce from the inequality

$$(1\text{-}68) \quad c \|A p_h u_h\|_F{}^2 \leq (J A p_h u_h, A p_h u_h) = (J f, A p_h u_h) \leq M \|f\|_F \|A p_h u_h\|_F$$

that the operators $s_h A p_h$ are stable.

Now let $A \in L(V, F)$ and choose J as the canonical isometry from F onto F'. Then $M = c = 1$ and Theorem 1-8 is proved.

Alternatively, let K be an isometry from V onto F' and $A \in K(c, M)$ and take $J = (A')^{-1}K'$. In this case, $\tilde{s}_h = L_h p_h' A' J = L_h p_h' K'$, and Theorem 1-9 follows from (1-66) if we prove that M is the norm of $(J')^{-1} = AK^{-1}$ and c is its constant of F'-ellipticity. But, since $A \in K(c, M)$, we obtain

$$|((J')^{-1}f, g) = |(AK^{-1}f, g)| = |(Au, Kv)| \leq M \|u\|_V \|v\|_V = M\|f\|_{F'} \|g\|_{F'}$$

and

$$((J')^{-1}f, f) = (Au, Ku) \geq c \|u\|_V^2 = c \|f\|_{F'}^2,$$

where $u = K^{-1}f$ and $v = K^{-1}g$. ∎

2. REGULARITY OF THE CONVERGENCE AND ESTIMATES OF ERROR IN TERMS OF n-WIDTH

We assume that $U \subset V \subset W$, $E \subset F \subset G$, and an operator A is an isomorphism from U onto E, from V onto F, and from W onto G, letting $s_h A p_h u_h = s_h f$ be an internal approximation of the equation $Au = f$.

Under what conditions does the regularity of the convergence hold? In other words, does $p_h u_h$ converge to u in U when $f \in E$ and does $p_h u_h$ converge to u in W when $f \in G$ when we know that $p_h u_h$ converges to u in V when $f \in F$?

The convergence in U holds if, for instance, we assume that the approximations $(V_h, p_h, \hat{r}_h)$ are quasi-optimal for the injection from U into V and convergent in U. In this case, we obtain the following error estimate:

$$\|u - p_h u_h\|_V \leq M E_U^V(n(h) - 1) \|u\|_U,$$

where $n(h)$ is the dimension of V_h and $E_U^V(n)$ is the n-width of the injection from U into V.

The convergence in W holds if, for instance, we assume that the approximations $(F_h, \hat{q}_h, s_h)$ are quasi-optimal for the injection from F into G and the approximations $(V_h, p_h, \hat{r}_h)$ are convergent in W. In this case, we obtain the following error estimate:

$$\|u - p_h u_h\|_W \leq M E_F^G(n(h) - 1) \|u\|_V.$$

2-1. Stability and Convergence in Smaller Spaces

Let A be an isomorphism from V onto F and suppose that we know a result of "regularity": if the data f belong to a smaller space $E \subset F$, then the solution u belongs to a smaller space $U \subset V$. We assume, in other words, that there exist Hilbert spaces U and E such that

$$(2\text{-}1) \qquad \begin{cases} \text{(i)} \ U \subset V, \ E \subset F, \\ \text{(ii)} \ A \text{ is an isomorphism from } U \text{ onto } E, \end{cases}$$

with the injections continuous in (i).

Knowing how to approximate in V solutions u of $Au = f$ when $f \in F$, can we approximate these solutions in U when the data belong to E? To begin with, assume that

$$(2\text{-}2) \qquad \begin{cases} \text{(i)} \ p_h \text{ maps } V_h \text{ into } U; \ P_h = p_h V_h \text{ is closed in } V \text{ (and thus in } U), \\ \text{(ii)} \ s_h \text{ maps } E \text{ and } F \text{ onto } F_h. \end{cases}$$

We thus can associate with p_h and s_h the following approximations: $(V_h, p_h, \hat{r}_h)$ of V, (V_h, p_h, r_h^U) of U, $(F_h, \hat{q}_h, s_h)$ of F, and (F_h, q_h^E, s_h) of E, where r_h^U (resp. $\hat{r}_h$) is the optimal restriction associated with p_h in U (resp. in V) and where q_h^E (resp. $\hat{q}_h$) is the optimal prolongation associated with s_h in E (resp. in F). (See Section 2.4-1.)

THEOREM 2-1   Let us assume that (2-1) and (2-2) hold and that the internal approximations $s_h A p_h$ of A are stable for the norms of V and F. In that case,

$$(2\text{-}3) \qquad \|p_h u_h\|_V \leq M \|\hat{q}_h s_h A p_h u_h\|_F, \qquad M \text{ is independent of } h.$$

If the approximations associated with p_h are quasi-optimal,

$$(2\text{-}4) \qquad \sup_h s_U^V(p_h) e_U^V(p_h) = M < + \infty,$$

and if there exist restrictions r_h from U onto V_h such that

$$(2\text{-}5) \qquad r_h p_h = 1; \ \sup_h e_U^U(p_h r_h) < + \infty; \ \sup_h e_V^V(p_h r_h) < + \infty,$$

then the internal approximation $s_h A p_h$ of A are stable for the norms of U and E:

$$(2\text{-}6) \qquad \|p_h u_h\|_U \leq M \|q_h^E s_h A p_h u_h\|_E, \qquad M \text{ is independent of } h.$$

Therefore, if the approximations (V_h, p_h, r_h^U) of U are convergent, the solutions u_h of the internal approximate equation

$$(2\text{-}7) \qquad s_h A p_h u_h = s_h f$$

converge in U to the solution u of $Au = f$ whenever $f \in E$:

$$(2\text{-}8) \qquad \|u - p_h u_h\|_U \leq M \|u - p_h r_h^U u\|_U.$$

If $n(h)$ is the dimension of the spaces V_h and F_h, the global error behaves like the $[n(h) - 1]$-width from U into V:

$$(2\text{-}9) \qquad \|u - p_h u_h\|_V \leq M E_U^V[n(h) - 1] \|u\|_U. \qquad \blacktriangle$$

Proof Since $Au = f$ and $s_h A p_h u_h = s_h f = s_h A u$, we deduce from (2-3) that

$$\|p_h(u_h - r_h u)\|_V \leq M \|\hat{q}_h s_h A(u - p_h r_h u)\|_F \leq M_1 \|u - p_h r_h u\|_V$$

By Theorem 2.2-4, we know that

$$\|u - p_h r_h u\|_V \leq e_U^V(p_h r_h) \|u\|_U \leq e_U^V(p_h) e_V^V(p_h r_h) \|u\|_U.$$

By the very definition of stability functions, we obtain

$$\|p_h(u_h - r_h u)\|_U \leq s_U^V(p_h) \|p_h(u_h - r_h u)\|_V.$$

Finally, $\|u - p_h r_h u\|_U \leq e_U^U(p_h r_h) \|u\|_U$ by definition of the truncation error. From these inequalities, therefore, we deduce that

$$(2\text{-}10) \quad \|u - p_h u_h\|_U \leq [e_U^U(p_h r_h) + M_1 s_U^V(p_h) e_U^V(p_h) e_V^V(p_h r_h)] \|u\|_U.$$

Then, by the assumptions (2-4) and (2-5), we claim that there exists a constant M independent of h such that

$$\|u - p_h u_h\|_U = \|(A^{-1} - p_h A_h^{-1} s_h)f\|_U \leq M \|f\|_E \qquad \text{for any} \quad f \in E,$$

where $A_h = s_h A p_h$.

Thus we deduce that the operators $p_h A_h^{-1} s_h$ are bounded in $L(E, U)$, and we obtain the inequality

$$\|p_h A_h^{-1} f_h\|_U = \|p_h A_h^{-1} s_h q_h^E f_h\|_U \leq M \|q_h^E f_h\|_E,$$

which is the inequality (2-6) with $u_h = A_h^{-1} f_h$.

Using Theorem 1-3 (the theorem of equivalence), we deduce (2-8) and the convergence in U of $p_h u_h$ to u. On the other hand, since the approximations are quasi-optimal, we deduce from (1-21) and Theorem 2.2-3 the estimate (2-9). ∎

In Theorem 2-1, we assume both the regularity of equation $Au = f$ and the regularity of the approximants of P_h (by assuming that p_h maps V_h into U). We can obtain weaker results regarding the regularity of the convergence by relaxing the assumption that p_h maps V_h into U, however.

Let us introduce a space K between U and V and a prolongation p_h^1 mapping V_h into K, related to the prolongation p_h by the stability function

$$s_K^V(p_h^1, p_h) = \sup_{v_h \in V_h} \frac{\|p_h^1 v_h\|_K}{\|p_h v_h\|_V}.$$

THEOREM 2-2   Let us assume (2-1) and the stability assumptions (2-3), calling K a Hilbert space such that $U \subset K \subset V$ and p_h^1 a prolongation mapping V_h into K. Then the norm in K of the error satisfies the following inequality:

$$(2\text{-}11) \quad \|u - p_h^1 u_h\|_U \leq M[e_U^K(p_h^1 r_h) + s_K^V(p_h^1, p_h)e_U^V(p_h r_h)] \|u\|_U,$$

for any restriction r_h.   ▲

Proof The proof, analogous to the proof of inequality (2-10), is left as an exercise.   ■

Theorem 2-2 implies the convergence of $p_h^1 u_h$ to the solution u whenever the right-hand side of inequality (2-11) converges to 0 for a convenient r_h.

We often encounter the following situation: the right-hand side of inequality (2-11) converges to 0 when $K = V$ and is bounded when $K = U$. In this case, we deduce that $p_h^1 u_h$ converges weakly to u in U. Indeed, since $p_h^1 u_h$ is bounded, a subsequence converges weakly to an element u_* of U. But since this subsequence converges to u in V, $u_* = u$ and $p_h^1 u_h$ converge weakly to u in U.

2-2. Stability and Convergence in Larger Spaces

Let A be an isomorphism from V onto F. If W and G are two Hilbert spaces such that:

$$(2\text{-}12) \quad \begin{cases} \text{(i) } V \subset W, F \subset G, \text{ the injections are continuous and dense,} \\ \text{(ii) } A \text{ can be extended to an isomorphism from } W \text{ onto } G, \end{cases}$$

with the injections continuous and dense in (i).

Knowing how to approximate in V solutions u of $Au = f$ when $f \in F$, can be approximate these solutions in W when the data belong to the larger space G? Let us assume

$$(2\text{-}13) \quad \begin{cases} \text{(i) } P_h = p_h V_h \text{ is closed in } W \text{ (and thus in } V), \\ \text{(ii) } s_h \text{ maps } F \text{ and } G \text{ onto } F_h. \end{cases}$$

Thus we can associate with p_h and s_h the following approximations $(V_h, p_h, \hat{r}_h)$ of V, (V_h, p_h, r_h^W) of W, $(F_h, \hat{q}_h, s_h)$ of F, and (F_h, q_h^G, s_h) of G where r_h^W is the optimal restriction associated with p_h in W and q_h^G is the optimal prolongation associated with s_h in G. (See Section 2.4-1.)

THEOREM 2-3   Let us assume (2-12) and (2-13) and that the internal approximations $s_h A p_h$ of A are stable for the norms of V and F:

$$\|p_h u_h\|_V \leq M \|\hat{q}_h s_h A p_h u_h\|_F, \qquad M \text{ is independent of } h.$$

If the approximations associated with $q_h{}^G$ are quasi-optimal,

(2-14)
$$\sup_h s_F{}^G(q_h{}^G)e_F{}^G(q_h{}^G) = M < +\infty;$$

then the internal approximations $s_h A p_h$ of A are stable for the norms of W and G:

(2-15) $\|p_h u_h\|_W \le M \|q_h{}^G s_h A p_h u_h\|_G$, M is independent of h.

Therefore, if the approximations $(V_h, p_h, r_h{}^W)$ are convergent in W, the solutions u_h of the internal approximate equation

(2-16)
$$s_h A p_h u_h = s_h f$$

converge in W to the solution u of $Au = f$ whenever $f \in G$:

(2-17) $\|u - p_h u_h\|_W \le M \|u - p_h r_h{}^W u\|_W.$

If $n(h)$ denotes the dimension of V_h and F_h and if f belongs to F, the global error in W behaves like the $[n(h) - 1]$-width of the injection from F into G

(2-18) $\|u - p_h u_h\|_W \le M E_F{}^G[n(h) - 1] \|u\|_V.$ ▲

Proof Let us assume for awhile that we have proved that

(2-19) $\|A^{-1} - p_h A_h{}^{-1} s_h\|_{L(G, W)} \le M_0$, M_0 independent of h,

where $A_h = s_h A p_h$. Then

$$\|p_h A_h{}^{-1} s_h\|_{L(G, W)} \le M_1,   M_1 \text{ independent of } h.$$

We deduce from this that

$$\|p_h A_h{}^{-1} f_h\|_W = \|p_h A_h{}^{-1} s_h q_h{}^G f_h\|_W \le M \|q_h{}^G f_h\|_G,$$

which is the inequality (2-15) with $u_h = A_h{}^{-1} f_h$.

Then, by Theorem 1-3 (the theorem of equivalence), we obtain the inequality (2-17) and the convergence of $p_h u_h$ to u in W.

To prove (2-19) we must first transpose the problem. The condition [(2-12)(i)] is equivalent to

(2-20) $$G' \subset F', \quad W' \subset V',$$

the injections are continuous and dense. Since A is an isomorphism from V onto F and from W onto G, its transpose A' is an isomorphism from F' onto V' and from $G' \subset F'$ onto $W' \subset V'$. However, the transpose of $A_h = s_h A p_h$ is equal to

(2-21) $$A_h' = p_h' A' s_h'.$$

Since the norm of an operator is equal to the norm of its transpose, we deduce from the stability of the A_h the stability of the A_h'; namely, we can write (2-3) in the form

$$\|A_h^{-1}f_h\|_{V_h} \le M \|f_h\|_{F_h}$$

where

$$\|u_h\|_{V_h} = \|p_h u_h\|_V \quad \text{and} \quad \|f_h\|_{F_h} = \|\hat{q}_h f_h\|_F$$

Then $\|(A_h')^{-1}f_h\|_{F_{h}'} \le M \|f_h\|_{V_{h}'}$. But, by the results of Section 2.4-2, the dual norm $\|u_h\|_{F_{h}'}$ of $\|\hat{q}_h f_h\|_F$ is equal to $\|s_h' u_h\|_{F'}$ and the dual norm $\|f_h\|_{V_{h}'}$ of $\|p_h u_h\|_V$ is equal to $\|\hat{r}_h' f_h\|_{V'}$. Therefore the stability of A_h from V onto F implies the inequality

$$(2\text{-}22) \quad \|s_h'(A_h')^{-1}f_h\|_{F'} \le M \|\hat{r}_h' f_h\|_{V'} \quad \text{or} \quad \|s_h' u_h\|_{F'} \le M \|\hat{r}_h' A_h' u_h\|_{V'}.$$

Now, we furnish a proof analogous to the one of Theorem 2-1. In so doing, we need the following estimate:

$$(2\text{-}23) \quad s_{G'}^{F'}(s_h') = \sup_{f_h \in F_h} \frac{\|s_h' f_h\|_{G'}}{\|s_h' f_h\|_{F'}} \le s_F^G(q_h^G).$$

Indeed, since $\|s_h' f_h\|_{G'}$ is the dual norm of $\|q_h^G u_h\|_G$ and $\|s_h' f_h\|_{F'}$ is the dual norm of $\|\hat{q}_h u_h\|_F$, we deduce that

$$s_{G'}^{F'}(s_h') \le \sup_{u_h \in V_h} \frac{\|\hat{q}_h u_h\|_F}{\|q_h^G u_h\|_G}.$$

On the other hand, $\|\hat{q}_h u_h\|_F = \|\hat{q}_h s_h q_h^G u_h\|_F \le \|q_h^G u_h\|_F$. These two inequalities imply (2-23). We need also the following identity:

$$(2\text{-}24) \quad e_F^G(q_h^G) = \|1 - q_h^G s_h\|_{L(F,G)} = \|1 - s_h'(q_h^G)'\|_{L(G',F')}.$$

Let u and u_h be the solutions of the transposed problems

$$(2\text{-}25) \quad A'u = f; \; A_h' u_h = p_h' A' s_h' u_h = p_h' f \quad \text{when } f \in G'.$$

Applying successively (2-23), (2-22), (2-25), (2-24), (2-14), and (2-12), we obtain the following inequalities:

$$\|s_h'(u_h - (q_h^G)'u)\|_{G'} \le s_F^G(q_h^G) \|s_h'(u_h - (q_h^G)'u)\|_{F'}$$
$$\le M s_F^G(q_h^G) \|\hat{r}_h' p_h' A'(u - s_h'(q_h^G)'u)\|_{V'} \le M s_F^G(q_h^G) \|u - s_h'(q_h^G)'u\|_{F'}$$
$$\le M s_F^G(q_h^G) e_F^G(q_h^G) \|u\|_F = M \|u\|_F \le M \|f\|_{W'}.$$

However, $\|u - s_h'(q_h^G)'u\|_{G'} \le \|u\|_{G'} \le M \|f\|_{W'}$. Thus we get

$$(2\text{-}26) \quad \|u - s_h' u_h\|_{G'} = \|((A')^{-1} - s_h' A_h'^{-1} p_h')f\|_{G'} \le M \|f\|_{W'},$$

and this implies (2-19), since

$$\|(A')^{-1} - s_h'(A_h')^{-1}p_h'\|_{L(W',G')} = \|A^{-1} - p_h A_h^{-1} s_h\|_{L(G,W)}.$$

Finally, we have to prove (2-18). Although we know by (2-14) and Theorem 2.2-3 that $e_F{}^G(q_h{}^G) \leq ME_F{}^G[n(h) - 1]$, we deduce from (2-22) and (2-25) that

$$(2\text{-}27) \quad \begin{cases} \|u - s_h'u_h\|_{F'} \leq M \|u - s_h'(q_h{}^G)'u\|_{F'} \\ \qquad \leq M \|1 - s_h'(q_h{}^G)'\|_{L(G',F')} \|u\|_{G'} \\ \qquad \leq Me_F{}^G(q_h{}^G) \|f\|_{W'} \leq ME_F{}^G[n(h) - 1] \|f\|_{W'}. \end{cases}$$

This amounts to saying that

$$\|(A')^{-1} - s_h'(A_h')^{-1}p_h'\|_{L(W',F')} = \|A^{-1} - p_h(A_h)^{-1}s_h\|_{L(F,W)}$$
$$\leq ME_F{}^G[n(h) - 1].$$

Therefore, if u is the solution of $Au = f$ and u_h the solution of $A_hu_h = s_hf$, the error is estimated by

$$\|u - p_hu_h\|_W = \|(A^{-1} - p_h(A_h^{-1})s_h)f\|_W \leq ME_F{}^G[n(h) - 1] \|f\|_F. \ \blacksquare$$

Remark 2-1 We can prove analogous statements when the restriction s_h, mapping F onto F_h, cannot be extended to a continuous operator from G onto F_h. One result of this kind is obtained by transposing Theorem 2-2. (See also Section 8.2-2.)

2-3. Approximation of the Value of a Functional at a Solution

In several problems, we are not interested in knowing the solution of the equation $Au = f$, but only in the value (j, u) of a functional j at u. Then it is more economical to determine once and for all the functional k, which is the solution of:

$$(2\text{-}28) \qquad\qquad A'k = j;$$

where A is an isomorphism from V onto F, j is given in V', and k belongs to F', and to apply k to the data f because

$$(2\text{-}29) \qquad (j, u) = (j, A^{-1}f) = ((A')^{-1}j, f) = (k, f).$$

Therefore, instead of approximating the solution u of $Au = f$ and computing $(j_h, u_h)_h$, we approximate (j, u) by $(k_h, f_h)_h$ where k_h is an approximate solution of (2-28).

THEOREM 2-4 Let A be an isomorphism from V onto F, $j \in V'$, and $k = (A')^{-1}j \in F'$. Let k_h be the solution of

$$(2\text{-}30) \qquad A_h'k_h = p_h'j \qquad \text{where} \quad A_h = s_hAp_h.$$

Then if the A_h are stable—that is, if (2-3) holds—and if u is the solution of $Au = f$, we obtain the inequality

$$(2\text{-}31) \qquad |(j, u) - (k_h, s_h f)_h| \leq M \, \|k - s_h' \hat{q}_h' k\|_{F'} \, \|u\|_V. \qquad \blacktriangle$$

Remark 2-2 Since we are only interested in the approximation of (j, u), a "pointwise" estimate of the error $k - s_h' \hat{q}_h' k$ is sufficient for our purpose.

Proof If we consider the dual approximations $(V_h', \hat{r}_h', p_h')$ and $(F_h', s_h', \hat{q}_h')$ of V' and F', we see that the transposes $A_h' = p_h' A' s_h'$ of $A_h = s_h A p_h$ are the internal approximations of A' for these dual approximations and that (2-30) is the internal approximation of $A'k = j$ [see (1-5)]. Deducing from (2-3) the stability of the A_h' [see (2-22)], by Theorem (1-3), we can further deduce that

$$\|k - s_h' k_h\|_{F'} \leq M \, \|k - s_h' \hat{q}_h' k\|_{F'}.$$

Then

$$\begin{aligned} |(j, u) - (k_h, s_h f)_h| &= |(k - s_h' k_h, f)| \\ &\leq \|k - s_h' k_h\|_{F'} \, \|f\|_F \\ &\leq M \, \|k - s_h' \hat{q}_h' k\|_{F'} \, \|u\|_V. \qquad \blacksquare \end{aligned}$$

Remark 2-3 We can deduce from Theorem 2-3 estimates of the global error. For instance, if we assume that $j \in W' \subset V'$, also adopting the hypotheses (2-3), (2-12), (2-13), and (2-14) of Theorem 2-3, we obtain, by (2-27), the following inequality:

$$(2\text{-}32) \qquad |(j, u) - (k_h, s_h f)_h| \leq M E_F{}^G [n(h) - 1] \, \|j\|_{W'} \, \|u\|_V. \qquad \blacksquare$$

Remark 2-4 We can also write

$$(2\text{-}33) \qquad (j, u) - (k_h, s_h f)_h = (j, u - p_h u_h),$$

where u_h is the solution of $s_h A p_h u_h = s_h f$ and deduce from (2-34) other estimates of error. $\qquad \blacksquare$

3. DISCRETE CONVERGENCE, CONSISTENCY, AND OPTIMAL APPROXIMATION OF LINEAR OPERATORS

Let $A \in L(V, F)$ and $f \in F$ be given and $u \in V$ be the solution of $Au = f$. Assume also that V_h and F_h are discrete spaces supplied with norms $\|v_h\|_{V_h}$ and $\|f_h\|_{F_h}$, $A_h \in L(V_h, F_h)$, $f_h \in F_h$ is given, and $u_h \in V_h$ is the solution of the discrete equation $A_h u_h = f_h$. Calling r_h and s_h restrictions mapping V onto

V_h and F onto F_h, respectively, we define

- the discrete errors $\|u_h - r_h u\|_{V_h}$ and $\|f_h - s_h f\|_{F_h}$,
- the lack of consistency $\|A_h r_h u - s_h A u\|_F$,
- the stability of the operators A_h: $\|u_h\|_{V_h} \leq S \|A_h u_h\|_{F_h}$.

Then the discrete convergence of u_h to u is implied by (1) the stability of the operators A_h, (2) the consistency of A_h to A (i.e., the fact that the lack of consistency converges to 0), and (3) the discrete convergence of f_h to f (i.e., the fact that the discrete error converges to 0).

Under convenient assumptions, we can prove a theorem of equivalence stating that the stability and the consistency are not only sufficient conditions, but also are necessary.

If $U \subset V$, let us set $\Phi_U^{F_h}(A_h) = \sup_{\|u\|_U \leq 1} \|A_h r_h u - s_h A u\|_{F_h}$. Then there exists a constant M such that

$$\|u_h - r_h u\|_V \leq M(\|f_h - s_h f\|_{F_h} + \Phi_U^{F_h}(A_h) \|u\|_U).$$

By taking $f_h = s_h f$, we minimize $\sup_{\|u\|_U \leq 1} \|u_h - r_h u\|_{V_h}$ whenever we minimize $\Phi_U^{F_h}(A_h)$ among the operators $A_h \in L(V_h, F_h)$.

If p_h is the optimal prolongation associated with r_h in U, we can prove that $\Phi_U^{F_h}(s_h A p_h) \leq \Phi_U^{F_h}(A_h)$ for any $A_h \in L(V_h, F_h)$, and by so doing, motivate the use of internal approximations.

3-1. Discrete Convergence and Consistency

If A is an isomorphism from V onto F and A_h an isomorphism from V_h onto F_h, we can furnish sufficient (and necessary) conditions for the "convergence" of the solutions $u_h \in V_h$ of

$$(3\text{-}1) \qquad A_h u_h = f_h; \qquad f_h \in F_h$$

to the solution u of

$$(3\text{-}2) \qquad A u = f; \qquad f \in F.$$

First we must introduce a way to compare f with f_h, u with u_h, and A with A_h. We can do that by introducing

$$(3\text{-}3) \qquad \begin{cases} \text{(i) a restriction } r_h \text{ mapping } V \text{ onto } V_h, \\ \text{(ii) a restriction } s_h \text{ mapping } F \text{ onto } F_h, \end{cases}$$

and by supplying the discrete spaces V_h and F_h with discrete norms $\|v_h\|_{V_h}$

Definition 3-1 We say that

$$(3-4) \quad \begin{cases} \text{(i)} \ \|u_h - r_h u\|_{V_h} \text{ is the } \textit{discrete error between } u \in V \text{ and } u_h \in V_h, \\ \text{(ii)} \ \|f_h - s_h f\|_{F_h} \text{ is the } \textit{discrete error between } f \text{ and } f_h, \\ \text{(iii)} \ \|A_h r_h u - s_h A u\|_{F_h} \text{ is the } \textit{lack of consistency} \text{ between } A \text{ and } A_h. \end{cases}$$

▲

Now we extend the notion of stability to this new situation:

Definition 3-2 The operators A_h are stable from V_h onto F_h (supplied with the discrete norms $\|v_h\|_{V_h}$ and $\|f_h\|_{F_h}$) if there exists a constant S independent of h such that

$$(3-5) \qquad \|u_h\|_{V_h} \leq S \|A_h u_h\|_{F_h} \qquad \text{for any} \quad u_h \in V_h.$$

▲

THEOREM 3-1 If we assume that A and A_h are isomorphisms from V onto F and from V_h onto F_h, then the stability of A_h and the consistency of A_h to A

$$(3-6) \quad \begin{cases} \text{(i) the operators } A_h \text{ are stable,} \\ \text{(ii)} \ \lim_{h \to 0} \|A_h r_h u - s_h A u\|_{F_h} = 0 \qquad \text{for any} \quad u \in V \text{ (consistency),} \end{cases}$$

imply the convergence property

$$(3-7) \quad \begin{cases} \text{the discrete convergence of } f_h \text{ to } f \text{ implies} \\ \qquad\qquad\qquad\qquad\qquad \text{the discrete convergence of } u_h \text{ to } u. \end{cases}$$

Conversely, (3-7) implies the stability of the A_h, and if, moreover

$$(3-8) \qquad \|A_h u_h\|_{F_h} \leq M \|u_h\|_{V_h}; \ M \text{ independent of } h,$$

(3-7) implies also the consistency of A_h to A.

▲

Proof Writing $A_h(u_h - r_h u) = f_h - s_h f + s_h A u - A_h r_h u$, the stability assumption, implies

$$(3-9) \quad \begin{cases} \|u_h - r_h u\|_{V_h} \leq M \|A_h(u_h - r_h u)\|_{F_h} \\ \qquad\qquad \leq M(\|f_h - s_h f\|_{F_h} + \|s_h A u - A_h r_h u\|_{F_h}) \end{cases}$$

Therefore, (3-6) implies (3-7). Conversely, let us assume (3-7) and prove the stability of the A_h, for if this were not the case, there would exist $f_h \in F_h$ such that

$$\|f_h\|_{F_h} = 1; \ \|(A_h^{-1})f_h\|_{V_h} = a(h); a(h) \to \infty.$$

Let $g_h = f_h/\sqrt{a(h)}$ and $v_h = A_h^{-1} g_h$. Then g_h converges discretely to 0 since $\|g_h\|_{F_h} = 1/\sqrt{a(h)}$ and v_h does not converge to 0 due to the convergence to infinity of $\|v_h\|_{V_h} = \sqrt{a(h)}$. We thus obtain a contradiction. Now, let us assume (3-7) and (3-8) and prove the consistency; that is, let us prove [(3-6)(ii)]. Let $u \in V$ and u_h be the solution of $A_h u_h = s_h Au$. Then $f_h = s_h Au$ converges discretely to $f = Au$, and, by (3-7), u_h converges discretely to u. Therefore, since the A_h are stable,

$$\|A_h r_h u - s_h Au\|_{F_h} \leq M \|r_h u - A_h^{-1} s_h Au\|_{V_h} = M \|u_h - r_h u\|_{V_h} \to 0. \quad \blacksquare$$

The inequality (3-9) is a "pointwise" estimate of the discrete error. To obtain an estimate of the global discrete error, we are led to introduce the global lack of consistency $\Phi_U{}^{F_h}(A_h)$:

$$(3\text{-}10) \quad \Phi_U{}^{F_h}(A_h) = \|A_h r_h - s_h A\|_{L(U, F_h)} = \sup_{u \in U} \frac{\|A_h r_h u - s_h Au\|_{F_h}}{\|u\|_U}.$$

We therefore deduce the following corollary:

COROLLARY 3-1   Assume that A and A_h are isomorphisms from V onto F and from V_h onto F_h, respectively, and that the operators A_h are stable. Then, if u is the solution of $Au = f$ and u_h the solution of $A_h u_h = s_h f$, the discrete error obeys the following estimate:

$$(3\text{-}11) \quad \|u_h - r_h u\|_{V_h} \leq M \Phi_U{}^{F_h}(A_h) \|u\|_U. \quad \blacktriangle$$

Now let us glance at the relations between discrete convergence and the notion of convergence we have used until now.

LEMMA 3-1   Let p_h be a prolongation mapping V_h onto V. The discrete convergence of u_h to u implies the convergence of $p_h u_h$ to u if and only if

$$(3\text{-}12) \quad \begin{cases} \text{(i) } \|p_h u_h\| \leq M \|u_h\|_{V_h}; \ M \text{ independent of } h, \\ \text{(ii) } \lim_{h \to 0} \|u - p_h r_h u\|_V = 0 \quad \text{for any} \quad u \in V. \end{cases} \quad \blacktriangle$$

Proof   The sufficiency follows from the decomposition $u - p_h u_h = u - p_h r_h u + p_h(r_h u - u_h)$. To prove the necessity, let us assume that

$$\|p_h u_h\|_V = M(h) \to \infty; \quad \|u_h\|_{V_h} = 1.$$

Therefore, if we set $v_h = u_h/\sqrt{M(h)}$, we see that v_h converges discretely to 0 and that $p_h v_h$ does not converge. This is impossible and thus, $M(h) \leq M$ and [(3-12)(ii)] holds. $\blacksquare$

COROLLARY 3-2 Let us adopt the assumptions of Corollary 3-1 and Lemma 3-1, from which the estimates of error can be deduced. The error $u - p_h u_h$ obeys the estimate

$$(3-13) \qquad \|u - p_h u_h\|_V \leq [e_U{}^V(p_h r_h) + M\phi_U{}^{F_h}(A_h)] \|u\|_U. \qquad \blacktriangle$$

We consider in Section 3-3 the relations between estimates of the error $u - p_h u_h$ and estimates of discrete errors.

3-2. Optimal Approximation of Operators and Internal Approximations

By Corollary 3-1, the discrete error between the solution u of $Au = f$ and the solution u_h of $A_h u_h = s_h f$ is estimated by

$$(3-14) \qquad \|u_h - r_h u\|_{V_h} \leq M\Phi_U{}^{F_h}(A_h) \|u\|_U$$

whenever the A_h are stable and $u \in U \subset V$.

Now we must look for the operator A_h mapping V_h onto F_h that minimizes the global lack of consistency. Such an operator A_h exists and is an internal approximation of A.

THEOREM 3-2 Let A be an operator from V onto F, r_h and s_h restrictions mapping V onto V_h and F onto F_h, and U a space satisfying

$$(3-15) \qquad\qquad U \subset V;$$

the injection is continuous and dense; r_h maps U onto V_h, and finally let $p_h{}^U$ be the optimal prolongation associated with r_h in U.

Then the internal approximation $s_h A p_h{}^U$ of A minimizes the global lack of consistency

$$(3-16) \qquad \Phi_U{}^{F_h}(s_h A p_h{}^U) \leq \Phi_U{}^{F_h}(A_h) \qquad \text{for any} \quad A_h \in L(V_h, F_h). \qquad \blacktriangle$$

Proof By (3-10), and since the norm of an operator is equal to the norm of its transpose, we get

$$(3-17) \quad \Phi_U{}^{F_h}(A_h) = \|A's_h' - r_h'A_h'\|_{L(F_h', U')} = \sup_{f_h \in F_h'} \frac{\|A's_h'f_h - r_h'A_h'f_h\|_{U'}}{\|f_h\|_{F_h'}}.$$

But, by Theorem 2.4-2, $r_h'(p_h{}^U)'$ is the orthogonal projector onto $r_h'V_h'$ in U' because $p_h{}^U$ is the optimal prolongation associated with r_h in U. Therefore, we deduce the following inequality for any $f_h \in F_h'$ and any A_h:

$$\|A's_h'f_h - r_h'(p_h{}^U)'A's_h'f_h\|_{U'} \leq \|A's_h'f_h - r_h'A_h'f_h\|_{U'}.$$

Thus taking the supremum on the unit ball of F_h', we obtain

$$\|A's_h' - r_h'(p_h{}^U)'A's_h'\|_{L(F_{h'},U')} \leq \|A's_h' - r_h'A_h'\|_{L(F_{h'},U')}.$$

Since the norm of an operator is equal to the norm of its transpose, this last inequality is equivalent to (3-16).   ∎

3-3. Estimates of Error and Discrete Errors

For the problem of choosing an approximate operator A_h, we can use as criteria either estimates of the error $u - p_h u_h$ or estimates of the discrete error $\|u_h - r_h u\|_{V_h}$. We must be careful, however, because we cannot compare these two kinds of estimates.

If we let u be the solution of $Au = f$ and u_h the solution of $A_h u_h = s_h f$, assuming that u belongs to a smaller space $U \subset V$, then, by the very definition of the n-width, the global error cannot be better than the $n(h)$-width

$$(3\text{-}18) \qquad \sup_{u \in U} \frac{\|u - p_h u_h\|_V}{\|u\|_U} \geq E_U{}^V(n(h)).$$

On the other hand, there are examples in which the global discrete error is 0:

$$(3\text{-}19) \qquad \sup_{u \in U} \frac{\|u_h - r_h u\|_{V_h}}{\|u\|_U} = 0.$$

For example let us take $F = V'$, $A = J$ (the canonical isometry from V onto V'), and $A_h = \hat{p}_h'J\hat{p}_h$ (where $\hat{p}_h$ is the optimal prolongation associated with r_h in V). By Theorem 2.4-2, the solution u_h of $A_h u_h = \hat{p}_h Au$ is no more than $r_h u$. Then the discrete error $\|u_h - r_h u\|_{V_h} = 0$ for any $u \in V$ and therefore (3-19) hold. On the other hand, the global error $\sup_{\|u\|_U \leq 1} \|u - \hat{p}_h u_h\|_V$ is equal to $e_U{}^V(\hat{p}_h)$, and the error function of $\hat{p}_h$ does not necessarily converge.

If we want to know whether the choice of an approximate operator A_h is relevant, we encounter the problem of finding lower estimates of the global discrete error. In other words, such estimates will play the role of the n-width for estimates of global errors.

THEOREM 3-3 Let A be an isomorphism from V onto F and from $U \subset V$ onto $E \subset F$ (the injections being continuous and dense). Let $q_h{}^E$ be the optimal prolongation associated with s_h in E. The global discrete error between solutions of $Au = f$ and solutions of any approximate problems $A_h u_h = s_h f$ is estimated in the following way:

$$(3\text{-}20) \qquad \sup_{u \in U} \frac{\|u_h - r_h u\|_{V_h}}{\|u\|_U} \geq M \|r_h A^{-1}(1 - q_h{}^E s_h)\|_{L(E,V_h)},$$

where M depends only on A.   ▲

Proof Since A is an isomorphism from U onto E, there exists a constant M such that

$$(3-21) \quad \begin{cases} \displaystyle\sup_{u \in U} \frac{\|u_h - r_h u\|_{V_h}}{\|u\|_U} \geq M \sup_{f \in E} \frac{\|A_h^{-1} s_h f - r_h A^{-1} f\|_{V_h}}{\|f\|_E} \\ \qquad = M \|A_h^{-1} s_h - r_h A^{-1}\|_{L(E, V_h)} \\ \qquad = M \|(A')^{-1} r_h' - s_h'(A_h')^{-1}\|_{L(V_h', E')}. \end{cases}$$

Since q_h^E is the optimal prolongation associated with s_h in E, we deduce from Theorem 2.4-2 that $s_h'(q_h^E)'$ is the orthogonal projector from E' onto $s_h' F_h'$. Therefore, for any f_h in V_h', we have the inequality

$$\|(A')^{-1} r_h' f_h - s_h'(A_h')^{-1} f_h\|_{E'} \geq \|(A')^{-1} r_h' f_h - s_h'(q_h^E)'(A')^{-1} r_h' f_h\|_{E'}.$$

From this, after taking the supremum on the unit ball of V_h', we deduce

$$\|(A')^{-1} r_h' - s_h'(A_h')^{-1}\|_{L(V_h', E')} \geq \|(1 - s_h'(q_h^E)')(A')^{-1} r_h'\|_{L(V_h', E')}.$$

Then, since the norm of an operator is equal to the norm of its transpose, we obtain

$$(3-22) \quad \|A_h^{-1} s_h - r_h A^{-1}\|_{L(E, V_h)} \geq \|r_h A^{-1}(1 - q_h^E s_h)\|_{L(E, V_h)},$$

which holds for any invertible operator A_h from V_h onto F_h.

Therefore, (3-21) and (3-22) imply (3-20). ∎

CHAPTER 4

Finite-Element Approximation of Functions of One Variable

For the sake of simplicity, we introduce and study the finite-element approximations of the Sobolev spaces $H^m(R)$ of functions of one variable; the results of this chapter are extended to Sobolev spaces $H^m(R^n)$, $H^m(\Omega)$, and $H_0^m(\Omega)$ in Chapter 5.

We begin by constructing approximations of $L^2(R)$ by steps functions and, in the second Section, we construct and study piecewise-polynomial approximations of Sobolev spaces $H^m(R)$.

In Section 3 we show that these piecewise-polynomial approximations are examples of finite-element approximations, and we estimate the error functions in this more general framework. [It is appropriate to note here that convergent finite-element approximations of the Sobolev spaces $H^m(R)$ are characterized.]

1. APPROXIMATION OF FUNCTIONS OF L^2 BY STEP FUNCTIONS AND BY CONVOLUTION

Step-function approximations of the space $L^2(R)$ are defined by the discrete space $L_h^2(R) = l^2(Z)$ of square summable sequences defined on Z, by p_h^0, where $p_h^0 u_h = \sum_{j \in Z} u_h^j \theta_{jh}(x)$, with $\theta_{jh}(x)$ denoting the characteristic function of $(jh, (j+1)h)$, and by r_h, where $r_h u = \{h^{-1} \int \lambda(x/h - j) u(x) \, dx\}_{j \in Z}$, $\lambda \in L^\infty(R)$ being a function with compact support satisfying $\int \lambda(x) \, dx = 1$. We prove that these approximations are convergent in L^2.

The rest of this section deals with the product of convolution $u * v = \int u(x-t) v(t) \, dt$ of two functions u and v, and we prove the properties that are needed subsequently. In particular, we shall study the approximation of a function u by the convolution product $u * \lambda_h$, where $\lambda \in L^1(R)$ is a function with compact support satisfying $\int \lambda(x) \, dx = 1$ and $\lambda_h(x) = h^{-1} \lambda(x/h)$.

1-1.   The Space L^2 and the Discrete Space $L_h{}^2$

Let $L^2 = L^2(R)$ be the space of measurable and square integrable (classes of) functions on R, supplied with the norm

$$(1\text{-}1) \qquad |u| = \left(\int |u(x)|^2 \, dx \right)^{1/2} ; (u, v) = \int u(x)v(x) \, dx.$$

Let h be a positive number associated with the grid of points jh where j ranges over the ring Z of positive or negative integers. We denote by $L_h{}^2 = L_h{}^2(R)$ the space of square summable sequences $u_h = (u_h{}^j)_{j \in Z}$ defined on Z supplied with the discrete norm

$$(1\text{-}2) \qquad |u_h|_h = \sqrt{h} \left(\sum_{j \in Z} |u_h{}^j|^2 \right)^{1/2} ; (u_h, v_h)_h = h \sum_{j \in Z} u_h{}^j v_h{}^j.$$

This (infinite dimensional) space plays the role of the discrete space of the approximations when the prolongations and restrictions are defined in the next sections.

1-2.   The Prolongations $p_h{}^0$

Let us denote by θ_{jh} the characteristic function of the interval $(jh, (j+1)h)$, that is, the function defined by

$$(1\text{-}3) \qquad \theta_{jh}(x) = \begin{cases} 1 & \text{if} \quad jh \leq x < (j+1)h \\ 0 & \text{if} \quad x < jh \quad \text{or} \quad (j+1)h \leq x. \end{cases}$$

We define $p_h{}^0$ to be the following operator:

$$(1\text{-}4) \qquad p_h{}^0 u_h = \sum_{j \in Z} u_h{}^j \theta_{jh} \qquad \text{where} \quad u_h = (u_h{}^j)_{j \in Z}.$$

1-3.   The Restrictions r_h

A family of restrictions r_h must be defined in order to provide some flexibility later in choosing the convenient restriction adapted to given requirements. First, we denote by λ or μ a function in L^∞ satisfying

$$(1\text{-}5) \qquad \lambda \text{ has a compact support;} \int \lambda(x) \, dx = 1.$$

With such a function λ, we associate the restriction r_h defined by

$$(1\text{-}6) \qquad (r_h u)^j = \int u(x) \lambda_h{}^j(x) \, dx \; ; j \in Z \; (u \in L^2),$$

where

(1-7) $\quad \lambda_h(x) = h^{-1}\lambda\left(\dfrac{x}{h}\right) \quad$ and $\quad \lambda_h{}^j(x) = h^{-1}\lambda\left(\dfrac{x}{h} - j\right); j \in Z.$

1-4. The Theorem of Convergence

Before stating this theorem, we must introduce the *oscillation* of a function u of L^2:

(1-8) $\qquad \omega(u, h) = \sup_{|v| \leq h} \left(\int |u(x - y) - u(x)|^2\, dx\right)^{\frac{1}{2}}.$

It is clear that $\omega(u, h)$ converges to 0 with h and that we can associate with any constant a a constant b such that

(1-9) $\qquad\qquad\qquad \omega(u, ah) \leq b\omega(u, h).$

On the other hand, by using the Taylor formula, we deduce

(1-10) $\qquad\qquad \omega(u, h) \leq h\,|Du| \qquad$ whenever $\quad Du \in L^2.$

THEOREM 1-1   When $p_h{}^0$ and r_h are defined by (1-4) and (1-6), they satisfy the following properties:

(1-11) $\quad \begin{cases} \text{(i) } |p_h{}^0 u_h| = |u_h|_h;\ u_h \in L_h{}^2, \\ \text{(ii) } |r_h u|_h \leq c\,|u|;\ c \text{ independent of } h;\ u \in L^2, \\ \text{(iii) } |u - p_h{}^0 r_h u| \leq c\omega(u, h);\ c \text{ independent of } h. \end{cases}$

Then the triples $(L_h{}^2, p_h{}^0, r_h)$ are convergent approximations of the space L^2. If $V = H^1$ and $H = L^2$, the error function is estimated by

(1-12) $\qquad\qquad\qquad e_V{}^H(p_h{}^0) \leq 2h$

and the optimal restriction associated with $p_h{}^0$ is the one defined by the function $\lambda = \pi$ where π is the characteristic function of $(0, 1)$.   ▲

Proof   We first notice that

$$|p_h{}^0 u_h(x)|^2 = \sum_{j \in Z} |u_h{}^j|^2\, \theta_{jh}(x).$$

Thus we deduce [(1-11(i)] after integration, since $\int \theta_{jh}(x)\, dx = h$. On the other hand, by the Cauchy-Schwarz inequality, we obtain

$$\left|\int u(x)\lambda_h{}^j(x)\, dx\right|^2 \leq \left(\int |\lambda_h{}^j(x)|\, dx\right) \cdot \int |u(x)|^2\, |\lambda_h{}^j(x)|\, dx.$$

Let $c_1 = \int |\lambda|$ and $c_2 = \sup_{x \in R} \sum_{j \in Z} |\lambda(x-j)|$; c_2 is finite because $\lambda \in L^\infty$ and has a compact support. Then

$$|r_h u|_h^2 \leq c_1 h \int \sum_{j \in Z} h^{-1} |u(x)|^2 \left| \lambda \left(\frac{x}{h} - j \right) \right| dx \leq c_1 c_2 \int |u(x)|^2 dx,$$

and this inequality implies [(1-11)(ii)].

Since $\sum_j \theta_{jh}(x) = 1$ and $u(x) = \int u(x) \lambda_h{}^j(y) \, dy$, we can write

$$(1\text{-}13) \quad \begin{cases} |u(x) - p_h{}^0 r_h u(x)|^2 = \sum_{j \in Z} \theta_{jh}(x) \left| \int (u(x) - u(y)) \lambda_h{}^j(y) \, dy \right|^2 \\ \qquad\qquad\qquad \leq c_1 \sum_j \theta_{jh}(x) \int |u(x) - u(y)|^2 \, |\lambda_h{}^j(y)| \, dy. \end{cases}$$

To prove [(1-11)(iii)], we can assume that u is continuous because the space of continuous functions is dense in L^2. Then, if $x \in (jh, (j+1)h)$ and $x/h - z - j$ belongs to the support of λ,

$$(1\text{-}14) \qquad |u(x) - u(x - zh)|^2 \leq c_2 \sup_{|z| \leq ah} |u(x) - u(x-z)|^2,$$

where $a - 1$ is the measure of the support of λ.

Therefore, if we integrate inequality (1-13) with respect to x, perform a change of variable, and use (1-14), we have

$$|u - p_h{}^0 r_h u|^2 \leq c_1 c_2 \sup_{|z| \leq ah} \int \sum_j \theta_{jh}(x) \, |u(x) - u(z)|^2 \, dx$$
$$= c_1 c_2 (\omega(u, ah))^2.$$

We thus have proved (1-11). The last sentence of the theorem is obvious, when $\lambda = \pi$, $c_1 = c_2 = 1$, and $a = 2$. The inequality (1-12) follows from the definition of the error function (Section 1.2-5) and of the Sobolev space H^1 (Section 1.1-6). ∎

1-5. Convolution of Functions and Measures

If u and v are two functions with compact support, the convolution of u and v is the function

$$(1\text{-}15) \qquad (u * v)(x) = \int u(x-y)v(y) \, dy = \int u(y)v(x-y) \, dy.$$

This definition extends to less smooth functions.

LEMMA 1-1   If u and v belong to L^2, then the convolution of u and v exists and is a bounded continuous function vanishing at ∞

$$(1\text{-}16) \qquad |(u * v)(x)| \leq |u| \cdot |v| \qquad \text{for any} \quad u, v \in L^2. \qquad ▲$$

Proof We first denote by $\delta(y) * u$ the translate of u:

$$(1\text{-}17) \qquad (\delta(y) * u)(x) = u(x - y).$$

It is clear that the translation $\delta(y) *$ is a continuous operator of L^2 and that $y \to \delta(y) * u$ is continuous from R into L^2. Indeed, we can first approximate u by a continuous function φ with compact support. Thus $|u - \varphi| \le \varepsilon$ and $|\delta(y) * \varphi - \delta(z) * \varphi| \le \varepsilon$ if $|z - y| \le \alpha$ (since φ is uniformly continuous). Therefore, $|\delta(y) * u - \delta(z) * u| \le 3\varepsilon$ for $|z - y| \le \alpha$.

In particular, $\delta(y) * u$ is continuous from R into L^2 supplied with the weak topology, and then the function

$$(1\text{-}18) \qquad \alpha(y) = (\delta(y) * u, v) = \int u(x - y)v(x)\, dx$$

is a bounded continuous function of y. This is true because the Cauchy-Schwarz inequality yields

$$(1\text{-}19) \qquad |\alpha(y)| \le |u| \cdot |v|.$$

But we see that $\alpha(y) = (\tilde{u} * v)(y)$ where $\tilde{u}(x) = u(-x)$. ∎

We deduce from the relation

$$(1\text{-}20) \quad \int (u * v)(x)\, dx = \iint u(x - y)v(y)\, dx \cdot dy = \int u(x)\, dx \int v(y)\, dy$$

the following Lemma:

LEMMA 1-2 If the functions λ and μ belong to L^1, the convolution $\lambda * \mu$ also belongs to L^1 and

$$(1\text{-}21) \qquad \int |\lambda * \mu| \le \int |\lambda| \cdot \int |\mu|.$$

Let us now prove Lemma 1-3.

LEMMA 1-3 If λ, μ belong to L^1 and u belongs to L^2, the convolution of λ, μ and u has a meaning and is an associative and commutative operation

$$(1\text{-}22) \qquad \lambda * \mu * u = \lambda * (\mu * u) = (\lambda * \mu) * u.$$

The operator of convolution by λ is a continuous operator of L^2 and its norm is equal to $\int |\lambda|$. Its transpose is the operator of convolution by $\tilde{\lambda}$ where $\tilde{\lambda}(x) = \lambda(-x)$. ▲

Proof We simply have to prove that if $\lambda \in L^1$ and $u \in L^2$, the convolution has a meaning and

$$(1\text{-}23) \qquad |\lambda * u| \leq \left(\int |\lambda| \right) \cdot |u| .$$

Indeed, let v be any function of L^2. By Lemma 1-1, the function $\alpha(y)$ defined by (1-18) is bounded and continuous. Then the integral $\int \lambda(y)\alpha(y)\, dy$ has a meaning; however, we can check by Fubini's theorem that

$$\int \lambda(y)\alpha(y)\, dy = \int \lambda(y)\left(\int u(x-y)v(x)\, dx \right) dy$$
$$= \int \left(\int u(x-y)\lambda(y)\, dy \right) v(x)\, dx.$$

Therefore the convolution of λ and u is a continuous linear functional on L^2, and thus belongs to L^2. Inequality (1-23) follows from (1-19). ∎

We can finally define operators of convolution by bounded measures. Indeed, by Lemma 1-1, if $u \in L^2$, the operator of convolution by $\tilde{u}$ is a continuous linear operator mapping L^2 into the Banach space B of bounded continuous functions vanishing at ∞.

The dual B' of B is the space of *bounded Radon measures*, and although the usual Lebesgue measure does not belong to this space, B' does contain all the measures with compact support. Since the operator of convolution by $\tilde{u}$ maps L^2 into B, its transpose maps B' into L^2. Then if λ is a bounded measure, we shall set

$$(1\text{-}24) \qquad (\tilde{u}\,*)'\lambda = u * \lambda = \lambda * u \in L^2.$$

Then we can state Lemma 1-4.

LEMMA 1-4 Let λ be a bounded Radon measure and u a function of L^2. Then the convolution of λ and u defined by

$$(1\text{-}25) \qquad \int (\lambda * u)(x)v(x)\, dx = \int \left(\int u(x-y)v(x)\, dx \right) d\lambda(y) \quad \text{for any} \quad v \in L^2$$

belongs to L^2 and the operator of convolution by λ is a continuous linear operator of L^2. ▲

Since the space L^1 is contained in B', we see that the definitions are consistent. For instance, the Dirac measure $\delta(y)$ at the point y is a bounded measure and it is possible to check that

$$(1\text{-}26) \qquad (\delta(y) * u)(x) = u(x-y).$$

1-6. Approximation by Convolution

Let λ be a measure satisfying

(1-27) $\qquad\qquad \lambda$ has a compact support; $\int d\lambda(x) = 1$

and use λ_h to denote the measure defined by

(1-28)
$$
\begin{cases}
(\lambda_h, u) = \int u(x)\, d\lambda_h(x) \\[2mm]
\qquad\quad = \int u(xh)\, d\lambda(x) \quad \text{for any continuous function } u.
\end{cases}
$$

If $\lambda \in L^1$, this definition of λ_h coincides with (1-7).

If $u \in L^2$, we say that $\lambda_h * u$ is the *approximation of u by convolution.*
Indeed, the following theorem holds:

THEOREM 1-2 If λ satisfies (1-27), the operators of convolution by
λ_h are bounded and converge pointwise to the identity

(1-29)
$$
\begin{cases}
\text{(i) } |\lambda_h * u| \le c\,|u|;\ c \text{ is a constant independent of } h. \\[2mm]
\text{(ii) } |u - \lambda_h * u| \le c\omega(u, h) \text{ and converges to 0 with } h.
\end{cases}
\qquad \blacktriangle
$$

Proof We first remark that the norm of λ_h in B' is bounded because

$$
|(\lambda_h, u)| = \left| \int u(xh)\, d\lambda(x) \right| \le \|\lambda\|_{B'} \sup_{x \in R} |u(xh)| = \|\lambda\|_{B'} \sup_{x \in R} |u(x)|.
$$

Therefore, since the norm of the operator $\lambda_h *$ is less than or equal to
$\|\lambda\|_{B'}$, we deduce that $|\lambda_h * u| \le \|\lambda\|_{B'} |u|$. On the other hand,

$$
(\lambda_h * u)(x) = \int u(x - y)\, d\lambda_h(y) = \int u(x - yh)\, d\lambda(y)
$$
$$
\text{and} \qquad u(x) = \int u(x)\, d\lambda(y).
$$

Then, if u is a continuous function with compact support,

(1-30)
$$
\begin{cases}
|(u - \lambda_h * u)(x)|^2 = \left| \int (u(x) - u(x - yh))\, d\lambda(y) \right|^2 \\[3mm]
\qquad\qquad \le \sup_{|z| \le dh} |u(x) - u(x - z)|^2\, \|\lambda\|_{B'}^2,
\end{cases}
$$

where d is the measure of the support of λ. Therefore, integrating this inequality with respect to x, we deduce that

$$(1\text{-}31) \qquad |u - \lambda_h * u|^2 \leq \|\lambda\|_{B'}{}^2 \omega(u, dh)$$

for any continuous function u with compact support.

Since the space of these functions is dense in L^2, we obtain the second inequality of the theorem. ∎

2. PIECEWISE-POLYNOMIAL APPROXIMATIONS OF SOBOLEV SPACES H^m

Let us denote by π the characteristic function of $(0, 1)$ and by π_m its m-fold convolution. Then the operator $p_h{}^m$ defined by $p_h{}^m u_h = \pi_m * p_h{}^0 u_h$ maps L_h^2 into H^m, and we can prove that the approximations $(V_h, p_h{}^m, r_h)$ are convergent in H^m.

Furthermore, the restriction of $p_h{}^m u_h$ to each interval $(jh, (j+1)h)$ is equal to $\sum_{0 \leq k \leq m} u_h{}^{j-k} \alpha_m{}^k(x/h - j)$, where $(\alpha_m{}^k(x))_{0 \leq k \leq m}$ is a basis of the space of polynomials of degree m.

The prolongations $p_h{}^m$ satisfy the commutation formulas $D^k p_h{}^m u_h = p_h{}^{m-k} \nabla_h{}^k u_h$, where $\nabla_h{}^k$ is the finite-difference operator defined by

$$(\nabla_h{}^k u_h)^j = h^{-k} \sum_{0 \leq i \leq k} (-1)^i \binom{k}{i} u_h{}^{j-i}.$$

We estimate the error functions of the piecewise-polynomial approximations in the next Section, proving that $e_s{}^k(p_h{}^m) \leq ch^{s-k}$, where $0 \leq k \leq s \leq m+1$, $k \leq m$. We prove in Section 2-6 that $s_s{}^k(p_h{}^m) \leq ch^{-(s-k)}$; in other words, the piecewise-polynomial approximations are quasi-optimal for the injection from H^s into H^k for $0 \leq k \leq s \leq m$.

2-1. Finite-Difference Operators

Let us denote by ∇_h the finite-difference operator defined by

$$(2\text{-}1) \qquad \nabla_h u(x) = h^{-1}(u(x) - u(x - h)) = h^{-1}\left[(\delta - \delta(h)) * u\right](x).$$

and by $\nabla_h{}^m$ its m-fold power.

The translation and the finite-difference operating on sequences are defined by

$$(2\text{-}2) \qquad (\delta(ih) * u_h)^j = u_h{}^{j-i} \; ; \; (\nabla_h u_h)^j = h^{-1}(u_h{}^j - u_h{}^{j-1}).$$

Let us consider now the characteristic function π of the interval $(0, 1)$ and $\pi_h(x) = h^{-1}\pi(x/h)$ equal to h^{-1} on $(0, h)$ and to 0 elsewhere. The (weak) derivative of π_h is equal to $h^{-1}(\delta - \delta(h))$ since, for any φ infinitely differentiable with compact support, we have

$$(D\pi_h, \varphi) = -(\pi_h, D\varphi) = -h^{-1}\int_0^h D\varphi(x)\, dx = -h^{-1}(\varphi(h) - \varphi(0)).$$

Therefore, we have an often-used relation for the operator of convolution by $D\pi_h$:

$$(2\text{-}3) \qquad\qquad \nabla_h = D\pi_h *.$$

First we deduce that if $u \in L^2$, the convolution $\pi_h * u \in H^1$, since $D(\pi_h * u) = D\pi * u = \nabla_h u \in L^2$. To have analogous formulas for $\nabla_h{}^m$, we are led to introduce π_m:

$$(2\text{-}4)\quad \pi_m = \underbrace{\pi * \pi * \cdots * \pi}_{m\ \text{times}};\ \pi_{m,h}(x) = h^{-1}\pi_m\left(\frac{x}{h}\right) = (\pi_h * \cdots * \pi_h)(x)$$

since

$$\pi_h * \pi_h = \int h^{-1}\pi\left(\frac{x}{h} - \frac{y}{h}\right)\pi\left(\frac{y}{h}\right)h^{-1}\, dy = h^{-1}\int \pi\left(\frac{x}{h} - z\right)\pi(z)\, dz.$$

We deduce Lemma 2-1 from this equation.

LEMMA 2-1 If $\pi_{m,h}$ denotes the m-fold convolution of the function $\pi_h(x) = h^{-1}\pi(x/h)$, where π is the characteristic function of $(0, 1)$, then the operator of convolution $\pi_{m,h} *$ by $\pi_{m,h}$ maps L^2 into the Sobolev space H^m and satisfies the following commutation properties:

$$(2\text{-}5) \qquad D^k\pi_{m,h} * u = \pi_{m-k,h} * \nabla_h{}^k u \qquad \text{for any}\ \ k \le m.$$

The function π_m is positive, $\int \pi_m(x)\, dx = 1$, and its support is the interval $(0, m)$. ▲

2-2. Construction of Approximations of the Space H^m

Let $H^m = H^m(R)$ be the Sobolev space of order m on R, and use the same discrete space and restriction of the approximations that were employed for the approximations of L^2 (see Sections 1-1 and 1-3). However, we must introduce new prolongations. Since $p_h{}^0 u_h$ belongs to L^2, we deduce from Lemma 2-1 that $\pi_{m,h} * p_h{}^0 u_h$ belongs to H^m. Therefore, we shall set

$$(2\text{-}6) \qquad p_h{}^m u_h = \pi_{m,h} * p_h{}^0 u_h = \sum_{j \in Z} u_h{}^j \theta_{jh} * \pi_{m,h}$$

We deduce from Lemma 2-2:

THEOREM 2-1   The prolongations $p_h{}^m$ and the restrictions r_h defined by (2-6) and (1-6) satisfy the following commutation formulas:

$$(2.7) \quad \begin{cases} \text{(i)} \;\; D^k p_h{}^m u_h = p_h{}^{m-k} \nabla_h{}^k u_h, \\ \text{(ii)} \;\; \nabla_h{}^k r_h u = r_h \nabla_h{}^k u = r_h(D^k u * \pi_{k,h}). \end{cases} \qquad \blacktriangle$$

Proof   It is sufficient to prove that the translation operators commute with $p_h{}^0$ and r_h. Indeed, if $j \in Z$,

$$\delta(jh) * p_h{}^0 u_h = \sum u_h{}^i \, \delta(jh) * \theta_{ih} = \sum_i u_h{}^i \theta_{(j+i)h} = \sum_i u_h{}^{i-j} \theta_{ih} = \sum_i (\delta(jh) * u_h)^i \theta_{ih}$$

since $\delta(jh) * \theta_{ih} = \theta_{(j+i)h}$. In the same way, since $\delta(jh) * \lambda_h{}^i = \lambda_h{}^{i+j}$, we obtain

$$\delta(jh) * (r_h u)^i = (u, \lambda_h{}^{i-j}) = (u, \delta(-jh) * \lambda_h{}^i) = (\delta(jh) * u, \lambda_h{}^i).$$

Therefore, since the finite difference operators are linear combinations of translations, they commute with $p_h{}^0$ and r_h:

$$\nabla_h{}^k p_h{}^0 u_h = p_h{}^0 \nabla_h{}^k u_h; \; \nabla_h{}^k r_h u = r_h \nabla_h{}^k u.$$

Finally, by Lemma 2-1

$$D^k p_h{}^m = D^k \pi_{m,h} * p_h{}^0 = \pi_{m-k,h} * \nabla_h{}^k p_h{}^0 = \pi_{m-k,h} * p_h{}^0 \nabla_h{}^k = p_h{}^{m-k} \nabla_h{}^k.$$

On the other hand, since $u \in H^m$, $D^k u \in L^2$, and thus

$$\nabla_h{}^k u = (D^k \pi_{k,h}) * u = \pi_{k,h} * D^k u. \qquad \blacksquare$$

Definition 2-1   The approximations $(L_h^2, p_h{}^m, r_h)$, where $p_h{}^m$ is defined by (2-6) and r_h is defined by (1-5) and (1-6), are the piecewise-polynomial approximations of degree m of the Sobolev spaces H^k for $k \leq m$. $\qquad \blacktriangle$

2-3.   Convergence Theorem

We deduce from Theorem 1-1 and 1-2 and the properties of convolution the following:

THEOREM 2-2   For any restrictions r_h defined by (1-5) and (1-6), $p_h{}^m r_h u$ converges to u in H^m. More precisely, the following estimates hold:

$$(2\text{-}8) \quad |D^k(u - p_h{}^m r_h u)| \leq c\omega(D^k u, h); \; c \text{ independent of } h; \; k \leq m. \qquad \blacktriangle$$

Proof   By Theorem 2-1, we can write

$$D^k(u - p_h{}^m r_h u) = D^k u - p_h{}^{m-k} r_h(D^k u * \pi_{k,h}) = A_h + B_h + C_h$$

where

(2-9)
$$\begin{cases} \text{(i) } A_h = D^k u - \pi_{m-k,h} * D^k u, \\ \text{(ii) } B_h = \pi_{m-k,h} * (D^k u - p_h{}^0 r_h D^k u), \\ \text{(iii) } C_h = \pi_{m-k,h} * p_h{}^0 r_h (D^k u - \pi_{k,h} * D^k u). \end{cases}$$

By Theorems 1-1 and 1-2, we deduce

(2-10)
$$\begin{cases} \text{(i) } |A_h| = |D^k u - \pi_{m-k,h} * D^k u| \leq c_1 \omega(D^k u, h), \\ \text{(ii) } |B_h| \leq |D^k u - p_h{}^0 r_h D^k u| \leq c_2 \omega(D^k u, h) \\ \text{(iii) } |C_h| \leq m |D^k u - \pi_{k,h} * D^k u| \leq m c_3 \omega(D^k u, h). \end{cases}$$

This follows because π_k and π_{m-k} are functions with compact support and integral equal to 1. ∎

2-4. Explicit Form of Function π_m

To obtain further properties of the prolongations $p_h{}^m$, we need the explicit form of π_m.

THEOREM 2-3   The restriction of π_{m+1} to each interval $(k, k + 1)$ is a polynomial $\alpha_m{}^k(x)$ of degree m:

(2-11) $$\pi_{m+1}(x) = \sum_{k=0}^{m} \alpha_m{}^k(x - k)\theta_k(x); \qquad \alpha_m{}^k(x) = \sum_{j=0}^{m} a_m(k, j) \frac{x^j}{j!},$$

where θ_k is the characteristic function of $(k, k + 1)$ and the coefficients $a_m(k,j)$ of $\alpha_m^k(x)$ are given by

(2-12) $$a_m(k, j) = \sum_{0 \leq i \leq k} (-1)^i \binom{m + 1}{i} \frac{(k - i)^{m-j}}{(m - j)!} \qquad 0 \leq k, j \leq m. \quad ▲$$

Proof   We prove Theorem 2-3 by induction. The statement is obviously true if $m = 0$, so let us assume it true for $m - 1$ and prove it for m. Writing $\pi_{m+1} = \pi * \pi_m = \sum_{0 \leq k \leq m-1} \delta(k) * \pi * (\alpha_{m-1}^k \theta)$, we see that

$$\pi * (\alpha_{m-1}^k \theta) = \begin{cases} \displaystyle\int_0^x \alpha_{m-1}^k(y)\, dy & \text{for } 0 \leq x \leq 1, \\ \displaystyle\int_{x-1}^1 \alpha_{m-1}^k(y)\, dy & \text{for } 1 \leq x \leq 2. \end{cases}$$

In other words, we can write

(2-13) $$\pi * (\alpha_{m-1}^k \theta) = \left(\int_0^x \alpha_{m-1}^k(y)\, dy \right)\theta + \delta(1) * \left(\int_x^1 \alpha_{m-1}^k(y)\, dy \right)\theta.$$

We therefore deduce that

$$(2\text{-}14)\begin{cases} \text{(i)} \quad \alpha_m{}^0(x) = \int_0^x \alpha_{m-1}^0(y)\, dy, \\[2ex] \text{(ii)} \quad \alpha_m{}^k(x) = \int_0^x (\alpha_{m-1}^k(y) - \alpha_{m-1}^{k-1}(y))\, dy + \int_0^1 \alpha_{m-1}^{k-1}(y)\, dy; \\[1ex] \hspace{6cm} 1 \le k \le m-1, \\[2ex] \text{(iii)} \quad \alpha_m{}^m(x) = -\int_0^x \alpha_{m-1}^{m-1}(y)\, dy + \int_0^1 \alpha_{m-1}^{m-1}(y)\, dy. \end{cases}$$

Therefore, the coefficients $a_m(k, j)$ obey the following recursive formulas:

$$(2\text{-}15)\begin{cases} \text{(i)} \quad a_m(0, 0) = 0; \qquad a_m(k, 0) = \sum_{0 \le j \le m-1} \dfrac{a_{m-1}(k-1, j)}{(j+1)!} \\[1ex] \hspace{7cm} 1 \le k \le m. \\[2ex] \text{(ii)} \quad a_m(k, j) = a_{m-1}(k, j-1) - a_{m-1}(k-1, j-1); \\[1ex] \hspace{4cm} 1 \le k \le m-1; \qquad 1 \le j \le m, \\[2ex] \text{(iii)} \quad a_m(0, j) = a_{m-1}(0, j-1); \\[1ex] \hspace{2cm} a_m(m, j) = -a_{m-1}(m-1, j-1); \quad 1 \le j \le m. \end{cases}$$

Now using the binomial formula and the relation $\binom{m+1}{i} = \binom{m}{i} - \binom{m}{i-1}$, we replace the coefficients $a_{m-1}(k, j)$ by their values given by the inductive hypothesis and, ultimately, obtain the theorem. In addition, we have shown the following result. ∎

COROLLARY 2-1 The polynomials $\alpha_m{}^k(x)$ and their coefficients $a_m(k, j)$ obey the recursive formulas (2-14) and (2-15). Moreover,

$$(2\text{-}16)\begin{cases} \text{(i)} \quad a_m(k, j) = (D^j \alpha_m{}^k)(0) = (D^j \pi_{m+1})(k). \\[2ex] \text{(ii)} \quad \sum_{0 \le k \le m} \alpha_m{}^k(x) = 1 \\[2ex] \text{(iii)} \quad \int_0^1 \alpha_m{}^k(y)\, dy = \alpha_{m+1}^{k+1}(0). \end{cases}$$

▲

Proof The statements [(2 − 16)(i)] and [(2 − 16)(iii)] are obvious, the function $\sum_{j \in Z} \delta(j) * \pi_{m+1} = \pi_m * 1 = \int \pi_m(x)\, dx = 1$ and its restriction to $(0, 1)$ is equal to $\sum_{k=0}^m \alpha_m{}^k(x)$. This implies [(2-16)(ii)]. ∎

Numerical values of some coefficients $a_m(k,j)$ include

$m = 1$

k \ j	0	1
0	0	1
1	1	-1

$m = 2$

k \ j	0	1	2
0	0	0	1
1	$\frac{1}{2}$	1	-2
2	$\frac{1}{2}$	-1	1

$m = 3$

k \ j	0	1	2	3
0	0	0	0	1
1	$\frac{1}{6}$	$\frac{1}{2}$	1	-3
2	$\frac{4}{6}$	0	-2	3
3	$\frac{1}{6}$	$-\frac{1}{2}$	1	-1

$m = 4$

k \ j	0	1	2	3	4
0	0	0	0	0	1
1	$\frac{1}{24}$	$\frac{1}{6}$	$\frac{1}{2}$	1	-4
2	$\frac{11}{24}$	$\frac{3}{6}$	$-\frac{1}{2}$	-3	6
3	$\frac{11}{24}$	$-\frac{3}{6}$	$-\frac{1}{2}$	3	-4
4	$\frac{1}{24}$	$-\frac{1}{6}$	$\frac{1}{2}$	-1	1

COROLLARY 2-2   The polynomials $\alpha_m{}^k(x)$ are linearly independent $(0 \leq k \leq m)$.   ▲

Proof   We prove Corollary 2-2 by induction, assuming that the statement is true for $m - 1$ and proving it for m. Let us suppose that

$$(2\text{-}17) \qquad \sum_{k=0}^{m} a_k \alpha_m{}^k(x) = 0.$$

Therefore, by differentiating, we deduce from (2-14) that

$$(2\text{-}18) \qquad \sum_{k=0}^{m-1} (a_k - a_{k+1}) \alpha_{m-1}^k(x) = 0$$

and thus, by the induction hypothesis, $a_0 = a_1 = \cdots = a_k = \cdots = a_m$. Then $a_0 \sum_{k=0}^{m} \alpha_m{}^k(x) = a_0 = 0$ and all the coefficients $a_k = 0$.   ■

2-5. Properties of the Prolongations $p_h{}^m$

We now apply the results of the previous section.

THEOREM 2-4   The restriction of the function $p_h{}^m u_h$ to each interval $(jh, (j + 1)h)$ is a polynomial of degree m whose coefficients involve only the m components $(u_h^{j-k})_{0 \leq k \leq m}$:

$$(2\text{-}19) \qquad p_h{}^m u_h = \sum_{j \in Z} \theta_{jh}(x) \sum_{0 \leq k \leq m} u_h^{j-k} \alpha_m{}^k \left(\frac{x}{h} - j \right)$$

and the value of $D^k p_h{}^m u_h$ at the point jh is equal to

$$(2\text{-}20) \qquad (Q_{m,h}^k u_h)^j = (D^k p_h{}^m u_h)(jh) = h^{-k} \sum_{0 \le i \le m} a_m(i, k) u_h^{j-i}.$$

The operator $p_h{}^m$ is an isomorphism from $L_h{}^2$ onto its closed range in H^m and there exists a left inverse r_h of $p_h{}^m$ (i.e., satisfying $r_h p_h{}^m = 1$) among the restrictions r_h defined by (1-6).

For such restrictions, the triples $(L_h{}^2, p_h{}^m, r_h)$ are convergent approximations of H^m. Moreover, if we set $V = H^k$ and $U = H^s$ with $0 \le k \le s \le m$, we have

$$(2\text{-}21) \qquad e_U{}^V(p_h{}^m r_h) \le c e_U{}^V(p_h{}^m); \ c \text{ is independent of } h. \qquad \blacktriangle$$

Proof The first statement of Theorem 2-4 follows from the definition of $p_h{}^m$ by (2-6) and from Theorem 2-3, since

$$(2\text{-}22) \qquad \theta_{jh} * \pi_{m,h} = \pi_{m+1}\left(\frac{x}{h} - j\right) = \sum_{0 \le k \le m} \alpha_m{}^k\left(\frac{x}{h} - k - j\right)\theta_{(j+k)h}(x)$$

and the second statement follows from Corollary 2-1 [(2-16)(i)]. A restriction r_h defined by (1-6) is a left inverse of $p_h{}^m$ if the function λ satisfies

$$(2\text{-}23) \qquad (h\pi_{m+1,h}^j, \lambda_h{}^k) = (\pi_{m+1}, \delta(k - j) * \lambda) = \begin{cases} 0 & \text{if } j \ne k, \\ 1 & \text{if } j = k. \end{cases}$$

For instance, we can try to find λ of the form

$$(2\text{-}24) \qquad \lambda(x) = \pi(x)\left(\sum_{0 \le j \le m} b_j \frac{x^j}{j!}\right).$$

In this case, the b_j's are solutions of the linear system

$$(2\text{-}25) \qquad \sum_{0 \le j \le m} b_j \int_0^1 \alpha_m{}^k(x) \frac{x^j}{j!}\, dx = \begin{cases} 0 & \text{if } 1 \le k \le m \\ 1 & \text{if } k = 0. \end{cases}$$

There exists a unique solution of this system because the polynomials $\alpha_m{}^k$ and $x^j/j!$ are linearly independent (see Corollary 2-2). The integral of such a function λ is equal to 1, since

$$(2\text{-}26) \qquad \int \lambda(x)\, dx = \sum_{j \in Z} \int \lambda(x)\, \delta(j) * \pi_m(x)\, dx = \int \lambda(x)\alpha_m{}^0(x)\, dx = 1.$$

Thus the last statement follows from Theorem 2.2-4 and from Theorem 2-2, which implies that:

$$(2\text{-}27) \qquad e_V{}^V(p_h{}^m r_h) = \|1 - p_h{}^m r_h\|_{L(H^k, H^k)} \le c; \ c \text{ independent of } h. \qquad \blacksquare$$

We estimate the error functions in Section 3-1 (Theorem 3-1).

2-6.   Estimates of the Stability Functions

If we view the triple (L_h^2, p_h^m, r_h) as an approximation of the space H^k for $0 \leq k \leq m$, we implicitly supply the discrete space with the norm $\|p_h^m u_h\|_k$. The problem of how these norms behave arises, and, since the discrete space plays the role of a "discrete Sobolev space," it is natural to introduce also "discrete Sobolev norms," defined as follows:

$$(2\text{-}28) \qquad \|u_h\|_{k,h} = \left(\sum_{0 \leq j \leq k} |\nabla_h^j u_h|_h^2 \right)^{1/2},$$

where the discrete L^2-norm $|u_h|_h$ is defined by (1-2).

The relations among these norms are given in Theorem 2-5.

THEOREM **2-5**   There exists a constant c independent of h such that

$$(2\text{-}29) \qquad c\,\|u_h\|_{k,h} \leq \|p_h^m u_h\|_k \leq \|u_h\|_{k,h} \qquad \text{for any} \quad k \leq m.$$

If we set $U = H^s$ and $V = H^k$ with $0 \leq k \leq s \leq m$, the stability functions are estimated in the following way:

$$(2\text{-}30) \qquad s_U^V(p_h^m) \leq ch^{-(s-k)}; \ c \text{ is independent of } h.$$

If $U = H^{m+1}$ and $V = H^m$,

$$(2\text{-}31) \qquad s_U^V(p_h^{m+1}, p_h^m) = \sup_{u_h \in L_h^2} \frac{\|p_h^{m+1} u_h\|_{m+1}}{\|p_h^m u_h\|_m} \leq ch^{-1}. \qquad \blacktriangle$$

Proof   It is enough to prove that there exists a constant c such that

$$(2\text{-}32) \qquad c\,|\nabla_h^j u_h|_h \leq |D^j p_h^m u_h| \leq |\nabla_h^j u_h|_h.$$

Since $D^j p_h^m u_h = \pi_{m-j,h} * p_h^0 \nabla_h^j u_h$, we deduce from Theorems 1-1 and 1-2 that $|D^j p_h^m u_h| \leq |\nabla_h^j u_h|_h$. On the other hand, if r_h is a left inverse of p_h^{m-j}, we can write

$$(2\text{-}33) \qquad \nabla_h^j u_h = \nabla_h^j r_h p_h^{m-j} u_h = r_h \nabla_h^j p_h^{m-j} u_h = r_h D^j p_h^m u_h.$$

Thus the inequality

$$(2\text{-}34) \qquad |\nabla_h^j u_h|_h = |r_h D^j p_h^m u_h|_h \leq c\,|D^j p_h^m u_h|$$

follows from Theorem 1-1, and the estimates of the stability functions follow from the estimates of

$$\sup_{u_h} \frac{\|u_h\|_{s,h}}{\|u_h\|_{m,h}} \qquad \text{for} \quad m \leq s.$$

But $\|u_h\|_{s,h} = \sum_{m+1 \le j \le s} |\nabla_h^{j-m} \nabla_h^m u_h|_h^2 + \|u_h\|_{m,h}^2$ and

$$|\nabla_h^k u_h|_h \le h^{-k} 2^k |u_h|_h.$$

Therefore

$$(2\text{-}35) \qquad \|u_h\|_{s,h}^2 \le (ch^{-2(s-m)} + 1) \|u_h\|_{m,h}^2 \le ch^{-2(s-m)} \|u_h\|_{m,h}^2. \qquad \blacksquare$$

2-7.   Optimal Properties of Prolongations $p_h{}^m$

The prolongation $p_h{}^m$ satisfies the property

$$(2\text{-}36) \qquad \nabla_h{}^m = r_h{}^0 D^m p_h{}^m,$$

where $r_h{}^0$ is the prolongation defined by

$$(2\text{-}37) \qquad r_h{}^0 u = \left(h^{-1} \int_{jh}^{(j+1)h} u(x)\, dx \right)_{j \in Z} = ((u, \pi_h{}^j))_{j \in Z}.$$

Indeed, $r_h{}^0 D^m p_h{}^m = r_h{}^0 p_h{}^0 \nabla_h{}^m = \nabla_h{}^m$, since $r_h{}^0 p_h{}^0 u_h = u_h$ for any u_h. In other words, the prolongation $p_h{}^m$ is such that the finite-difference operator $\nabla_h{}^m$ is the internal restriction of D^m. It is the optimal prolongation satisfying the property described by Theorem 2-6.

THEOREM 2-6   If a prolongation p_h mapping $L_h{}^2$ into $H^m(R)$ satisfies

$$(2\text{-}38) \qquad r_h{}^0 D^m p_h u_h = \nabla_h{}^m u_h \qquad \text{for any } u_h,$$

then the prolongation $p_h{}^m$ satisfies

$$(2\text{-}39) \qquad |D^m p_h{}^m u_h| \le |D^m p_h u_h| \qquad \text{for any } u_h. \qquad \blacktriangle$$

Proof   Indeed, by the commutation formulas, we can write

$$(2\text{-}40) \qquad \begin{cases} |D^m p_h{}^m u_h|^2 = (D^m p_h{}^m u_h, p_h{}^0 \nabla_h{}^m u_h) = (D^m p_h{}^m u_h, p_h{}^0 r_h{}^0 D^m p_h u_h) \\ \le |D^m p_h{}^m u_h| \cdot |p_h{}^0 r_h{}^0 D^m p_h u_h| \le |D^m p_h{}^m u_h| \cdot |D^m p_h u_h| \end{cases}$$

since, by the last part of Theorem 1-1, $p_h{}^0 r_h{}^0$ is an orthogonal projector of L^2. $\qquad \blacksquare$

3.   FINITE-ELEMENT APPROXIMATIONS OF SOBOLEV SPACES H^m

We associate with functions $\mu \in H^m(R)$ and $\lambda \in L^\infty(R)$ with compact support satisfying $\int \mu(x)\, dx = \int \lambda(x)\, dx = 1$—finite-element approximations $(L_h{}^2, p_h, r_h)$, where,

$$p_h u_h = \sum_{j \in Z} u_h{}^j \mu\left(\frac{x}{h} - j\right) \qquad \text{and} \qquad r_h u = \left\{ h^{-1} \int \lambda\left(\frac{x}{h} - j\right) u(x)\, dx \right\}_{j \in Z}.$$

These approximations are convergent if and only if the function μ satisfies the following criterion of m-convergence:

$$\sum_{k \in Z} \frac{k^j}{j!} \mu(x - k) = \sum_{0 \le i \le j} b^{j-i} \frac{x^i}{i!} \quad \text{for} \quad 0 \le j \le m.$$

In this case, we prove that $e_s^k(p_h) \le ch^{s-k}$ for $0 \le k \le s \le m + 1, k \le m$. By using an equivalent form of the criterion of m-convergence, we verify that the piecewise-polynomial approximations of degree m satisfy the criterion of m-convergence.

3-1. Finite-Element Approximations

Let us introduce

(3-1)
$$\begin{cases} \text{(i) a function } \mu \in H^m(R) \text{ with compact support, } \int \mu(x)\, dx = 1, \\ \text{(ii) a function } \lambda \in L^\infty(R) \text{ with compact support, } \int \lambda(x)\, dx = 1. \end{cases}$$

In considering the discrete space $V_h = L_h^2$, we associate with μ and λ the following prolongations and restrictions:

(3-2)
$$\begin{cases} \text{(i)} \quad p_h u_h = \sum_{j \in Z} u_h^j \mu\left(\frac{x}{h} - j\right), \\ \text{(ii)} \quad r_h u = \left\{ h^{-1} \int \lambda\left(\frac{x}{h} - j\right) u(x)\, dx \right\}_{j \in Z}. \end{cases}$$

Definition 3-1 Approximations (L_h^2, p_h, r_h) associated with μ and λ by (3-2) are finite-element approximations of the Sobolev space $H^m = H^m(R)$. ▲

We prove that these approximations are convergent if and only if μ satisfies the following criterion of m-convergence:

Definition 3-2 A function $\mu \in H^m$ with compact support satisfies the criterion of m-convergence if there exists $b^0 = 1, b^1, \ldots, b^m$ such that

(3-3)
$$\sum_{k \in Z} \frac{k^j}{j!} \mu(x - k) = \sum_{0 \le p \le j} b^{j-p} \frac{x^p}{p!} \quad \text{for} \quad 0 \le j \le m. \quad ▲$$

We learn later that these scalars b^k are the kth moments of $\tilde{\mu}(x) = \mu(-x)$:

(3-4)
$$b^k = (-1)^k \int \mu(x) \frac{x^k}{k!}\, dx = \int \mu(-x) \frac{x^k}{k!}\, dx.$$

We begin by proving that these approximations are convergent if λ and μ are related by

$$(3\text{-}5) \qquad \iint \mu(x)\lambda(y)(x-y)^k \, dx \, dy = \begin{cases} 1 & \text{if } k = 0, \\ 0 & \text{if } 1 \le k \le m, \end{cases}$$

that is, if the moments $d^k = \int \lambda(x)x^k/k! \, dx$ of λ are defined by the relations

$$(3\text{-}6) \qquad d^k + \sum_{1 \le j \le k} d^{k-j}b^j = 0 \qquad \text{for } 1 \le k \le m, \, d^0 = 1.$$

We use the Taylor expansion formula and write a function $u \in H^{m+1}$ in the following form: on each interval $(jh, (j+1)h)$

$$(3\text{-}7) \qquad u(x) = X_{jh}{}^s(u)(x) + Y_{jh}{}^s(D^s u)(x),$$

where

$$(3\text{-}8) \qquad \begin{cases} \text{(i)} \quad X_{jh}{}^s(u)(x) = \sum_{0 \le q \le s-1} h^q (D^q u)(jh)\left(\frac{x}{h} - j\right)^q \Big/ q!, \\[2mm] \text{(ii)} \quad Y_{jh}{}^s(u)(x) = \int_{jh}^x \frac{(x-t)^{s-1}}{(s-1)!} u(t) \, dt. \end{cases}$$

PROPOSITION 3-1   Let us assume that μ satisfies the criterion of m-convergence and that λ is related to μ by (3-5). Then on each interval $(jh, (j+1)h)$

$$(3\text{-}9) \qquad (1 - p_h r_h)u = (1 - p_h r_h)Y_{jh}{}^s(D^s u) \quad \text{for } 0 \le s \le m+1. \qquad \blacktriangle$$

Proof   We have to prove that

$$(3\text{-}10) \qquad \begin{cases} p_h r_h \left(\frac{x}{h} - j\right)^q \Big/ q! = \left(\frac{x}{h} - j\right)^q \Big/ q! \\ \qquad \text{for any } j \in Z \qquad \text{and} \qquad \text{for any } q \text{ with } 0 \le q \le m. \end{cases}$$

By definition of p_h and r_h, we obtain

$$(3\text{-}11) \qquad p_h r_h \left(\frac{x}{h} - j\right)^q \Big/ q! = \sum_{k \in Z} \left(\int \lambda(x) \frac{(x+k-j)^q}{q!} \, dx \right) \mu\left(\frac{x}{h} - k\right).$$

Using the Newton formula, we can write

$$(3\text{-}12) \qquad \int \lambda(x) \frac{(x+k-j)^q}{q!} \, dx = \sum_{0 \le p \le q} d^{q-p} \frac{(k-j)^p}{p!}.$$

Then the criterion of m-convergence implies that

$$(3\text{-}13) \qquad \sum_{k \in Z} \frac{(k-j)^p}{p!} \mu\left(\frac{x}{h} - k\right) = \sum_{k \in Z} \frac{k^p}{p!}\mu\left(\frac{x}{h} - k - j\right) = \sum_{0 \le i \le p} b^{p-i}\left(\frac{x}{h} - j\right)^i \Big/ i!.$$

Therefore, we deduce from (3-11) through (3-13) and from assumption (3-5) that

$$(3\text{-}14) \begin{cases} p_h r_h \left(\dfrac{x}{h} - j\right)^q \Big/ q! = \displaystyle\sum_{0 \le p \le q} \sum_{0 \le i \le p} d^{q-p} b^{p-i} \left(\dfrac{x}{h} - j\right)^i \Big/ i! \\[2mm] \quad = \displaystyle\sum_{0 \le i \le q} \left(\dfrac{x}{h} - j\right)^i \Big/ i! \sum_{0 \le p \le q} d^{q-p} b^{p-i} \\[2mm] \quad = \displaystyle\sum_{0 \le i \le q} \left(\dfrac{x}{h} - j\right)^i \Big/ i! \sum_{0 \le r \le q-i} d^{q-i-r} b^r = \left(\dfrac{x}{h} - j\right)^q \Big/ q! \quad \blacksquare \end{cases}$$

Remark 3-1   If a function u satisfies $Y_{jh}{}^{s+1}(D^{s+1}u) = 0$—that is, if the restriction of u to each interval $(jh, (j+1)h)$ is a polynomial of degree $s \le m$—Proposition 3-1 implies that $p_h r_h u = u$. In other words, the kernel of $(1 - p_h r_h)$ contains the piecewise polynomial of degree m. $\blacksquare$

THEOREM 3-1   Let us assume that μ satisfies the criterion of m-convergence and that λ is related to μ by (3-5). Then there exists a constant c such that

$$(3\text{-}15) \qquad |D^k(u - p_h r_h u)| \le c h^{s-k} |D^s u| \qquad \text{for } 0 \le k \le s \le m+1, k \le m. \qquad \blacktriangle$$

Proof   By Proposition 3-1, we can write

$$(3\text{-}16) \qquad |D^k(u - p_h r_h u)|^2 = \sum_{j \in Z} U_{jh}{}^2,$$

where

$$(3\text{-}17) \begin{cases} U_{jh} = \displaystyle\int_{jh}^{(j+1)h} |D^k(1 - p_h r_h) Y_{jh}{}^s (D^s u)|^2 \, dx \\[2mm] \quad \le \displaystyle\int_{jh}^{(j+1)h} |Y_{jh}{}^{s-k}(D^s u)|^2 \, dx + \int_{jh}^{(j+1)h} |D^k p_h r_h Y_{jh}{}^s (D^s u)|^2 \, dx, \end{cases}$$

since $D^k Y_{jh}{}^s(u) = Y_{jh}{}^{s-k}(u)$.

Let us begin by estimating the first term of the right-hand side of (3-17). By the Cauchy-Schwarz inequality, we obtain

$$(3\text{-}18) \qquad |Y_{jh}{}^{s-k}(v)(x)|^2 \le \frac{(x - jh)^{2(s-k)-1}}{(s-k-1)! \, (2(s-k)-1)} \int_{jh}^x |v(t)|^2 \, dt.$$

Integrating this inequality on $(jh, (j+1)h)$, we obtain

$$(3\text{-}19) \qquad \int_{jh}^{(j+1)h} |Y_{jh}{}^{s-k}(v)(x)|^2 \, dx \le c h^{2(s-k)} \int_{jh}^{(j+1)h} |v(t)|^2 \, dt.$$

It remains to estimate the second term of the right-hand side of (3-17). We notice that

$$(3\text{-}20) \qquad D^k p_h v_h = h^{-k} \sum_{p \in Z} v_h{}^p (D^k \mu)\left(\frac{x}{h} - p\right).$$

Therefore,

$$(3\text{-}21) \qquad \begin{cases} |(D^k p_h r_h v)(x)|^2 \\ \quad \le h^{-2k} \left(\sum_{p \in Z} |(\lambda_h{}^p, v)|^2 \left|(D^k \mu)\left(\frac{x}{h} - p\right)\right|\right)\left(\sum_{p \in Z} \left|(D^k \mu)\left(\frac{x}{h} - p\right)\right|\right). \end{cases}$$

Since μ has a compact support, there exists a constant c such that

$$\sup_{x \in R} \sum_{p \in Z} \left|(D^k \mu)\left(\frac{x}{h} - p\right)\right| \le c.$$

On the other hand, if (a, b) denotes the support of λ, the Cauchy-Schwarz inequality implies that $|(\lambda_h{}^p, v)|^2 \le ch^{-1}\int_{(a+p)h}^{(b+p)h} |v(x)|^2 \, dx$. Therefore, integrating $D^k p_h r_h v$ on $(jh, (j+1)h)$, we deduce from (3-21) that

$$(3\text{-}22) \qquad \int_{jh}^{(j+1)h} |(D^k p_h r_h v)(x)|^2 \, dx \le ch^{-2k} \sum_{|p-j| \le r} \int_{(a+p)h}^{(b+p)h} |v(x)|^2 \, dx,$$

where $2r + 1$ is the number of integers p such that the measure of the intersection of the supports of $\mu(x)$ and $\mu(x - p)$ is positive.

If we replace v by $Y_{jh}{}^s(D^s u)$ in (3-22), we deduce from (3-19) (with $s - k$ replaced by s) that

$$(3\text{-}23) \qquad \int_{jh}^{(j+1)h} |D^k p_h r_h Y_{jh}{}^s (D^s u)|^2 \, dx \le ch^{2(s-k)} \int_{(j-c)h}^{(j+c)h} |D^s u(t)|^2 \, dt,$$

where $c = r + b - a$.

Thus we deduce from (3-19) (with v replaced by $D^s u$) and (3-23), that

$$(3\text{-}24) \qquad U_{jh}{}^2 \le ch^{2(s-k)} \int_{(j-c)h}^{(j+c)h} |D^s u(t)|^2 \, dt.$$

This implies that

$$(3\text{-}25) \qquad |D^k(u - p_h r_h u)|^2 = \sum_{j \in Z} U_{jh}{}^2 \le ch^{2(s-k)} \int |D^s u(t)|^2 \, dt. \qquad \blacksquare$$

3-2. The Criterion of m-Convergence

Before proving the converse of Theorem 3-1, we prove another formulation of the criterion of m-convergence involving the Fourier transform $\hat{\mu}$ of the function μ and show that the function π_{m+1} satisfies the criterion of m-convergence. For this purpose, we denote the Fourier transform by

$$(3\text{-}26) \qquad \hat{u}(y) = (Fu)(y) = \int e^{ixy} u(x) \, dx$$

and recall the Poisson formula

(3-27) $$\sum_{k \in Z} u(k) = \sum_{j \in Z} \hat{u}(2\pi j).$$

THEOREM 3-2 If a function $\mu \in H^m$ satisfies the criterion of m-convergence, then it satisfies

(3-28) $$\begin{cases} (D^j \hat{\mu})(2\pi k) = 0 & \text{for any} \quad k \in Z, k \neq 0 \\ & \text{and} \quad \text{for any} \ \ j \text{ with } 0 \leq j \leq m. \end{cases}$$

Conversely, if μ satisfies (3-28), then it satisfies the criterion of m-convergence with

(3-29) $$b^j = \int \mu(-x) \frac{x^j}{j!} \, dx = (\partial^j \hat{\mu})(0) \qquad \text{where} \quad \partial^j = \frac{(i)^j D^j}{j!}. \qquad \blacktriangle$$

Proof The Fourier transform of the function

$$\varphi(x) = \frac{x^j}{j!} \mu(t - x)$$

is equal to

(3-30) $$\begin{cases} \hat{\varphi}(y) = e^{iyt} \sum_{0 \leq k \leq j} \frac{t^{j-k}}{(j-k)!} \int e^{iyx} \frac{x^k}{k!} \mu(-x) \, dx \\ \\ = e^{iyt} \sum_{0 \leq k \leq j} \frac{t^{j-k}}{(j-k)!} \partial^k \hat{\mu}(-y), \end{cases}$$

where we set

(3-31) $$\partial^k \hat{\mu}(-y) = \left(\frac{(i)^k D^k \hat{\mu}}{k!} \right)(-y) = \int \frac{x^k}{k!} e^{ixy} \mu(-x) \, dx.$$

Therefore, by applying the Poisson formula, we obtain

(3-32) $$\begin{cases} \sum_{k \in Z} \varphi(k) = \sum_{k \in Z} \frac{k^j}{j!} \mu(t-k) = \sum_{p \in Z} \hat{\varphi}(2\pi p) \\ \\ = \sum_{0 \leq k \leq j} \frac{t^{j-k}}{(j-k)!} \int \frac{x^k}{k!} \mu(-x) \, dx \\ \\ + \sum_{p \neq 0} \sum_{0 \leq k \leq j} e^{2i\pi pt} \frac{t^{j-k}}{(j-k)!} (\partial^k \hat{\mu})(-2\pi p). \end{cases}$$

Since the functions $e^{2i\pi pt}(t^k/k!)$ are linearly independent, the decomposition (3-32) is unique. Therefore, the criterion of m-convergence holds if and only if (3-28) and (3-29) are satisfied. $\blacksquare$

Now we verify that the piecewise-polynomial function $\mu = \pi_{m+1}$ of degree m satisfies the criterion of m-convergence.

THEOREM 3-3 The piecewise-polynomial function π_{m+1} of degree m satisfies the criterion of m-convergence

$$(3\text{-}33) \qquad \sum_{k \in Z} \frac{k^j}{j!} \, \pi_{m+1}(x - k) = \sum_{0 \le k \le j} b_m^{k-j} \frac{x^j}{j!},$$

where

$$(3\text{-}34) \qquad b_m{}^j = \sum_{j_1 + \cdots + j_{m+1} = j} \frac{(-1)^j}{(j_1 + 1)! \cdots (j_{m+1} + 1)!}. \qquad \blacktriangle$$

Proof The Fourier transform $\hat\pi(y)$ of the characteristic function $\pi(x)$ of $(0, 1)$ is equal to

$$(3\text{-}35) \qquad \hat\pi(y) = \frac{e^{iy} - 1}{iy} = \sum_{0 \le n \le \infty} \frac{(iy)^n}{(n + 1)!}.$$

Then the Fourier transform of π_{m+1} is equal to

$$(3\text{-}36) \qquad \hat\pi_{m+1}(y) = (\pi(y))^{m+1}.$$

This implies that

$$(3\text{-}37) \qquad \partial^j (\hat\pi(y))^{m+1} = \sum_{j_1 + \cdots + j_{m+1} = j} \partial^{j_1} \hat\pi(y) \cdots \partial^{j_{m+1}} \hat\pi(y).$$

When $j \le m$, the decomposition $j = j_1 + \cdots + j_{m+1}$ contains at least an integer j_k equal to 0. Then, since $\hat\pi(2\pi k) = 0$ for any $k \ne 0$, $k \in Z$, we deduce that $(D^j \hat\pi_{m+1})(2\pi k) = 0$ for any $k \ne 0$. Then Theorem 3-2 implies that π_{m+1} satisfies the criterion of m-convergence.

On the other hand, by (3-29) and (3-37), we obtain (3-34), since $(\partial^j \hat\pi)(0) = (-1)^j/(j + 1)!$. $\blacksquare$

Remark 3-2 Theorem 3-2 implies also that if μ satisfies the criterion of m-convergence, so does the functions $\mu * \lambda$ when λ is a function with compact support satisfying $\int \lambda(x) \, dx = 1$; indeed, the Fourier transform of $\mu * \lambda$ is the product $\hat\mu(y)\hat\lambda(y)$ and satisfies (3-28) whenever μ does. $\blacksquare$

3-3. Characterization of Convergent Finite-Element Approximations

We prove the converse of Theorem 3-1; namely, if finite-element approximations are convergent, then the function μ satisfies the criterion of m-convergence. Actually, we prove an even stronger result.

THEOREM 3-4   Let μ be a function of H^m with compact support. Let $u(x) = \sin x/x$. If there exists a family of sequences $u_h \in L_h^2$ satisfying

(3-38)
$$\begin{cases} \text{(i)} \ \sup_h |u_h|_h \leq M, \\ \text{(ii)} \ \lim \|u - p_h u_h\|_m = 0, \end{cases}$$

then μ satisfies the criterion of m-convergence.   ▲

Proof   Actually, we use Theorem 3-2 and show that (3-38) implies that the Fourier transform $\hat{\mu}(y)$ of μ satisfies (3-28).

Let us recall that the Fourier transform is an isomorphism from $L^2(R_x)$ onto $L^2(R_y)$, that its inverse is defined by

(3-39)
$$u(x) = (F^{-1}\hat{u})(x) = (2\pi)^{-1} \int e^{-ixy} \hat{u}(y) \, dy,$$

and that it is an isomorphism from $H^m(R_x)$ onto the space $\hat{H}^m(R_y)$ defined by

(3-40)
$$\begin{cases} \hat{H}^m(R_y) = \left\{ \hat{u} \in L^2(R_y) \right. \\ \qquad \text{such that} \qquad \left. \left(\int (1 + |y|^2)^m |\hat{u}(y)|^2 \, dy \right)^{1/4} < + \infty \right\}. \end{cases}$$

The Fourier transform $\hat{u}(y)$ of the function $u(x) = \sin x/x$ is the characteristic function of the interval $(-\pi, +\pi)$, and thus u belongs to the Sobolev spaces $H^m(R)$ for all m [by (3-40)]. However, we can write the Fourier transform of $p_h u_h$ in the form

(3-41)
$$F(p_h u_h)(y) = \hat{u}_h(y)\mu(yh),$$

where

(3-42)
$$\hat{u}_h(y) = h \sum_{k \in Z} u_h^k \exp(iykh)$$

can be viewed as the discrete Fourier transform of the sequence u_h. Let us note that $\hat{u}_h(y)$ has period $2\pi/h$.

First we see that

(3-43)
$$\int_{-\pi/h}^{+\pi/h} |\hat{u}_h(y)|^2 \, dy = |u_h|_h^2,$$

since

$$\int_{-\pi/h}^{+\pi/h} |\hat{u}_h(y)|^2 \, dy = h \int_{-\pi}^{+\pi} \left| \sum_{k \in Z} u_h^k e^{iky} \right|^2 dy = h \sum_{k \in Z} |u_h^k|^2 = |u_h|_h^2.$$

Therefore, assumption [(3-38)(i)] implies that the functions $\hat{u}_h(y)$ are bounded in $L^2(-\pi, +\pi)$. Thus there exists a subsequence of h (again denoted by h) such that

(3-44) $\hat{u}_h(y)$ converges weakly to a function $u_*(y)$ in $L^2(-\pi, +\pi)$.

Second, assumption [(3-38)(ii)] amounts to saying that

$$(3-45) \quad \lim_{h \to 0} \int_{-\infty}^{+\infty} |\hat{u}(y) - \hat{u}_h(y)\hat{\mu}(yh)|^2 (1 + |y|^2)^p \, dy = 0 \quad \text{for} \quad 0 \le p \le m,$$

by (3-40).

Since $\hat{u}(y) = 1$ on $(-\pi, +\pi)$, we deduce from (3-45) with $p = 0$ that

$$(3-46) \qquad \int_{-\pi}^{+\pi} |1 - \hat{u}_h(y)\hat{\mu}(yh)|^2 \, dy \text{ converges to 0.}$$

Therefore, $\hat{u}_h(y)\hat{\mu}(yh)$ converges to 1 in $L^2(-\pi, +\pi)$. Since $\hat{\mu}(y)$ is continuous, we deduce that

$$(3-47) \quad \hat{u}_h(y) = \frac{(\hat{u}_h(y)\hat{\mu}(yh))}{\hat{\mu}(yh)} \quad \text{converges to} \quad \frac{1}{\hat{\mu}(0)} \text{ in } L^2(-\pi, +\pi).$$

Comparing with (3-44), we deduce that $\hat{\mu}(0) \ne 0$. By multiplying μ by a constant if needed, we have proved that $\hat{\mu}(0) = 1$ and that $\hat{u}_h(y)$ converges to 1 in $L^2(-\pi, +\pi)$.

Since $\hat{u}(y) = 0$ on the intervals $[(2j - 1)\pi/h, (2j + 1)\pi/h]$ for any $j \ne 0$, we deduce from (3-45) that the integrals

$$(3-48) \quad \left\{ I_h{}^j = \int_{(2j-1)\pi/h}^{(2j+1)\pi/h} |\hat{u}_h(y)\hat{\mu}(yh)|^2 (1 + |y|^2)^p \, dy \text{ converge to 0} \right.$$
$$\text{for} \quad 0 \le p \le m.$$

By setting $z = y - 2j\pi/h$ and using the periodicity of $\hat{u}_h(y)$, we can write

$$(3-49) \quad I_h{}^j = \int_{-\pi/h}^{+\pi/h} |\hat{u}_h(z)\hat{\mu}(zh + 2j\pi)|^2 \, h^{-2p}(h^2 + |zh + 2\pi j|^2)^p \, dz.$$

Therefore, we deduce that

$$(3-50) \quad \left\{ h^{-2p}|\hat{u}(yh + 2\pi j)|^2 \, |\hat{u}_h(y)|^2 \, (h^2 + |yh + 2\pi j|^2)^p \text{ converges to 0 in} \right.$$
$$L^2(-\pi, +\pi) \quad \text{for any} \quad 0 \le p \le m.$$

By using the Taylor formula, we arrive at the implication that

$$(3-51) \quad (D^p\hat{\mu})(2\pi j) = 0 \quad \text{for any} \quad j \ne 0 \quad \text{and} \quad \text{for any} \quad 0 \le p \le m. \quad \blacksquare$$

Remark 3-3 If finite-element approximations associated with a function μ are convergent in H^m, then the function μ satisfies the condition (3-4); thus we recognize the crucial role played by polynomials.

We note also that the rate of convergence of the error function depends on the degree of the polynomial $\sum_{0 \leq k \leq j} b^{j-k}(x^k/k!)$ manufactured from the translates of the function μ.

The simplest functions satisfying the criterion of m-convergence are the functions $\pi_{m+1}(x)$, and moreover, they are the functions with minimal support. ∎

3-4. Stability Properties of Finite-Element Approximations

Definition 3-3 If $\mu \in H^m(R)$ is a function with compact support, then μ satisfies the "stability property" if there is no

$$(3\text{-}52) \quad y \in (-\pi, +\pi) \quad \text{such that} \quad \hat{\mu}(y + 2\pi j) = 0 \quad \text{for any } j \in Z. \quad ▲$$

THEOREM 3-5 Let us assume that μ satisfies the stability property. Then there exists a constant M such that

$$(3\text{-}53) \quad \begin{cases} \text{(i)} \ M^{-1} |p_h u_h| \leq |u_h|_h \leq M |p_h u_h|, \\ \text{(ii)} \ \|p_h u_h\|_s \leq M h^{-(s-k)} \|p_h u_h\|_k \quad \text{for} \ \ 0 \leq k \leq s \leq m. \end{cases} \quad ▲$$

Proof By using the Fourier transform, we can write [see (3-40)]

$$(3\text{-}54) \quad \begin{cases} \|p_h u_h\|_s^2 = \sum_{j \in Z} \int_{(2j-1)\pi/h}^{(2j+1)\pi/h} |\hat{u}_h(y)|^2 |\hat{\mu}(yh)|^2 (1 + |y|^2)^s \, dy \\ \qquad = h \int_{-\pi}^{+\pi} \left| \sum_{j \in Z} u_h{}^j e^{ijz} \right|^2 \sum_{j \in Z} |\hat{\mu}(z + 2\pi j)|^2 \left(1 + \frac{|z + 2\pi j|^2}{h^2} \right)^s dz \end{cases}$$

by setting $z = yh - 2\pi j$.

Let us begin by proving [(3-53)(ii)]. We estimate $\|p_h u_h\|_s$ by

$$(3\text{-}55) \quad \begin{cases} \|p_h u_h\|_s^2 \leq h^{-2(s-k)+1} \int_{-\pi}^{+\pi} \left| \sum_{j \in Z} u_h{}^j e^{ijz} \right|^2 \\ \qquad \times \left\{ \sum_{j \in Z} |\hat{\mu}(z + 2\pi j)|^2 \left(1 + \frac{|z + 2\pi j|^2}{h^2} \right)^k \right\} \frac{\alpha_s(z)}{\beta_k(z)} \, dz, \end{cases}$$

where

$$(3\text{-}56) \quad \begin{cases} \text{(i)} \ \alpha_s(z) = \sum_{j \in Z} |\hat{\mu}(z + 2\pi j)|^2 (1 + |z + 2\pi j|^2)^s, \\ \text{(ii)} \ \beta_k(z) = \sum_{j \in Z} |\hat{\mu}(z + 2\pi j)|^2 |z + 2\pi j|^{2k}. \end{cases}$$

But $\beta_k(z) = 0$ if and only if $\hat{\mu}(z + 2\pi j) |z + 2\pi j| = 0$ for any $j \in Z$. Since μ satisfies the stability property, then $\beta_k(z) > 0$ on the compact

$(-\pi, +\pi)$, and thus there exists a constant $c_k > 0$ such that

$$(3\text{-}57) \qquad\qquad 0 < c_k = \inf_{z \in (-\pi, +\pi)} \beta_k(z).$$

On the other hand, $\alpha_s(z)$ is bounded on $(-\pi, +\pi)$. Indeed, if we set $\hat{\gamma}(y) = |\hat{\mu}(y)|^2 (1 + |y|^2)^s$, we deduce that

$$(3\text{-}58) \qquad\qquad \alpha_s(z) = \sum_{j \in Z} \hat{\gamma}(z + 2\pi j) = \sum_{j \in Z} \gamma(j) e^{ijz}$$

by applying the Poisson formula to the function $\gamma(x) e^{izx}$. Since $\hat{\gamma}(y)$ is the Fourier transform of the function $\gamma(x) = (1 + D^2)^k (\mu * \bar{\mu})$, which has a compact support, then $\gamma(j) = 0$ except for a finite number of values of j. Therefore

$$(3\text{-}59) \qquad\qquad \sup_{z \in (-\pi, +\pi)} |\alpha_s(z)| = M_s \quad \text{and} \quad \frac{\alpha_s(z)}{\beta_k(z)} \le \frac{M_s}{c_k}.$$

Then inequality (3-55) implies that

$$\|p_h u_h\|_s^2 \le \frac{M_s}{c_k} h^{-2(s-k)} \|p_h u_h\|_k^2.$$

We have yet to prove [(3-53)(i)]. For $s = k = 0$, we deduce from (3-54) and (3-58) that

$$(3\text{-}60) \qquad |p_h u_h| = \|p_h u_h\|_0 \le M_0 h \int_{-\pi}^{+\pi} \left| \sum_{j \in Z} u_h^{\,j} e^{ijz} \right|^2 dz = M_0 |u_h|_h^2.$$

On the other hand, we deduce from (3-57) and (3-54) (for $k = 0$) that

$$(3\text{-}61) \qquad \left\{ \begin{aligned} c_0 |u_h|_h^2 &= c_0 \int_{-\pi}^{+\pi} h \left| \sum_{j \in Z} u_h^{\,j} e^{ijz} \right|^2 dz \\ &\le h \int_{-\pi}^{+\pi} \left| \sum_{j \in Z} u_h^{\,j} e^{ijz} \right|^2 \sum_{j \in Z} |\hat{\mu}(z + 2\pi j)|^2 \, dz \\ &= |p_h u_h|. \end{aligned} \right.$$

CHAPTER 5

Finite-Element Approximation of Functions of Several Variables

This chapter is devoted to the study of finite-element approximations of the Sobolev spaces $H^m(R^n)$, $H^m(\Omega)$, and $H_0^m(\Omega)$, where Ω is a smooth bounded open subset of R^n.

The prolongations p_h of finite-element approximations of $H^m(R^n)$ are associated with a function $\mu \in H^m(R^n)$ with compact support by

$$p_h u_h = \sum_{j \in Z^n} u_h^j \mu\left(\frac{x}{h} - j\right).$$

We prove that these approximations are convergent if and only if μ satisfies the criterion of m-convergence

$$\sum_{k \in Z^n} \frac{k^j}{j!} \mu(x - k) = \sum_{0 \le p \le j} b^{j-p} \frac{x^p}{p!} \qquad \text{for} \quad 0 < |p| \le m.$$

In this case, the error function satisfies

$$e_s^k(p_h) = \sup \frac{\|u - p_h r_h u\|_k}{\|u\|_s} \le c \, |h|^{s-k} \qquad \text{for} \quad 0 \le k \le s \le m + 1, \, k \le m,$$

We shall prove also that if μ satisfies a stability property, then there exists a constant c such that

$$s_s^k(p_h) = \sup \frac{\|p_h v_h\|_s}{\|p_h v_h\|_k} \le c \, |h|^{-(s-k)} \qquad \text{for} \quad 0 \le k \le s \le m$$

so that these approximations are quasi-optimal for the injection from H^s into H^k for $0 \le k \le s \le m$.

We introduce also the number of levels of finite-element approximations, which is the number of multi-integers k such that the measure of the intersection of the supports of $\mu(x)$ and $\mu(x - k)$ is positive.

Two examples of convergent piecewise-polynomial approximations of the Sobolev space $H^m(R^n)$ are provided, and these have, respectively, $(2m + 1)^n$ and $2(2m)^n - (2m - 1)^n$ levels. Now it is possible to construct finite-element approximations of the Sobolev spaces $H^m(\Omega)$ and $H_0^m(\Omega)$. In the first case, the convergence properties are easily obtained, but the requirement that Ω satisfy a "property of μ-stability" must be met if we are to obtain quasi-optimal approximations of $H^m(\Omega)$. In the second case, the stability properties of the finite-element approximations of $H_0^m(\Omega)$ are straightforward, but we have to assume that Ω satisfies a "property of μ-convergence" in order to estimate the error functions.

1. APPROXIMATIONS OF THE SOBOLEV SPACES $H^m(R^n)$

In this section we study finite-element approximations of the Sobolev spaces $H^m(R^n)$ and construct two examples of piecewise-polynomial approximations of multidegree m which are convergent in $H^m(R^n)$.

The first is associated with the function $\mu = \pi_{(m+1)}$, which is the $(m + 1)$th-fold convolution of the characteristic function $\pi(x)$ of the cube $(0, 1)^n$. The second approximation is associated with the function

$$\mu(x) = \int_0^1 \pi_{(m)}(x_1 - t, \dots, x_n - t)\, dt.$$

The number of levels of these approximation equals, respectively, $(2m + 1)^n$ and $2(2m)^n - (2m - 1)^n$.

These piecewise-polynomial approximations satisfy commutation formulas of the form

$$D^p p_h u_h = q_h \nabla_h^p u_h \qquad \text{for} \quad |p| \leq m.$$

1-1. Notations

We use $x = (x_1, \dots, x_n)$, $y = (y_1, \dots, y_n)$ to denote vectors of R^n. The symbol h denotes either the parameter $h = (h_1, \dots, h_n) \in R^n$ with positive components or the product $h = h_1 \dots h_n$ of its components. Also, we set $x/h = (x_1/h_1, \dots, x_n/h_n)$, $xh = (x_1 h_1, \dots, x_n h_n)$ and $|h| = \max(h_1, \dots, h_n)$.

We denote the multi-integers of Z^n by i, j, k, p, q. In particular, we use the following multi-integers:

$$(1\text{-}1) \qquad \begin{cases} \text{(i)} \ \ (m) = (m, \dots, m, \dots, m), \\ \text{(ii)} \ \ \ \varepsilon_j = (0, \dots, 0, 1, 0, \dots, 0), \end{cases}$$

and the following notations

$$(1\text{-}2) \quad \begin{cases} \text{(i)} & j \leq k \quad \text{if } j_1 \leq k_1, \ldots, j_n \leq k_n, \\ \text{(ii)} & |j| = j_1 + \cdots + j_n, \\ \text{(iii)} & j! = j_1! \cdots j_n!, \\ \text{(iv)} & \binom{j}{k} = \binom{j_1}{k_1} \cdots \binom{j_n}{k_n}, \\ \text{(v)} & x^j = x_1^{j_1} \cdots x_n^{j_n}, \end{cases}$$

and so on,

If d and b are two vectors of R^n, we can set

$$(1\text{-}3) \qquad (a, b) = (a_1, b_1) \times \cdots \times (a_n, b_n).$$

We associate with a given function $u(x)$ the following:

$$(1\text{-}4) \quad \begin{cases} \text{(i)} & \tilde{u}(x) = u(-x), \\ \text{(ii)} & u_h(x) = h^{-1}u\left(\dfrac{x}{h}\right) = \dfrac{1}{h_1 \cdots h_n} u\left(\dfrac{x_1}{h_1}, \ldots, \dfrac{x_n}{h_n}\right), \\ \text{(iii)} & u_h{}^j(x) = h^{-1}u\left(\dfrac{x}{h} - j\right) = \dfrac{1}{h_1 \cdots h_n} u\left(\dfrac{x_1}{h_1} - j_1, \ldots, \dfrac{x_n}{h_n} - j_n\right), \\ \text{(iv)} & \hat{u}(y) = \displaystyle\int e^{iyx}u(x)\, dx \\ & \qquad = \displaystyle\int e^{i(x_1y_1 + \cdots + x_ny_n)}u(x_1, \ldots, x_n)\, dx_1 \cdots dx_n, \\ \text{(v)} & (u, v) = \displaystyle\int u(x)v(x)\, dx \\ & \qquad = \displaystyle\int u(x_1, \ldots, x_n)v(x_1, \ldots, x_n)\, dx_1 \cdots dx_n. \end{cases}$$

If u and v are two functions, we denote the convolution product by

$$(1\text{-}5) \qquad u * v(x) = \int u(x - t)v(t)\, dt = \int u(t)v(x - t)\, dt,$$

and the results of Sections 4.1-5 and 4.1-6 can be easily extended to the case of functions of n variables. If u and v are measures, we define the convolution product by

$$(1\text{-}6) \qquad (u * v, \varphi) = \int u(x)v(y)\varphi(x + y)\, dx\, dy.$$

If $u_1, \ldots, u_n$ are n functions of one variable, we associate with them their tensor product $u(x)$; that is, the function

$$(1\text{-}7) \qquad u(x) = u_1(x_1) \cdots u_n(x_n)$$

of n variables.

In particular, if π_m denotes the m-fold convolution of the characteristic function of the interval $(0, 1)$ and $k = (k_1, \ldots, k_n)$ is a multi-integer, we set

$$(1\text{-}8) \qquad \pi_k(x) = \pi_{k_1}(x_1) \cdots \pi_{k_n}(x_n),$$

where $\pi_0(x_i)$ is the Dirac measure.

If k is a multi-integer, we use the classical notations

$$(1\text{-}9) \quad \begin{cases} \text{(i)} \;\; D^k = D_1^{k_1} \cdots D_n^{k_n} = \dfrac{\partial^{|k|}}{\partial x_1^{k_1} \cdots \partial x_n^{k_n}} \; ; \quad D_i = D^{\varepsilon_i} = \dfrac{\partial}{\partial x_i} \, , \\[2ex] \text{(ii)} \;\; \nabla_h^k = \nabla_{h_1}^{k_1} \cdots \nabla_{h_n}^{k_n}; \;\; \nabla_{h_i} = \nabla_h^{\varepsilon_i} = \dfrac{\delta - \delta(h\varepsilon_i)}{h_i} \, . \end{cases}$$

Definition 1-1 The Sobolev space $H^m = H^m(R^n)$ is the space

$$(1\text{-}10) \quad H^m = \{ u \in L^2(R^n) \qquad \text{such that} \qquad D^k u \in L^2(R^n) \qquad \text{for} \;\; |k| \le m \}$$

supplied with the norm

$$(1\text{-}11) \qquad \|u\|_m = \left(\sum_{|k| \le m} |D^k u|^2 \right)^{1/2}$$

where

$$|v| = \|v\|_0 = \sqrt{(v, v)} = \left(\int |v(x)|^2 \, dx \right)^{1/2} . \qquad \blacktriangle$$

Sometimes, we shall set $H^0(R^n) = L^2(R^n)$.

1-2. Finite-Element Approximations

We associate with the Sobolev spaces $H^m(R^n)$ the discrete spaces

$$(1\text{-}12) \quad L_h^2 = L_h^2(R^n) = \{ u_h = \{u_h^{\,j}\}_{j \in Z^n} \qquad \text{such that} \qquad |u_h|_h < +\infty \},$$

where

$$(1\text{-}13) \quad |u_h|_h = h^{1/2} \left[\sum_{j \in Z^n} |u_h^{\,j}|^2 \right]^{1/2} = \sqrt{(u_h, u_h)_h}; \quad (u_h, v_h)_h = h \sum_{j \in Z^n} u_h^{\,j} v_h^{\,j}.$$

Let $\mu \in H^m(R^n)$ and $\lambda \in L^\infty(R^n)$ be functions with compact support such that $\int \mu(x) \, dx = \int \lambda(x) \, dx = 1$, and associate with them the prolongations p_h

and the restrictions r_h defined by

$$(1\text{-}14) \quad \begin{cases} \text{(i)} \ \ p_h u_h = \sum_{j \in Z^n} u_h{}^j \mu\left(\frac{x}{h} - j\right), \\[2mm] \text{(ii)} \ \ r_h u \ = \{(\lambda_h{}^j, u)\}_{j \in Z^n} = \left\{ h^{-1} \int \lambda\left(\frac{x}{h} - j\right) u(x) \, dx \right\}_{j \in Z^n}. \end{cases}$$

Definition 1-2   Let $\mu \in H^m(R^n)$ and $\lambda \in L^\infty(R^n)$ be functions with compact support satisfying $\int \mu(x) \, dx = \int \lambda(x) \, dx = 1$. We say that the approximations $(L_h{}^2, p_h, r_h)$ associated with μ and λ by (1-14) are finite-element approximations of the Sobolev spaces $H^k(R^n)$ for $k \le m$. ▲

Definition 1-3   A function $\mu \in H^m(R^n)$ with compact support satisfying $\int \mu(x) \, dx = 1$ satisfies the criterion of m-convergence if one of the two following equivalent properties holds:

$$(1\text{-}15) \quad \begin{cases} \sum_{k \in Z^n} \frac{k^j}{j!} \mu(x - k) = \sum_{0 \le p \le j} b^{j-p} \frac{x^p}{p!} \quad \text{for} \ \ |j| \le m; \\[3mm] \hspace{4cm} b^p = \int \mu(-x) \frac{x^p}{p!} \, dx, \end{cases}$$

and

$$(1\text{-}16) \qquad (D^j \hat{\mu})(2\pi k) = 0 \qquad \text{for any} \ \ k \in Z^n, k \ne 0.$$

We say that μ satisfies the stability property if there is no

$$(1\text{-}17) \quad y \in (-\pi, +\pi)^n \quad \text{such that} \quad \hat{\mu}(y + 2\pi j) = 0 \quad \text{for any} \ \ j \in Z^n. \ \ ▲$$

Remark 1-1   The equivalence of properties (1-15) and (1-16) follows from a trivial extension of Theorem 4.3-2 to functions of several variables. In the same way, the extension of Theorems 4.3-1, 4.3-4, and 4.3-5 to the case of functions of several variables implies Theorem 1-1. ■

THEOREM 1-1   Let $\mu \in H^m(R^n)$ and $\lambda \in L^\infty(R^n)$ be functions with compact support satisfying $\int \mu(x) \, dx = \int \lambda(x) \, dx = 1$. Finite-element approximations $(L_h{}^2, p_h, \hat{r}_h)$ (where $\hat{r}_h$ is the optimal restriction associated with p_h in H^m) are convergent in H^m if and only if μ satisfies the criterion of m-convergence. Furthermore, if λ is related to μ by

$$(1\text{-}18) \qquad \int \mu(x)\lambda(y)(x - y)^k \, dx \, dy = \begin{cases} 1 & \text{if} \ \ k = 0, \\ 0 & \text{if} \ \ 0 < |k| \le m, \end{cases}$$

then the finite-element approximations are convergent and there exists a constant c such that

$$(1\text{-}19) \quad e_s{}^k(p_h) \leq e_s{}^k(p_h r_h) \leq c \, |h|^{s-k} \quad \text{for} \quad 0 \leq s \leq k \leq m + 1, \quad k \leq m,$$

where we set $e_s{}^k(p_h r_h) = e_U{}^V(p_h r_h)$ with $U = H^s$ and $V = H^k$.

If the function μ satisfies the stability property, then there exists a constant c such that

$$(1\text{-}20) \quad \begin{cases} \text{(i)} \quad s_s{}^k(p_h) \leq c \, |h|^{-(s-k)} \quad \text{for} \quad 0 \leq k \leq s \leq m, \\ \text{(ii)} \quad c^{-1} \, |p_h u_h| \leq |u_h|_h \leq c \, |p_h u_h|, \end{cases}$$

where we set $s_s{}^k(p_h) = s_U{}^V(p_h)$ with $U = H^s$ and $V = H^k$. ▲

In other words, if μ satisfies the criterion of m-convergence, then *the prolongations p_h associated with μ are quasi-optimal for the injections from H^s into H^k for $0 \leq k \leq s \leq m$.*

Definition 1-4 Let $\mu \in H^m$ be a function with compact support satisfying $\int \mu(x) \, dx = 1$ and p_h the prolongations associated with μ. We define the "number of levels" of μ (or of the prolongations p_h) as the number of multi-integers k such that the measure of the intersection of the supports of $\mu(x)$ and $\mu(x - k)$ is positive. ▲

As we see in Chapter 8, this number is related to the sparsity of the matrices of internal approximations of boundary-value problems involving these prolongations. The following proposition characterizes the number of levels. ▲

PROPOSITION 1-1 The number of levels of a positive function μ with compact support is equal to the number of multi-integers k belonging to the interior of the support of the function $\mu * \bar{\mu}$ (where $\bar{\mu}(x) = \mu(-x)$).

Proof Proposition 1-1 follows from the formula

$$\mu * \bar{\mu}(k) = \int \mu(x)\mu(x - k) \, dx.$$

If the measure of the intersection of the supports of $\mu(x)$ and $\mu(x - k)$ is equal to 0, then $\mu * \bar{\mu}(k) = 0$. Conversely, if $\mu * \bar{\mu}(k) = 0$, then $\mu(x)\mu(x - k) = 0$, since μ is positive and therefore $\mu(x)\mu(x - k)$ is a positive continuous function. Then the measure of the intersection of the supports of $\mu(x)$ and $\mu(x - k)$ is equal to 0. ■

Next we study two examples of piecewise-polynomial approximations of multidegree m which are convergent finite-element approximations of H^m; their numbers of levels are, respectively, $(2m + 1)^n$ and $2(2m)^n - (2m - 1)^n$.

1-3. $(2m + 1)^n$-Level Piecewise-Polynomial Approximations

Definition 1-5   The $(2m + 1)^n$-level piecewise-polynomial approximations of the Sobolev spaces $H^k(R^n)$ (for $k \leq m$) are the finite-element approximations associated with the function

$$(1\text{-}21) \qquad \mu(x) = \pi_{(m+1)}(x) = \pi_{m+1}(x_1) \cdots \pi_{m+1}(x_n). \qquad \blacktriangle$$

Since the Fourier transform of $\pi_{(m+1)}(x)$ is equal to $\hat{\pi}_{(m+1)}(y) = \hat{\pi}(y)^{m+1} = [\hat{\pi}(y_1) \cdots \hat{\pi}(y_n)]^{m+1}$, it is clear that $\pi_{(m+1)}$ satisfies the criterion of m-convergence (see Theorem 4.3-3).

Furthermore, the function $\pi_{(m+1),h}(x)$ satisfies the following commutation properties:

$$(1\text{-}22) \qquad D^p \pi_{(m+1),h} = \nabla_h{}^p \pi_{(m+1)-p,h} \qquad \text{for any } \ p \leq (m)$$

(see Lemma 4.2-1). Then the properties of piecewise-polynomial approximations of functions of one variable can be easily extended to the case of $(2m + 1)^n$-level piecewise approximations of functions of n variables.

We set

$$(1\text{-}23) \quad
\begin{cases}
\text{(i)} \quad p_h{}^k u_h = \displaystyle\sum_{j \in Z^n} u_h{}^j \pi_{k+(1)}\left(\frac{x}{h} - j\right) = \pi_{k,h} * p_h{}^0 u_h, \\[3mm]
\text{(ii)} \quad p_h^{(m)} u_h = \displaystyle\sum_{j \in Z^n} u_h{}^j \pi_{(m+1)}\left(\frac{x}{h} - j\right) = \pi_{(m+1),h} * p_h{}^0 u_h \\[3mm]
\qquad\qquad = \displaystyle\sum_{j \in Z^n} \theta_{jh}(x) \sum_{0 \leq k \leq (m)} u_h^{j-k} \alpha_{(m)}^k \left(\frac{x}{h} - j\right),
\end{cases}$$

where $\theta_{jh}(x)$ denotes the characteristic function of $[jh, (j + (1))h]$, where $\alpha_{(m)}(x) = \alpha_m{}^{k_1}(x_1) \cdots \alpha_m{}^{k_n}(x_n)$ is the tensor product of the polynomials of degree m defined in Theorem 4.2-3.

We summarize the properties of the $(2m + 1)^n$-level piecewise-polynomial approximations $(L_h{}^2, p_h^{(m)}, r_h)$ in Theorem 1-2.

THEOREM 1-2   The $(2m + 1)^n$-level piecewise-polynomial approximations $(L_h{}^2, p_h^{(m)}, r_h)$ have the following properties:

1. Commutation formulas

$$(1\text{-}24) \quad
\begin{cases}
\text{(i)} \quad D^p p_h^{(m)} u_h = p_h^{(m)-p} \nabla_h{}^p u_h & \text{for any } \ p \leq (m), \\[2mm]
\text{(ii)} \quad \nabla_h{}^p r_h u = r_h \nabla_h{}^p u = r_h(\pi_{p,h} * D^p u).
\end{cases}$$

2. Convergence properties

For any function $\lambda \in L^\infty(R^n)$ with compact support satisfying $\int \lambda(x)\, dx = 1$, the approximations $(L_h^2, p_h^{(m)}, r_h)$ (where r_h is associated with λ) are convergent in $H^k(R^n)$ for $k \leq m$. Furthermore, if λ satisfies

$$(1\text{-}25) \qquad \int \pi_{(m+1)}(x)\lambda(y)(x - y)^k\, dx\, dy = 0 \qquad \text{for} \quad 0 < |k| \leq m,$$

then there exists a constant c such that

$$(1\text{-}26) \quad \begin{cases} e_s^{\,k}(p_h^{(m)}) \leq e_s^{\,k}(p_h^{(m)} r_h) \leq c\,|h|^{s-k} & \text{for} \\ & 0 \leq k \leq s \leq m + 1, \quad k \leq m. \end{cases}$$

3. Stability properties

There exists a constant c such that

$$(1\text{-}27) \quad \begin{cases} \text{(i)} & |D^p p_h^{(m)} u_h| \leq |\nabla_h^{\,p} u_h|_h \leq c\,|D^p p_h^{(m)} u_h| & \text{for any} \\ & & p \leq (m), \\ \text{(ii)} & s_s^{\,k}(p_h^{(m)}) \leq c\,|h|^{-(s-k)} & \text{for} \quad 0 \leq k \leq s \leq m, \\ \text{(iii)} & s_{m+1}^{\,k}(p_h^{(m+1)}, p_h^{(m)}) \leq c\,|h|^{-(m+1-k)} & \text{for} \quad k \leq m. \end{cases}$$

4. Number of levels

The number of levels of the $(2m + 1)^n$-level piecewise-polynomial approximations is equal to $(2m + 1)^n$. ▲

Proof The proofs of statements 1 through 3 are straightforward extensions of the results stated in Section 4.2.

It remains to compute the number of levels. By Proposition 1-1, the number of levels is equal to the number of multi-integers k belonging to the support of the function $\pi_{(m+1)} * \tilde{\pi}_{(m+1)}$, which is the cube $(-(m + 1), m + 1)^n$. Then the multi-integers belonging to the interior of this cube are defined by the inequalities $-m \leq k_i \leq m$ for $i = 1, \ldots, n$. There are $(2m + 1)^n$ such multi-integers k. ■

1-4. $[2(2m)^n - (2m - 1)^n]$-Level Piecewise-Polynomial Approximations

Definition 1-6 The $[2(2m)^n - (2m - 1)^n]$-level piecewise-polynomial approximations of the Sobolev spaces $H^k(R^n)$ (for $k \leq m$) are the finite-element approximations associated with the function

$$(1\text{-}28) \qquad \mu(x) = \mu_m(x) = \psi * \pi_{(m)}(x)$$

where ψ is the measure defined by

$$(1\text{-}29) \qquad (\psi, u) = \int_0^1 u(x, \ldots, x)\, dx$$

whose support is the line $x_1 = t, \ldots, x_n = t$ when t ranges over $(0, 1)$.   ▲

We have to study the properties of the function $\mu_{(m)}$.

LEMMA 1-1   The function $\mu_{(m)} = \psi * \pi_{(m)}$ belongs to $H^m(R^n)$; it has a compact support, and satisfies $\int \mu_{(m)}(x)\, dx = 1$ and the criterion of m-convergence.   ▲

Proof   It is clear that the support of $\mu_{(m)}$ is compact and that

$$\int \mu_{(m)}(x)\, dx = 1.$$

The Fourier transform $\hat{\mu}_{(m)}(y)$ of $\mu_{(m)}(x)$ is equal to

$$(1\text{-}30) \qquad \hat{\mu}_{(m)}(y) = \hat{\pi}(y)^m \hat{\pi}(y_1 + \cdots + y_n),$$

since $\hat{\mu}_{(m)}(y) = \hat{\pi}_{(m)}(y)\hat{\psi}(y)$ and since $\hat{\psi}(y) = \int_0^1 e^{ix(y_1 + \cdots + y_n)}\, dx = \hat{\pi}(|y|)$, where we set $|y| = y_1 + \cdots + y_n$.

The function $\mu_{(m)}$ belongs to H^m if and only if $y^p \hat{\mu}_{(m)}(y)$ belongs to $L^2(R_y^n)$ for $|p| \leq m$.

If $|p| \leq m - 1$, then the components p_i of p are at most equal to $m - 1$ and

$$y^p \hat{\mu}_{(m)}(y) = y_1^{p_1}\hat{\pi}(y_1)^m \cdots y_n^{p_n}\hat{\pi}(y_n)^m \hat{\pi}(y_1 + \cdots + y_n) \in L^2(R_y^n).$$

If $|p| = m$, then $p = m\varepsilon_j = (0, \ldots, 0, m, 0, \ldots, 0)$. If we take $j = n$ for simplicity, we can write

$$y^p \hat{\mu}_{(m)}(y) = \hat{\pi}(y_1)^m \cdots \hat{\pi}(y_{n-1})^m y_n^{m-1}\hat{\pi}(y_n)^m y_n \hat{\pi}(y_1 + \cdots + y_n),$$

and we deduce that $y^p \hat{\mu}_{(m)}$ belongs to $L^2(R_y^n)$ when $|p| = m$.

Now let us prove that $\mu_{(m)}$ satisfies the criterion of m-convergence. If we set $\partial^p = (i)^p D^p/p! = \partial_1^{p_1} \cdots \partial_n^{p_n}$, we obtain, by (1-30)

$$\partial^p \hat{\mu}_{(m)}(y) = \sum_{0 \leq j \leq p} \partial_1^{p_1-j_1}\hat{\pi}(y_1)^m \cdots \partial_n^{p_n-j_n}\hat{\pi}(y_n)^m \partial^{|j|}\hat{\pi}(y_1 + \cdots + y_n).$$

Since $\partial^j \hat{\pi}^m(2\pi k) = 0$ for $j \leq m - 1$ and for $k \neq 0$, we deduce that $\mu_{(m)}$ satisfies the criterion of m-convergence.   ■

The simplexes ω_σ^j defined by

$$(1\text{-}31) \quad \omega_\sigma^j = \{j, j + \varepsilon_{\sigma(1)}, j + \varepsilon_{\sigma(1)} + \varepsilon_{\sigma(2)}, \ldots, j + \varepsilon_{\sigma(1)} + \cdots + \varepsilon_{\sigma(n)}\}$$

where σ ranges over the $n!$ permutations of $(1, 2, \ldots, n)$ are a partition of the cube $(j, j + (1))$.

LEMMA 1-2 The restriction of $\mu_{(m)}$ to each simplex $\omega_\sigma{}^j$ is a polynomial of multidegree m. ▲

Proof Since $\pi_{(m)}(x) = \sum_{0 \le k \le (m-1)} \alpha^k_{(m-1)}(x - k)\theta(x - k)$, where θ is the characteristic function of $(0, 1)^n$ and where $\alpha^k_{(m-1)}(x)$ is a polynomial of multidegree $m - 1$, the function $\mu_{(m)}(x)$ is the sum of translates of functions of the form

$$(1\text{-}32) \qquad \psi * \theta\alpha(x) = \int_0^1 \alpha(x - t)\theta(x - t)\, dt,$$

where $\alpha(x)$ is a polnomial of multidegree $m - 1$.

For any fixed x, we actually integrate $\alpha(x - t)$ on the intersection $(a(x), b(x))$ of the intervals $(0, 1)$ and $(x_j - 1, x_j)$ for $j = 1, \ldots, n$. The values of $a(x)$ and $b(x)$ range over $0, 1, x_j, x_j - 1$. Then such an interval $(a(x), b(x))$ does not depend on x when x ranges over a simplex $\omega_\sigma{}^j$. Therefore, on any simplex $\omega_\sigma{}^j$ contained in the support of $\psi * \theta\alpha$, this function is a polynomial of multidegree m. ■

LEMMA 1-3 The function $\mu_{(m)}$ satisfies the following properties of commutation:

$$(1\text{-}33) \quad \begin{cases} \text{(i)} \ D^p\mu_{(m),h} = \nabla_h{}^p\mu_{(m)-p,h} \quad \text{for} \ |p| \le m, \text{ where } \mu_k = \psi * \pi_k, \\[2mm] \text{(ii)} \ \dfrac{h_1 D_1 + \cdots + h_n D_n}{h_1 + \cdots + h_n} \mu_{(m),h} = \hat{\nabla}_h \pi_{(m),h} \quad \text{where} \\[4mm] \qquad\qquad\qquad\qquad \hat{\nabla}_h = \dfrac{\delta - \delta_{(h)}}{h_1 + \cdots + h_n} * . \end{cases}$$
▲

Proof Equation 1-33 (i) follows from the commutation properties of the function $\pi_{(m)}$. Equation 1-33(ii) follows from

$$(1\text{-}34) \quad \begin{cases} ((h_1 D_1 + \cdots + h_n D_n)\psi_h, u) \\[1mm] \qquad = -(\psi_h, (h_1 D_1 + \cdots + h_n D_n)u) \\[1mm] \qquad = \displaystyle\int_0^1 (h_1 D_1 + \cdots + h_n D_n)u(th_1, \ldots, th_n)\, dt \\[2mm] \qquad = -\displaystyle\int_0^1 \dfrac{d}{dt} u(th_1, \ldots, th_n)\, dt \\[2mm] \qquad = u(0, \ldots, 0) - u(h_1, \ldots, h_n) \\[1mm] \qquad = (\delta - \delta(h), u). \end{cases}$$
■

LEMMA 1-4 If k is the multi-integer $(0, 1, \ldots, 1)$, the function $\mu_k = \psi * \pi_k$ is the characteristic function of the subset ω_k defined by

(1-35) $\quad \begin{cases} \omega_k = \{x \in R^n \quad \text{such that} \quad 0 \leq x_1 \leq 1; \, x_1 \leq x_j \leq x_1 + 1 \\ \qquad\qquad\qquad\qquad\qquad\qquad\qquad \text{for} \quad j = 2, \ldots, n\}. \quad \blacktriangle \end{cases}$

Proof We know that $\pi_k(x) = \pi_0(x_1)\pi(x_2) \cdots \pi(x_n)$ is a measure because we denoted by $\pi_0(x_1)$ the Dirac measure with respect to the first variable. Then, for any function $u \in L^2$, we obtain

$$(\psi * \pi_k, u) = \int \psi(y)\pi_k(x)u(x + y) \, dx \, dy$$

$$= \int_0^1 dt \int_0^1 dx_2 \cdots \int_0^1 dx_n \, u(t, x_2 + t, \ldots, x_n + t)$$

$$= \int_0^1 dx_1 \int_{x_1}^{x_1+1} dx_2 \cdots \int_{x_1}^{x_1+1} dx_n \, u(x_1, x_2, \ldots, x_n)$$

$$= \int_{\omega_k} u(x) \, dx.$$

Therefore $\mu_k = \psi * \pi_k$ is the characteristic function of ω_k and thus belongs to $L^2(R^n)$. $\quad\blacksquare$

LEMMA 1-5 The functions $\mu_{(m)}(x - k)$ are linearly independent. $\quad \blacktriangle$

Proof Let us assume that $\alpha(x) = \sum_{j \in Z^n} u^j \mu_{(m)}(x - j) = 0$. By Lemma 2-3 [with $h = (1, \ldots, 1)$], we obtain

$$(D_1 + \cdots + D_n)\alpha(x) = \sum_{j \in Z^n} (u^j - u^{j-(1)})\pi_{(m)}(x - j) = 0.$$

Since the functions $\pi_{(m)}(x - j)$ are linearly independent, we deduce that $u^k - u^{k-(1)} = 0$ for any $k \in Z^n$. Then all the u^j are equal to the same number u and thus $\alpha(x) = u \sum_{j \in Z^n} \mu_{(m)}(x - j) = u = 0$. This implies that all the u^j are equal to 0, and thus that the functions $\mu_{(m)}(x - j)$ are linearly independent. $\quad\blacksquare$

We denote by $\bar{p}_h^{\,k}$ the prolongations associated with the functions $\mu_k = \mu * \pi_k$:

(1-36) $\quad \bar{p}_h^{\,k} u_h = \psi * p_h^{k-(1)} u_h = \sum_{j \in Z^n} u_h^j \mu_k \left(\frac{x}{h} - j \right) \quad$ where $\quad \mu_k = \psi * \pi_k(x).$

THEOREM 1-3 The $[2(2m)^n - (2m - 1)^n]$-level piecewise-polynomial approximations $(L_h{}^2, \bar{p}_h^{(m)}, r_h)$ satisfy the following properties:

1. Commutation formulas

$(1\text{-}37)$
$$\begin{cases} \text{(i)} \ D^p \bar{p}_h^{(m)} u_h = \bar{p}_h^{(m)-p} \nabla_h{}^p u_h & \text{for } |p| \leq m, \\ \text{(ii)} \ \dfrac{h_1 D_1 + \cdots + h_n D_n}{h_1 + \cdots + h_n} \bar{p}_h^{(m)} u_h = p_h^{(m)} \hat{\nabla}_h u_h. \end{cases}$$

2. Convergence properties

For any function $\lambda \in L^\infty(R^n)$ with compact support satisfying $\int \lambda(x) \, dx = 1$, the approximations $(L_h{}^2, \bar{p}_h^{(m)}, r_h)$ are convergent approximations of the Sobolev spaces $H^k(R^n)$ for $k \leq m$. Furthermore, if λ satisfies

$(1\text{-}38)$
$$\int \mu_{(m)}(x)\lambda(y)(x - y)^p \, dx \, dy = 0 \qquad \text{for } 0 < |p| \leq m,$$

then there exists a constant c such that

$(1\text{-}39)$ $e_s{}^k(\bar{p}_h^{(m)}) \leq e_s{}^k(\bar{p}_h^{(m)} r_h) \leq c \, |h|^{s-k}$ for $0 \leq k \leq s \leq m + 1, k \leq m$

3. Stability properties

There exists a constant c such that

$(1\text{-}40)$
$$\begin{cases} \text{(i)} \ c^{-1} |D^p \bar{p}_h^{(m)} u_h| \leq |\nabla_h{}^p u_h|_h \leq c \, |D^p \bar{p}_h^{(m)} u_h| & \text{for } |p| \leq m, \\ \text{(ii)} \ s_s{}^k(\bar{p}_h^{(m)}) \leq c \, |h|^{-(s-k)} & \text{for } 0 \leq k \leq s \leq m, \\ \text{(iii)} \ s_{m+1}^k(\bar{p}_h^{(m+1)}, \bar{p}_h^{(m)}) \leq c \, |h|^{-(m+1-k)} & \text{for } 0 \leq k \leq m. \end{cases}$$

4. Number of levels

The number of levels of the $[2(2m)^n - (2m - 1)^n]$-level piecewise-polynomial approximations is equal to $2(2m)^n - (2m - 1)^n$. ▲

Remark 1-2 The difference between the numbers of levels of these two examples of piecewise-polynomial approximations of multidegree m is equal to

$(1\text{-}41)$ $(2m + 1)^n - 2(2m)^m + (2m + 1)^n = \sum_{0 \leq k \leq n/2} \binom{n}{2k} (2m)^{n-2k},$

and, in particular, to 0 when $n = 1$, to 2 when $n = 2$, to $12m$ when $n = 3$. ■

Proof The proofs of the foregoing statements are analogous to the proofs of the results of Chapter 4. Let us give only a few hints.

- the commutation formulas follow from Lemma 1-3.
- the convergence properties for any function λ follow from the following decomposition of $D^p \bar{p}_h^{(m)} r_h u$:

  If $|p| \leq m - 1$, we write

(1-42) $\quad D^p \bar{p}_h^{(m)} r_h u = \varphi_h * p_h^0 r_h(\pi_{p,h} * D^p u)$ where $\varphi = \psi * \pi_{(m-1)-p}$.

  If $|p| = m$, then $p = m\varepsilon_j$ for an index $j = 1, \ldots, n$. Taking $j = 1$, for instance, we would write

(1-43) $\quad D_1^m \bar{p}_h^{(m)} r_h u = \varphi_h * \bar{p}_h^k r_h(\pi_{m\varepsilon_1,h} * D_1^m u)$ where $\varphi = \pi_{(m)-m\varepsilon_1-k}$

and $k = (0, 1, 1, \ldots, 1)$. But by Lemma 1-4, $\bar{p}_h^k u_h$ is a step function equal to u_h^j on each subset $\omega_{k,h}^j = \{x \in R^n$ such that $x/h - j \in \omega_k\}$, where ω_k is defined by (1-35). Since the subsets $\omega_{k,h}^j$ are a partition of R^n, the proof of Theorem 4.1-1 implies that $u - p_h^k r_h u$ converges to 0 in $L^2(R^n)$ for any $u \in L^2(R^n)$.

Using the decompositions (1-42) and (1-43), the proof of Theorem 4.2-2 implies the first convergence properties. The second convergence properties follow from Theorem 1-1 and Lemma 1-1.

It is clear that stability inequalities [(1-40)(ii)] and [(1-41)(iii)] follow from inequalities [(1-40)(i)]. The latter inequalities can be proved as in Theorem 4.2-5: we need the existence of a function λ such that $r_h \bar{p}_h^{(m)} = 1$, that is, such that $\int \mu_{(m)}(x - j)\lambda(x)\,dx = 1$ if $j = 0$, $= 0$ if $j \neq 0$. The existence of such a function λ follows from Lemma 1-5 as in Theorem 4.2-4.

The rest of the proof is devoted to the computation of the number of levels.

By Proposition 1-1, we must compute the number of multi-integers k belonging to the support of

$$\mu_{(m)} * \tilde{\mu}_{(m)}(x) = \int_{-1}^{1} \pi_2(t + 1)\pi_{(2m)}(x - t + (m))\,dt.$$

Indeed,

$$\pi_{(m)} * \tilde{\pi}_{(m)} = \pi_{(2m)}(x + (m)) \text{ and } (\psi * \tilde{\psi}, u) = \int_{-1}^{1} \pi_2(t + 1)u(t, \ldots, t)\,dt.$$

Therefore, t ranges over the intersection of the intervals $(-1, +1)$ and $(x_j - m, x_j + m)$ for $j = 1, \ldots, n$. Then the support of $\mu_{(m)} * \tilde{\mu}_{(m)}$ is the set of vectors x such that this intersection is not empty; that is, such that

(1-44) $\quad \begin{cases} -m - 1 \leq x_j \leq m + 1; \quad j = 1, \ldots, n, \\ -2m \leq x_i - x_j \leq 2m, \quad i > j = 1, \ldots, n. \end{cases}$

The multi-integers belonging to the interior of this support satisfy

(1-45) $\begin{cases} -m \leq k_j \leq m \\ -(2m-1) \leq k_i - k_j \leq 2m-1 \end{cases}$; $j = 1, \ldots, n; \quad i > j,$

and the number of these multi-integers equals the number of k such that

(1-46) $\begin{cases} 0 \leq k_j \leq 2m \\ -(2m-1) \leq k_i - k_j \leq 2m-1 \end{cases}$; $j = 1, \ldots, n; \quad i > j.$

But inequalities (1-46) are inequalities defining the support of $\mu_{(2m-1)}(x) = \int_0^1 \pi_{(2m-1)}(x-t)\,dt$. Indeed, this integral is different from 0 if t ranges over the nonempty intersection of intervals $(0, 1)$ and $(x_j - (2m-1), x_j)$; $j = 1, \ldots, n$. This support is also the sum of the cube $(0, 2m-1)^n$ and of the support of ψ. Therefore, a multi-integer k belongs to the support of $\mu_{(2m-1)}$ if either

(1-47) $\begin{cases} 0 \leq k_j \leq 2m-1 \ (j = 1, \ldots, n) \\ \qquad\qquad \text{or} \qquad 0 \leq k_j + 1 \leq 2m-1 \ (j = 1, \ldots, n). \end{cases}$

Let Q be the set of multi-integers k belonging to the cube $(0, 2m-1)^n$ and $\bar{Q}$ be the set of multi-integers k belonging to the cube $(-1, 2m-2)^n$. Then the number of levels is equal to the number of elements of $Q \cup \bar{Q}$.

But the number of elements of Q (or of $\bar{Q}$) is equal to $(2m)^n$ and the number of elements of $Q \cap \bar{Q} = (0, 2m-2)^n$ is equal to $(2m-1)^n$. We deduce that the number of elements of $Q \cup \bar{Q}$ is equal to $2(2m)^n - (2m-1)^n$. ∎

In particular, we have proved that the levels k belong to the support of $\mu_{(2m-1)}(x + (m))$. It is useful to represent these levels by Figures 1 and 2.

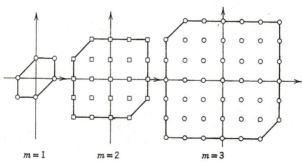

$m = 1$ $m = 2$ $m = 3$

FIGURE 1   Levels of $\bar{p}_h^{(m)}$ for $n = 2$ and $m = 1$ (7 levels), $m = 2$ (23 levels), $m = 3$ (47 levels).

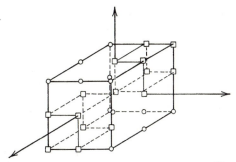

FIGURE 2   Levels of $p_h^{(1)}$ and $\bar{p}_h^{(1)}$ for $n = 3$: squares, 15 levels of $\bar{p}_h^{(1)}$; squares and circles, 27 levels of $p_h^{(1)}$.

2. APPROXIMATIONS OF THE SOBOLEV SPACES $H^m(\Omega)$

When Ω is a smooth bounded open subset of R^n, the operator of restriction ρ, which associates with a function u of $H^m(R^n)$ its restriction to Ω, maps $H^m(R^n)$ onto $H^m(\Omega)$; thus it has a continuous right inverse ω.

Then we construct the prolongations and the restrictions of the approximations of $H^m(\Omega)$ by taking p_h and $r_h\omega$ where p_h and r_h are the prolongations and the restrictions of approximations of $H^m(R^n)$. We deduce that the approximations of $H^m(\Omega)$ are convergent whenever the approximations of $H^m(R^n)$ are convergent and satisfy $e_s^k(p_h) \leq c\,|h|^{s-k}$ for $0 \leq k \leq s \leq m+1$, $k \leq m$.

In order to obtain the estimates $s_s^k(p_h) \leq c\,|h|^{-(s-k)}$, we must assume that Ω satisfies a property of μ-stability (when μ satisfies the stability property).

2-1. Sobolev Spaces $H^m(\Omega)$

Let Ω be a smooth bounded open subset of R^n and Γ its boundary.

Definition 2-1   The Sobolev space $H^m(\Omega)$ is the space of functions $u \in L^2(\Omega)$ such that $D^p u \in L^2(\Omega)$ for any $|p| \leq m$.   ▲

Now we set

$$(2\text{-}1)\quad (u, v) = \int_\Omega u(x)v(x)\,dx,\ |u| = \sqrt{(u, u)},\ \|u\|_{m,\Omega} = \left(\sum_{|p|\leq m} |D^p u|^2\right)^{1/2},$$

and sometimes we set $H^0(\Omega) = L^2(\Omega)$, $\|u\|_{0,\Omega} = |u|$.

We can prove that $H^m(\Omega)$ is a Hilbert space as in Section 1.1-6. A more comprehensive study of the Sobolev spaces $H^m(\Omega)$ appears in Section 6.3; in particular, we use the following result (see Theorem 6.3-2):

LEMMA 2-1 If Ω is a smooth bounded open subset of R^n, the operator of restriction ρ which associates with a function $u \in H^m(R^n)$ its restriction $\rho u = u|_\Omega$ to Ω is a continuous linear operator from $H^m(R^n)$ onto $H^m(\Omega)$. In particular, there exists a continuous right inverse ω of ρ satisfying

$$(2\text{-}2) \quad \begin{cases} \text{(i)} \ \omega \in L(H^k(\Omega), H^k(R^n)) & \text{for } 0 \le k \le m, \\ \text{(ii)} \ \rho\omega u = u & \text{for any } u \in H^m(\Omega). \end{cases} \quad \blacktriangle$$

In other words, Lemma 2-1 implies that $H^m(\Omega)$ is the space of restrictions to Ω of the functions u of $H^m(R^n)$. This further implies that we can identify the norm of $H^m(\Omega)$ to the norm (again denoted $\|u\|_{m,\Omega}$) defined by

$$(2\text{-}3) \quad \|u\|_{m,\Omega} = \inf_{\rho v = u} \|v\|_{m,R^n}.$$

2-2. Finite-Element Approximations of $H^m(\Omega)$

Let $\mu \in H^m(R^n)$ and $\lambda \in L^\infty(R^n)$ be functions with compact support satisfying $\int \mu(x)\, dx = \int \lambda(x)\, dx = 1$. We may associate with μ and λ approximations of $H^m(\Omega)$ in the following way.

1. Discrete space $H_h^\mu(\Omega)$

We associate with μ the following grids:

$$(2\text{-}4) \quad \begin{cases} \text{(i)} \ \mathcal{R}_h^\mu(\Omega) = \left\{ j \in Z^n \ \text{ such that } \ \Omega \cap \text{support}\, \mu\left(\dfrac{x}{h} - j\right) \ne \varnothing \right\}, \\[2mm] \text{(ii)} \ \mathcal{R}_h^\mu(\Gamma) = \left\{ j \in Z^n \ \text{ such that } \ \Gamma \cap \text{support}\, \mu\left(\dfrac{x}{h} - j\right) \ne \varnothing \right\}, \\[2mm] \text{(iii)} \ \mathcal{R}_{0,h}^\mu(\Omega) = \mathcal{R}_h^\mu(\Omega) - \mathcal{R}_h^\mu(\Gamma) \\[2mm] \qquad\qquad\quad = \left\{ j \in Z^n \ \text{ such that } \ \text{support}\, \mu\left(\dfrac{x}{h} - j\right) \subset \Omega \right\}. \end{cases}$$

Now we define the discrete spaces $H_h^\mu(\Omega)$ and $H_{0,h}^\mu(\Omega)$ by

$$(2\text{-}5) \quad \begin{cases} \text{(i)} \ H_h^\mu(\Omega) \text{ is the space of sequences } u_h = \{u_h{}^j\}_j \\ \qquad\qquad\qquad\qquad \text{when } j \text{ ranges over } \mathcal{R}_h^\mu(\Omega). \\ \text{(ii)} \ H_{0,h}^\mu(\Omega) \text{ is the space of sequences } u_h = \{u_h{}^j\}_j \\ \qquad\qquad\qquad\qquad \text{when } j \text{ ranges over } \mathcal{R}_{0,h}^\mu(\Omega). \end{cases}$$

Since Ω is bounded, these spaces are finite-dimensional spaces $R^{n(h)}$, where $n(h)$ is the number of multi-integers belonging to the grids $\mathcal{R}_h^\mu(\Omega)$ or $\mathcal{R}_{0,h}^\mu(\Omega)$.

We supply these spaces with the following inner product and norms:

$$(2\text{-}6) \quad (u_h, v_h)_h = h \sum_{j \in \mathcal{R}_h^{\mu(\Omega)}} u_h{}^j v_h{}^j, \quad |u_h|_h = \sqrt{(u_h, u_h)_h}$$

2. Prolongations p_h

Let u_h belong to $H_h^\mu(\Omega)$ and associate with μ the prolongation p_h defined by

$$(2\text{-}7) \qquad p_h u_h = \sum_{j \in \mathcal{R}_h^\mu(\Omega)} u_h^j \mu\left(\frac{x}{h} - j\right) \qquad \text{when} \quad x \in \Omega.$$

This prolongation p_h maps $H_h^\mu(\Omega)$ into $H^m(\Omega)$.

3. Restrictions $r_{h,\Omega}$

Let $\omega \in L(H^m(\Omega), H^m(R^n))$ be a continuous right inverse of ρ and r_h be the restriction associated with λ by

$$r_h u = \left\{ \int \lambda_h^j(x) u(x)\, dx \right\}_{j \in Z^n}.$$

If $u \in H^m(\Omega)$, se set

$$(2\text{-}8) \quad r_{h,\Omega} u = \{r_h(\omega u)^j\}_{j \in \mathcal{R}_h^\mu(\Omega)} \qquad \text{where} \quad r_h v^j = \int \lambda_h^j(x) v(x)\, dx.$$

Definition 2-2 The approximations $(H_h^\mu(\Omega), p_h, r_{h,\Omega})$ associated with μ and λ by (2-5), (2-7), and (2-8) are finite-element approximations of the Sobolev spaces $H^k(\Omega)$ for $k \leq m$. ▲

THEOREM 2-1 Let us suppose that μ satisfies the criterion of m-convergence and that λ is related to μ by $\iint \mu(x)\lambda(y)(x-y)^p\, dx\, dy = 0$ for $0 < |p| \leq m$. Then

$$(2\text{-}9) \qquad \lim_{h \to 0} \|u - p_h r_{h,\Omega} u\|_{k,\Omega} = 0 \qquad \text{for any} \quad k \leq m,$$

and there exists a constant c such that

$$(2\text{-}10) \quad \begin{cases} e_s^k(p_h) \leq e_s^k(p_h r_{h,\ \Omega}) \leq c\, |h|^{s-k} \\ \qquad\qquad \text{for}\ \ 0 \leq k \leq s \leq m+1,\ \ k \leq m, \end{cases}$$

where we set $e_s^k(p_h r_{h,\Omega}) = e_U^V(p_h r_{h,\Omega})$ with $U = H^s(\Omega)$ and $V = H^k(\Omega)$. ▲

Proof We can write $u = \rho\omega u$ and $p_h r_{h,\Omega} u = \rho p_h r_h \omega u$ when $u \in H^m(\Omega)$. Therefore $u - p_h r_{h,\Omega} u = \rho(1 - p_h r_h)\omega u$ and, by Theorem 1-1,

$$(2\text{-}11) \quad \begin{cases} \|u - p_h r_{h,\Omega} u\|_{k,\Omega} \leq \|(1 - p_h r_h)\omega u\|_{k,R^n} \leq c\, |h|^{s-k}\, \|\omega u\|_{s,R^n} \\ \qquad\qquad\qquad\qquad\qquad\qquad\qquad\qquad\qquad \leq c\, |h|^{s-k}\, \|u\|_{s,\Omega}. \end{cases} \blacksquare$$

2-3. Quasi-Optimal Finite-Element Approximations of $H^m(\Omega)$

By Theorem 1-1, we know that if μ satisfies the stability property then there exists a constant c such that $\|p_h u_h\|_s \leq c\, |h|^{-(s-k)}\, \|p_h u_h\|_k$.

This section is devoted to the extension of these inequalities to the case of Sobolev spaces $H^m(\Omega)$, that is to say, to the proof of inequalities

$$(2\text{-}12) \qquad \|p_h u_h\|_{s,\Omega} \le c\,|h|^{-(s-k)}\,\|p_h u_h\|_{k,\Omega} \qquad \text{for} \quad 0 \le k \le s \le m.$$

First, we show that such inequalities hold when we supply the discrete spaces $H_h{}^\mu(\Omega)$ with discrete norms different from the discrete norms $\|p_h u_h\|_m$.

Let us denote by ρ_h the operator that associates with a sequence $u_h = \{u_h{}^j\}_{j\in Z^n}$ defined on Z^n its restriction $\rho_h u_h = \{u_h{}^j\}_{j\in \mathscr{R}_h{}^\mu(\Omega)}$ to the grid $\mathscr{R}_h{}^\mu(\Omega)$. This operator ρ_h maps $L_h{}^2$ onto $H_h{}^\mu(\Omega)$. Therefore it is natural to introduce the following discrete norms:

Definition 2-3 We use $\|u_h\|_{s,\Omega,h}$ to denote the discrete norms defined by

$$(2\text{-}13) \qquad \|u_h\|_{s,\Omega,h} = \inf_{\rho_h v_h = u_h} \|p_h v_h\|_{s,R^n}. \qquad\qquad \blacktriangle$$

THEOREM 2-2 Let u_h belong to $H_h{}^\mu(\Omega)$. Then

$$(2\text{-}14) \qquad \|p_h u_h\|_{s,\Omega} \le \|u_h\|_{s,\Omega,h} \qquad \text{for} \quad 0 \le s \le m.$$

Furthermore, if μ satisfies the stability property, then there exists a constant c such that

$$(2\text{-}15) \qquad \begin{cases} \text{(i)} \ \ \|u_h\|_{s,\Omega,h} \le c\,|h|^{-(s-k)}\,\|u_h\|_{k,\Omega,h} & \text{for} \quad 0 \le k \le s \le m, \\[2mm] \text{(ii)} \ \ c^{-1}\,\|u_h\|_{0,\Omega,h} \le |u_h|_h \le c\,\|u_h\|_{0,\Omega,h}. \end{cases} \qquad \blacktriangle$$

Proof First, since $\rho_h v_h = u_h$ amounts to saying that $\rho p_h v_h = \rho p_h u_h$, we deduce that, by using (2-3)

$$(2\text{-}16) \quad \|p_h u_h\|_{s,\Omega} = \inf_{\rho v = \rho p_h u_h} \|v\|_{s,R^n} \le \inf_{\rho p_h v_h = \rho p_h u_h} \|p_h v_h\|_{s,R^n} = \|u_h\|_{s,\Omega,h}.$$

Inequalities 2-15 follow from (2-16) and Theorem 2.1-7 if we take $U = L_h{}^2$ supplied with the norm $\|p_h v_h\|_{k,R^n}$, $V = L_h{}^2$ supplied with the norm $\|p_h v_h\|_{s,R^n}$ and P to be the kernel of ρ_h: indeed, if $u_h \in L_h{}^2$, Theorem 2.1-7 states that inequality

$$(2\text{-}17) \qquad \begin{cases} \|\rho_h u_h\|_{s,\Omega,h} = \inf_{\rho_h v_h = 0} \|p_h(u_h - v_h)\|_{s,R^n} \\[2mm] \qquad \le \|p_h u_h\|_{s,R^n} \\[2mm] \qquad \le c\,|h|^{-(s-k)}\,\|p_h u_h\|_{k,R^n} \end{cases}$$

implies the inequality

$$(2\text{-}18) \quad \|\rho_h u_h\|_{s,\Omega,h} \le c\,|h|^{-(s-k)} \inf_{\rho_h v_h = 0} \|p_h(u_h - v_h)\|_{k,R^n} = c\,|h|^{-(s-k)}\,\|u_h\|_{k,\Omega,h}.$$

Then inequality [(2-15)(i)] is proved.

Now we notice that

$$(2\text{-}19) \quad |u_h|_h^2 = h \sum_{j \in \mathscr{R}_h^\mu(\Omega)} |u_h{}^j|^2 = \inf_{p_h v_h = u_h} h \sum_{j \in Z^n} |v_h{}^j|^2 = \inf_{p_h v_h = u_h} |v_h|_{h,Z^n}.$$

But since there exists a constant c such that $c^{-1} \|p_h u_h\|_{0,R^n} \leq |u_h|_{h,Z^n} \leq \|p_h u_h\|_{0,R^n}$, (2-19) implies [(2-15)(ii)]. ∎

Definition 2-4 We say that Ω has the property of μ-stability for a subsequence of h if there exists a constant c such that inequalities

$$(2\text{-}20) \quad \|u_h\|_{s,\Omega,h} \leq c \|p_h u_h\|_{s,\Omega} \quad \text{for} \quad 0 \leq s \leq m$$

hold for this subsequence. ▲

Theorem 2-2 and Definition 2-4 imply the following result:

THEOREM 2-3 Let $\mu \in H^m(R^n)$ be a function with compact support satisfying the stability property and Ω a smooth bounded open subset satisfying the property of μ-stability for a subsequence of h. Then there exists a constant c such that inequalities

$$(2\text{-}21) \quad \begin{cases} \text{(i)} \ s_s^k(p_h) = \sup \dfrac{\|p_h v_h\|_{s,\Omega}}{\|p_h v_h\|_{k,\Omega}} \leq c \, |h|^{-(s-k)} & \text{for} \quad 0 \leq s \leq k \leq m, \\ \text{(ii)} \ c^{-1} |p_h u_h| \leq |u_h|_h \leq c \, |p_h u_h| \end{cases}$$

hold for this subsequence.

Then if μ satisfies the criterion of m-convergence, the finite-element approximations $(H_h^\mu(\Omega), p_h, \hat{r}_{h,\Omega}^k)$ [where $\hat{r}_{h,\Omega}^k$ denotes the optimal restriction associated with p_h in $H^k(\Omega)$] are quasi-optimal for the injection from $H^s(\Omega)$ into $H^k(\Omega)$ when $0 \leq k \leq s \leq m$ and h ranges over the subsequence.

Furthermore, there exists a constant M such that inequalities

$$(2\text{-}22) \quad \|u - p_h \hat{r}_{h,\Omega}^k u\|_{k,\Omega} \leq M E_s^k[n(h) - 1], \quad 0 \leq k \leq s \leq m$$

hold for this subsequence, where $n(h)$ is the dimension of $H_h^\mu(\Omega)$ and $E_s^k(n)$ is the n-width of the injection from $H^s(\Omega)$ into $H^k(\Omega)$. ▲

To state a sufficient condition of μ-stability, we say that if $u_h \in H_h^\mu(\Omega)$, $\tilde{u}_h$ denotes the sequence defined by

$$(2\text{-}23) \quad \tilde{u}_h{}^j = \begin{cases} 0 & \text{if} \quad j \notin \mathscr{R}_h^\mu(\Omega), \\ u_h{}^j & \text{if} \quad j \in \mathscr{R}_h^\mu(\Omega). \end{cases}$$

PROPOSITION 2-1 Let us assume that there exists a subsequence of h and a constant c such that inequalities

$$(2\text{-}24) \qquad \|p_h \tilde{u}_h\|_{s,R^n} \leq c \, \|p_h u_h\|_{s,\Omega} \qquad \text{for} \quad 0 \leq s \leq m$$

hold for this subsequence.

Then Ω satisfies the property of μ-stability for this subsequence. ▲

Proof Since $\rho_h \tilde{u}_h = u_h$, we obtain

$$(2\text{-}25) \qquad \|u_h\|_{s,\Omega,h} = \inf_{\rho_h v_h = u_h} \|p_h v_h\|_{s,R^n} \leq \|p_h \tilde{u}_h\|_{s,R^n} \leq c \, \|p_h u_h\|_{s,\Omega}. \qquad ■$$

2-4. Piecewise-Polynomial Approximations of $H^m(\Omega)$

The finite-element approximations associated either with the function $\mu = \pi_{(m+1)}$ or with the function $\mu = \mu_{(m)} = \psi * \pi_{(m)}$ are piecewise-polynomial approximations that are convergent approximations by Theorem 2-1, 1-2, and 1-3. In this case we set

$$(2\text{-}26) \quad \begin{cases} \text{(i)} \;\; H_h^{(m)}(\Omega) = H_h^{\mu}(\Omega) \quad \text{if} \quad \mu = \pi_{(m+1)}, \qquad \bar{H}_h^{(m)}(\Omega) = H_h^{\mu}(\Omega) \\ \hspace{8.5cm} \text{if} \quad \mu = \mu_{(m)}, \\[2mm] \text{(ii)} \;\; p_h^{(m)} u_h = p_h u_h \quad \text{if} \quad \mu = \pi_{(m+1)}, \qquad \bar{p}_h^{(m)} u_h = p_h u_h \\ \hspace{8.5cm} \text{if} \quad \mu = \mu_{(m)} \end{cases}$$

Definition 2-5 The approximation $(H_h^{(m)}(\Omega), p_h^{(m)}, r_{h,\Omega})$ are the $(2m+1)^n$-level piecewise-polynomial approximations of the Sobolev spaces $H^k(\Omega)$ (for $k \leq m$) and that the approximations $(\bar{H}_h^{(m)}(\Omega), \bar{p}_h^{(m)}, r_{h,\Omega})$ are the $[2(2m)^n - (2m-1)^n]$-level piecewise-polynomial approximations of these Sobolev spaces. ▲

THEOREM 2-4 Let Ω be a smooth bounded open subset of R^n. The $(2m+1)^n$-level and the $[2(2m)^n - (2m-1)^n]$-level piecewise-polynomial approximations have the following properties:

1. Commutation formulas

$$(2\text{-}27) \quad D^p p_h^{(m)} u_h = p_h^{(m)-p} \nabla_h{}^p u_h, \quad D^p \bar{p}_h^{(m)} u_h = \bar{p}_h^{(m)-p} \nabla_h{}^p u_h \quad \text{for} \;\; |p| \leq m.$$

2. Convergence properties

For any restriction $r_{h,\Omega} = r_h \omega$ associated with a function $\lambda \in L^\infty(\Omega)$ with compact support satisfying $\int \lambda(x)\,dx = 1$, and for any k such that $0 \leq k \leq m$, we have

$$(2\text{-}28) \qquad \lim_{h \to 0} \|u - p_h^{(m)} r_{h,\Omega} u\|_{k,\Omega} = 0, \; \lim_{h \to 0} \|u - \bar{p}_h^{(m)} r_{h,\Omega} u\|_{k,\Omega} = 0. \qquad ▲$$

Proof The commutation formulas follow from the commutation formulas of Theorems 1-2 and 1-3. The convergence properties stated in Theorems 1-2 and 1-3 imply that $u - p_h^{(m)} r_h u$ and $u - \bar{p}_h^{(m)} r_h u$ converge to 0 in $H^k(R^n)$. Therefore, if $u \in H^k(\Omega)$, we deduce that

$$u - p_h^{(m)} r_{h,\Omega} u = \rho(1 - p_h^{(m)} r_h)\omega u \quad \text{and} \quad u - \bar{p}_h^{(m)} r_{h,\Omega} u = \rho(1 - \bar{p}_h^{(m)} r_h)\omega u$$

converge to 0 in $H^k(\Omega)$.   ∎

To achieve a sufficient condition of μ-stability in the case of piecewise-polynomial approximations, let us assume that there exists a constant c such that either inequalities

$$(2\text{-}29) \quad \begin{cases} \text{(i)} \quad \int_{\omega_h^j} |\varphi(x)|^2 \, dx \leq c \int_{\omega_h^j \cap \Omega} |\varphi(x)|^2 \, dx \quad \text{or} \\[2ex] \text{(ii)} \quad \int_{\omega_{\sigma,h}^j} |\varphi(x)|^2 \, dx \leq c \int_{\omega_{\sigma,h}^j \cap \Omega} |\varphi(x)|^2 \, dx \end{cases}$$

hold when h ranges over a subsequence of h, j ranges over Z^n, σ ranges over the permutations of $(1, \ldots, n)$ and φ ranges over the space of polynomials of multidegree $m + 1$, where $\omega_h^j = (jh, (j + (1))h)$ and $\omega_{\sigma,h}^j$ is the simplex equal to $\{x \text{ such that } x/h \in \omega_\sigma^j\}$ [see (1-31)].   ∎

THEOREM 2-5 Let us assume (2-29). Then either Ω is $\pi_{(m+1)}$-stable or Ω is $\mu_{(m)}$-stable. Therefore, there exists a constant c such that either inequalities

$$(2\text{-}30) \quad \begin{cases} \text{(i)} \quad s_s^k(p_h^{(m)}) \leq c \, |h|^{-(s-k)} \quad \text{for} \quad 0 \leq k \leq s \leq m, \\[1ex] \text{(ii)} \quad s_{m+1}^k(p_h^{(m+1)}, p_h^{(m)}) \leq c \, |h|^{-(m+1-k)} \quad \text{for} \quad 0 \leq k \leq m \end{cases}$$

or inequalities

$$(2\text{-}31) \quad \begin{cases} \text{(i)} \quad s_s^k(\bar{p}_h^{(m)}) \leq c \, |h|^{-(s-k)} \quad \text{for} \quad 0 \leq k \leq s \leq m, \\[1ex] \text{(ii)} \quad s_{m+1}^k(\bar{p}_h^{(m+1)}, \bar{p}_h^{(m)}) \leq c \, |h|^{-(m+1-k)} \quad \text{for} \quad 0 \leq k \leq m \end{cases}$$

hold for this subsequence.   ▲

Proof We prove this theorem only in the case of $(2m + 1)^n$-level piecewise-polynomial approximations. Let us consider the extension $\tilde{u}_h$ of $u_h \in H_h^{(m)}(\Omega)$, where $\tilde{u}_h^j = u_h^j$ when $j \in \mathcal{R}_h^{(m)}(\Omega)$ and $\tilde{u}_h^j = 0$ when $j \notin \mathcal{R}_h^{(m)}(\Omega)$.

Theorem 1-2 implies that

$$(2\text{-}32) \quad \|D^p p_h^{(m)} u_h\|_{0,\Omega} \leq \|D^p p_h^{(m)} \tilde{u}_h\|_{0,R^n} \leq |\nabla_h^p u_h|_{h,Z^n} \quad \text{for} \quad |p| \leq m.$$

Now we prove that inequalities [(2-29)(i)] imply that

$$(2\text{-}33) \qquad \|D^p p_h^{(m)} \tilde{u}_h\|_{0,R^n} \leq c \, \|D^p p_h^{(m)} u_h\|_{0,\Omega}.$$

Indeed, we can write

$$(2\text{-}34) \qquad D^p p_h^{(m)} \tilde{u}_h = \sum_{j \in Z^n} \theta_{jh}.x) \varphi_{jh}(x),$$

where $\varphi_{jh}(x)$ is the polynomial $\sum_{k \leq (m)-p} (\nabla_h{}^p u_h)^{j-k} \alpha_{(m)-p}^k(x/h - j)$.
Therefore

$$(2\text{-}35) \qquad \begin{cases} \displaystyle\int_{R^n} |D^p p_h^{(m)} \tilde{u}_h|^2 \, dx = \sum_{j \in Z^n} \int_{\omega_h{}^j} |\varphi_{jh}(x)|^2 \, dx \leq c \sum_{j \in \Omega_h} \int_{\omega_h{}^j \cap \Omega} |\varphi_{jk}(x)|^2 \, dx \\ \displaystyle\qquad\qquad\qquad\qquad\qquad\qquad = c \int_\Omega |D^p p_h^{(m)} u_h|^2 \, dx, \end{cases}$$

where Ω_h is the subset of $j \in Z^n$ such that $\Omega \cap \omega_h{}^j \neq \varnothing$.
Therefore Theorem 1-2 and (2-33) imply that

$$(2\text{-}36) \qquad c^{-1} |\nabla_h{}^p \tilde{u}_h|_{h,Z^n} \leq \|D^p p_h^{(m)} \tilde{u}_h\|_{0,R^n} \leq c \, \|D^p p_h^{(m)} u_h\|_{0,\Omega}.$$

We thus have proved that for any $|p| \leq m$, inequalities

$$(2\text{-}37) \qquad c^{-1} |\nabla_h{}^p \tilde{u}_h|_{h,Z^n} \leq \|D^p p_h^{(m)} u_h\|_{0,\Omega} \leq c \, |\nabla_h{}^p \tilde{u}_h|_{h,Z^n}$$

hold when h ranges over the subsequence for which (2-29) holds. And, of course we can deduce easily inequalities (2-30) from (2-37). ∎

Remark 2-1 If $\Omega = (a, b) = (a_1, b_1) \times \cdots \times (a_n, b_n)$, then (2-29) holds if we choose the sequence h_p where $h_{i_p} = (b_i - a_i)/p$ ∎

We need the following corollary.

COROLLARY 2-1 Let us assume that Ω is a smooth bounded open subset of R^n satisfying (2-29) for $m = 2k$. Then there exists a restriction $r_{h,\Omega}$ such that

$$(2\text{-}38) \qquad \begin{cases} \text{(i)} \quad s_{2k}^k(p_h^{(2k)}, p_h^{(2k-1)}) e_{2k}^k(p_h^{(2k-1)} r_{h,\Omega}) \leq M, \\ \text{(ii)} \quad s_k^k(p_h^{(2k)}, p_h^{(2k-1)}) \leq M, \\ \text{(iii)} \quad \lim_{h \to 0} \|u - p_h^{(2k)} r_{h,\Omega} u\|_{2k,\Omega} = 0, \\ \text{(iv)} \quad e_{2k}^k(p_h^{(2k-1)} r_{h,\Omega}) \leq c \, |h|^k \\ \text{(v)} \quad \lim_{h \to 0} e_{2k}^k(p_h^{(2k)} r_{h,\Omega}) = 0, \end{cases}$$

hold when h ranges over the subsequence for which (2-29) holds. The same statements hold when we replace $p_h^{(2k)}$ and $p_h^{(2k-1)}$ by $\bar{p}_h^{(2k)}$ and $\bar{p}_h^{(2k-1)}$. ▲

Proof Let $\lambda \in L^\infty(\Omega)$ be a function with compact support related to $\mu = \pi_{(2k)}$ by $\iint \pi_{(2k)}(x)\lambda(y)\lambda(x-y)^p \, dx \, dy = 0$ for $0 < |p| \leq 2k - 1$. Then Theorems 2-1 and 2-5 imply [(2-38)(i) and (ii)]. Moreover, Theorem 2-4 implies [(2-38)(iii)] and Theorem 2-5 implies [(2-38)(ii)]. Finally, it remains to prove [(2-38)(v)]. Since the injection from $H^{2k}(\Omega)$ into $H^k(\Omega)$ is compact (see Theorem 6.3-10) and since $u - p_h^{(2k)} r_{h,\Omega} u$ converges to 0 in $H^k(\Omega)$ by Theorem 2-4, then Theorem 1.2-3 implies [(2-38)(v)]. ∎

3. APPROXIMATION OF THE SOBOLEV SPACES $H_0{}^m(\Omega)$

We construct finite-element approximations of the Sobolev spaces $H_0{}^m(\Omega)$ by taking the discrete space $H_{0,h}^\mu(\Omega)$ to be the space of sequences $u_h \in L_h^2$ such that the support of $p_h u_h$ is contained in Ω. By doing so, we can prove that the stability functions satisfy $s_s^k(p_h) \leq c \, |h|^{-(s-k)}$ when μ satisfies the stability property. In order to prove that the error functions satisfy $e_s^k(p_h) \leq c \, |h|^{s-k}$ for $0 \leq k \leq s \leq m + 1$, $k \leq m$, we must assume that Ω satisfies a property of μ-convergence (when μ satisfies the criterion of m-convergence).

3-1. Sobolev Spaces $H_0{}^m(\Omega)$

Definition 3-1 We use $H_0{}^m(\Omega)$ to denote the closure in $H^m(\Omega)$ of the space $\Phi_0(\Omega)$ of infinitely differentiable functions with compact support in Ω and $H^{-m}(\Omega)$ for the dual of $H_0{}^m(\Omega)$. ▲

We see that if $\Omega = R^n$, then $H_0{}^m(R^n) = H^m(R^n)$. If Ω is bounded, these spaces are distinct (see Theorem 6.3-2).

PROPOSITION 3-1 The operator which associates with $u \in H_0{}^m(\Omega)$ its extension $\tilde{u}(x)$ by 0 outside of Ω

$$(3\text{-}1) \qquad \tilde{u}(x) = \begin{cases} u(x) & \text{if } x \in \Omega, \\ 0 & \text{if } x \notin \Omega, \end{cases}$$

is an isometry from $H_0{}^m(\Omega)$ into $H^m(R^n)$. ▲

Proof For any function $u \in \Phi_0(\Omega)$, we have $\|u\|_{H_0{}^m(\Omega)} = \|\tilde{u}\|_{m, R^n}$. Therefore, since $\Phi_0(\Omega)$ is dense in $H_0{}^m(\Omega)$, this equality holds for any function $u \in H_0{}^m(\Omega)$. ∎

Remark 3-1 This operator of extension by 0 outside of Ω is a right inverse of the operator of restriction ρ to Ω, which is continuous from $H_0{}^m(\Omega)$ into $H^m(R^n)$, not continuous from $H^m(\Omega)$ into $H^m(R^n)$. For instance, if $\Omega = (0, 1)$ and u is the characteristic function of $(0, 1)$, then u belongs to $H^m(\Omega)$ for any m, but its extension by 0 outside Ω belongs only to $L^2(R)$, since its derivative $D\tilde{u} = \delta - \delta(1)$ does not belong to $L^2(R)$. This, incidentally implies that $H_0{}^m(\Omega)$ is different from $H^m(\Omega)$. ∎

We now prove the Poincaré inequalities.

PROPOSITION 3-2 Let Ω be a bounded open subset of R^n. Then there exists a constant c such that

$$(3\text{-}2) \quad \begin{cases} \|u\|_{0,\Omega} \leq c \|D^p u\|_{0,\Omega} \quad \text{for any} \quad u \in H_0{}^m(\Omega) \quad \text{and} \quad \text{for any} \\ \hspace{9cm} |p| \leq m. \quad \blacktriangle \end{cases}$$

Proof Since Ω is bounded, there exists a vector $a \in R^n$ such that

$$(3\text{-}3) \quad \Omega - ap \cap \Omega = \varnothing \quad \text{for any} \quad p, \quad \text{such that} \quad 0 < |p| \leq m.$$

Let $\chi_i(x_i)$ be the characteristic function of $(0, a_i)$ and define

$$(3\text{-}4) \qquad\qquad \chi_p(x) = \chi_{p_1}(x_1) \cdots \chi_{p_n}(x_n),$$

where $\chi_{p_i}(x_i)$ is the p_ith-fold convolution of $\chi_i(x_i)$ if $p_i > 0$ and is the Dirac measure if $p_i = 0$.

Let $\tilde{u}(x)$ be the extension of u by 0 outside Ω. Therefore

$$(3\text{-}5) \quad D^p(\chi_p * \tilde{u}) = a^{-p} \sum_{k \leq p} (-1)^{|k|} \binom{p}{k} \tilde{u}(x - ak) = a^{-p} u(x) \quad \text{when} \quad x \in \Omega.$$

In other words, we have proved that we can write

$$(3\text{-}6) \quad u(x) = a^p \chi_p * D^p \tilde{u} \quad \text{when} \quad x \in \Omega, u \in H_0{}^m(\Omega), \text{ and } |p| \leq m,$$

which in turn implies (3-2). ∎

Remark 3-2 Inequalities 3-2 imply that the norms $\|u\|_m$ and

$$\left(\sum_{|p|=m} |D^p u|^2 \right)^{1/2}$$

are equivalent on $H_0{}^m(\Omega)$ [but not on $H^m(\Omega)$!]. On the other hand, the proof of Proposition 3-2 shows that inequalities (3-2) hold if there exists a vector $a \in R^n$ such that (3-3) is satisfied. ∎

We continue the study of Sobolev spaces $H_0{}^m(\Omega)$ in Section 6.3.

3-2.   Finite-Element Approximations of $H_0^m(\Omega)$

Let $\mu \in H^m(R^n)$ and $\lambda \in L^\infty(\Omega)$ be functions with compact support satisfying $\int \mu(x)\, dx = \int \lambda(x)\, dx = 1$. We associate with μ

(3-7)
$$
\begin{cases}
\text{(i)} & \text{the grid } \mathscr{R}_{0,h}^\mu(\Omega) \\
& \qquad = \left\{ j \in Z^n \quad \text{such that} \quad \text{support } \mu\left(\dfrac{x}{h} - j\right) \subset \Omega \right\}, \\[2mm]
\text{(ii)} & \text{the discrete space } H_{0,h}^\mu(\Omega) \\
& \qquad = \{ u_h \in L_h^2 \quad \text{such that} \quad u_h^{\ j} = 0 \quad \text{for } j \notin \mathscr{R}_{0,h}^\mu(\Omega)\}, \\[2mm]
\text{(iii)} & \text{the prolongation } p_h \text{ defined by } p_h u_h = \displaystyle\sum_{j \in Z^n} u_h^{\ j} \mu\left(\dfrac{x}{h} - j\right),
\end{cases}
$$

and we associate with λ the restriction $r_{0,h}$ defined by

(3-8)
$$
(r_{0,h} u)^j =
\begin{cases}
0 & \text{if } j \notin \mathscr{R}_{0,h}^\mu(\Omega), \\[2mm]
(r_h u)^j = \displaystyle\int \lambda_h^{\ j}(x)\tilde{u}(x)\, dx & \text{if } j \in \mathscr{R}_{0,h}^\mu(\Omega).
\end{cases}
$$

Definition 3-2   Approximations $(H_{0,h}^\mu(\Omega), p_h, r_{0,h})$ associated with μ and λ by (3-7) and (3-8) are finite approximations of the Sobolev spaces $H_0^k(\Omega)$ for $k \leq m$.   ▲

THEOREM 3-1   If we assume that μ satisfies the stability property, then there exists a constant c such that

(3-9)
$$
s_s^k(p_h) \leq c\, |h|^{-(s-k)} \quad \text{for} \quad 0 \leq k \leq s \leq m.
$$ ▲

Proof   Theorem 3-1 follows from Theorem 1-1 because $\|p_h u_h\|_{k,\Omega} = \|p_h \tilde{u}_h\|_{k,R^n}$ for any $u_h \in H_{0,h}^\mu(\Omega)$, since the support of $p_h u_h$ is contained in Ω.   ■

Now let us study the convergence properties of the finite-element approximations of the Sobolev spaces $H_0^m(\Omega)$.

3-3.   Convergent Finite-Element Approximations of $H_0^m(\Omega)$

Definition 3-3   A smooth bounded open subset of R^n satisfies the property of μ-convergence if there exist vectors $a_{jh} \in R^n$ and a constant c such that

(3-10)
$$
\begin{cases}
\text{(i)} & \text{for any } j \in \mathscr{R}_h^\mu(\Gamma),\ x - a_{jh} ph \notin \Omega \\
& \qquad \text{when} \quad x \in \Omega \cap \text{support } \mu\left(\dfrac{x}{h} - j\right) \\
& \qquad\qquad\qquad\qquad \text{for any } p \text{ such that } 0 \leq |p| \leq m, \\[2mm]
\text{(ii)} & \displaystyle\sup_h \sup_{j \in Z^n} |a_{jh}| \leq c.
\end{cases}
$$ ▲

THEOREM 3-2 Let $\mu \in H^m(R^n)$ be a function with compact support satisfying the criterion of m-convergence, assuming that Ω satisfies the property of μ-convergence. Let $\lambda \in L^\infty(\Omega)$ be a function with compact support satisfying $\int \lambda(x)\,dx = 1$ related to μ by $\int\int \mu(x)\lambda(y)(x-y)^p\,dx\,dy = 0$ for $0 < |p| \leq m$. Then

$$(3\text{-}11) \quad \begin{cases} \lim_{h \to 0} \|u - p_h r_{0,h} u\|_{k,\Omega} = 0 \\ \qquad\qquad \text{for any} \quad u \in H_0^k(\Omega) \quad \text{and} \quad \text{for} \quad 0 \leq k \leq m, \end{cases}$$

and there exists a constant c such that

$$(3\text{-}12) \quad \begin{cases} \|u - p_h r_{0,h} u\|_{k,\Omega} \leq c\,|h|^{(s-k)}\,\|u\|_{s,\Omega} \quad \text{for any} \quad u \in H_0^k(\Omega), \\ \qquad\qquad\qquad\qquad\qquad 0 \leq k \leq s \leq m+1, \quad k \leq m. \quad \blacktriangle \end{cases}$$

Proof Let u belong to $H_0^{m+1}(\Omega)$. We write

$$(3\text{-}13) \qquad u - p_h r_{0,h} u = \tilde{u} - p_h r_h \tilde{u} + p_h(r_h \tilde{u} - r_{0,h} u).$$

Theorem 1-1 implies that

$$(3\text{-}14) \qquad \|\tilde{u} - p_h r_h \tilde{u}\|_{k,\Omega} \leq c\,|h|^{s-k}\,\|u\|_{s,\Omega}.$$

It remains to estimate for any $|p| \leq m$ the norms of X_h defined by

$$(3\text{-}15) \quad \begin{cases} X_h = D^p p_h(r_h \tilde{u} - r_{0,h}\tilde{u}) = \sum_{j \in \mathscr{R}_h{}^\mu(\Gamma)} (\lambda_h{}^j, \tilde{u}) D^p \mu\left(\dfrac{x}{h} - j\right) \\ \qquad\qquad = h^{-p} \sum_{j \in \mathscr{R}_h{}^\mu(\Gamma)} (\lambda_h{}^j, \tilde{u})(D^p \mu)\left(\dfrac{x}{h} - j\right). \end{cases}$$

Since there exists a constant c such that $\sum_{j \in Z^n} |D^p \mu(x - j)| \leq c$, we deduce from the Cauchy-Schwarz inequality that

$$(3\text{-}16) \quad X_h{}^2 \leq c\,|h|^{-2k} \sum_{j \in \mathscr{R}_h{}^\mu(\Gamma)} |(\lambda_h{}^j, \tilde{u})|^2 \left|(D^p\mu)\left(\dfrac{x}{h} - j\right)\right| \quad \text{where} \quad k = |p|.$$

If we choose λ such that its support is contained in the support of μ (this is always possible), the property of μ-convergence implies the possibility of writing

$$(3\text{-}17) \quad \begin{cases} \tilde{u}(x) = h^q a_{jh}{}^q \chi_{jh,q} * D^q u \\ \qquad\qquad \text{when} \quad x \in \Omega \cap \text{support } \mu\left(\dfrac{x}{h} - j\right), j \in \mathscr{R}_h{}^\mu(\Gamma), \end{cases}$$

where we set $\chi_{jh,q}(x) = \chi_{j_1 h_1, q_1}(x_1) \cdots \chi_{j_n h_n, q_n}(x_n)$ with

$$
(3\text{-}18) \quad
\begin{cases}
\text{(i)} \ \chi_{j_k h_k}(x_k) =
\begin{cases}
0 & \text{if either } x_k \leq 0 \quad \text{or} \quad x_k \geq a_{j_k h_k} h_k \\[4pt]
\dfrac{1}{h_k} & \text{if } 0 \leq x_k \leq a_{j_k h_k} h_k,
\end{cases} \\[24pt]
\text{(ii)} \ \chi_{j_k h_k, q_k} =
\begin{cases}
\text{the } q_k\text{th-fold convolution of } \chi_{j_k h_k} & \text{if } q_k > 0, \\[4pt]
\text{the Dirac measure} & \text{if } q_k = 0.
\end{cases}
\end{cases}
$$

Indeed

$$
(3\text{-}19) \quad D^p(\chi_{jh,q} * \tilde{u}) = (a_{jh}h)^{-q} \sum_{0 \leq k \leq q} (-1)^k \binom{q}{k} \tilde{u}(x - a_{jh}hk) = (a_{jh}h)^{-q} \tilde{u}(x)
$$

when x belongs to $\Omega \cap$ support $\mu(x/h - j)$.

Since $\sup_h \sup_{j \in \mathbf{Z}^n} |a_{jh}| \leq c$, there exists a compact K such that the support of $\lambda_h{}^j * \chi_{jh,q}$ is contained in $K + jh$. Then if we set $s = |q|$, we deduce from (3-16) and (3-17) that there exists a constant c such that

$$
(3\text{-}20) \quad
\begin{cases}
X_h{}^2 \leq c\,|h|^{2(s-k)} \sum_{j \in \mathcal{R}_h{}^\mu(\Gamma)} |(\lambda_h{}^j, \chi_{jh,q} * D^q\tilde{u})|^2 \left|(D^p\mu)\!\left(\dfrac{x}{h} - j\right)\right| \\[14pt]
\qquad \leq c\,|h|^{2(s-k)} h^{-1} \sum_{j \in \mathcal{R}_h{}^\mu(\Gamma)} \left(\displaystyle\int_{K+jh} |D^q\tilde{u}(x)|^2\,dx\right) \left|(D^p\mu)\!\left(\dfrac{x}{h} - j\right)\right|,
\end{cases}
$$

since

$$
(3\text{-}21) \quad
\begin{cases}
|(\lambda_h{}^j, \chi_{jh,q} * D^q\tilde{u})| = |(\lambda_h{}^j * \tilde{\chi}_{jh,q}, D^q\tilde{u})| \\[8pt]
\qquad \leq |\lambda_h{}^j * \tilde{\chi}_{jh,q}| \left(\displaystyle\int_{K+jh} |D^q\tilde{u}|^2\,dx\right)^{1/2} \leq h^{-1/2} \left(\displaystyle\int_{K+jh} |D^q\tilde{u}|^2\,dx\right)^{1/2}.
\end{cases}
$$

By integrating inequality (3-20), we obtain

$$
(3\text{-}22) \quad
\begin{cases}
\|D^p p_h(r_h\tilde{u} - r_{0,h}\tilde{u})\|_0^2 \leq c\,|h|^{2(s-k)} \sum_{j \in \mathcal{R}_h{}^\mu(\Gamma)} \displaystyle\int_{K+jh} |D^q\tilde{u}|^2\,dx \\[14pt]
\qquad\qquad\qquad\qquad\qquad \leq c\,|h|^{2(s-k)} \|D^q u\|_{0,\Omega}^2.
\end{cases}
$$

Then inequalities (3-21) and (3-14) imply (3-12). ∎

CHAPTER 6

Boundary-Value Problems and the Trace Theorem

In this chapter we begin an elementary study of variational boundary-value problems; that is, boundary-value problems that are equivalent to a variational equation of the form

$$u \in V \; ; \; a(u, v) = l(v) \quad \text{for any } v \in V,$$

where $a(u, v)$ and $l(v)$ are, respectively, bilinear and linear forms continuous on V.

Besides characterizing these variational boundary-value problems, we study the existence and uniqueness of the solutions and some regularity properties.

In order to motivate and illustrate the abstract study of boundary-value problems made in Section 2, we construct variational formulations of boundary-value problems for the Laplacian.

In the theory of boundary-value problems, the so-called trace theorems play a crucial role: we devote the third Section of this chapter to the proof of these theorems, as well as other properties of Sobolev spaces that are subsequently required.

We gather examples of boundary-value problems for elliptic operators of order $2k$ in Chapter 7.

1. SOME VARIATIONAL BOUNDARY-VALUE PROBLEMS FOR THE LAPLACIAN

In this section we assume the trace theorem (see Theorem 6.3-1) and the Green formula (see Theorem 6.2-1), and we construct the variational formulations of the Dirichlet problem, the Neumann problem, and the mixed and oblique problems for the Laplacian.

1-1. The Laplacian

Here we study boundary-value problems associated with the operator $\Lambda = -\Delta + \lambda$, where Δ is the Laplacian:

$$(1\text{-}1) \qquad\qquad \Delta = \sum_{1 \le i \le n} D_i^2.$$

First, Λ maps the space $\Phi_0(\Omega)$ into itself, and, by transposition, maps the space $\Phi_0(\Omega)$ of distributions into itself; if $u \in \Phi_0'(\Omega)$ is a distribution, Λu is defined by

$$(1\text{-}2) \qquad (\Lambda u, \phi) = (u, \Lambda^*\phi) = a(u, \phi) \qquad \text{for any} \quad \phi \in \Phi(\Omega),$$

where, by definition of weak derivatives of distributions:

$$(1\text{-}3) \qquad \begin{cases} \text{(i)} \quad \Lambda^*\phi = \Lambda\phi = -\Delta\phi + \lambda\phi, \\[2mm] \text{(ii)} \quad a(u, \phi) = \sum_{1 \le i \le n} \int_\Omega D_i\, u D_i\, \phi\, dx + \lambda \int_\Omega u \cdot \phi \cdot dx. \end{cases}$$

Then we consider the bilinear form $a(u, \phi)$ defined by [(1-3)(ii)]. It is defined on $\Phi_0'(\Omega) \times \Phi_0(\Omega)$. But it is clear that it is also a continuous bilinear form on $V \times V$, where $V = H^1(\Omega)$.

Now, if we view $a(u, v)$ as a bilinear form on $V \times V$, we can no longer write (1-2), since $\Phi_0(\Omega)$ is not dense in $V = H^1(\Omega)$ in general. Nevertheless, $V_0 = H_0^1(\Omega)$ is the closure of $\Phi_0(\Omega)$ in $V = H^1(\Omega)$, and therefore (1-2) holds when $\phi \in H_0^1(\Omega)$ because we can extend it by continuity.

We summarize these two points in Lemma 1-1.

LEMMA 1-1   Let Λ and $a(u, v)$ be defined by (1-2) and (1-3). If u ranges over $V = H^1(\Omega)$ and v ranges over $V_0 = H_0^1(\Omega)$, then

$$(1\text{-}4) \qquad\qquad (\Lambda u, v) = a(u, v). \qquad\qquad \blacktriangle$$

The relation (1-4) *does not hold when both u and v range over* $V = H^1(\Omega)$.

At this point, we must state some fundamental properties of Sobolev spaces and introduce "boundary operators."

1-2. Characterization of Sobolev Spaces $H_0^1(\Omega)$

We have defined $H_0^1(\Omega)$ to be the closure of $\Phi_0(\Omega)$ in $H^1(\Omega)$. Actually, if the bounded open subset Ω of R^n and its boundary Γ are smooth enough, we can characterize a function u of $H_0^1(\Omega)$ to be a function of $H^1(\Omega)$ whose restriction to Γ vanishes.

Let us introduce some notations. If v is an infinitely differentiable function, we set

$$(1\text{-}5) \quad \begin{cases} \text{(i) } \gamma_0 v \text{ is the restriction of } v \text{ to the boundary } \Gamma \text{ of } \Omega, \\[2mm] \text{(ii) } \gamma_j v = \dfrac{\partial^j}{\partial n^j}\, v \text{ is the normal derivative of order } j \text{ (defined on } \Gamma). \end{cases}$$

Calling them "trace" operators or "boundary" operators, we deduce from Theorem 3-1 the following result:

LEMMA 1-2   Let us assume that Ω and Γ are smooth.

1. The operator γ_0 can be extended to a continuous linear operator mapping $H^1(\Omega)$ into $L^2(\Gamma)$, but the operator γ_1 *cannot* be extended to $H^1(\Omega)$.

2. The range of γ_0, denoted by $T = H^{1/2}(\Gamma)$, is a Hilbert space when it is supplied with the strongest norm for which γ_0 is continuous. It is dense in $L^2(\Gamma)$.

3. The space $H_0^1(\Omega)$ is characterized by

$$(1\text{-}6) \quad \begin{cases} V_0 = H_0^1(\Omega) \text{ is the kernel of } \gamma_0 \text{ mapping} \\[2mm] \hspace{3cm} V = H^1(\Omega) \text{ onto } T = H^{1/2}(\Gamma). \end{cases}$$

4. Finally, $H^1(\Omega)$ and $H_0^1(\Omega)$ satisfy:

$$(1\text{-}7) \quad V = H^1(\Omega) \quad \text{and} \quad V_0 = H_0^1(\Omega) \quad \text{are dense in } H = L^2(\Omega). \quad \blacktriangle$$

The last statement of Lemma 1-2 is obvious (but important!), since $\Phi_0(\Omega)$, which is contained in $H_0^1(\Omega)$, is dense in $L^2(\Omega)$.

From now on, we assume that Ω is a bounded open subset of R^n and that it is smooth enough for Lemma 1-2 to be true.

1-3.   The Green Formula

We have extended the operator γ_0 to $H^1(\Omega)$. By the Green formula, we may extend the operator γ_1 to a suitable subspace of $H^1(\Omega)$.

When the functions u and v are infinitely differentiable, we can write the (classical) Green formula

$$(1\text{-}8) \qquad \sum_{1 \le j \le n} \int_\Omega D_j u\, D_j v\, dx = -\int_\Omega \Delta u \cdot v\, dx + \int_\Gamma \frac{\partial u}{\partial n}\, v\, d\sigma.$$

We deduce from (1-8) the formula

$$(1\text{-}9) \qquad a(u, v) = (\Lambda u, v) + \langle \gamma_1 u, \gamma_0 v \rangle,$$

where

(1-10) $\begin{cases}\text{(i) } (u, \phi) \text{ is the duality pairing on } \Phi'(\Omega) \times \Phi(\Omega) \text{ extending} \\ \qquad\qquad\qquad\qquad\qquad\qquad\qquad\qquad \displaystyle\int_\Omega u\phi \, dx \\ \text{(ii) } \langle t, s \rangle \text{ is a duality pairing extending } \displaystyle\int_\Gamma ts \, d\sigma.\end{cases}$

(Section 3 provides details of the spaces of functions on Γ.)

When v ranges over $H^1(\Omega)$, $\gamma_0 v$ ranges over $H^{1/2}(\Gamma)$ and (1-9) will have a meaning whenever u ranges over $H^1(\Omega)$ and $\gamma_1 u$ ranges over the dual T' of $T = H^{1/2}(\Gamma)$.

Let us introduce the following space:

(1-11) $\begin{cases}V(\Lambda) = H^1(\Omega, \Delta) \text{ is the space of functions } u \text{ belonging to} \\ V = H^1(\Omega) \qquad \text{such that} \quad \Lambda u \text{ (or, equivalently, } \Delta u) \text{ belongs to} \\ L^2(\Omega), \text{ supplied with the norm } \|u\| = (\|u\|_1{}^2 + |\Delta u|^2)^{1/2},\end{cases}$

and let us set

(1-12) $$T' = (H^{1/2}(\Gamma))' = H^{-1/2}(\Gamma).$$

We deduce from Theorem 2-1 the following consequence:

LEMMA 1-3 The operator γ_1 can be extended to a continuous operator mapping the space $V(\Lambda) = H^1(\Omega, \Delta)$ into $T' = H^{-1/2}(\Gamma)$. Then, when u ranges over $V(\Lambda) = H^1(\Omega, \Delta)$ and v ranges over $V = H^1(\Omega)$, the Green formula [(1-9)] has meaning.   ▲

We have at hand everything necessary for constructing boundary-value problems associated with the Laplacian.

1-4. The Dirichlet Problem for the Laplacian

To prove that the Dirichlet problem for the Laplacian is equivalent to a variational equation, we need Lemma 1-4.

LEMMA 1-4 Let f belong to $H = L^2(\Omega)$. The two following problems are equivalent:

1. Dirichlet problem for $-\Delta + \lambda$
Look for u satisfying

(1-13) $\begin{cases}\text{(i) } \quad u \in V(\Lambda) = H^1(\Omega, \Delta), \\ \text{(ii) } \gamma_0 u = u|_\Gamma = 0 \qquad \text{on} \quad \Gamma, \\ \text{(iii) } \Lambda u = -\Delta u + \lambda u = f \qquad \text{on} \quad \Omega.\end{cases}$

2. Variational Dirichlet problem for $-\Delta + \lambda$

Look for u satisfying

(1-14) $\begin{cases} \text{(i)} \quad u \in V_0 \doteq H_0^1(\Omega), \\ \text{(ii)} \; a(u, v) = (f, v) \quad \text{for any} \quad v \in H_0^1(\Omega). \end{cases}$ ▲

Proof Indeed, it is clear that (1-13) implies (1-14). First, $u \in H_0^1(\Omega)$, since $H^1(\Omega, \Delta) \subset H^1(\Omega)$ and $\gamma_0 u = 0$ (see Lemma 1-2). Second, multiplying [(1-13)(iii)] by $v \in H_0^1(\Omega)$ and integrating on Ω, we obtain [(1-14)(ii)] by Lemma 1-1.

Conversely, let u be a solution of (1-14). The variational equation [(1-14)(ii)] holds for any $v \in H_0^1(\Omega)$, and we deduce from Lemma 1-1 that it satisfies $\Lambda u = -\Delta u + \lambda u = f$. Since $u \in H_0^1(\Omega)$, $\gamma_0 u = 0$ by Lemma 1-2. Finally, since $f \in L^2(\Omega)$, $\Delta u = f - \lambda u$ belongs to $L^2(\Omega)$ and thus, $u \in H^1(\Omega, \Delta)$.

1-5.   The Neumann Problem for the Laplacian

LEMMA 1-5 If we let f belong to $L^2(\Omega)$ and t to $H^{-\frac{1}{2}}(\Gamma)$, the two following problems are equivalent:

1. Unhomogeneous Neumann problem for $-\Delta + \lambda$

Look for u satisfying

(1-15) $\begin{cases} \text{(i)} \; u \in V(\Lambda) = H^1(\Omega, \Lambda), \\ \text{(ii)} \qquad \Lambda u = -\Delta u + \lambda u = f \quad \text{on} \quad \Omega, \\ \text{(iii)} \qquad \dfrac{\partial u}{\partial n} = \gamma_1 u = t \quad \text{on} \quad \Gamma. \end{cases}$

2. Variational Neumann problem for $-\Delta + \lambda$

Look for u satisfying

(1-16) $\begin{cases} \text{(i)} \qquad u \in H^1(\Omega), \\ \text{(ii)} \; a(u, v) = (f, v) + \langle t, \gamma_0 v \rangle \quad \text{for any} \quad v \in H^1(\Omega). \end{cases}$ ▲

Proof First, we must recognize that (1-16) has a meaning since $t \in T'$ and $\gamma_0 v \in T = H^{\frac{1}{2}}(\Gamma)$ when v ranges over $H^1(\Omega)$.

It is clear that (1-15) implies (1-16): we multiply [(1-15)(ii)] by $v \in H^1(\Omega)$ and we integrate on Ω; using the Green formula and [(1-15)(iii)], we find [(1-16)(ii)].

Conversely, the variational equation [(1-16)(ii)] holds in particular for any $v \in H_0^1(\Omega)$. Then we deduce from Lemma 1-1 that $\Lambda u = f$ (since $\langle t, \gamma_0 v \rangle = 0$

when $v \in H_0^1(\Omega)$). Since f belongs to $L^2(\Omega)$ we deduce that $u \in H^1(\Omega, \Delta)$. Then, by Lemma 1-3, we can use Green formula

$$a(u, v) = (\Lambda u, v) + \langle \gamma_1 u, \gamma_0 v \rangle = (f, v) + \langle t, \gamma_0 v \rangle = (\Lambda u, v) + \langle t, \gamma_0 v \rangle.$$

Therefore, $\langle \gamma_1 u - t, \gamma_0 v \rangle = 0$ for any $v \in H^1(\Omega)$. Since γ_0 maps $H^1(\Omega)$ *onto* $T = H^{1/2}(\Gamma)$, we obtain [(1-15)(iii)].

1-6. A Mixed Problem for the Laplacian

We would like to associate with the differential equation $\Lambda u = f$ the boundary conditions

$$(1\text{-}17) \quad \begin{cases} \text{(i)} \ u|_{\Gamma_1} = 0 & \text{where} \ \Gamma_1 \text{ is a smooth open subset of } \Gamma, \\ \text{(ii)} \ \dfrac{\partial u}{\partial n}\bigg|_{\Gamma_2} = 0 & \text{where} \ \Gamma_2 \text{ is the interior of } \Gamma - \Gamma_1. \end{cases}$$

First, we have to define a generalized meaning for these boundary conditions. We prove in Corollary 3-1 (Section 3-5) that there exists a continuous projector σ_1 of $H^{1/2}(\Gamma)$ such that

$$(1\text{-}18) \quad \begin{cases} \text{(i)} \ t \in H^{1/2}(\Gamma) & \text{and} \quad \sigma_1 t = 0 \quad \text{if and only if} \ t|_{\Gamma_1} = 0, \\ \text{(ii)} \ t \in H^{-1/2}(\Gamma) & \text{and} \quad \sigma_2' t = 0 \quad \text{if and only if} \ t|_{\Gamma_2} = 0, \end{cases}$$

where $\sigma_2 = 1 - \sigma_1$.

Now we can define the mixed problems.

LEMMA 1-6 Let f belong to $L^2(\Omega)$ and t to $H^{1/2}(\Gamma)$. The two following problems then become equivalent.

1. Mixed problem for $-\Delta + \lambda$
Look for u satisfying

$$(1\text{-}19) \quad \begin{cases} \text{(i)} \ u \in H^1(\Omega, \Delta), \\ \text{(ii)} \ -\Delta u + \lambda u = f & \text{on} \ \Omega, \\ \text{(iii)} \ \sigma_1 \gamma_0 u = u|_{\Gamma_1} = 0 & \text{on} \ \Gamma_1, \\ \text{(iv)} \ \sigma_2' \gamma_1 u = \dfrac{\partial u}{\partial n}\bigg|_{\Gamma_2} = t|_{\Gamma_2} & \text{on} \ \Gamma_2. \end{cases}$$

2. Variational mixed problem for $-\Delta + \lambda$
Look for u satisfying

$$(1\text{-}20) \quad \begin{cases} \text{(i)} \ u \in W = H_{\Gamma_1}^1(\Omega), \text{ space of } u \in H^1(\Omega) \quad \text{such that} \\ \qquad\qquad\qquad\qquad\qquad\qquad\qquad\qquad\qquad\qquad \sigma_1 \gamma_0 u = 0, \\ \text{(ii)} \ a(u, v) = (f, v) + \langle t, \gamma_0 v \rangle \quad \text{for any} \ v \in W = H_{\Gamma_1}^1(\Omega). \end{cases} \quad \blacktriangle$$

Proof Lemma 1-6 follows from the Green formula written in the following way:

(1-21) $a(u, v) = (\Lambda u, v) + \langle \sigma'_1 \gamma_1 u, \sigma_1 \gamma_0 v \rangle + \langle \sigma'_2 \gamma_1 u, \sigma_2 \gamma_0 v \rangle.$

Indeed, σ_1 is a projector of $H^{1/2}(\Gamma)$ and $1 - \sigma_1 = \sigma_2$. Then σ'_1 and σ'_2 are projectors of $H^{-1/2}$, and we can write

$$\langle t, s \rangle = \langle t, (\sigma_1^2 + \sigma_2^2)s \rangle = \langle \sigma'_1 t, \sigma_1 s \rangle + \langle \sigma'_2 t, \sigma_2 s \rangle,$$

and thus deduce (1-21) from the Green formula (1-9). Now, let us prove the equivalence of the two problems in Lemma 1-6. It is clear that (1-19) implies (1-20), Indeed, [(1-19)(i) and (iii)] imply [(1-20)(i)]. On the other hand, we multiply [(1-19)(ii)] by v, integrate on Ω, and use (1-21). Thus we find [(1-20)(ii)].

Conversely, [(1-20)(ii)] holds for any v in $H_0^1(\Omega)$, whereupon Lemma 1-1 implies that $\Lambda u = f$. Since f belongs to $L^2(\Omega)$, we deduce that $u \in H^1(\Omega, \Delta)$. Since $u \in H_{\Gamma_1}^1(\Omega)$, $\sigma_1 \gamma_0 u = 0$.

Now we can write [(1-20)(ii)] and use the Green formula (by Lemma 1-3). We deduce that

$\langle \sigma'_2(\gamma_1 u - t), \sigma_2 \gamma_0 v \rangle = 0$ for any $u \in H^1(\Omega)$ such that $\sigma_1 \gamma_0 u = 0.$

This relation implies [(1-19)(iv)].

1-7.   An Oblique Problem for the Laplacian

We have seen that the three last examples of boundary-value problems for the Laplacian were variational problems of the form

(1-22) $u \in W \; ; a(u, v) = (f, v) + \langle t, \gamma_0 v \rangle$ for any $v \in W,$

where W is a closed subspace of $V = H^1(\Omega)$ such that

(1-23) $V_0 = H_0^1(\Omega) \subset W \subset V = H^1(\Omega).$

(We took W to be successively $H_0^1(\Omega)$, $H^1(\Omega)$, and $H_{\Gamma_1}^1(\Omega)$.)

To obtain more examples of boundary-value problems for the Laplacian we can choose other closed subspace W of $H^1(\Omega)$ satisfying (1-23), we can perturb the form $a(u, v)$ by adding to it $m(\gamma_0 u, \gamma_0 v)$, where $m(t, s)$ is any continuous bilinear form on $T \times T$ ($T = H^{1/2}(\Gamma)$), or we can do both. As an illustration, let us choose $W = H^1(\Omega)$ and perturb $a(u, v)$.

LEMMA 1-7   Let M be an operator mapping $T = H^{1/2}(\Gamma)$ into T' and $m(t, s) = \langle Mt, s \rangle$ the associated bilinear form. Let f belong to $L^2(\Omega)$ and t to $H^{-1/2}(\Gamma)$. The two following problems are equivalent.

1. Oblique problem for $-\Delta + \lambda$
Look for u satisfying

$$(1\text{-}24) \quad \begin{cases} \text{(i)} \quad u \in H^1(\Omega, \Delta), \\ \text{(ii)} \quad \Lambda u = -\Delta u + \lambda u = f \quad \text{on} \quad \Omega, \\ \text{(iii)} \quad \gamma_1 u + Mu = t \quad \text{on} \quad \Gamma. \end{cases}$$

2. Variational Oblique Problem for the Laplacian
Look for u satisfying

$$(1\text{-}25) \quad \begin{cases} \text{(i)} \quad u \in H^1(\Omega), \\ \text{(ii)} \quad a(u, v) + m(\gamma_0 u, \gamma_0 v) = (f, v) + \langle t, \gamma_0 v \rangle \quad \text{for any} \\ \qquad\qquad\qquad\qquad\qquad\qquad\qquad\qquad\qquad v \in H^1(\Omega). \quad \blacktriangle \end{cases}$$

Proof The proof of Lemma 1-7 is analogous to the proof of Lemma 1-5 and is left as an exercise. ∎

Let us give examples of operators M from T into T'.

EXAMPLE 1-1 Let $m(x)$ be a function belonging to $L^\infty(\Gamma)$. Let us set

$$(1\text{-}26) \qquad\qquad m(t,s) = \int_\Gamma m(x)t(x)s(x) \, d\sigma.$$

Then the operator M defined by $m(t, s)$ is the operator of multiplication by $m(x)$ and the boundary condition $[(1\text{-}24)(\text{iii})]$ can be written

$$(1\text{-}27) \qquad\qquad \gamma_1 u + m(x)u = t \quad \text{on} \quad \Gamma.$$

EXAMPLE 1-2 Let $\alpha(x, y)$ be a function of $L^2(\Gamma \times \Gamma)$ and define $m(t, s)$ as

$$(1\text{-}28) \qquad m(t,s) = \int_\Gamma d\sigma(x) \int_\Gamma d\sigma(y)\alpha(x, y)t(y)s(x) \quad \text{on} \quad \Gamma.$$

Then the operator M is an integral operator and the boundary condition $[(1\text{-}24)(\text{iii})]$ can be written

$$(1\text{-}29) \qquad\qquad \gamma_1 u + \int_\Gamma \alpha(x, y)u(y) \, d\sigma(y) = t \quad \text{on} \quad \Gamma.$$

EXAMPLE 1-3 Let $Mu = \partial u/\partial \sigma$ be the "tangential" derivative of u. We can prove that M is a continuous linear operator from $H^{1/2}(\Gamma)$ into its dual

$H^{-\frac{1}{2}}(\Gamma)$. The boundary condition [(1-24)(iii)] can be written

$$(1\text{-}30) \qquad\qquad \gamma_1 u + \frac{\partial}{\partial \sigma} u = t \quad \text{on} \ \ \Gamma.$$

1-8.   Existence and Uniqueness of the Solutions

Since we have proved the equivalence of the foregoing boundary-value problems to variational equations, we can use the Lax-Milgram theorem for proving the existence and uniqueness of the solutions.

Indeed, it is clear that the form $a(u, v)$ is $H^1(\Omega)$-elliptic when $\lambda > 0$. We thus deduce Lemma 1-8.

LEMMA 1-8   Let us assume that λ is positive. Then there exists a unique solution of the Dirichlet (respectively Neumann, mixed) problem for $-\Delta + \lambda$.

If $m(t, t) = \langle Mt, t \rangle \geq 0$ for any $t \in H^{\frac{1}{2}}(\Gamma)$, there exists a unique solution of the oblique problem (1-24) for $-\Delta + \lambda$ whenever λ is positive.   ▲

Remark 1-1   If we consider the Dirichlet problem, the Poincaré inequality (see Section 5.3-1) implies the $H_0^1(\Omega)$-ellipticity for $\lambda = 0$, and then the existence and the uniqueness of a solution of the Dirichlet problem for $-\Delta$.   ■

Remark 1-2   More generally, by using Theorem 2.1-16, we can deduce the existence and uniqueness of solutions of the foregoing boundary-value problems for the operator $-\Delta + \lambda$ for negative values of λ which do not belong to the spectrum of the boundary-value problem.   ■

2.   *VARIATIONAL BOUNDARY-VALUE PROBLEMS AND THEIR ADJOINTS*

We start with a Hilbert space V and a "trace theorem": there exists a pivot space H, a Hilbert space T, and an operator $\gamma \in L(V, T)$ satisfying

$$\begin{cases} \text{(i) } \gamma \text{ maps } V \text{ onto } T, \\ \text{(ii) } V \text{ is contained in } H \text{ with a stronger topology,} \\ \text{(iii) the kernel } V_0 \text{ of } \gamma \text{ is dense in } H. \end{cases}$$

Second, we associate with a continuous bilinear form $a(u, v)$ on V the formal operator Λ defined by $a(u, v) = (\Lambda u, v)$ for any $u \in V$, $v \in V_0$, which maps its domain $V(\Lambda) = \{u \in V \text{ such that } \Lambda u \in H\}$ into H.

Then we prove the Green formula: there exists a unique operator $\delta \in L(V, T')$ such that

$$a(u, v) = (\Lambda u, v) + \langle \delta u, \gamma v \rangle \qquad \text{for any} \quad u \in V(\Lambda), v \in V.$$

This is our basic tool for characterizing variational boundary-value problems associated with $a(u, v)$ and a projector σ_1 of T:

$$\begin{cases} \text{(i)} \ u \in V(\Lambda), \\ \text{(ii)} \ \Lambda u = f, \qquad \text{where } f \text{ is given in } H, \\ \text{(iii)} \ \gamma_1 u = \sigma_1 \gamma u = t_1 \qquad \text{where } t_1 \text{ is given in } T_1 = \sigma_1 T, \\ \text{(iv)} \ \delta_2 u = \sigma_2' \delta u = t_2 \quad \text{where } t_2 \text{ is given in } T_2' = \sigma_2' T', \sigma_2 = 1 - \sigma_1, \end{cases}$$

which are equivalent to a variational equation on a closed subspace W of V.

When the form $a(u, v)$ is coercive, we employ the Lax-Milgram theorem and the Riesz-Fredholm alternative to deduce the existence and uniqueness of solutions of such boundary-value problems.

Finally, we study regularity properties: roughly speaking, if we know that (Λ, γ) is an isomorphism from U onto $H \times S$, under what assumptions can we deduce that (Λ, γ) is also an isomorphism from

$$H(\Lambda) = \{u \in H \qquad \text{such that} \quad \Lambda u \in H\} \qquad \text{onto} \quad H \times R,$$

where $R \supset T$, $U \subset V(\Lambda)$ and $S \subset T$. For that purpose, we need the introduction of the formal adjoint Λ^* of Λ and a Green formula relating Λ and its formal adjoint.

2-1. Spaces V, H and Operator γ

To study variational equations on a Hilbert space V satisfying an abstract "trace property," we need to know that a pivot space H, a Hilbert space T, and an operator $\gamma \in L(V, T)$ exist, satisfying

$$\text{(2-1)} \qquad \begin{cases} \text{(i)} \ \gamma \text{ maps } V \text{ onto } T, \\ \text{(ii)} \ V \text{ is contained in } H \text{ with a stronger topology}, \\ \text{(iii)} \ \text{the kernel } V_0 \text{ of } \gamma \text{ is } dense \text{ in } H. \end{cases}$$

The trace property (2-1) and theorem 2.1-4 imply the following inclusions:

$$\text{(2-2)} \qquad \begin{cases} \text{(i)} \ V \subset H = H' \subset V', \\ \text{(ii)} \ V_0 \subset H = H' \subset V_0', \end{cases}$$

where the injections are continuous and dense.

EXAMPLE   By Lemma 1-2, the assumption (2-1) is satisfied when $V = H^1(\Omega)$, $\gamma = \gamma_0$, $T = H^{1/2}(\Gamma)$, and $H = L^2(\Omega)$.

2-2.   Formal Operator Λ Associated with $a(u, v)$

Let us consider now a continuous bilinear form $a(u, v)$ on $V \times V$.

For any fixed u in V, the map that associates the value $a(u, v)$ with $v \in V_0$ is continuous on V_0. We denote by Λu this linear functional on V_0:

(2-3) $\Lambda u \in V_0'$ is the functional defined by $(\Lambda u, v) = a(u, v)$ when $v \in V_0$

Then the operator Λ associating $\Lambda u \in V_0'$ with any $u \in V$ is a continuous linear operator from V into V_0'.

On the other hand, we can identify H as a dense subspace of V_0'—by [(2-2)(ii)]. Then we introduce the space $V(\Lambda)$ defined by

(2-4) $V(\Lambda)$ is the space of $u \in V$ such that $\Lambda u \in H$,

supplied with the norm

(2-5) $\|u\|_{V(\Lambda)} = (\|u\|_{V}^2 + \|\Lambda u\|_{H}^2)^{1/2}$.

It is a Hilbert space (see Section 2.1-10).

Then the operator Λ associated with $a(u, v)$ by (2-3) satisfies

$$\Lambda \in L(V, V_0') \cap L(V(\Lambda), H).$$

Definition 2-1   We say that Λ is the "formal operator" associated with the form $a(u, v)$.   ▲

EXAMPLE   If $V = H^1(\Omega), a(u, v) = \sum_{1 \leq i \leq n} \int_\Omega D_i u D_i v \, dx + \lambda \int_\Omega uv \, dx$, the associated formal operator Λ is $-\Delta + \lambda$ (see Lemma 1-1). The space $V(\Lambda)$ is the space $H^1(\Omega, \Delta)$ defined by (1-11).

2-3.   The Green Formula

The equivalence between boundary-value problems and variational equations is based on the existence of a Green's formula as in Theorem 2-1.

THEOREM 2-1   Let us assume (2-1) and let $a(u, v)$ be a continuous bilinear form on $V \times V$. Then there exists a *unique* operator δ mapping $V(\Lambda)$ into T' such that the Green formula holds:

(2-6) $a(u, v) = (\Lambda u, v) + \langle \delta u, \gamma v \rangle$ for any $u \in V(\Lambda), v \in V$,

where $(\cdot, \cdot)$ is the inner product on $H \times H$ and $\langle \cdot, \cdot \rangle$ is the duality pairing on $T' \times T$.   ▲

Proof Let us assume that u ranges over $V(\Lambda)$. Since $\Lambda u \in H$, the bilinear form $(\Lambda u, v)$ is continuous on $V(\Lambda) \times H$, and then is continuous on $V(\Lambda) \times V$. On the other hand, $a(u, v)$ is continuous on $V(\Lambda) \times V \subset V \times V$.

Then the bilinear form $(Bu, v) = a(u, v) - (\Lambda u, v)$ is continuous on $V(\Lambda) \times V$ and B is a continuous linear operator mapping $V(\Lambda)$ into V'. Actually,

(2-7) B maps $V(\Lambda)$ into the orthogonal $V_0^{\perp}$ of V_0 in V'

since, by the very definition of Λ, $(Bu, v) = 0$ for any $v \in V_0$. (See Section 2.1-7.)

On the other hand, since γ maps V onto T, its transpose γ' is an isomorphism from T' onto its closed range (by Theorem 2.1-14). Since V_0 is the kernel of γ, the range of γ' is $V_0^{\perp}$. Let μ be the inverse of γ' mapping $V_0^{\perp}$ onto T' and set $\delta = \mu B$. The operator δ maps $V(\Lambda)$ into T'. Since $Bu \in V_0^{\perp}$ to (2-7), we can write

(2-8) $Bu = \gamma' \mu Bu = \gamma' \, \delta u.$

Then, applying Bu to $v \in V$, we obtain

$$a(u, v) - (\Lambda u, v) = (Bu, v) = (\gamma' \, \delta u, v) = \langle \delta u, \gamma v \rangle.$$

Let us show that δ is unique. If there exists δ_1 such that (2-6) holds, we deduce that $Bu = \gamma' \, \delta_1 u$ belongs to $V_0^{\perp}$. Applying μ, we find

$$\delta u = \mu Bu = \mu \gamma' \, \delta_1 u = \delta_1 u \qquad \text{for any} \quad u \in V(\Lambda). \qquad \blacksquare$$

Definition 2-2 We say that γ is the Dirichlet operator and δ is the Neumann operator associated with $a(u, v)$, γ, and H. $\blacktriangle$

EXAMPLES Theorem 2-1 and Lemma 1-2 imply Lemma 1-3. Indeed, by comparing the Green formulas (1-9) and (2-6) and by the uniqueness of the operator δ, we see that δ is the unique operator from $H^1(\Omega, \Lambda)$ into $H^{-\frac{1}{2}}(\Gamma)$ extending γ_1 and such that (1-9) holds.

Theorem 2-1 implies also Proposition 1.1-1. We choose $V = H^1(I)$, $T = R^2$ and $\gamma u = (u(0), u(1))$. Assumption 2-1 is satisfied. We take $a(u, v) = \int_I (a(x) \, Du \, Dv + b(x) uv) \, dx$. Then $\Lambda u = -D(a(x) Du) + b(x)u$. By Theorem 2-1, there exists a map δ from $V(\Lambda)$ into R^2 such that the Green formula holds. In other words, there exist two linear functionals $a_1(u)$ and $a_0(u)$ such that

$$\int_I (a(x) \, Du \, Dv + D(a(x) \, Du)v) \, dx = a_1(u)v(1) - a_0(u)v(0).$$

By comparing with the integration by parts formula, we can write that $a_i(u)$ is the extension of the functional $a_i(u) = a(i)Du(i)$ $(i = 0, 1)$.

2-4. Abstract Neumann and Dirichlet Problems Associated with $a(u, v)$

We can extend the results of Sections 1-4 and 1-5 to a more general situation.

PROPOSITION 2-1 Let V, H, γ, and T satisfy the trace property (2-1) and $a(u, v)$ be a continuous bilinear form on $V \times V$. Let Λ and δ be the formal and Neumann operators associated with $a(u, v)$. If $f \in H$ and $t \in T'$ are given, then the following two problems are equivalent:

1. The Neumann problem for Λ
Look for $u \in V(\Lambda)$ satisfying

$$(2-9) \qquad \begin{cases} \text{(i) } \Lambda u = f, \\ \text{(ii) } \delta u = t. \end{cases}$$

2. The variational Neumann problem for Λ
Look for $u \in V$ such that

$$(2-10) \qquad a(u, v) = (f, v) + \langle t, \gamma v \rangle \qquad \text{for any} \quad v \in V. \qquad \blacktriangle$$

Proof The proof is the same as the proof of Lemma 1-5. Let u be a solution of the variational Neumann problem. Since this equation holds for any $v \in V_0$, we deduce that $\Lambda u = f$ and since, furthermore $f \in H$, this solution u belongs to $V(\Lambda)$. Then we can apply the Green formula, which implies that

$$(\Lambda u - f, v) = \langle t - \delta u, \gamma v \rangle = 0 \qquad \text{for any} \quad v \in V.$$

Since γ maps V onto T, this implies that $\delta u = t$. Then u is a solution of the Neumann problem for Λ. Conversely, if u is a solution of the Neumann problem, it satisfies, by the Green formula,

$$a(u, v) = (\Lambda u, v) + \langle \delta u, \gamma v \rangle = (f, v) + \langle t, \gamma v \rangle \qquad \text{for any} \quad v \in V. \qquad \blacksquare$$

Now let us consider the Dirichlet problem. Since γ is onto, there exists at least one continuous right inverse $\mu \in L(T, V)$ of γ (see Theorem 2.1-13).

PROPOSITION 2-2 Let V, H, γ, and T satisfying (2-1) and $a(u, v)$ be a continuous bilinear form on $V \times V$, and let Λ be the formal operator associated with $a(u, v)$, with $f \in H$ and $t \in T$ given. Now we are in a position to consider the two following problems.

1. The Dirichlet problem for Λ
Look for $u \in V(\Lambda)$ satisfying

$$(2-11) \qquad \begin{cases} \text{(i) } \Lambda u = f, \\ \text{(ii) } \gamma u = t. \end{cases}$$

2. The variational Dirichlet problem for Λ

Look for $w \in V_0 = \ker \gamma$ satisfying

$$(2\text{-}12) \qquad a(w, v) = (f, v) - a(\mu t, v) \qquad \text{for any} \quad v \in V_0.$$

Then if u is the solution of the Dirichlet problem, $w = u - \mu t$ is the solution of the variational Dirichlet problem and, conversely, if w is the solution of the variational Dirichlet problem, $u = w + \mu t$ is the solution of the Dirichlet problem. ▲

Proof. Indeed, let w be a solution of the variational Dirichlet problem. Then $a(u, v) = (\Lambda u, v) = a(w + \mu t, v) = (f, v)$ for any $v \in V_0$. This implies that $\Lambda u = f$ and that $u \in V(\Lambda)$ (since $f \in H$). On the other hand, $\gamma u = \gamma w + \gamma \mu t = t$.

Conversely, if u is a solution of the Dirichlet problem, then $w = u - \mu t$ satisfies $\gamma w = 0$ and $a(w, v) = (\Lambda u, v) - a(\mu t, v) = (f, v) - a(\mu t, v)$ for any $v \in V_0$. ■

Besides the Neumann and Dirichlet problems, we can construct a whole family of boundary-value problems for Λ which are equivalent to variational equations.

2-5. Mixed Type Boundary-Value Problems Associated with $a(u, v)$

Starting with a bilinear form $a(u, v)$ continuous on $V \times V$, we can construct variational equations of the form

$$(2\text{-}13) \qquad \begin{cases} \text{(i) } u \in W, \\ \text{(ii) } a(u, v) + m(\gamma u, \gamma v) = (f, v) \qquad \text{for any} \quad v \in W \end{cases}$$

where

$$(2\text{-}14) \qquad \begin{cases} \text{(i) } W \text{ is a closed subspace of } V \qquad \text{such that} \qquad V_0 \subset W \subset V, \\ \text{(ii) } m(t, s) \text{ is a continuous bilinear form on } T \times T. \end{cases}$$

We learn in this section that such a variational equation is equivalent to a boundary-value problem for the formal operator Λ associated with $a(u, v)$.

Let σ_1 be a continuous projector of T, $\sigma_2 = 1 - \sigma_1$. Then the kernel $W = \ker \sigma_1 \gamma$ of the operator $\sigma_1 \gamma$ is a closed subspace of V containing V_0. Conversely, we can show that it is possible to associate with a closed subspace W satisfying [(2-14)(i)] a projector σ_1 such that $W = \ker \sigma_1 \gamma$.

Then $\gamma_1 = \sigma_1 \gamma$ maps V onto $T_1 = \sigma_1 T$ and $\gamma_2 = \sigma_2 \gamma$ maps V(and W) onto $T_2 = \sigma_2 T$. Furthermore, if μ is a right inverse of γ, then $\mu_1 = \mu \sigma_1 \in L(T_1, V)$ is a right inverse of γ_1.

On the other hand, the continuous bilinear form $m(t, s)$ on $T \times T$ defines a continuous linear operator $M \in L(T, T')$ by

$$\langle Mt, s \rangle = m(t, s) \qquad \text{for any} \quad t, s \in T,$$

(and conversely).

We associate with the continuous projector σ_1 of T and the continuous bilinear form $m(t, s)$ on $T \times T$ the so-called (σ_1, m) boundary-value problem for Λ which is equivalent to a variational equation.

THEOREM 2-2   Let us introduce the following items:

1. The space V, H, and T and the Dirichlet operator γ satisfying the trace property (2-1).
2. A continuous bilinear form $a(u, v)$ on $V \times V$; its associated formal operator Λ, and its associated Neumann operator δ.
3. A continuous bilinear form $m(t, s)$ on $T \times T$ and its associated operator $M \in L(T, T')$.
4. A continuous projector σ_1 of T, $\sigma_2 = 1 - \sigma_1$, $T_j = \sigma_j T$, $\gamma_j = \sigma_j \gamma$, $\delta_j = \sigma'_j \delta$ and $\mu_j = \mu \sigma_j$, where $j = 1, 2$ and μ is a right inverse of γ.
5. Data $f \in H$, $t_1 \in T_1$, and $t_2 \in T'_2$.

Now we can consider the two following problems.

1. The (σ_1, m) boundary-value problem for Λ
Look for $u \in V(\Lambda)$ satisfying

(2-15)
$$\begin{cases} \text{(i)} \ \Lambda u = f, \\ \text{(ii)} \ \gamma_1 u = t_1, \\ \text{(iii)} \ \delta_2 u + \sigma'_2 M \gamma u = t_2. \end{cases}$$

2. The variational (σ_1, m) boundary-value problem for Λ
Look for $w \in W = \ker \sigma_1 \gamma$ satisfying

(2-16)
$$\begin{cases} a(w, v) + m(\gamma w, \gamma v) = (f, v) + \langle t_2, \gamma_2 v \rangle - a(\mu_1 t_1, v) - \\ \qquad\qquad m(\gamma \mu_1 t, \gamma v) \qquad \text{for any} \quad v \in W. \end{cases}$$

Then if u is a solution of the (σ_1, m) boundary-value problem, $w = u - \mu_1 t_1$ is the solution of the variational (σ_1, m) boundary-value problem. Conversely, if w is a solution of the variational (σ_1, m) boundary-value problem, then $u = w + \mu_1 t_1$ is a solution of the variational (σ_1, m) boundary-value problem.   ▲

Proof Since σ_1 is a continuous projector, we can write the Green formula in the following form:

$$(2\text{-}17) \quad \begin{cases} a(u, v) = (\Lambda u, v) + \langle \delta_1 u, \gamma_1 v \rangle + \langle \delta_2 u, \gamma_2 v \rangle \\ \qquad\qquad\qquad\qquad \text{when} \quad u \in V(\Lambda), \quad v \in V. \end{cases}$$

On the other hand, if $v \in W = \ker \gamma_1$, then $\gamma v = \gamma_2 v$, since $\gamma = \gamma_1 + \gamma_2$.

Let w be a solution of the variational problem. Since the variational equation holds for $v \in V_0$, we deduce that $\Lambda u = \Lambda(w + \mu_1 t_1) = f$, and then that $u \in V(\Lambda)$. However, since $w \in W$, $\gamma_1 u = \gamma_1 w + \gamma_1 \mu_1 t_1 = t_1$. In order to prove that u satisfies [(2-15)(iii)], we write the Green formula in the form (2-17), thus deducing that

$$(2\text{-}18) \quad (\Lambda u - f, v) = \langle t_2 - \delta_2 u - \sigma_2' M \gamma u, \gamma_2 v \rangle = 0 \quad \text{for any} \quad v \in W.$$

Since γ_2 maps W onto T_2, we have proved that u is a solution of the (σ_1, m) boundary-value problem. Conversely, if u is a solution of the boundary-value problem, $w = u - \mu_1 t_1$ is a solution of the variational equation. Indeed, $\gamma_1 w = \gamma_1 u - \gamma_1 \mu_1 t_1 = 0$, and then $w \in W$, although when v ranges over w, $\gamma v = \gamma_2 v$ and we deduce from the Green formula that (2-16) holds. ∎

Definition 2-3 The boundary conditions of a (σ_1, m) boundary-value problem may be broken down into two parts:

1. The "stable conditions" or "forced conditions" $\gamma_1 u = t_1$, which are defined on the space V itself.
2. The "natural conditions" $\delta_2 u + \sigma_2' M \gamma u = t_2$, which are defined on the subspace $V(\Lambda)$ by the Green formula. ▲

When we use the variational formulation of a (σ_1, m) boundary-value problem, *we do not deal explicitly with the "natural" conditions that are involved in the form $a(u, v)$*. Under these conditions, it is convenient to introduce the following definitions.

Definition 2-4 Under the assumptions of Theorem 2-2, we say that

$$(2\text{-}19) \quad \begin{cases} \text{a } (0, m)\text{-problem is a } \textit{Neumann type} \text{ boundary-value problem,} \\ \text{a } (\sigma_1, m)\text{-problem is a } \textit{mixed type} \text{ boundary-value problem} \\ \text{a } (1, 0)\text{-problem is the } \textit{Dirichlet} \text{ problem} \end{cases}$$

associated with the bilinear form $a(u, v)$ continuous on $V \times V$. ▲

Even if a boundary-value problem is defined by the data of a bilinear form $a(u, v)$ continuous on a space $V \times V$ satisfying the assumptions (2-1), there are other ways to construct equivalent variational equations. (See, e.g., the conjugate problems in Chapter 10). *Each variational formulation of a given boundary-value problem furnishes a way of constructing approximate problems by using Theorem* 3.1-6.

Remark 2-1   Let us denote by $A_W \in L(W, W')$ the operator defined by

$$(2\text{-}20) \qquad a(u, v) = (A_W u, v) \qquad \text{for any} \quad u \in W, v \in W.$$

Since V_0 is dense in H, then W is dense in H and we deduce from Theorem 2.1-5 that

$$(2\text{-}21) \qquad\qquad\qquad W \subset H \subset W',$$

the injections being continuous and dense.

We introduce the domain $D(A_W)$ of A_W defined by

$$(2\text{-}22) \qquad D(A_W) = \{u \in W \qquad \text{such that} \qquad A_W u \in H\}$$

supplied with the graph norm.   ∎

COROLLARY **2-1**   If we make the assumptions of Theorem 2-2 and let $D(A_W)$ be the domain of the operator $A_W \in L(W, W')$ defined by the restriction to W of the continuous bilinear form $a(u, v)$, then

$$(2\text{-}23) \qquad D(A_W) = \{u \in V(\Lambda) \qquad \text{such that} \qquad \gamma_1 u = \delta_2 u = 0\}. \qquad ▲$$

Proof   The proof of Corollary 2-1 is left as an exercise.   ∎

2-6.   Existence and Uniqueness of the Solutions of Boundary-Value Problems

We study here the existence and uniqueness of the boundary-value problems associated with a (V, H)-coercive continuous bilinear form $a(u, v)$.

Let V, H, T be Hilbert spaces satisfying the trace property

$$(2\text{-}24) \quad \begin{cases} \text{(i)} \ \gamma \text{ maps } V \text{ onto } T, \\ \text{(ii)} \ H \text{ is a pivot space,} \\ \text{(iii)} \ V \text{ and } V_0 = \ker \gamma \text{ are dense in } H \text{ with a stronger topology.} \end{cases}$$

Furthermore, we assume that

$$(2\text{-}25) \quad \begin{cases} \text{(i) the injection from } V \text{ into } H \text{ is compact,} \\ \text{(ii) } a(u, v) \text{ is } (V, H)\text{-coercive.} \end{cases}$$

Let $\Lambda \in L(V, V_0') \cap L(V(\Lambda), H)$ and $\delta \in L(V(\Lambda), T')$ be, respectively, the formal operator and the Neumann operator associated with $a(u, v)$, and call Λ^* and δ^* the analogous operators associated with $a_*(u, v) = a(v, u)$.

Letting σ_1 be a projector of T, $\gamma_1 = \sigma_1\gamma$ and $W = \ker \gamma_1$, we consider the operators $A_W \in L(W, W')$ defined by the restriction of $a(u, v)$ to W, $A_W{}^* \in L(W, W')$ defined by the restriction of $a_*(u, v)$ to W.

By Theorem 2.1-16, $A_W + \lambda$ is an isomorphism from W onto W' whenever λ does not belong to the spectrum $S(A_W)$ of the operator A_W, which is a countable subset of isolated points.

If $f \in H$, $t_1 \in T_1 = \sigma_1 T$, and $t_2 \in T_2' = (1 - \sigma_1')T'$ are given, we can study the mixed boundary-value problems

$$(2\text{-}26) \quad \begin{cases} \text{(i)} \;\; u \in V(\Lambda), \Lambda u = f, \gamma_1 u = t_1, \delta_2 u = t_2, \\ \text{(ii)} \;\; u \in V(\Lambda), \Lambda^* u = f, \gamma_1 u = t_1, \delta_2^* u = t_2. \end{cases}$$

THEOREM 2-3 Let us assume (2-24) and (2-25). If λ does not belong to $S(A_W)$, then the mixed boundary-value problems [(2-26)(i) and (ii)] have unique solutions for any $f \in H$, $t_1 \in T_1$, and $t_2 \in T_2'$. If λ does belong to $S(A_W)$, then λ is an eigenvalue of the same problems. The dimension $n(\lambda)$ of the subspaces $N_\lambda(A_W)$ and $N_\lambda(A_W^*)$ defined by

$$(2\text{-}27) \quad \begin{cases} \text{(i)} \;\; N_\lambda(A_W) = \{u \in V(\Lambda) \;\; \text{such that} \quad \Lambda u + \lambda u = 0 \;\; \text{and} \\ \qquad\qquad\qquad\qquad\qquad\qquad\qquad\qquad \gamma_1 u = \delta_2 u = 0\}, \\ \text{(ii)} \;\; N_\lambda(A_W^*) = \{u \in V(\Lambda) \;\; \text{such that} \quad \Lambda^* u + \lambda u = 0 \;\; \text{and} \\ \qquad\qquad\qquad\qquad\qquad\qquad\qquad\qquad \gamma_1 u = \delta_2^* u = 0\} \end{cases}$$

is finite. In this case, the boundary-value problem [(2-26)(i)] has solutions if and only if the data $f \in H$, $t_1 \in T_1$, and $t_2 \in T_2'$ satisfy the compatibility conditions

$$(2\text{-}28) \qquad (f, v) + \langle t_2, \gamma_2 v \rangle - \langle t_1, \delta_1^* v \rangle = 0 \quad \text{for any} \quad v \in N_\lambda(A_W^*). \qquad \blacktriangle$$

Proof This theorem follows from Theorem 2.1-16, and from Theorem 2-2.

Indeed, if $\lambda \notin S(A_W)$, the operators $A_W + \lambda$ and $A_W^* + \lambda$ are isomorphisms from W onto W'. If μ is a continuous right inverse of γ and if $\mu_1 = \mu\sigma_1$, then the linear form l defined by

$$(2\text{-}29) \qquad l(v) = (f, v) + \langle t_2, \gamma_2 v \rangle - a(\mu_1 t_1, v) - \lambda(\mu_1 t_1, v)$$

is continuous on W.

Therefore, there exists a unique solution $w \in W$ of $(A_W + \lambda)w = l$. Theorem 2-2 implies that $u = w + \mu_1 t_1$ is a solution of [(2-26)(i)], and conversely.

Now, if $\lambda \in S(A_W)$, then the kernels of $A_W + \lambda$ and $A_W^* + \lambda$ have the same finite-dimension $n(\lambda)$; moreover, Theorem 2-2 implies that these kernels are the subspaces $N_\lambda(A_W)$ and $N_\lambda(A_W^*)$ defined by (2-27).

Finally, the closed range of $A_W + \lambda$ is the orthogonal in W' of the kernel $N_\lambda(A_W^*)$ of its transpose $(A_W + \lambda)' = A_W^* + \lambda$.

On the other hand, if $f \in H$, $t_1 \in T_1$, and $t_2 \in T_2'$ satisfy the compatibility conditions (2-28), then the form l defined by (2-29) belongs to the orthogonal of $N_\lambda(A_W^*)$. Indeed, we can write

$$l(v) = (f, v) + \langle t_2, \gamma_2 v \rangle - a_*(v, \mu_1 t_1) - \lambda(\mu_1 t_1, v).$$

Since $N_\lambda(A_W^*)$ is contained in $V(\Lambda^*)$, we can use the Green formula and we obtain

$$l(v) = (f, v) + \langle t_2, \gamma_2 v \rangle - (\Lambda^* v, \mu_1 t_1) - \lambda(v, \mu_1 t_1) - \langle \delta_1^* v, t_1 \rangle$$
$$- \langle \delta_2^* v, \gamma_2 \mu_1 t_1 \rangle$$
$$= (f, v) + \langle t_2, \gamma_2 v \rangle - \langle t_1, \delta_1^* v \rangle,$$

when v ranges over $N_\lambda(A_W^*)$.

This means that there are solutions w of the equation $(A_W + \lambda)w = l$, and Theorem 2-2 implies that $u = w + \mu_1 t_1$ is a solution of [(2-26)(i)].

Conversely, the Green formula implies that if u is the solution of [(2-26)(i)], then the compatibility conditions (2-28) hold. ∎

Remark 2-2   We can reformulate Theorem 2-3 by first setting

(2-30) $R_\lambda(A_W) = \{(f, t_1, t_2) \in H \times T_1 \times T_2' \quad$ satisfying (2-28)$\}$.

Then, if we assume (2-24) and (2-25), Theorem 2-3 amounts to saying that the operator $(\Lambda + \lambda, \gamma_1, \delta_2)$ defines an isomorphism from $V(\Lambda)/N_\lambda(A_W)$ onto $R_\lambda(A_W)$. When λ does not belong to the spectrum $S(A_W)$, then $N_\lambda(A_W) = 0$ and $R_\lambda(A_W) = H \times T_1 \times T_2'$. ∎

Remark 2-3   If we assume that $a(u, v)$ is V-elliptic instead of assuming (2-25), we deduce from the Lax-Milgram theorem that the operators $(\Lambda, \gamma_1, \delta_2)$ are isomorphisms from $V(\Lambda)$ onto $H \times T_1 \times T_2'$. ∎

Remark 2-4   We can use also Theorem 2.1-16 in order to study the existence and uniqueness of (σ_1, m)-boundary-value 'problems. For simplicity, let us consider the case of a Neumann type problem in which we assume that

(2-31) $\begin{cases} \text{(i) } a(u, v) \text{ is a } V\text{-elliptic continuous bilinear form on } V. \\ \text{(ii) } M \text{ is a compact operator from } T \text{ into } T'. \end{cases}$

Then the operator $A = A_V \in L(V, V')$ defined by $a(u, v)$ is an isomorphism and the operator $\gamma' M \gamma \in L(V, V')$ is a compact operator.

Let us denote by $S(A, M)$ the spectrum of the operator $\gamma' M \gamma$ relative to A (see Section 2.1-15).

THEOREM 2-4 Let us assume (2-24) and (2-31). If λ does not belong to $S(A, M)$, then there exists a unique solution $u \in V(\Lambda)$ of the Neumann-type problem

$$(2\text{-}32) \quad \Lambda u = f, \, \delta u + \lambda M \gamma u = t \quad \text{for any} \quad f \in H \quad \text{and} \quad t \in T'.$$

If λ does belong to $S(A, M)$, then λ is an eigenvalue of the Neumann-type problem: the proper subspace $N_\lambda(A, M)$ and $N_\lambda(A^*, M^*)$ are defined by

$$(2\text{-}33) \quad \begin{cases} \text{(i)} \ N_\lambda(A, M) = \{u \in V(\Lambda) \quad \text{such that} \quad \Lambda u = 0 \quad \text{and} \\ \qquad\qquad\qquad\qquad\qquad\qquad\qquad\qquad\qquad \delta u + \lambda M \gamma u = 0\}, \\ \text{(ii)} \ N_\lambda(A^*, M^*) = \{u \in V(\Lambda^*) \quad \text{such that} \quad \Lambda^* u = 0 \quad \text{and} \\ \qquad\qquad\qquad\qquad\qquad\qquad\qquad\qquad\qquad \delta^* u + \lambda M^* \gamma u = 0\} \end{cases}$$

have the same positive dimension $n(\lambda)$. If $f \in H$ and $t \in T'$ satisfy the compatibility conditions

$$(2\text{-}34) \quad (f, v) + \langle t, \gamma v \rangle = 0 \quad \text{for any} \quad v \in N_\lambda(A^*, M^*),$$

then there exist solutions u of the Neumann-type problem (2-32), and conversely. ▲

Proof The proof, which is analogous to that for Theorem 2.3, is left as an exercise. ■

Remark 2-5 It is clear that the domains $V(\Lambda + \lambda) = V(\Lambda)$ do not depend on λ and that the Neumann operator δ associated with the form $a(u, v) + \lambda(u, v)$ does not depend on λ. Therefore, we deduce from Theorem 2-3 that if $a(u, v)$ is a (V, H)-coercive bilinear form on $V \times V$, then

$$(2\text{-}35) \quad \begin{cases} \text{(i)} \ \Lambda \text{ maps } V(\Lambda) \text{ onto } H, \\ \text{(ii)} \ \gamma \text{ maps } V(\Lambda) \text{ onto } T, \\ \text{(iii)} \ \delta \text{ maps } V(\Lambda) \text{ onto } T'. \end{cases}$$

It is useful to introduce the following subspaces of $V(\Lambda)$ and $V(\Lambda^*)$ defined by homogeneous conditions: we set

$$(2\text{-}36) \quad \begin{cases} \text{(i)} \ D_0 = D(A_{V_0}) = \{u \in V(\Lambda) \text{ such that } \gamma u = 0\}, \, D_0^* = D(A_{V_0}^*), \\ \text{(ii)} \ D = D(A_V) = \{u \in V(\Lambda) \text{ such that } \delta u = 0\}, \, D^* = D(A_V^*) \end{cases}$$

(see Remark 2-1). ■

THEOREM 2-5 We suppose the assumptions of Theorem 2-3. The subspaces D and D^* are dense in V and H and the subspaces D_0 and D_0^* are dense in V_0 and H. In addition the images $\gamma(D)$ and $\gamma(D^*)$ are dense in T, and the images $\delta(D_0)$ and $\delta^*(D_0^*)$ are dense in T'.   ▲

Proof By replacing $a(u, v) + \lambda(u, v)$ by $a(u, v)$ for λ large enough, we can assume that $a(u, v)$ is V-elliptic.

Let us prove, for instance, that D is dense in V. By Theorem 2.1-2, we must prove that if $f \in V'$ vanishes on D, then f is equal to 0.

Let $u \in V$ be the solution of $A^*u = f$. Therefore $(f, v) = a(v, u) = (Av, u) = 0$ for any $v \in D$. Since A is an isomorphism from $D = D(A_V)$ onto H, we deduce that $(u, w) = 0$ for any $w \in H$. Then $u = 0$ and $f = A^*u = 0$.

Now let us prove that $\gamma(D)$ is dense in T. Let $t \in T'$ vanish on $\gamma(D)$, and prove that $t = 0$. Indeed, let $u \in V(\Lambda^*)$ be the solution of $\Lambda^*u = 0$ and $\delta^*u = t$. We deduce from the Green formula that

$$0 = \langle t, \gamma v \rangle = \langle \delta^*u, \gamma v \rangle = a_*(u, v) = a(v, u)$$
$$= (\Lambda v, u) + \langle \delta v, \gamma u \rangle = (\Lambda v, u) \qquad \text{for any} \quad v \in D.$$

Since $\Lambda = A$ is an isomorphism from D onto H, we deduce that $u = 0$, and thus that $t = \delta^*u = 0$.

The proofs of the other statements are analogous and are left as an exercise.   ■

2-7. Formal Adjoint of an Operator and Green's Formula

Let D be a continuous linear operator from a Hilbert space U into a Hilbert space E'. Its transpose D' is the continuous linear operator from E onto U' defined by

$$(2\text{-}37) \qquad (D'u, v) = (u, Dv) \qquad \text{for any} \quad u \in E \qquad \text{and} \qquad v \in U.$$

It is useful to associate with D its formal adjoint D^* and its domain $E(D^*)$ when the following trace properties hold: we suppose that there exist an operator α mapping U into a Hilbert space Q and a Hilbert space K' such that

$$(2\text{-}38) \qquad \begin{cases} \text{(i) } \alpha \text{ maps } U \text{ onto } Q, \\ \text{(ii) } U \text{ is contained in } K' \text{ with a stronger topology,} \\ \text{(iii) the kernel } U_0 = \ker \alpha \text{ of } \alpha \text{ is dense in } K'. \end{cases}$$

Definition 2-5 Under the assumptions (2-38), we define the formal adjoint D^* of D by

$$(2\text{-}39) \qquad (D^*u, v) = (u, Dv) \qquad \text{for any} \quad u \in E, v \in U_0,$$

and its domain $E(D^*)$ by

(2-40) $\qquad E(D^*) = \{u \in E \qquad \text{such that} \qquad D^*u \in K\}$

supplied with the graph norm $\|u\|_{E(D^*)} = (\|u\|_E^2 + \|D^*u\|_K^2)^{1/2}$. ▲

This definition has meaning because K can be identified to a dense subspace of U_0' by [(2-38)(ii)] and Theorem 2.1-5.

The operators D and D^* are related by a Green's formula:

THEOREM 2-6 If we assume (2-38), then there exists a *unique* operator $\alpha^* \in L(E(D^*), Q')$ such that

(2-41) $\begin{cases} (D^*u, v) - (u, Dv) = \langle \alpha^*u, \alpha v \rangle \qquad \text{for any} \quad u \in U \quad \text{and} \\ \qquad\qquad\qquad\qquad\qquad\qquad\qquad\qquad\qquad v \in E(D^*) \end{cases}$ ▲

Proof The proof is analogous to the proof of Theorem 2-1. Let $u \in E(D^*)$ and j be the injection from U into K'. Then $jD^*u - D'u$ belongs to the orthogonal $U_0^\perp$ of U_0 in U' by (2-37) and (2-39). Since $U_0 = \ker \alpha$, we deduce from [(2-38)(i)] and Theorem 2.1-14 that the transpose α' of α is an isomorphism from Q' onto its closed range $U_0^\perp$. Then there exists a unique element α^*u of Q' such that $jD^*u - D'u = \alpha'\alpha^*u$ [i.e., such that the Green formula (2-41) holds]. It is clear that α^* is a continuous linear operator from $E(D^*)$ into Q'. ∎

Now we apply Theorem 2-5 in the following situation. Let us assume the hypotheses (2-24) and let $a(u, v)$ be a continuous bilinear form on $V \times V$. Let $\Lambda, \Lambda^*, \delta, \delta^*$ be the formal operators and Neumann operators associated with $a(u, v)$ and $a_*(u, v) = a(v, u)$, respectively.

By applying Theorem 2-1 to $a(u, v)$ and to $a_*(u, v)$, we deduce the following Green formula:

(2-42) $\begin{cases} (\Lambda^*u, v) - (u, \Lambda v) = \langle \gamma u, \delta v \rangle - \langle \delta^*u, \gamma v \rangle \qquad \text{for any} \\ \qquad\qquad\qquad\qquad\qquad u \in V(\Lambda^*) \qquad \text{and} \qquad v \in V(\Lambda). \end{cases}$

Furthermore, we assume that there exist Hilbert spaces U, S, and R such that

(2-43) $\begin{cases} \text{(i)} \;\; U \subset V(\Lambda) \cap V(\Lambda^*), \text{ the injections being continuous,} \\ \text{(ii)} \;\; S \subset T \subset R, \text{ the injections being continuous and dense,} \\ \text{(iii)} \;\; (\gamma, \delta) \quad \text{and} \quad (\gamma, \delta^*) \text{ map } U \text{ onto } S \times R', \\ \text{(iv)} \;\; U_0 = D \cap D_0 = D^* \cap D_0^* \text{ is dense in } H, \\ \text{(v)} \;\; D, D_0, D^*, \text{ and } D_0^* \text{ are contained in } U. \end{cases}$

We thus can use Theorem 2-6 with $E = K = H$, $D = \Lambda$, $Q = R' \times S$ and $\alpha = (\delta, \gamma)$.

We deduce that the formal adjoint $\Lambda^* \in L(H, U_0')$ of Λ is the unique extension of the formal operator $\Lambda^* \in L(V(\Lambda^*), H)$ associated with the form $a_*(u, v)$, and this is why we identify these two operators.

Let $H(\Lambda^*) = \{u \in H$ such that $\Lambda^*u \in H\}$ be the domain of the formal adjoint Λ^* of Λ. Theorem 2-6 implies that there exists a unique operator $\alpha^* \in L(H(\Lambda^*), R \times S')$ such that the Green formula holds. We can write such an operator α^* in the form $\alpha^*u = (\gamma u, -\delta^*u)$ and the Green formula (2-41) in the form

$$(2\text{-}44) \quad \begin{cases} (\Lambda^*u, v) - (u, \Lambda v) = \langle \gamma u, \delta v \rangle - \langle \delta^*u, \gamma v \rangle & \text{for any} \quad v \in U \\ & \text{and} \quad u \in H(\Lambda^*). \end{cases}$$

By comparing (2-42) and (2-44), we deduce that we can extend $\gamma \in L(V(\Lambda^*), T)$ to $L(H(\Lambda^*), R)$ and $\delta^* \in L(V(\Lambda^*), T')$ to $L(H(\Lambda^*), S')$.

By exchanging the roles of $a(u, v)$ and $a_*(u, v)$, we can extend Λ to an operator $\Lambda \in L(H, U_0') \cap L(H(\Lambda), H)$, which plays the role of the formal adjoint of Λ^*; also, we can extend γ and δ to $L(H(\Lambda), R)$ and $L(H(\Lambda), S')$, respectively.

2-8. Theorems of Regularity

After the operators γ and δ have been extended to the space $H(\Lambda)$, it is possible to prove some results of regularity of the solutions of the Neumann and Dirichlet problems.

THEOREM 2-7   Let us assume (2-24), (2-25), and (2-43).

If λ does not belong to the spectrum $S(A) = S(A_V)$, then the operator $(\Lambda + \lambda, \delta)$ is an isomorphism from U onto $H \times R'$, from $V(\Lambda)$ onto $H \times T'$, and from $H(\Lambda)$ onto $H \times S'$.

If λ does not belong to the spectrum $S(\Lambda) = S(A_{V_0})$, the operator $(\Lambda + \lambda, \gamma)$ is an isomorphism from U onto $H \times S$, from $V(\Lambda)$ onto $H \times T$, and from $H(\Lambda)$ onto $H \times R$.

The same statements hold when we replace Λ by Λ^* and δ by δ^*.   ▲

Proof   For simplicity, when $\lambda \notin S(A)$, we replace $\Lambda + \lambda$ by Λ. We have already proved that (Λ, δ) and (Λ^*, δ^*) are isomorphisms from $V(\Lambda)$ and $V(\Lambda^*)$ onto $H \times T'$ (see Theorem 2-3 with $W = V$). In particular, this implies that Λ^* is an isomorphism from D^* onto H. Then its transpose is an isomorphism from H onto $(D^*)'$. If $f \in H$ and $t \in S'$, then the form $l = f + \gamma t$ is continuous on D^*. Therefore, there exists a unique solution u of the equation $(\Lambda^*)' u = l$, which can also be stated

$$(2\text{-}45) \quad \begin{cases} \text{(i)} \ u \in H, \\ \text{(ii)} \ (u, \Lambda^*v) = (f, v) + \langle t, \gamma v \rangle \quad \text{for any} \quad v \in D^*. \end{cases}$$

When v ranges over $U_0 = D^* \cap D_0^*$, we deduce that $\Lambda u = f$, and thus that $u \in H(\Lambda)$, since Λ is the formal adjoint of Λ^*. By using the Green formula (2-44), we deduce from (2-45) that $\langle \delta u - t, \gamma v \rangle = 0$ for any $v \in D^*$. Since γ maps D^* onto S—by [(2-43)(iii)]—u is the solution of the Neumann problem

$$(2\text{-}46) \qquad \Lambda u = f \quad \text{and} \quad \delta u = t$$

and belongs to $H(\Lambda)$.

We have proved that (Λ, δ) maps $H(\Lambda)$ onto $H \times S'$; by Theorem 2.1-12, it is an isomorphism. Now, to prove that (Λ, δ) is an isomorphism from U onto $H \times R'$, it is enough to show that (Λ, δ) maps U onto $H \times R'$. For that purpose, we must demonstrate that its range is both closed and dense in $H \times R'$.

First we show that the range of (Λ, δ) is closed: indeed, let $u_n \in U$ be a sequence such that

$$(2\text{-}47) \quad \Lambda u_n \quad \text{converges to } f \text{ in } H \quad \text{and} \quad \delta u_n \text{ converges to } t \text{ in } R'.$$

Since δ maps U onto R' by assumption, there exists a continuous right inverse ρ of δ. Then $v_n = u_n - \rho \delta u_n$ belongs to D, and Λv_n converges to $f - \Lambda \rho \delta t$ in H. Since Λ is an isomorphism from D onto H, this implies that v_n converges in D to an element v satisfying $\Lambda v = f - \Lambda \rho \delta t$. Therefore $u = v + \rho \delta t$ belongs to U and satisfies $f = \Lambda u$ and $t = \delta u$. Then the range of (Λ, δ) is closed.

By using Theorem 2.1-2, we show that the range of (Λ, δ) is dense by proving that any continuous linear form $(u, t) \in H \times S$ on $H \times S'$ satisfying

$$(2\text{-}48) \qquad (u, \Lambda v) + \langle t, \delta v \rangle = 0 \quad \text{for any} \quad v \in U$$

is equal to 0.

When v ranges over U_0, we deduce that $\Lambda^* u = 0$, and thus that $u \in H(\Lambda^*)$. Applying the Green formula (2-44), we obtain

$$(2\text{-}49) \qquad \delta^* u = 0 \quad \text{and} \quad t = -\gamma u.$$

Since (Λ^*, δ^*) is an isomorphism from $H(\Lambda^*)$ onto $H \times S'$, we deduce that $u = 0$ and that $t = -\gamma u = 0$.

The proof of the second statement is analogous, since in this theorem the roles played by γ and δ are symmetric. ∎

Remark 2-6 Let us consider the case where $\lambda \in S(A)$. By [(2-43)(v)], the subspace $N_\lambda(A)$ is contained in U. We can show that if $f \in H$ and t satisfy the compatibility conditions

(2-50) $\qquad (f, v) + \langle t, \gamma v \rangle = 0 \qquad$ for any $\quad v \in N_\lambda(A^*)$,

then there exist solutions u of the Neumann problem

$$\Lambda u + \lambda u = f, \quad \delta u = t,$$

which belong to U if $t \in R'$, to $V(\Lambda)$ if $t \in T'$, and to $H(\Lambda)$ if $t \in S'$.

In the same way, if $\lambda \in S(\Lambda)$ and if $f \in H$ and t satisfy the compatibility conditions

(2-51) $\qquad (f, v) - \langle t, \delta^* v \rangle = 0 \qquad$ for any $\quad v \in N_\lambda(\Lambda^*) = N_\lambda(A_{V_0}^*)$,

then there exist solutions u of the Dirichlet problem

$$\Lambda u + \lambda u = f, \quad \gamma u = t$$

which belong to U if $t \in S$, to $V(\Lambda)$ if $t \in T$, and to $H(\Lambda)$ if $t \in R$. The proof of these statements is left as an exercise. ∎

THEOREM 2-8 Let us assume (2-24), (2-25), and (2-43). Then the space U is dense in the spaces $V(\Lambda)$ and $H(\Lambda)$. ▲

Proof Let us prove that U is dense in $V(\Lambda)$. By Theorem 2.1-2 we must show that if a linear form $l(v)$ on $V(\Lambda)$ vanishes on U, then l is equal to 0.

Since $V(\Lambda) = \{u \in V$ such that $\Lambda u \in H\}$, we deduce from Theorem 2.1-8 that $l(v)$ can be written in the form

(2-52) $\quad l(v) = (f, v) - (u, \Lambda v) \qquad$ where $\quad f \in V' \qquad$ and $\qquad u \in H$.

Replacing Λ by $\Lambda + \lambda$ if needed, we can assume that $a(u, v)$ is V-elliptic. Then there exists $u_0 \in V$ such that $(f, v) = a_*(u_0, v)$ for any $v \in V$.

We can write

(2-53) $\begin{cases} l(v) = a_*(u_0, v) - (u_1, \Lambda v) = a(v, u_0) - (u_1, \Lambda v) & \text{where} \quad u_0 \\ & \text{and} \qquad u \in V. \end{cases}$

Since $v \in V(\Lambda)$, we can apply the Green formula and, finally, we can write $l(v)$ in the following form:

(2-54) $\qquad l(v) = (u_0 - u, \Lambda v) + \langle \delta v, \gamma u_0 \rangle, \ \gamma u_0 \in T'$.

Since (Λ, δ) is an isomorphism from U onto $H \times R'$, there exists a unique $v \in U$ satisfying $\Lambda v = w$, $\delta v = t$ when w ranges over H and t ranges over R'. Since $l(v)$ vanishes on U, we deduce that

(2-55) $\quad (u_0 - u, w) + \langle t, \gamma u_0 \rangle = 0 \qquad$ for any $\quad w \in H \qquad$ and $\qquad t \in R'$.

This implies first, that $u_0 = u$ and second that $\gamma u_0 = 0$ (since R' is dense in T'). Therefore $l(v)$ vanishes on $V(\Lambda)$, and thus U is dense in $V(\Lambda)$. The proof of the density of U in $H(\Lambda)$ is left as an exercise. ∎

3. THE TRACE THEOREM AND PROPERTIES OF SOBOLEV SPACES

The study of abstract boundary-value problems we made in the previous section was based on the assumption that trace properties held. Such trace properties do hold for Sobolev spaces, and if Ω is a smooth bounded open subset of R^n, Γ its boundary, $\gamma_j u$ the normal derivative of order j of a function u, we can prove that

$$\gamma = (\gamma_0, \ldots, \gamma_{m-1}) \text{ maps } V = H^m(\Omega) \text{ onto } T = \prod_{0 \leq j \leq m-1} H^{m-j-\frac{1}{2}}(\Gamma)$$

and

$$\text{the kernel } V_0 \text{ of } \gamma \text{ is } H_0^m(\Omega), \text{ dense in } L^2(\Omega),$$

where the spaces $H^s(\Gamma)$ are defined in a convenient way.

We also prove the following properties of Sobolev spaces:

- Sobolev inequalities: the space $H^s(\Omega)$ is a space of continuous functions if $s > n/2$,
- a theorem of compactness: if $s > k$, the injection from $H^s(\Omega)$ into $H^k(\Omega)$ is compact,
- the space $H^m(\Omega)$ is a space of order $\theta = (s - m)/(s - k)$ between $H^s(\Omega)$ and $H^k(\Omega)$ if $0 \leq k \leq m \leq s$.

3-1. Statement of the Trace Theorem

We define Sobolev space $H^s(\Gamma)$ on the boundary Γ of a bounded open subset Ω of R^n when s is a real number, and we call the trace operators γ_j. This being done, we prove the trace Theorem 3-1.

THEOREM 3-1 Let us assume that Ω is a smooth bounded open subset of R^n and that Γ is its boundary. Then the trace operators γ_j can be extended to continuous linear operators mapping $H^m(\Omega)$ *onto* $T_j = H^{m-j-\frac{1}{2}}(\Gamma)$ for $0 \leq j \leq m - 1$. Moreover, the operator $\gamma = (\gamma_0, \gamma_1, \ldots, \gamma_j, \ldots, \gamma_{m-1})$ maps $H^m(\Omega)$ *onto* the product space $\prod_{0 \leq j \leq m-1} T_j = \prod_{0 \leq j \leq m-1} H^{m-j-\frac{1}{2}}(\Gamma)$.

Finally, the space $H_0^m(\Omega)$, the closure of $\Phi_0(\Omega)$ in $H^m(\Omega)$, is the kernel of the operator γ mapping $H^m(\Omega)$ onto $\prod_{0 \leq j \leq m-1} H^{m-j-\frac{1}{2}}(\Gamma)$. ▲

Then if $V = H^m(\Omega)$, this theorem implies assumptions (2-1) and (2-38) (and thus, the results of Section 2), since we can choose $T = \prod_{0 \leq j \leq m-1} T_j$

$$\gamma = (\gamma_0, \ldots, \gamma_{m-1}), \quad H = L^2(\Omega), \text{ and } V_0 = H_0^m(\Omega).$$

Therefore, variational equations on $H^m(\Omega)$ are equivalent to boundary-value problems associated with projectors of $\prod_{0 \leq j \leq m-1} T_j$ and bilinear forms on this space.

3-2.   Change of Coordinates

In order to define and study the spaces $H^s(\Gamma)$ and the trace operators, we must use the equations of Ω and its boundary Γ.

Definition 3-1   We say that Ω is "smooth" if it is a bounded open subset of R^n, if its boundary Γ is an infinitely differentiable manifold of dimension $n - 1$, and if Ω lies in a same side of Γ.   ▲

Therefore, there exists a covering of Γ by a finite number of bounded open subsets $\Theta_j (1 \leq j \leq J)$ of R^n and maps ϕ_j and ψ_j such that

(3-1)

> (i) ϕ_j maps Θ_j into $Q = \{y = (y', y_n) \in R^n$ such that $|y'| < 1, -1 < y_n < 1\}$,
>
> (ii) ϕ_j maps $\Theta_j \cap \Omega$ into $Q_+ = \{y \in Q$ such that $y_n > 0\}$,
>
> (iii) ϕ_j maps $\Theta_j \cap \Gamma$ into $Q_0 = \{y \in Q$ such that $y_n = 0\}$,
>
> (iv) ϕ_j and ψ_j are infinitely differentiable, their Jacobians are positive,
>
> (v) ψ_j is the inverse of ϕ_j.

There exists also an open subset Θ_0 of Ω, contained in a compact of Ω, such that $(\Theta_0, \ldots, \Theta_j, \ldots)$ is a covering of Ω (see Figure 3).

Let $\alpha_0, \ldots, \alpha_j, \ldots$ be a partition of unity such that

(3-2) $\alpha_j \in \Phi(R^n)$, support $\alpha_j \subset \Theta_j$, $0 \leq \alpha_j(x) \leq 1$; $\sum_j \alpha_j(x) = 1$.

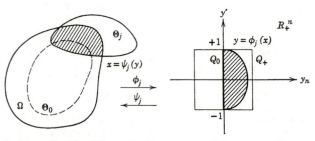

FIGURE 3   Open subset Θ_0 of Ω contained in a compact of Ω such that $(\Theta_0, \ldots, \Theta_j, \ldots)$ is a covering of Ω.

and introduce the following notations:

$$(3\text{-}3) \qquad \phi_j^*(\sqrt{\alpha_j}\,u)(y) = (\sqrt{\alpha_j}\,u)(\psi_j(y)) \, ; \ \psi_j^* v(x) = v(\phi_j(x)),$$

when u is defined on Ω and v is defined on Q.

Therefore, we can write a function u of $H^m(\Omega)$ in the form

$$(3\text{-}4) \qquad u(x) = \sqrt{\alpha_0}\,\sqrt{\alpha_0}u + \sum_{1 \leq j \leq J} \sqrt{\alpha_j}\,\psi_j^* \phi_j^*(\sqrt{\alpha_j}\,u).$$

Then a function u belongs to $H^m(\Omega)$ if and only if $\sqrt{\alpha_0}u$ belongs to $H^m = H^m(R^n)$ and $\phi_j^*(\sqrt{\alpha_j}u)$ belongs to $H^m(R_+^n)$ with compact support

$$\text{in } \bar{Q}_+ = \{y = (y', y_n) \in Q \qquad \text{such that} \qquad y_n \geq 0\}.$$

In other words, the operator θ defined by

$$(3\text{-}5) \qquad \theta u = (\sqrt{\alpha_0}\,u, \ldots, \phi_j^*(\sqrt{\alpha_j}\,u), \ldots)$$

is an isomorphism from $H^m(\Omega)$ onto its range, closed in $H^m(R^n) \times (H^m(R_+^n))^J$. Indeed, the map that associates with $(v_0, v_1, \ldots, v_J)$ the function $\alpha_0 v_0 + \sum_{1 \leq j \leq J} \sqrt{\alpha_j}\psi_j^* v_j$ is a right inverse of θ.

Using this isomorphism θ, we can deduce the properties of $H^m(\Omega)$ from the properties of $H^m(R_+^n)$.

3-3. Sobolev Spaces $H^s(R^n)$ for Real Numbers s

Since the Fourier transform is an isomorphism from $L^2(R_x^n)$ onto $L^2(R_y^n)$, it maps the space $H^m = H^m(R_x^n)$ onto the space $\hat{H}^m = \hat{H}^m(R_y^n)$ defined by

$$(3\text{-}6) \qquad \left\{ \hat{H}^m \text{ is the space of } \hat{u} \in L^2(R_y^n) \qquad \text{such that} \qquad (1 + |y|^2)^{m/2}\hat{u}(y) \in L^2(R_y^n) \right.$$

supplied with the norm

$$(3\text{-}7) \qquad \|v\|_m = \left(\int (1 + |y|^2)^m |v(y)|^2 \, dy \right)^{1/2}.$$

Therefore, we can define $\hat{H}^s$ for any real number s by replacing m by s in (3-6). Such a space is a Hilbert space because $\hat{H}^s$ is the L^2-space for the measure $(1 + |y|^2)^s \, dy$. For the duality pairing extending $\int u(y)v(y)\,dy$, we can identify the dual of $\hat{H}^s$ with the space $\hat{H}^{-s}$.

Definition 3-2 If s is positive, we define $H^s = H^s(R^n)$ to be the space of functions $u \in L^2$ whose Fourier transform belongs to $\hat{H}^s$, supplied with the norm

$$(3\text{-}8) \qquad \|u\|_s = \left(\int (1 + |y|^2)^s |\hat{u}(y)|^2 \, dy \right)^{1/2}.$$

If s is negative, we set

$$(3\text{-}9) \qquad\qquad H^s = (H^{-s})'; \; s \leq 0. \qquad\qquad \blacktriangle$$

Since the Fourier transform is an isomorphism from H^s onto $\hat{H}^s$ its transpose is an isomorphism from $\hat{H}^{-s}$ onto H^{-s}. But this transpose is an extension of the Fourier transform, since

$$\int dy\, v(y) \int e^{ix\cdot y} u(x)\, dx = \int dx\, u(x) \int e^{ix\cdot y} v(y)\, dy,$$

when $u \in L^2(R_x^n)$ and $v \in L^2(R_y^n)$. We call this transpose the Fourier transform.

We thus have proved the result of Lemma 3-1.

LEMMA 3-1   The spaces H^s are the Fourier transforms of the spaces $\hat{H}^s$ for any real number s and the dual of H^s is identified with the space H^{-s}.

$$\blacktriangle$$

3-4.   Sobolev Spaces $H^s(\Gamma)$ and $H^s(\Omega)$

Let us consider the covering of Γ and the system of local coordinates defined by (3-1). The operator θ defined by (3-5) is a one to one operator associating with a function u defined on Γ the sequence $(\ldots, \phi_j^*(\sqrt{\alpha_j}u)(y', 0), \ldots)$ of J functions with compact support defined on R^{n-1}.

Definition 3-3   We say that u belongs to $H^s(\Gamma)$ if the functions $\phi_j^*(\sqrt{\alpha_j}u)(y', 0)$ belongs to $H^s(R^{n-1})$ for $1 \leq j \leq J$. The space $H^s(\Gamma)$ is supplied with the norm

$$(3\text{-}10) \qquad \|u\|_{H^s(\Gamma)} = \left(\sum_{1 \leq j \leq J} \|\phi_j^*\left(\sqrt{\alpha_j}u\right)(y', 0)\|_{H^s(R^{n-1})}^2 \right)^{1/2}. \qquad \blacktriangle$$

In other words, we identify $H^s(\Gamma)$ with a closed subspace of $(H^s(R^{n-1}))^J$ by the isomorphism θ. Naturally, this definition seems to depend on the choice of the system of local coordinates defining the manifold Γ. Actually, it is possible to check that this definition does not depend on such a system: the different norms (3-10) are all equivalent.

Definition 3-3 implies that the spaces $H^s(\Gamma)$ "behave" like the spaces $H^s(R^{n-1})$.

Definition 3-4   We define the space $H^s(\Omega)$ as the space of the restrictions to Ω of functions u of $H^s = H^s(R^n)$. $\qquad\qquad \blacktriangle$

3-5. Trace Operators and Operators of Extension: Theorems of Density

Let us begin by defining the notions of trace operators and operators of extension when $\Omega = R_+^n$ is the half-space and $\Gamma = R^{n-1}$ is its boundary. In this case, the trace operators γ_j are defined by

$$(3\text{-}11) \qquad (\gamma_j u)(x') = (D_{x_n}^j u)(x', x_n)\big|_{x_n=0} \text{ defined on } \Gamma = R^{n-1}.$$

Let ρ be the operator of restriction associating with a function u defined on R^n its restriction ρu to R_+^n. It is a continuous linear operator from $H^m(R^n)$ into $H^m(R_+^n)$.

We shall say that ω is an "operator of extension" if ω is a continuous right inverse of ρ. Such an operator exists: let us consider, for instance, ω defined by

$$(3\text{-}12) \qquad \omega(u)(x', x_n) = \begin{cases} u(x', x_n) & \text{if } x_n > 0 \\ \displaystyle\sum_{1 \le j \le m} \alpha_j u(x', -jx_n) & \text{if } x_n < 0, \end{cases}$$

where the α_j's are solution of the Vandermonde system

$$(3\text{-}13) \qquad \sum_{1 \le j \le m} (-j)^k \alpha_j = 1 \qquad \text{for } 0 \le k \le m - 1.$$

Then (3-13) implies that

$$(3\text{-}14) \qquad (\gamma_j u)(x') = (\gamma_j \omega u)(x') \qquad \text{for } 0 \le j \le m - 1$$

and ωu belongs to $H^m(R^n)$, since

$$(D^p u)(x', x_n) = \begin{cases} (D^p u)(x', x_n) = (D_{x_n}^{p_n} D^{p-p_n} u)(x', x_n) & \text{if } x_n > 0 \\ \displaystyle\sum_{1 \le j \le m} (-j)^{p_n} \alpha_j (D_{x_n}^{p_n} D^{p-p_n} u)(x', -jx_n) & \text{if } x_n < 0, \end{cases}$$

and the limits of these expressions when x_n converges to 0 are equal.

Moreover,

$$(3\text{-}15) \qquad \|\omega u\|_{H^m(R^n)} \le c\, \|u\|_{H^m(R_+^n)} \qquad \text{for any } u \in H^m(R_+^n).$$

Now let Ω be a smooth bounded open subset of R^n and Γ its boundary. Let us consider the system of local coordinates defined by (3-1). We define the trace operators γ_j and the extension operator ω in the following way:

$$(3\text{-}16) \qquad \gamma_i u = \sum_{1 \le j \le J} \sqrt{\alpha_j}\; \psi_j^* [\gamma_i \phi_j^* (\sqrt{\alpha_j}\, u)]$$

and

$$(3\text{-}17) \qquad \omega u = \sum_{1 \le j \le J} \sqrt{\alpha_j}\; \psi_j^* \omega(\phi_j^* (\sqrt{\alpha_j}\, u)).$$

Thus we have proved Theorem 3-2.

THEOREM 3-2   If Ω is a smooth bounded open subset of R^n, there exists an operator of extension from $H^m(\Omega)$ into $H^m(R^n)$.   ▲

We need the following density theorem:

THEOREM 3-3   The space $\Phi_0 = \Phi_0(R^n)$ is dense in the spaces $H^s = H^s(R^n)$. The space $\Phi_0(\Gamma)$ is dense in the spaces $H^s(\Gamma)$ and the space $\Phi_0(\bar{\Omega})$ is dense in the spaces $H^m(\Omega)$ when Ω is a smooth bounded subset of R^n and $\bar{\Omega}$ denotes the closure of Ω.   ▲

Proof   1. Φ_0 is dense in H^m for m is a positive integer
Let μ be a positive function with compact support, and assume that

$$(3\text{-}18) \qquad \mu \in \Phi_0 \; ; \; \int \mu(x) \, dx = 1.$$

Let $\theta \in \Phi_0$, $\theta(x) = 1$ for $|x| \leq 1$ and $\theta(x) = 0$ for $|x| \geq 2$. If $h > 0$ converges to 0, we associate with u the functions ϕ_h where

$$(3\text{-}19) \qquad \phi_h(x) = \theta(xh) \cdot (\mu_h * u)(x).$$

The function ϕ_h is infinitely differentiable with compact support.

We proved that $\mu_h * u$ converges to u in H^m in Theorem 4.1-2. Now it is easy to check that $\theta(xh)u(x)$ converges to u in H^m by using the Leibniz formula. Indeed, we can write

$$(3\text{-}20) \qquad \begin{cases} \dfrac{D^p}{p!}(\theta(xh)u - u) = (\theta(xh) - 1)\dfrac{D^p u}{p!} \\[2ex] \qquad + \sum_{\substack{j \leq p \\ j \neq p}} \dfrac{h^{p-j}}{(p-j)!}(D^{p-j}\theta)(xh)\dfrac{D^j u(x)}{j!} \end{cases}$$

Since

$$\int |(\theta(xh) - 1)u(x)|^2 \, dx \leq \int_{|x| \geq h^{-1}} |u(x)|^2 \, dx$$

converges to 0, and since

$$\int |(D^k\theta)(xh)v(x)|^2 \, dx \leq \sup_{x \in R^n} |D^k\theta(x)|^2 \int |v(x)|^2 \, dx,$$

we deduce from (3-20) that

$$(3\text{-}21) \qquad |D^p(\theta(xh)u - u)|_{L^2} \text{ converges to 0 for any } |p| \leq m.$$

Therefore, the functions ϕ_h defined by (3-19) converge to u in H^m.

2. Φ_0 is dense in H_0 for s real and positive

This statement follows because Φ_0 is dense in H^m and H^m is dense in H^s if m is an integer $\geq s$. In this case, we can write

$$(3\text{-}22) \qquad \|u - \phi_h\|_s \leq \|u - u_h\|_s + c\,\|u_h - \phi_h\|_m,$$

where $u_h \in H^m$ converges to u in H^s and ϕ_h is defined by (3-19) with u replaced by u_n.

It is equivalent to prove that H^m is dense in H^s or to prove that $\hat{H}^m$ is dense in $\hat{H}^s$. Since the operator of multiplication by $(1 + |y|^2)^{s/2}$ is an isometry from $\hat{H}^t$ onto $\hat{H}^{t+s}$ (which is obvious by the very definition of spaces $\hat{H}^s$), it is in particular an isomorphism from $\hat{L}^2 = \hat{H}^0$ onto $\hat{H}^s$. Since $\hat{H}^t$ is dense in $\hat{L}^2$ for any $t > 0$, we deduce that $\hat{H}^{t+s}$ is dense in $\hat{H}^s$ for any $t > 0$. In particular, if m is an integer $\geq s$, H^m is dense in H^s.

3. Φ_0 is dense in the spaces H^s

Since the injection from H^s into Φ_0' is continuous and dense (for $s \geq 0$), we deduce from Theorem 2.1-4 that Φ_0 is a dense subspace of H^{-s}.

4. $\Phi_0(\Gamma)$ is dense in $H^s(\Gamma)$

Using the system of local coordinates defined by (3-1), we see that the function $v_j = \phi_j^*(\sqrt{\alpha_j} u)$ belongs to $H^s(R^{n-1})$ and their supports are contained in Q_0. There exist functions v_{jh} of $\Phi_0(R^{n-1})$ with compact support in Q_0 converging to v_j in $H^s(R^{n-1})$. Therefore, the functions $u_h = \sum_{1 \leq j \leq J} \sqrt{\alpha_j}\,\psi_j^* v_{jh}$ belong to $\Phi_0(\Gamma)$ and converge to u in $H^s(\Gamma)$.

5. $\Phi_0(\bar{\Omega})$ is dense in $H^m(\Omega)$ (for m positive integer)

If $u \in H^m(\Omega)$, the function $\omega u \in H^m$ by Theorem 3-2. But ωu is the limit of functions $v_h \in \Phi$. Then the functions $u_h = \rho v_h$ belong to $\Phi_0(\bar{\Omega})$ and converge to u in $H^m(\Omega)$, since $u = \rho \omega u$ is the limit of $\rho v_h = u_h$. ∎

We say that Γ_1 is a "smooth" open subset of Γ if $\phi_j(\Theta_j \cap \Gamma_1)$ is a smooth bounded open subset of R^{n-1} for $1 \leq j \leq J$. In this case, we can define an operator of extension $\omega \in L(H^s(\Gamma_1), H^s(\Gamma))$, which is a right inverse of the operator of restriction $\rho \in L(H^s(\Gamma), H^s(\Gamma_1))$.

COROLLARY 3-1 Let Γ_1 be a smooth open subset of Γ, ρ, and ω, respectively, the operator of restriction to Γ_1 and the operator of extension, and Γ_2 the interior of $\Gamma - \Gamma_1$. Then, if $s \geq 0$, the projectors $\omega\rho$ and $1 - \rho'\omega'$ satisfy

$$\ker \omega\rho = \{u \in H^s(\Gamma) \quad \text{such that} \quad u|_{\Gamma_1} = 0\}$$
$$\ker (1 - \rho'\omega') = \{f \in H^{-s}(\Gamma) \quad \text{such that} \quad f|_{\Gamma_2} = 0\},$$

where $f|_{\Gamma_2} = 0$ means that f is the limit of functions of $L^2(\Gamma)$ vanishing almost everywhere on Γ_2. ▲

Proof Since Γ_1 is smooth, it is sufficient to prove Theorem 3-1 when Γ is replaced by R^{n-1}, Γ_1 by R_+^{n-1}, Γ_2 by R_-^{n-1} and ω is the operator defined by (3-12) with n replaced by $n - 1$ and $m \geq s$. Then the first statement is obvious.

Now, if $f \in L^2(R^{n-1})$, we can check that

$$\rho'\omega'f = \begin{cases} f(x', x_{n-1}) + \sum_{1 \leq j \leq m} c_j f(x', -\dfrac{x_{n-1}}{j}) & \text{if } x \in R_+^{n-1} \\ 0 & \text{if } x \in R_-^{n-1}. \end{cases}$$

Then a function $f \in L^2(R^{n-1})$ satisfies $f = \rho'\omega'f$ if and only if its restriction to R_-^{n-1} is equal to 0.

Since $(1 - \rho'\omega')$ is a continuous projector of both $L^2(R^{n-1})$ and $H^{-s}(R^{n-1})$, and since $L^2(R^{n-1})$ is dense in $H^{-s}(R^{n-1})$, the kernel of $(1 - \rho'\omega')$ in $L^2(R^{n-1})$ is dense in the kernel of $(1 - \rho'\omega')$ in $H^{-s}(R^{n-1})$.

Therefore we can approximate any $f \in \ker(1 - \rho'\omega')$ by functions of $L^2(R^{n-1})$ vanishing on R^{n-1}. ∎

3.6 Properties of the Spaces $H^m(R_+^n)$

We denote by (x, t) $(x \in R^{n-1}, t \in R)$ vectors of R^n, which had previously been called (x', x_{n-1}). We define the trace operator γ_j by $(\gamma_j u)(x) = (D_t^j u)(x, 0)$ when $u \in \Phi_0(R^n)$.

THEOREM 3-4 If $0 \leq j \leq m - 1$, the trace operators γ_j can be extended to continuous linear operators from $H^m(R^n)$ into $H^{m-j-1/2}(R^{n-1})$.

Proof: We use Fourier transforms; we denote by F_n, F_{n-1} and F the Fourier transforms of functions of n, $n - 1$ and one variables respectively. We can write

$$(3\text{-}23) \quad \begin{cases} (F_n u)(y, s) = \displaystyle\int_{R^{n-1} \times R} \int dx\, dt\, e^{i<n, y>} e^{i\, ts} u(x, t) \\ = F_{n-1}[Fu(\cdot, s)](x) = F[F_{n-1} u(x, \cdot)](t) \end{cases}$$

So $F = F_{n-1} F = F F_{n-1}$. Let us choose $u \in \Phi_0(R^n)$; we get

$$\begin{cases} (\gamma_j u)(x) = [F^{-1} F [D_t^j u(x, \cdot)]](0) \\ = F^{-1}[(is)^j (Fu)(x, s)](0) = \displaystyle\int_{-\infty}^{+\infty} (is)^j (Fu)(x, s)\, ds \end{cases}$$

We deduce the following formula.

(3-24) $$(F_{n-1}\,\gamma_j\,u)\,(y) = \int_{-\infty}^{+\infty} (is)^j\,(F_n\,u)\,(y, s)\,ds$$

We notice now that if we set

$$G_{m,j} = \int_{-\infty}^{+\infty} \frac{t^{2j}\,dt}{(1 + t^2)^m} < +\infty \quad \text{when } m > j + \tfrac{1}{2}$$

we have the following equality:

(3-25) $$\int_{-\infty}^{+\infty} \frac{s^{2j}\,ds}{(1 + \|y\|^2 + s^2)^m} = \frac{1}{(1 + \|y\|^2)^{m-j-1/2}}\,G_{m,j}$$

For proving (3-25), it suffices to make the change of variables $s = (1 + \|y\|^2)^{1/2}\,t$. Therefore, by writing formula (3-24) in the form

(3-26) $$\left\{\begin{array}{l} (F_{n-1}\,\gamma_j\,u)\,(y) = \\[2mm] = (i)^j \int_{-\infty}^{+\infty} (F_n\,u)\,(y, s)\,(1 + \|y\|^2 + s^2)^{m/2} \\[4mm] \qquad\qquad \dfrac{s^j\,ds}{(1 + \|y\|^2 + s^2)^{m/2}} \end{array}\right.$$

and using the Cauchy-Schwarz inequality and (3-25), we obtain:

(3-27) $$\left\{\begin{array}{l} |\,(F_{n-1}\,\gamma_j\,u)\,(y)\,|^2 \leq \\[2mm] c\left(\int_{-\infty}^{+\infty} |F_n\,u\,(y, s)|^2\,(1 + \|y\|^2 + s^2)^m\,ds\right) \\[4mm] \qquad\qquad\qquad \left(\int \dfrac{s^{2j}\,ds}{(1 + \|y\|^2 + s^2)^m}\right) \\[4mm] \leq c\,G_{m,j} \int_{-\infty}^{+\infty} |F_n\,u\,(y, s)\,|^2\,\dfrac{(1 + \|y\|^2 + s^2)^m}{(1 + \|y\|^2)^{m-j-1/2}}\,ds \end{array}\right.$$

Since

(3-28) $\|\gamma_j u\|^2_{m-j-1/2} = \int_{R^{n-1}} |(F_{n-1}\gamma_j u)|^2 (1 + \|y\|^2)^{m-j-1/2} dy$

we deduce from (3-27) that

(3-29)
$$\begin{cases} \|\gamma_j u\|^2_{m-j-1/2} \leq \\ c\, G_{m,j} \int_{R^{n-1} \times R} \int |F_n u\,(y,s)|^2 (1 + \|y\|^2 + s^2)^m \, ds\, dy \\ = c\, G_{m,j} \|u\|_m^2 \end{cases}$$

This inequality shows that γ_j is continuous from $\Phi_0(R^n)$ supplied with the topology of $H^m(R^n)$ into $\Phi_0(R^{n-1})$ supplied with the topology of $H^{m-j-1/2}(R^{n-1})$. ∎

THEOREM 3-5   If $0 \leq j \leq m-1$, the trace operators γ_j can be extended to continuous linear operators from $H^m(R^n_+)$ into $H^{m-j-1/2}(R^{n-1})$. ▲

Proof.   We set $\Omega = R^n_+$ and $\Gamma = R^{n-1}$. Then the operator of extension ω defined by the relation (3-12) is continuous from $H^m(R^n_+)$ into $H^m(R^{n-1})$ and satisfies the relation $\gamma_j \omega u = \gamma_j u$ (see Theorem 3-2). Hence Theorem 3-5 follows from Theorem 3-4. ∎

3-7. Proof of the Trace Theorem

We prove the trace theorem when $\Omega = R^n_+$. If Ω is smooth, we deduce Theorem 3-1 by using the system of local coordinates defined by (3-1).

1. γ_j maps $H^m(R^n_+)$ into $H^{m-j-\frac{1}{2}}(R^{n-1})$
Indeed, by Theorem 3-5, $D_t^j u(x, t)$ is a continuous function from $\bar{R}_+$ into $H^{m-j-\frac{1}{2}}(R^{n-1})$ whenever $u \in H^m(R^n_+)$. Therefore we can define $(\gamma_j u)(x) = D_t^j u(x, t)|_{t=0} \in H^{m-j-\frac{1}{2}}(R^{n-1})$.

2. γ_j maps $H^m(R^n_+)$ onto $H^{m-j-\frac{1}{2}}(R^{n-1})$
Let $a_j(x)$ belong to $H^{m-j-\frac{1}{2}}(R^{n-1})$ and let $\tilde{a}_j(y) \in \hat{H}^{m-j-\frac{1}{2}}(R^{n-1})$ be its Fourier transform with respect to $x \in R^{n-1}$. Then we can consider the function $\tilde{u}_j(y, t)$ defined by

(3-30) $\tilde{u}_j(y, t) = |y|^{-j}\tilde{a}_j(y)\phi_j(|y|\,t),$

where $\phi_j(t)$ belongs to $\Phi_0(R)$ and satisfies $\phi_j^{(j)}(0) = 1$.

Therefore

$\gamma_j \tilde{u}_j(y) = |y|^{j-j}\tilde{a}_j(y)D_t^j\phi_j(0) = \tilde{a}_j(y)$

and

$D_t^k \tilde{u}_j(y, t) \in L^2(R_+, H^{m-k}(R^{n-1})).$

Indeed, $D_t^k \tilde{u}_j(y, t) = |y|^{k-j}\tilde{a}_j(y)(D_t^k\phi_j)(|y|\,t),$ and

$$\int_{R_+} \int_{R^{n-1}} |y|^{2(m-k)} |D_t^k \tilde{u}_j(y, t)|^2 \, dy \, dt$$

$$= \int_{R_+} \int_{R^{n-1}} |y|^{2(m-j)} |\tilde{a}_j(y)|^2 |(D_t^k \phi_j(|y| t)|^2 \, dy \, dt$$

$$= \int_{R_+} |(D^k \phi_j(t)|^2 \, dt \int_{R^{n-1}} |y|^{2(m-j)-1} |\tilde{a}_j(y)|^2 \, dy$$

$$\leq c \, \|a_j\|_{H^{m-j-\frac{1}{2}}(R^{n-1})}^2 .$$

Then, by Theorem 3-4, the function $u_j(x, t)$ belongs to $H^m(R_+^n)$ and satisfies $\gamma_i u_j = a_j$.

3. The map $(\gamma_0, \ldots, \gamma_{m-1})$ maps $H^m(R_+^n)$ onto $\prod_{0 \leq j \leq m-1} H^{m-j-\frac{1}{2}}(R^{n-1})$

Let $(a_0(x), \ldots, a_{m-1}(x)) \in \prod_{0 \leq j \leq m-1} H^{m-j-\frac{1}{2}}(R^{n-1})$ and u_j be the inverse Fourier transform of the functions $\tilde{u}_j(y, t)$ defined by (3-30), and consider the functions $v_j(x, t)$ defined by

$$(3\text{-}31) \qquad v_j(x, t) = \sum_{1 \leq r \leq m} \alpha_r^j u_j(x, rt).$$

Then

$$(3\text{-}32) \quad (D_t^k v_j)(x, 0) = \sum_{1 \leq r \leq m} \alpha_r^j r^k (D_t^k u_j)(x, 0) = \begin{cases} 0 & \text{if } j \neq k \\ a_j(x) & \text{if } j = k \end{cases}$$

if $\sum_{1 \leq r \leq m} r^k \alpha_r^j = \delta jk$ for $0 \leq k \leq m-1$. (i.e., if the α_r^j are the elements of the inverse matrix of the Vandermonde matrix).

Therefore, the function $u(x, t) = \sum_{0 \leq j \leq m-1} v_j(x, t)$ belongs to $H^m(R_+^n)$ and satisfies $\gamma_j u(x) = a_j(x)$ for $0 \leq j \leq m-1$.

4. $H_0^m(R_+^n)$ equals the kernel of $(\gamma_0, \ldots, \gamma_{m-1})$.

Let us denote by V_0 the kernel of $(\gamma_0, \ldots, \gamma_{m-1})$. Since $H_0^m(R_+^n)$ is the closure of the space of functions with compact support, we deduce that $H_0^m(R_+^n) \subset V_0$. To prove the other inclusion, note that the functions $D_t^j u(x, t)$ are continuous from R_+ into $L^2(R^{n-1})$ (by Theorem 3-5), which enables us to write the Taylor formula. Therefore, if $u \in V_0$, we deduce that

$$(3\text{-}33) \qquad u(x, t) = \int_0^t \frac{(t-\tau)^{m-1}}{(m-1)!} D_t^m u(x, \tau) \, d\tau \in H^m(R_+^n).$$

Let $\theta(t) \in \Phi(R)$ satisfying

$$(3\text{-}34) \quad \begin{cases} \theta(t) = 1 & \text{if } |t| \leq 1, \, 0 \leq \theta(t) \leq 1 \quad \text{if } 1 \leq |t| \leq 2 \\ & \text{and } \theta(t) = 0 \quad \text{if } |t| \geq 2. \end{cases}$$

Let us admit for a while that

$$(3\text{-}35) \qquad u_\varepsilon(t) = \theta\left(\frac{t}{\varepsilon}\right) u(x, t)$$

converges to 0 in $H^m(R_+^n)$ when $u \in V_0$.

Therefore the function $u - u_\varepsilon$ is equal to 0 for $|t| \leq \varepsilon$ and converges to u. In other words, (3-35) implies that $V_0 \subset H_0^m(R_+^n)$. Let us prove (3-35). By Leibniz's formula,

$$D_t^k u_\varepsilon = \sum_{0 \leq j \leq k} \binom{k}{j} \varepsilon^{-j} \theta^{(j)} \left(\frac{t}{\varepsilon}\right) D_t^k u(x, t).$$

Then, by (3-33), we must prove that

$$(3\text{-}36) \qquad \phi_\varepsilon(t) = \varepsilon^{-j} \theta^{(j)} \left(\frac{t}{\varepsilon}\right) \int_0^t (t - \tau)^{m-k+j-1} v(x, \tau) \, d\tau$$

converges to 0 when $v(x, t) \in L^2(R_+^n)$ and $0 \leq j \leq k \leq m$. But integrating (3-36) on R_+^n and performing the change of variables $t = s\varepsilon$ and $\tau = \sigma\varepsilon$, we obtain

$$\int_{R^{n-1}} \int_{R_+} |\phi_\varepsilon(t)|^2 \, dt \, dx$$

$$= \varepsilon^{2(m-k)+1} \int_{R^{n-1}} dx \int_{R_+} ds \left| \theta^{(j)}(s) \int_0^s (s - \sigma)^{m-k+j-1} v(x, \sigma\varepsilon) \, d\sigma \right|^2,$$

which implies (3-35).

3-8.   Sobolev Inequalities and the Trace Theorem in Space $H^s(\Omega)$

THEOREM 3-6   Let Ω be a smooth bounded open subset of R^n and Γ its boundary. Then

$$(3\text{-}37) \qquad \begin{cases} \text{(i) } H^m(\Gamma) \text{ is a space of order } \theta = \dfrac{s - m}{s - k} \text{ between } H^s(\Gamma) \text{ and} \\ \qquad H^k(\Gamma) \quad \text{for } k \leq m \leq s, \\[2ex] \text{(ii) } H^m(\Omega) \text{ is a space of order } \theta = \dfrac{s - m}{s - k} \text{ between } H^s(\Omega) \text{ and} \\ \qquad H^k(\Omega) \quad \text{for } 0 \leq k \leq m \leq s, \\[2ex] \text{(iii) } H^m(\Omega)' \text{ is a space of order } \theta = \dfrac{m - k}{s - k} \text{ between } H^k(\Omega)' \text{ and} \\ \qquad H^s(\Omega)' \quad \text{for } 0 \leq k \leq m \leq s. \quad \blacktriangle \end{cases}$$

Proof   First, we prove that $H^m(R^n)$ is a space of order $\theta = (s - m)/(s - k)$ between $H^s(R^n)$ and $H^k(R^n)$: by the Fourier transform, we obtain

$$(3\text{-}38) \qquad \begin{cases} \|\hat{u}\|_m^2 = \int (1 + |y|^2)^{(1-\theta)s} |\hat{u}(y)|^{2(1-\theta)} (1 + |y|^2)^{k\theta} |\hat{u}(y)|^2 \, dy \\ \qquad\qquad\qquad\qquad\qquad\qquad\qquad \leq \|\hat{u}\|_s^{1-\theta} \|\hat{u}\|_k^\theta, \end{cases}$$

after applying the Hölder inequality with $p = 1/(1 - \theta)$ and $p' = 1/\theta$.

We deduce [(3-37)(i)] from (3-38) by the very definition of spaces $H^s(\Gamma)$. If Ω is smooth, we deduce [(3-37)(ii)] from (3-38) by writing $u = \rho\omega u$ where

ω is a right inverse of the operator ρ of restriction to Ω. In the same way, we deduce [(3-37)(iii)] from (3-38) by replacing m by $-m$, s by $-k$, and k by $-s$ and by writing that $f = \omega'\rho'f$ for any $f \in H^k(\Omega)'$. ∎

THEOREM 3-7 Let Ω be a smooth bounded open subset of R^n. If $s > n/2$, the Sobolev space $H^s(\Omega)$ is contained in the space $C(\bar{\Omega})$ of continuous functions on $\bar{\Omega}$ and the space $H^s(R^n)$ is contained in the space $B(R^n)$ of bounded continuous functions vanishing at ∞. ▲

Proof We prove the second statement first. Since the Fourier transform maps $L^1(R^n)$ into B, it is sufficient to prove that $\hat{H}^s(R^n)$ is contained in $L^1(R^n)$. This is a consequence of the Cauchy-Schwarz inequality, since

$$(3\text{-}39) \quad \begin{cases} \displaystyle\int_{R^n} |\hat{u}(y)|\, dy = \int_{R^n} |\hat{u}(y)|\, (1 + |y|^2)^{s/2}(1 + |y|^2)^{-s/2}\, dy \\[2mm] \displaystyle \qquad \leq \|\hat{u}\|_{H^s}\left(\int_{R^n}(1 + |y|^2)^{-s}\, dy\right)^{1/2} \leq c\, \|\hat{u}\|_{H^s} \quad \text{if } s > n/2. \end{cases}$$

Now, if Ω is smooth, we deduce the first statement by using the operators ρ and ω. ∎

Finally, let us extend Theorem 3-1 to the case where m is a positive real number.

THEOREM 3-8 Let Ω be a smooth bounded open subset of R^n and Γ its boundary. Then if m is a positive real number and the integer j satisfies $0 \leq j < m - \frac{1}{2}$, the trace operator γ_j maps $H^m(\Omega)$ onto $H^{m-j-1/2}(\Gamma)$, and the operator $\gamma = (\gamma_0, \ldots, \gamma_J)$ maps $H^m(\Omega)$ onto $\prod_{j=0}^{J} H^{m-j-1/2}(\Gamma)$, where J is the largest integer smaller than $m - \frac{1}{2}$. ▲

Proof If m is no longer an integer, we replace Theorem 3-4 by the following statement: the space $H^m(R^n_+)$ is the space of restrictions to R^n_+ of functions $u(x, t)$ of $L^2(R, H^m(R^{n-1}))$ such that $|s|^m \tilde{u}(y, s)$ belongs to $L^2(R_s, L^2(R^{n-1}_y))$. This statement, in turn, follows from the theorem of Fubini and from the fact that the Fourier transform is an isomorphism from $L^2(R, X)$ onto $L^2(R, X)$ when X is a Hilbert space.

Therefore, the proof of Theorems 3-5 and the proofs of statements through 3 of Section 3-7 can be extended to the case in which m is no longer an integer. This implies Theorem 3-8. ∎

Remark 3-1 We can prove also that the closure $H^m(\Omega)$ of the functions with compact support is the kernel of the operator $\gamma = (\gamma_0, \ldots, \gamma_J)$. ∎

3-9. Theorem of Compactness

Here we prove that the canonical injection from one Sobolev space into another is compact when Ω is bounded.

THEOREM 3-9   Let us assume that Ω is a smooth bounded open subset of R^n. Then the canonical injection from $H^s(\Omega)$ into $H^{s-\varepsilon}(\Omega)$ is compact for any $\varepsilon > 0$.   ▲

Proof The transpose of a compact operator being compact, we can assume $s > 0$. We are to prove that if a sequence u_n converges weakly to 0 in $H^s(\Omega)$, then it converges strongly in $H^{s-\varepsilon}(\Omega)$.

Let $\omega \in L(H^s(\Omega), H^s(R^n))$ be an operator of extension and $\theta \in \Phi_0$ be a function equal to 1 on Ω. Then the norms in $H^s = H^s(R^n)$ of the functions $v_n = \theta \omega u_n$ are bounded by a constant c, their supports remain in a fixed compact K, and $\rho v_n = v_n|_\Omega = u_n$.

Therefore, by using the Fourier transform, we must prove that $X_n = Y_n + Z_n$ converges to 0, where

$$(3\text{-}40) \quad \begin{cases} \text{(i)} \quad Y_n = \int_{|y| \geq M} (1 + |y|^2)^{-\varepsilon}(1 + |y|^2)^s \, |v_n(y)|^2 \, dy, \\[2mm] \text{(ii)} \quad Z_n = \int_{|y| \leq M} (1 + |y|^2)^{s-\varepsilon} |\hat{v}_n(y)|^2 \, dy. \end{cases}$$

With any $\delta > 0$, we associate the constant M involved in (3-40) in such a way that $c(1 + M^2)^{-\varepsilon} \leq \delta/2$. Then we deduce the inequality

$$(3\text{-}41) \quad Y_n \leq (1 + M^2)^{-\varepsilon} \int_{R^n} (1 + |y|^2)^s \, |\hat{v}_n(y)|^2 \, dy \leq (1 + M^2)^{-\varepsilon} \, \|v_n\|_s \leq \frac{\delta}{2}.$$

Second, we estimate Z_n in the following way:

$$(3\text{-}42) \quad Z_n \leq (1 + M^2)^{s-\varepsilon} \int_{|y| \leq M} |\hat{v}_n(y)|^2 \, dy.$$

It remains to show that

$$(3\text{-}43) \quad \int_{|y| \leq M} |\hat{v}_n(y)|^2 \, dy \leq \frac{\delta}{2} \, (1 + M^2)^{s-\varepsilon} \quad \text{for} \quad n \geq n(\delta).$$

This follows from the Lebesgue theorem. Indeed, by Lemma 3-1 we can write $\tilde{v}_n(y) = \int_{R^n} v_n(x) \exp{(ixy)} \, dx = \int v_n(x)\phi(x) \exp{(ixy)} \, dx$ where $\phi(x) \in \Phi_0(R^n)$ and is equal to 1 on K. Then, by the Cauchy-Schwarz inequality, we deduce that

$$(3\text{-}44) \quad |\hat{v}_n(y)| \leq \|v_n\|_{L^2(R^n)} \, \|\phi\|_{L^2(R^n)} \leq N \quad \text{where } N \text{ is a constant.}$$

On the other hand, since v_n converges weakly to 0 in $L^2(R^n)$, we deduce that $\hat{v}_n(y)$ converges almost everywhere to 0. Then the Lebesgue theorem implies (3-43) and the proof of Theorem 3-9 is completed.   ■

CHAPTER 7

Examples of Boundary-Value Problems

We gather in this chapter examples of boundary-value problems for second-order elliptic operators (Section 1) and for elliptic operators of order $2k$ (Section 2). Approximations of all these problems appear in the subsequent chapters.

1. BOUNDARY-VALUE PROBLEMS FOR SECOND-ORDER DIFFERENTIAL OPERATORS

1-1. Second-Order Linear Differential Operators

Let us consider the following integro-differential bilinear form:

$$(1\text{-}1) \qquad a(u, v) = \sum_{1 \leq i, j \leq n} \int_{\Omega} a_{ij}(x) D_i u D_j v \, dx + \int_{\Omega} a_0(x) uv \, dx,$$

where Γ is the boundary of

$$(1\text{-}2) \qquad \begin{cases} \text{(i)} \ \ \hat{\Omega}, \text{ a smooth bounded open subset of } R^n, \\ \text{(ii)} \ \ \text{the coefficients } a_{ij}(x) \text{ and } a_0(x) \in L^\infty(\Omega). \end{cases}$$

Then, if $V = H^1(\Omega)$, $a(u, v)$ is a continuous bilinear form. Indeed, if we set

$$(1\text{-}3) \qquad (u, v) = \int_{\Omega} u(x) v(x) dx \qquad \text{and} \qquad |u| = \sqrt{(u, v)},$$

we deduce from (1-1) and the Cauchy-Schwarz inequality that

$$(1\text{-}4) \qquad \begin{cases} |a(u, v)| \leq M \left(\sum_{1 \leq i, j \leq n} |D_i u| \, |D_j v| + |u| \, |v| \right) \\ \leq M \left(\sum_{1 \leq i \leq n} |D_i u|^2 + |u|^2 \right)^{\!\! 1/2} \left(\sum_{1 \leq j \leq n} |D_j v|^2 + |v|^2 \right)^{\!\! 1/2} = M \, \|u\|_1 \, \|v\|_1, \end{cases}$$

where $M = \max \left(|a_{ij}|_{L^\infty(\Omega)}, |a_0|_{L^\infty(\Omega)} \right)$.

On the other hand, by Theorem 6.3-1, assumption (2-1) of Section 6.2-1 is satisfied when we choose

(1-5) $\begin{cases} V = H^1(\Omega), \, T = H^{1/2}(\Gamma), \, \gamma u = \gamma_0 u = u|_\Gamma, \, V_0 = H_0^1(\Omega) \\ \qquad\qquad\qquad\qquad\qquad \text{and} \qquad H = L^2(\Omega). \end{cases}$

Then the formal differential operator Λ associated with $a(u, v)$ is

(1-6) $$\Lambda u = -\left[\sum_{1 \le i,j \le n} D_j(a_{ij}(x)D_i u) \right] + a_0(x)u.$$

Indeed, the space $\Phi_0(\Omega)$ is dense in $H_0^1(\Omega)$ and, for any v in this space we can write $a(u, v) = (\Lambda u, v)$ by the very definition of derivatives of distributions. Then Λ belongs to $L(H^1(\Omega), H^{-1}(\Omega))$, where $H^{-1}(\Omega) = (H_0^1(\Omega))'$ and to $L(H^1(\Omega, \Lambda), L^2(\Omega))$, where $H^1(\Omega, \Lambda) = V(\Lambda)$ is the domain of Λ defined by

(1-7) $\quad H^1(\Omega, \Lambda) = \{u \in H^1(\Omega) \quad$ such that $\quad \Lambda u \in L^2(\Omega)\}$

(see Section 6.2-2).

Therefore, by Theorem 6.2-1 there exists an operator

$$\delta \in L(H^1(\Omega, \Lambda), H^{-1/2}(\Gamma)),$$

where $H^{-1/2}(\Gamma) = (H^{1/2}(\Gamma))'$, such that the following Green formula holds:

(1-8) $\quad a(u, v) = (\Lambda u, v) + \langle \delta u, \gamma_0 v \rangle \qquad$ for any $\quad u \in H^1(\Omega, \Lambda), v \in H^1(\Omega).$

If the coefficients $a_{ij}(x)$ are smooth enough, (1-8) is the usual Green formula:

(1-9) $$a(u, v) = (\Lambda u, v) + \int_\Gamma \frac{\partial u}{\partial n_\Lambda} \cdot v \, d\sigma(x),$$

where $\partial u/\partial n_\Lambda$ is the "normal derivative of u with respect to Λ" (i.e., $\partial u/\partial n_\Lambda = \sum_{i,j=1}^n a_{ij}(x)D_j u \cos(n, x_i)$, $\cos(n, x_i)$ being the directional cosines).

This is why we set $\delta u = \partial u/\partial n_\Lambda$ in this section.

Now by Theorem 6.2-2 we can characterize the boundary-value problems for Λ [defined by (1-6)] which are equivalent to variational equations defined on closed subspaces W of $H^1(\Omega)$ such that

(1-10) $$H_0^1(\Omega) \subset W \subset H^1(\Omega).$$

1-2. Elliptic Second-Order Partial Differential Operators

Let Λ be the second-order differential operator defined by (1-6). This operator is "elliptic" if there exists a positive constant c such that

(1-11) $\quad \sum_{1 \le i,j \le n} a_{ij}(x)z_i z_j \ge c \left(\sum_{1 \le i \le n} |z_i|^2 \right)^{1/2} \qquad$ for almost all $\quad x \in \Omega.$

LEMMA 1-1 Let us assume that Λ [still as defined by (1-6)] is elliptic. Then if

$$(1-12) \qquad a_0(x) \geq c \qquad \text{for almost every} \quad x \in \Omega,$$

the form $a(u, v)$ defined by (1-1) is $H^1(\Omega)$-elliptic. Furthermore, if

$$(1-13) \qquad a_0(x) \geq 0 \qquad \text{for almost every} \quad x \in \Omega,$$

the form $a(u, v)$ is $H_0^1(\Omega)$-elliptic. ▲

Proof Indeed, by (1-11) and (1-12) we verify that

$$(1-14) \quad \left\{ a(v, v) \geq c \left(\sum_{1 \leq i \leq n} \int_\Omega |D_i v(x)|^2 \, dx + \int_\Omega |v(x)|^2 \, dx \right) = c \, \|v\|_1^2 \right.$$
$$\text{when} \quad v \in H^1(\Omega).$$

When $v \in H_0^1(\Omega)$, we deduce from (1-11), (1-13), and the Poincaré inequality (see Proposition 5.3-2) that

$$(1-15) \qquad a(v, v) \geq c \sum_{1 \leq i \leq n} \int_\Omega |D_i v(x)|^2 \, dx = c \, \|v\|_{0,1}^2 \geq c' \, \|v\|_1^2.$$

1-3. The Dirichlet Problem

The Dirichlet problem is associated with the choice $W = H_0^1(\Omega)$.

COROLLARY 1-1 Let $a(u, v)$ be defined by (1-1) and Λ by (1-6); let $f \in L^2(\Omega)$. Then the following two problems are equivalent:

1. *The Dirichlet problem for Λ*
Look for u satisfying

$$(1-16) \qquad \begin{cases} \text{(i)} & u \in H^1(\Omega, \Lambda), \\ \text{(ii)} & \Lambda u = f \quad \text{on} \quad \Omega, \\ \text{(iii)} & \gamma_0 u = 0 \quad \text{on} \quad \Gamma. \end{cases}$$

2. *The variational Dirichlet problem*
Look for u satisfying

$$(1-17) \qquad \begin{cases} \text{(i)} & u \in H_0^1(\Omega), \\ \text{(ii)} & a(u, v) = (f, v) \quad \text{for any} \quad v \in H_0^1(\Omega). \end{cases}$$

Moreover, if Λ is elliptic and 0 is not an eigenvalue of the Dirichlet problem, the operator (Λ, γ_0) is an isomorphism from $H^1(\Omega, \Lambda)$ onto $L^2(\Omega) \times H^{1/2}(\Gamma)$. ▲

1-4. The Neumann Problem

The Neumann problem is associated with the choice $W = H^1(\Omega)$.

COROLLARY 1-2 If we let $a(u, v)$ be defined by (1-1) and Λ by (1-6), and assume $f \in L^2(\Omega)$ and $t \in H^{-1/2}(\Gamma)$, then the two following problems are equivalent:

1. *The Neumann problem for Λ*
Look for u satisfying

$$(1\text{-}18) \qquad \begin{cases} \text{(i)} \ \ u \in H^1(\Omega, \Lambda), \\[4pt] \text{(ii)} \ \ \Lambda u = f \quad \text{on} \ \ \Omega, \\[4pt] \text{(iii)} \ \dfrac{\partial u}{\partial n_\Lambda} = t \quad \text{on} \ \ \Gamma. \end{cases}$$

2. *The variational Neumann problem for Λ*
Look for u satisfying

$$(1\text{-}19) \qquad \begin{cases} \text{(i)} \ \ u \in H^1(\Omega), \\[4pt] \text{(ii)} \ \ a(u, v) = (f, v) + \langle t, \gamma_0 v \rangle \quad \text{for any} \ \ v \in H^1(\Omega). \end{cases}$$

Furthermore, if Λ is elliptic and 0 is not an eigenvalue of the Neumann problem, the operator $(\Lambda, \partial u / \partial n_\Lambda)$ is an isomorphism from $H^1(\Omega, \Lambda)$ onto $L^2(\Omega) \times H^{-1/2}(\Gamma)$. ▲

1-5. Mixed Problems

Let Γ_1 be a smooth open subset of Γ and Γ_2 the interior of $\Gamma - \Gamma_1$. Let $\rho \in L(H^{1/2}(\Gamma), H^{1/2}(\Gamma_1))$ be the operator of restriction to Γ_1, $\omega \in L(H^{1/2}(\Gamma_1), H^{1/2}(\Gamma))$ the operator of extension, $\sigma_1 = \omega\rho$ the projector of $H^{1/2}(\Gamma)$—whose kernel is the space of functions t vanishing on Γ_1—and $\sigma_2' = 1 - \sigma_1'$ the projector of $H^{-1/2}(\Gamma)$—whose kernel is the subspace of functions t vanishing on Γ_2. (See Corollary 6.3-1.)

A mixed problem is associated with the choice $W = H_{\Gamma_1}^1(\Omega)$, which is the kernel of $\sigma_1 \gamma_0$, that is, the space of functions $u \in H^1(\Omega)$ vanishing on Γ_1.

COROLLARY 1-3 Let $a(u, v)$ be defined by (1-1) and Λ by (1-6) Let $f \in L^2(\Omega)$ and $t \in H^{-1/2}(\Gamma_2)$. Then the two following problems are equivalent:

1. *Mixed problem for* Λ
Look for u satisfying

$$(1\text{-}20) \quad \begin{cases} \text{(i)} \ \ u \in H^1(\Omega, \Lambda), \\ \text{(ii)} \ \ \Lambda u = f \quad \text{on} \quad \Omega, \\ \text{(iii)} \ \ u|_{\Gamma 1} = 0 \quad \text{on} \quad \Gamma_1, \\ \text{(iv)} \ \ \dfrac{\partial u}{\partial n_\Lambda}\bigg|_{\Gamma 2} = t \quad \text{on} \quad \Gamma_2. \end{cases}$$

2. *Variational mixed problem for* Λ
Look for u satisfying

$$(1\text{-}21) \quad \begin{cases} \text{(i)} \ \ u \in H_{\Gamma_1}^{\ 1}(\Omega), \\ \text{(ii)} \ \ a(u, v) = (f, v) + \langle t, \gamma_0 v \rangle \quad \text{for any} \quad v \in H_{\Gamma_1}^{\ 1}(\Omega). \end{cases}$$

Furthermore, if Λ is elliptic and 0 is not an eigenvalue of the mixed problem, the operator $(\Lambda u, u|_{\Gamma_1}, \partial u/\partial n_\Lambda|_{\Gamma_2})$ is an isomorphism from $H^1(\Omega, \Lambda)$ onto $L^2(\Omega) \times H^{1/2}(\Gamma_1) \times H^{-1/2}(\Gamma_2)$. ▲

1-6. Oblique Problems

Let $\alpha(x)$ be a function in $L^\infty(\Gamma)$. Since the bilinear form

$$\int_\Gamma \alpha(x) t(x) s(x) \, d\sigma \, (x)$$

is continuous on $H^{1/2}(\Gamma)$, the operator M which associates with t the function $\alpha t = Mt$ belongs to $L(H^{1/2}(\Gamma), H^{-1/2}(\Gamma))$.

This problem is associated with the form $a(u, v) + \langle \alpha \gamma_0 u, \gamma_0 v \rangle$ and $W = H^1(\Omega)$.

COROLLARY 1-4 Let $a(u, v)$ be defined by (1-1) and Λ by (1-6). Let $f \in L^2(\Omega)$, $t \in H^{-1/2}(\Gamma)$, and $\alpha \in L^\infty(\Gamma)$. Then the two following problems are equivalent:

1. *Oblique problem for* Λ
Look for u satisfying

$$(1\text{-}22) \quad \begin{cases} \text{(i)} \ \ u \in H^1(\Omega, \Lambda), \\ \text{(ii)} \ \ \Lambda u = f \quad \text{on} \quad \Omega, \\ \text{(iii)} \ \ \dfrac{\partial u}{\partial n_\Lambda} + \alpha u = t \quad \text{on} \quad \Gamma. \end{cases}$$

2. *Variational oblique problem for* Λ

Look for u satisfying

$$(1\text{-}23) \quad \begin{cases} \text{(i)} \ u \in H^1(\Omega), \\ \text{(ii)} \ a(u, v) + \langle \alpha\gamma_0 u, \gamma_0 v \rangle = (f, v) + \langle t, \gamma_0 v \rangle \quad \text{for any} \quad v \in H^1(\Omega). \end{cases}$$

Furthermore, if Λ is elliptic, and if 0 is not an eigenvalue of the oblique problem, there exists a unique solution of these problems, and the operator $(\Lambda u, \partial u/\partial n_\Lambda + \alpha u)$ is an isomorphism from $H^1(\Omega, \Lambda)$ onto $L^2(\Omega) \times H^{-\frac{1}{2}}(\Gamma)$.

▲

Naturally, we can choose other operators M mapping $H^{\frac{1}{2}}(\Gamma)$ into its dual; if M is a positive operator (i.e., if $\langle Mt, t \rangle \geq 0$), all the conclusions of Corollary 1-4 hold. We can choose, for instance,

$$(1\text{-}24) \quad \begin{cases} \text{(i)} \ Mt = \int_\Gamma \alpha(x, y)t(y) \, d\sigma(y); \quad \alpha \in L^2(\Gamma \times \Gamma); \quad \alpha \geq 0, \\ \text{(ii)} \ Mt = \dfrac{\partial t}{\partial \sigma} \end{cases}$$

is the tangential derivative on Γ.

(See another example of this type in Section 1-11.)

1-7.   Interface Problems

Assuming that Ω is halved in two parts Ω_1 and Ω_2 by a manifold Σ (see Figure 4), we now consider functions a_{ij}^k and a_0^k in $L^\infty(\Omega_k)(k = 1, 2)$ and the following operators Λ^k:

$$(1\text{-}25) \quad \Lambda^k u^k = -\left(\sum_{1 \leq i, j \leq n} D_j(a_{ij}^k(x)D_i u^k) \right) + a_0^k(x)u^k; \quad k = 1, 2.$$

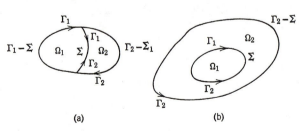

(a)   (b)

FIGURE 4   Examples of domains Ω halved in two parts Ω_1 and Ω_2 by a manifold Σ.

Let $f^k \in L^2(\Omega_k)$. We would like to solve the *interface problem:*
Look for u^1 and u^2 satisfying

(1-26)
$$\begin{cases} \text{(i) } u^1 \in H^1(\Omega_1, \Lambda^1); \qquad u^2 \in H^1(\Omega_2, \Lambda^2), \\[2mm] \text{(ii) } \Lambda^1 u^1 = f^1 \quad \text{on} \quad \Omega_1; \qquad \Lambda^2 u^2 = f^2 \quad \text{on} \quad \Omega_2, \\[2mm] \text{(iii) } \dfrac{\partial u^1}{\partial n_{\Lambda^1}} = 0 \quad \text{on} \quad \Gamma_1 - \Sigma; \qquad \dfrac{\partial u^2}{\partial n_{\Lambda^2}} = 0 \quad \text{on} \quad \Gamma_2 - \Sigma, \\[2mm] \text{(iv) } u^1 = u^2 \quad \text{on} \quad \Sigma; \qquad \dfrac{\partial u^1}{\partial n_{\Lambda^1}} + \dfrac{\partial u^2}{\partial n_{\Lambda^2}} = 0 \quad \text{on} \quad \Sigma. \end{cases}$$

In order to construct the equivalent variational problem, we have to introduce

(1-27)
$$\begin{cases} \text{(i) the space } V = H^1(\Omega_1) \times H^1(\Omega_2), \\ \text{(ii) the space } H = L^2(\Omega_1) \times L^2(\Omega_2) = L^2(\Omega), \\ \text{(iii) the space } T = H^{\frac{1}{2}}(\Gamma_1) \times H^{\frac{1}{2}}(\Gamma_2), \\ \text{(iv) the operator } \gamma = (\gamma_0^{\,1}, -\gamma_0^{\,2}) \text{ mapping } V \text{ onto } T, \\ \text{(v) the space } V_0 = H_0^1(\Omega_1) \times H_0^1(\Omega_2), \text{ kernel of } \gamma. \end{cases}$$

For this choice, assumption (2-1) of Section 6.2-1 holds, and we can present the following continuous bilinear form on $V \times V$:

(1-28)
$$a(u, v) = a^1(u^1, v^1) + a^2(u^2, v^2)$$

where

$$a^k(u^k, v^k) = \int_{\Omega_k} \left(\sum_{1 \le i, j \le n} a_{ij}^{\,k}(x) D_i u^k \, D_j v^k + a_0^{\,k}(x) u^k v^k \right) dx.$$

Then the formal operator associated with $a(u, v)$ is the operator Λ defined by $\Lambda u = (\Lambda^1 u^1, \Lambda^2 u^2)$. Its domain is the space

(1-29)
$$V(\Lambda) = H^1(\Omega_1, \Lambda^1) \times H^1(\Omega_2, \Lambda^2).$$

Then the interface problem is associated to the space W defined by

(1-30)
$$W = \{u = (u^1, u^2) \in V \quad \text{such that} \quad u^1|_\Sigma = u^2|_\Sigma\}.$$

We deduce Corollary 1-5 from the results of Chapter 6.

COROLLARY 1-5 Assume that Σ is a smooth open subset of both Γ_1 and Γ_2, and let $f^k \in L^2(\Omega_k) (k = 1, 2)$. Then the interface problem (1-26) is equivalent to the following variational problem, in which we look for u satisfying

(1-31) $\begin{cases} \text{(i) } u \in W \text{ defined by (1-30),} \\ \text{(ii) } a(u, v) = \displaystyle\int_\Omega f^1 v^1 \, dx + \int_\Omega f^2 v^2 \, dx \quad \text{for any} \quad v = (v^1, v^2) \in W, \end{cases}$

where $a(u, v)$ is defined by (1-28).

If the operators Λ^k are elliptic and if $a_0^k(x) \geq c$ almost everywhere in Ω_k, then there exists a unique solution of these problems. ▲

Proof Indeed, by Theorem 6.2-1, we can write the Green formula in the following way:

$$a(u, v) = (\Lambda u, v) + \langle \delta^1 u, \gamma_0^1 v \rangle - \langle \delta^2 u, \gamma_0^2 v \rangle \quad \text{for} \quad u \in V(\Lambda), v \in V,$$

where δ^1 is the extension of $\partial/\partial n_{\Lambda^1}$ and δ^2 the extension of $-(\partial/\partial n_{\Lambda^2})$.

Since Σ is smooth, we deduce from Corollary 6.3-1 there exist operators of restriction to Σ and operators of extension ρ^k and ω^k satisfying

$$\rho^k \in L(H^{1/2}(\Gamma_k), H^{1/2}(\Sigma)); \ \omega^k \in L(H^{1/2}(\Sigma), H^{1/2}(\Gamma_k))(k = 1, 2).$$

Then the operator $\sigma = (\omega^1 \rho^1, \omega^2 \rho^2)$ is a projector of T and the space W defined by (1-30) is the kernel of $\sigma\gamma$. Therefore, the first part of Corollary 1-5 follows from Theorem 6.2-1, since we can rewrite the Green formula in the form

$$\begin{aligned} a(u, v) = (\Lambda u, v) &+ \langle (1 - \sigma_1') \, \delta^1 u^1, (1 - \sigma_1)\gamma_0^1 v^1 \rangle \\ &- \langle (1 - \sigma_2') \, \delta^2 u^2, (1 - \sigma_2)\gamma_0^2 v^2 \rangle \\ &+ \langle \sigma_1' \, \delta^1 u^1, \sigma_1 \gamma_0^1 v^1 \rangle - \langle \sigma_2' \, \delta^2 u^2, \sigma_2 \gamma_0^2 v^2 \rangle \end{aligned}$$

where we have set $\sigma_1 = \omega^1 \rho^1$ and $\sigma_2 = \omega^2 \rho^2$. ■

1-8. The Regularity Theorem

We have proved that the operators (Λ, γ) and $(\Lambda, \partial/\partial n_\Lambda)$ are isomorphisms mapping $H^1(\Omega, \Lambda)$ onto $L^2(\Omega) \times H^{1/2}(\Gamma)$ and $L^2(\Omega) \times H^{-1/2}(\Gamma)$, respectively. We would like to "restrict" and "extend" these isomorphisms. In order to apply Theorem 6.2-7, we need the following theorem of regularity.

THEOREM 1-1 Let us assume that Ω is smooth and that the coefficients $a_{ij}(x)$ and $a_0(x)$ of the form $a(u, v)$ defined by (1-1) are continuously differentiable. Let us assume that the operator Λ is elliptic and that $a_0(x) \geq c > 0$. Let f belong to $L^2(\Omega)$.

Then the solution u of the homogeneous Neumann (respectively Dirichlet) problem belongs to the space $U = H^2(\Omega)$. Furthermore, the operator $(\gamma_0, \partial/\partial n_A)$ maps $H^2(\Omega)$ onto the space $H^{3/2}(\Gamma) \times H^{1/2}(\Gamma)$. ▲

Proof We prove this theorem in three steps.

1. Use of the change of coordinates.

We begin by replacing the smooth open subset Ω by R^n_+. In order to do that, we use the change of coordinates defined in Section 6.3-2. Let us set $u^j = \phi_j^*(\sqrt{\alpha_j}\, u)$ and $u^0 = \sqrt{\alpha_0}\, u$. Then we can write

$$\int_\Omega a_{ik}(x) D_i u D_k v\, dx = \int_\Omega \alpha_0 a_{ik} D_i u D_k v\, dx + \sum_{1 \leq j \leq J} \int_\Omega \alpha_j a_{ik} D_i u D_k v\, dx$$

$$= \int_\Omega \alpha_0 a_{ik} D_i u D_k v\, dx$$

$$+ \sum_{1 \leq j \leq J} \int_{R^n_+} \phi_j^*(\alpha_j a_{ik}) \phi_j^* D_i u \phi_j^* D_k v\, \frac{D\psi_j}{Dy}\, dy$$

$$= \int_{R^n} \sum_{|p|,|q| \leq 1} b_{pq}{}^0 D^p u^0 D^q v^0\, dx$$

$$+ \sum_{1 \leq j \leq J} \sum_{|q|,|q| \leq 1} \int_{R^n_+} b_{pq}{}^j(y) D^p u^j D^q v^j\, dy.$$

Then the form $a(u, v)$ is split in the following way:

$$a(u, v) = a^0(u^0, v^0) + \sum_{1 \leq j \leq J} a^j(u^j, v^j)$$

where

$$a^j(u, v) = \sum_{|p|,|q| \leq 1} c_{pq}{}^j(y) D^p u D^q v\, dy.$$

They are $H^1(R^n_+)$-elliptic forms if $1 \leq j \leq J$ and $H^1(R^n)$-elliptic if $j = 0$.

Therefore, the problem $a(u, v) = (f, v)$ for any $v \in H^1(\Omega)$ is equivalent to the problems

$$(1\text{-}32) \begin{cases} \text{(i)} \quad a^0(u^0, v) = (f^0, v) \quad \text{for any} \quad v \in H^1(R^n); \quad f^0 = \sqrt{\alpha_0}\, f \\ \text{(ii)} \quad a^j(n^j, v) = (f^j, v) \quad \text{for any} \quad v \in H^1(R^n_+); \\ \qquad\qquad\qquad\qquad\qquad\qquad f^j = \phi_j^*(\sqrt{\alpha_j}\, f) \dfrac{D\psi_j}{Dy} \end{cases}$$

Now we must prove that the solutions u^0 and u^j belong to $H^2(R^n)$ and $H^2(R^n_+)$, respectively.

2. Regularity of the solution when $\Omega = R_+^n$.

Let us prove that the solution $u = u^j$ of [(1-32)(ii)] belongs to $H^2(R_+^n)$. (We cancel the index j in this proof).

If $u \in H^1(R_+^n)$, the finite-differences $\nabla_{h_i} u$ belong to $H^1(R_+^n)$ for $i = 1, \ldots, n - 1$ (but not for $i = n$). Let us set $h = h_i$ and

$$a_h(u, v) = \sum_{|p|,|q| \leq 1} \int_{R_+^n} [\nabla_h c_{pq}(y)] D^p u D^q v \, dy.$$

Since the coefficients c_{pq} are continuously differentiable, the norms $|c_{pq}|_{L^\infty}$ are bounded and the norms of the forms $a_h(u, v)$ are bounded also.

On the other hand, since u is solution of [(1-32)(ii)], we obtain

$$a(u, \delta(-h) * v) = (f, \delta(-h) * v)$$

$$= \sum_{|p|,|q| \leq 1} \int_{R_+^n} (\delta(h) * c_{pq})(\delta(h) * D^p u) \cdot D^q v \, dy$$

$$= -h a_h(\delta(h) * u, v) + a(\delta(h) * u, v),$$

and thus we deduce the relation

(1-33) $\qquad a(\nabla_h u, v) = -(f, \tilde{\nabla}_h v) - a_h(\delta(h) * u, v).$

Replacing v by $\nabla_h u$ and using the $H^1(R_+^n)$-ellipticity, we get

(1-34) $\qquad \|\nabla_h u\|_1^2 \leq c^{-1} a(\nabla_h u, \nabla_h u) \leq M \|\nabla_h u\|_1,$

since $|(f, \tilde{\nabla}_h v)| \leq |f| \, \|v\|_1$ and $|a_h(\delta(h) * u, v)| \leq M \|u\|_1 \|v\|_1$.

Then the $\nabla_h u$ are bounded in $H^1(R_+^n)$ and we can extract from h a subsequence (again denoted by h) such that $\nabla_h u = \nabla_{h_i} u$ converges weakly to $D_i u$ in $H^1(R_+^n)$. Thus we have proved that

(1-35) $\quad D_i D^p u \in L^2(R_+^n) \qquad$ for $\quad i = 1, \ldots, n - 1 \qquad$ and $\qquad |p| \leq 1$.

It remains to prove that

(1-36) $\qquad\qquad D_n^2 u \qquad$ belongs to $\quad L^2(R_+^n)$.

Since $\sum_{|p|,|q| \leq 1} D^q(c_{pq}(y)D^p u) = -f$, we deduce from (1-36) that $c_{nn}(y)D_n^2 u$ belongs to $L^2(R_+^n)$. By the $H^1(R_+^n)$-ellipticity, we deduce that $c_{nn} \geq c > 0$ and that $D_n^2 u$ belongs to $L^2(R_+^n)$.

Analogous proofs show that the solution u^0 of [(1-32)(i)] belongs to $H^2(R^n)$. Then the solution of the homogeneous Neumann problem belongs to $H^2(\Omega)$. We can prove in the same way that the solution of the homogeneous Dirichlet problem belongs to $H^2(R_+^n)$.

$$\left(\gamma_0, \frac{\partial}{\partial n_A} \right) \text{ maps } H^2(\Omega) \text{ onto } H^{3/2}(\Gamma) \times H^{1/2}(\Gamma)$$

Again we replace Ω by R_+^n. In this case the integration by parts formula shows that

$$\frac{\partial u}{\partial n_\Lambda} = B\gamma_0 u(y') + c_{nn}(y')\gamma_1 u(y'),$$

where B is the "tangential" differential operator defined by

$$Bt = \sum_{\substack{|p| \leq 1 \\ p \neq n}} c_{pn}(y')D_{y'}{}^p t(y').$$

By Theorem 6.3-1 we know that $\gamma_0 u \in H^{3/2}(\Gamma)$ and $\gamma_1 u \in H^{1/2}(\Gamma)$. Since the coefficients c_{pq} are continuously differentiable, the operator B maps $H^{3/2}(R^{n-1})$ into $H^{1/2}(R^{n-1})$, and thus $\partial u/\partial n_\Lambda$ belongs to $H^{1/2}(R^{n-1})$. On the other hand, (γ_0, γ_1) maps $H^2(R_+^n)$ onto $H^{3/2}(\Gamma) \times H^{1/2}(\Gamma)$. Indeed, if (t_0, t_1) belongs to this latter space, the pair $(t_0, (t_1 - Bt_0)/c_{nn})$ belongs to this space as well. Thus there exists a function u of H^2 such that $\gamma_0 u = t_0$ and $\gamma_1 u = (t_1 - Bt_0)/c_{nn}$. In other words, this function u satisfies $\gamma_0 u = t_0$ and $\partial u/\partial n_\Lambda = t_1$. ∎

1-9. Theorems of Isomorphism

If $a(u, v)$ is the bilinear form defined by (1-1), its adjoint form is $a_*(u, v) = \sum_{1 \leq i, j \leq n} \int_\Omega a_{ji} D_i u D_j v \, dx + \int_\Omega a_0(x)uv \, dx$. We use

$$\Lambda^* u = - \sum_{1 \leq i, j \leq n} D_j(a_{ji} D_i u) + a_0 u$$

to denote its formal operator and $\delta^* u = \partial u/\partial n_{\Lambda^*}$ for its Neumann operator. Let us set:

(1-37) $U = H^2(\Omega)$, $H = L^2(\Omega)$, $S = H^{3/2}(\Gamma)$, and $R' = H^{1/2}(\Gamma)$,

so that Theorem 1-1 implies the assumptions (2-43) of Theorem 6.2-7.

If Λ is defined by (1-6), let us introduce $H(\Omega, \Lambda)$ defined by

(1-38) $H(\Omega, \Lambda) = \{u \in L^2(\Omega)$ such that $\Lambda u \in L^2(\Omega)\}$.

We deduce Theorem 1-2 from Theorems 6.2-7 and 1-1.

THEOREM 1-2 If we suppose that the assumptions of Theorem 1-1 are satisfied, then the operator $(\Lambda, \partial/\partial n_\Lambda)$ is

(1-39) an isomorphism from $\begin{cases} H^2(\Omega) & \text{onto} \quad L^2(\Omega) \times H^{1/2}(\Gamma), \\ H^1(\Omega, \Lambda) & \text{onto} \quad L^2(\Omega) \times H^{-1/2}(\Gamma), \\ H(\Omega, \Lambda) & \text{onto} \quad L^2(\Omega) \times H^{-3/2}(\Gamma), \end{cases}$

and the operator (Λ, γ_0) is

(1-40) an isomorphism from $\begin{cases} H^2(\Omega) & \text{onto} \quad L^2(\Omega) \times H^{3/2}(\Gamma), \\ H^1(\Omega, \Lambda) & \text{onto} \quad L^2(\Omega) \times H^{1/2}(\Gamma), \\ H(\Omega, \Lambda) & \text{onto} \quad L^2(\Omega) \times H^{-1/2}(\Gamma). \end{cases}$ ▲

1-10. Value of the Solution at a Point of the Boundary

Let us consider now a problem that arises in the theory of optimal control for instance. Let $t \in H^{\frac{1}{2}}(\Gamma)$ and $f \in L^2(\Omega)$, and let u be the solution of the problems

$$\Lambda^* u = f \quad \text{and} \quad \frac{\partial u}{\partial n_{\Lambda^*}} = t.$$

If a is a point of the boundary Γ of Ω, we would like to compute the value $u(a)$ of the solution u at the point a for any f and t. This has a meaning if $n \leq 3$; indeed, the solution u belongs to $H^2(\Omega)$ by Theorem 1-2 and the trace of u on Γ belongs to $H^{\frac{3}{2}}(\Gamma)$. By Theorem 6.3-8, it is a continuous function on Γ if $n \leq 3$, since $\frac{3}{2} > (n - 1)/2$.

COROLLARY 1-6   Let us suppose the assumptions of Theorem 1-1 hold and that $n \leq 3$, and let $a \in \Gamma$ be a point of the boundary. Then the value $u(a)$ of the solution u of $\Lambda^* u = f$ and $\partial u/\partial n_{\Lambda^*} = t$ is the functional k on $L^2(\Omega) \times H^{\frac{1}{2}}(\Gamma)$ defined by

$$(1\text{-}41) \qquad k(f, t) = \int_{\Omega} k(x) f(x) \, dx + \int_{\Gamma} k(x) t(x) \, d\sigma(x),$$

where $k \in H(\Omega, \Lambda)$ is the solution of the boundary-value problems

$$\Lambda k = 0 \quad \text{and} \quad \frac{\partial k}{\partial n_{\Lambda}} = \delta(a),$$

$\delta(a)$ being the Dirac measure on a.   ▲

Proof   We have seen that the Dirac measure $\delta(a)$—defined by $\langle \delta(a), u \rangle = u(a)$—is a continuous linear form on $H^{\frac{3}{2}}(\Gamma)$ and thus belongs to $H^{-\frac{3}{2}}(\Gamma)$. Then there exists a unique solution $k \in H(\Omega, \Lambda)$ satisfying $\Lambda k = 0$ and $\partial k/\partial n_{\Lambda} = \delta(a)$ (see Theorem 1-2).

Then, since $k \in H(\Omega, \Lambda)$ and u belongs to $H^2(\Omega)$, we can apply the Green formula, obtaining

$$u(a) = \langle \delta(a), \gamma_0 u \rangle = \left\langle \frac{\partial k}{\partial n_{\Lambda}}, \gamma_0 u \right\rangle$$

$$= -(\Lambda k, u) + (k, \Lambda^* u) + \left\langle \gamma_0 k, \frac{\partial u}{\partial n_{\Lambda^*}} \right\rangle$$

$$= (k, f) + \langle \gamma_0 k, t \rangle. \qquad ■$$

1-11. Problems with Elliptic Differential Boundary Conditions

For the sake of simplicity, let us take $\Omega = R_+^n$ and $\Gamma = R^{n-1}$, so that we may consider an elliptic operator Λ defined by (1-6), with $\tilde{\Delta}$ denoting the Laplacian on the boundary, defined by

$$(1\text{-}42) \qquad \tilde{\Delta}t = \sum_{1 \leq i \leq n-1} D_i^2 t(x'); \qquad x' \in \Gamma = R^{n-1}.$$

To prove that the following problem is equivalent to a variational equation and has a unique solution, let $f \in L^2(\Omega)$ and $t \in H^{-1}(\Gamma)$, and look for u satisfying

$$(1\text{-}43) \qquad \begin{cases} \text{(i)} \ u \in H(\Omega, \Lambda), \\[2mm] \text{(ii)} \ \Lambda u = f \quad \text{on} \quad \Omega, \\[2mm] \text{(iii)} \ \dfrac{\partial u}{\partial n_\Lambda} - \tilde{\Delta}\gamma_0 u + \gamma_0 u = t \quad \text{on} \quad \Gamma. \end{cases}$$

Let us consider the following space V,

$$(1\text{-}44) \qquad V = \{u \in H^1(\Omega) \quad \text{such that} \quad \gamma_0 u \in H^1(\Gamma)\}$$

supplied with the norm

$$(1\text{-}45) \qquad \|u\| = (\|u\|_1^2 + \|\gamma_0 u\|_{H^1(\Gamma)}^2)^{1/2}.$$

It is easy to check that V is a Hilbert space by using Theorem 6.3-1. Assumption (2-1) of section 6.2-1 is fulfilled if we choose

$$(1\text{-}46) \qquad T = H^1(\Omega), \gamma = \gamma_0, V_0 = H_0^1(\Omega), \text{ and } H = L^2(\Omega).$$

COROLLARY 1-7   Problem 1-43 is equivalent to the variational equation in which we look for u satisfying

$$(1\text{-}47) \qquad \begin{cases} \text{(i)} \ u \in V \\[2mm] \text{(ii)} \ a(u, v) + m(\gamma_0 u, \gamma_0 v) = (f, v) + \langle t, \gamma_0 v \rangle \quad \text{for any} \quad v \in V, \end{cases}$$

where $m(t, s)$ is defined by

$$(1\text{-}48) \qquad m(t, s) = \sum_{1 \leq j \leq n-1} \int_{R^{n-1}} D_j t D_j s \, dx' + \int_{R^{n-1}} t(x') s(x') \, dx'.$$

Furthermore, if Λ is elliptic and if $a_0(x) \geq c > 0$ almost everywhere on Ω, there exists a unique solution of these problems.   ▲

Proof It is clear that the formal operator associated with $a(u, v) + m(\gamma_0 u, \gamma_0 v)$ is Λ and that its Neumann operator is $\partial/\partial n_\Lambda - \tilde{\Delta}\gamma_0 + \gamma_0$.

However, when $a(u, v)$ is $H^1(\Omega)$-elliptic, the form $a(u, v) + m(\gamma_0 u, \gamma_0 v)$ is V-elliptic when V is defined by (1-44) and (1-45). Then Corollary 1-7 follows from Theorems 6.2-2 and 6.2-3. ∎

2. BOUNDARY-VALUE PROBLEMS FOR DIFFERENTIAL OPERATORS OF HIGHER ORDER

2-1. Linear Differential Operators of Orders $2k$

Let Ω be a smooth bounded open subset of R^n, Γ its boundary, and $a_{pq}(x)$ functions in $L^\infty(\Omega)$, and let us consider the following bilinear form $a(u, v)$ defined by

$$(2\text{-}1) \qquad a(u, v) = \sum_{|p|,|q|\le k} \int_\Omega a_{pq}(x) D^p u D^q v \, dx.$$

We can check that $a(u, v)$ is continuous on $V \times V$, where $V = H^k(\Omega)$. On the other hand, the assumption (2-1) of Section 6.2-1 is satisfied if we choose

$$(2\text{-}2) \qquad \begin{cases} V = H^k(\Omega), T = \displaystyle\prod_{0\le j\le k-1} H^{k-j-\frac{1}{2}}(\Gamma), \gamma = (\gamma_0, \dots, \gamma_{k-1}), \\ \qquad\qquad V_0 = H_0^k(\Omega), \text{ and } H = L^2(\Omega) \end{cases}$$

(see Theorem 6.3-1).

Then, since $\Phi_0(\Omega)$ is dense in $H_0^k(\Omega)$, the formal operator associated with $a(u, v)$ is defined by

$$(2\text{-}3) \qquad \Lambda u = \sum_{|p|,|q|\le k} (-1)^{|q|} D^q(a_{pq}(x) D^p u).$$

This differential operator Λ of order $2k$ belongs to $L(H^k(\Omega), H^{-k}(\Omega))$ and to $L(H^k(\Omega, \Lambda), L^2(\Omega))$, where $V(\Lambda) = H^k(\Omega, \Lambda)$ is the domain of Λ defined by

$$(2\text{-}4) \qquad H^k(\Omega, \Lambda) = \{u \in H^k(\Omega) \quad \text{such that} \quad \Lambda u \in L^2(\Omega)\}.$$

Then, by Theorem 6.2-1, there exists a continuous operator $\delta = (\delta_{2k-1}, \dots, \delta_j, \dots, \delta_k)$ mapping $H^k(\Omega, \Lambda)$ into $T' = \prod_{2k-1\ge j\ge k} H^{k-j-\frac{1}{2}}(\Gamma)$, such that the Green formula holds:

$$(2\text{-}5) \qquad \begin{cases} a(u, v) = (\Lambda u, v) + \displaystyle\sum_{0\le j\le k-1} \langle \delta_{2k-1-j} u, \gamma_j v \rangle & \text{for any} \quad u \in H^k(\Omega, \Lambda) \\ & \text{and} \qquad v \in H^k(\Omega). \end{cases}$$

If the coefficients $a_{pq}(x)$ are smooth enough, we can show that the operators δ_j are differential operators of order j ($k \le j \le 2k - 1$). In order to do that, we replace Ω by R_+^n, using the change of coordinates defined in Section 6.3-2 and finally integrating by parts with respect to x_n.

We assume the $(W, L^2(\Omega))$-coercivity of $a(u, v)$ when W is a closed subspace of $H^k(\Omega)$ such that

$$(2\text{-}6) \qquad\qquad H_0^k(\Omega) \subset W \subset H^k(\Omega).$$

This property depends on the regularity of Ω, on the regularity of the coefficients a_{pq}, and on the choice of W. (If $k > 1$, the ellipticity of the operator Λ does not imply necessarily the $(H^k(\Omega), L^2(\Omega))$-coercivity.) A sufficient condition of $H^k(\Omega)$-ellipticity is

$$\sum_{|p|,|q|\leq k} a_{pq}(x)z^p z^q \geq c \sum_{|p|\leq k} |z^p|^2$$

for $x \in \Omega$ almost everywhere and where $c > 0$.

2-2. The Dirichlet Problem

The Dirichlet problem is associated with the choice $W = H_0^k(\Omega)$.

COROLLARY 2-1 Let $a(u, v)$ be defined by (2-1) and Λ by (2-3). Let $f \in L^2(\Omega)$, as well, and then the two following problems are equivalent:

1. The Dirichlet problem for Λ '
Look for u satisfying

$$(2\text{-}7) \qquad \begin{cases} \text{(i) } u \in H^k(\Omega, \Lambda), \\[4pt] \text{(ii) } \Lambda u = f, \\[4pt] \text{(iii) } \gamma_j u = 0 \quad \text{for} \quad 0 \leq j \leq k - 1. \end{cases}$$

2. The variational Dirichlet problem
Look for u satisfying

$$(2\text{-}8) \qquad \begin{cases} \text{(i) } u \in H_0^k(\Omega), \\[4pt] \text{(ii) } a(u, v) = (f, v) \quad \text{for any} \quad v \in H_0^k(\Omega). \end{cases}$$

Furthermore, if $a(u, v)$ is $H_0^k(\Omega)$-elliptic, there exists a unique solution of these problems and the operator $(\Lambda, \gamma_0, \ldots, \gamma_{k-1})$ is an isomorphism from $H^k(\Omega, \Lambda)$ onto $L^2(\Omega) \times \prod_{0 \leq j \leq k-1} H^{k-j-\frac{1}{2}}(\Gamma)$.   ▲

2-3. The Neumann Problem

The Neumann problem is associated with the choice $W = H^k(\Omega)$.

COROLLARY 2-2 Let $a(u, v)$ be defined by (2-1) and Λ by (2-3). Let $f \in L^2(\Omega)$ and $t_j \in H^{k-j-\frac{1}{2}}(\Gamma)$ for $k \leq j \leq 2k - 1$. Then the two following problems are equivalent:

1. **The Neumann problem for Λ**
Look for u satisfying

$$(2\text{-}9) \quad \begin{cases} \text{(i) } u \in H^k(\Omega, \Lambda), \\ \text{(ii) } \Lambda u = f, \\ \text{(iii) } \delta_j u = t_j \quad \text{for} \quad k \leq j \leq 2k - 1. \end{cases}$$

2. **The variational Neumann problem**
Look for u satisfying

$$(2\text{-}10) \quad \begin{cases} \text{(i) } u \in H^k(\Omega), \\ \text{(ii) } a(u, v) = (f, v) + \sum_{0 \leq j \leq k-1} \langle t_{2k-1-j}, \gamma_j v \rangle \quad \text{for any} \quad v \in H^k(\Omega). \end{cases}$$

Furthermore, if $a(u, v)$ is $H^k(\Omega)$-elliptic, there exists a unique solution of these problems and $(\Lambda, \delta_k, \ldots, \delta_{2k-1})$ is an isomorphism from $H^k(\Omega, \Lambda)$ onto $L^2(\Omega) \times \prod_{k \leq j \leq 2k-1} H^{k-j-\frac{1}{2}}(\Gamma)$.   ▲

2-4. Regularity and Theorems of Isomorphism

The adjoint form of $a(u, v)$ is $a_*(u, v) = \sum_{|p|, |q| \leq k} \int_\Omega a_{pq}(x) D^p v D^q u \, dx$, its formal operator is defined by $\Lambda^* u = \sum_{|p|, |q| \leq k} (-1)^{|q|} D^q(a_{qp} D^p u)$, and we denote its Neumann operator by $\delta^* = (\delta^*_{2k-1}, \ldots, \delta^*_k)$, which maps $H^k(\Omega, \Lambda)$ into $\prod_{2k-1 \geq j \geq k} H^{k-j-1/2}(\Gamma)$.

On the other hand, it is clear that $H^{2k}(\Omega)$ is contained in $H^k(\Omega, \Lambda)$ if the coefficients $a_{pq}(x)$ belong to $C^k(\Omega)$.

Let us denote by D_0 (respectively D_0^*) the kernel of $\gamma = (\gamma_0, \ldots, \gamma_{k-1})$ in $H^k(\Omega, \Lambda)$ [respectively in $H^k(\Omega, \Lambda^*)$] and by D (respectively D^*) the kernel of $\delta = (\delta_{2k-1}, \ldots, \delta_k)$ in $H^k(\Omega, \Lambda)$ [respectively of $\delta^* = (\delta^*_{2k-1}, \ldots \delta^*_k)$ in $H^k(\Omega, \Lambda^*)$].

We assume the following properties of regularity:

$$(2\text{-}11) \quad \begin{cases} \text{(i) the coefficients } a_{pq}(x) \text{ belong to } C^k(\bar{\Omega}), \\ \text{(ii) the subspaces } D_0, D_0^*, D, \text{ and } D^* \text{ are contained in } H^{2k}(\Omega), \\ \text{(iii) } (\gamma, \delta) \text{ and } (\gamma, \delta^*) \text{ map } H^{2k}(\Omega) \text{ onto } \prod_{0 \leq j \leq 2k-1} H^{2k-j-\frac{1}{2}}(\Gamma), \\ \text{(iv) } H_0^{2k}(\Omega) = \ker(\gamma, \delta) = \ker(\gamma, \delta^*). \end{cases}$$

Let us set

$$(2\text{-}12) \quad \begin{cases} U = H^{2k}(\Omega), \ H = L^2(\Omega), \ S = \prod_{0 \geq j \geq k-1} H^{2k-j-\frac{1}{2}}(\Gamma), \\ \qquad\qquad\qquad \text{and} \quad R' = \prod_{2k-1 \geq j \geq k} H^{2k-j-\frac{1}{2}}(\Gamma). \end{cases}$$

Then it is clear that assumptions (2-11) imply the assumptions (6.2-43) of Theorem 6.2-7. Thus we deduce Theorem 2-1.

THEOREM 2-1 If we suppose that $a(u, v)$ is $H^k(\Omega)$-coercive and that (2-11) holds, then the operator $(\Lambda, \delta_k, \ldots, \delta_j, \ldots, \delta_{2k-1})$ is an isomorphism from

$$H^{2k}(\Omega) \quad \text{onto} \quad L^2(\Omega) \times \prod_{k \leq j \leq 2k-1} H^{2k-j-\frac{1}{2}}(\Gamma),$$

(2-13) $\qquad H^k(\Omega, \Lambda) \quad \text{onto} \quad L^2(\Omega) \times \prod_{k \leq j \leq 2k-1} H^{k-j-\frac{1}{2}}(\Gamma),$

$$H(\Omega, \Lambda) \quad \text{onto} \quad L^2(\Omega) \times \prod_{k \leq j \leq 2k-1} H^{-j-\frac{1}{2}}(\Gamma).$$

The operator $(\Lambda, \gamma_0, \ldots, \gamma_j, \ldots, \gamma_{k-1})$ is an isomorphism from

$$H^{2k}(\Omega) \quad \text{onto} \quad L^2(\Omega) \times \prod_{0 \leq j \leq k-1} H^{2k-j-\frac{1}{2}}(\Gamma),$$

(2-14) $\qquad H^k(\Omega, \Lambda) \quad \text{onto} \quad L^2(\Omega) \times \prod_{0 \leq j \leq k-1} H^{k-j-\frac{1}{2}}(\Gamma),$

$$H(\Omega, \Lambda) \quad \text{onto} \quad L^2(\Omega) \times \prod_{0 \leq j \leq k-1} H^{-j-\frac{1}{2}}(\Gamma). \qquad \blacktriangle$$

2-5. Other Boundary-Value Problems

We can associate with Λ variational boundary-value problems associated with a projector of T and an operator M from T into T'. For instance, we can construct mixed problems (see Section 1-5), oblique problems (Section 1-6), and interface problems (Section 1-7).

Let us give here another example of boundary conditions, where I and J are a partition of $(0, 1, \ldots, k-1)$ and σ_1 is the projector such that

(2-15) $\quad \sigma_1$ maps $T = \prod_{0 \leq j \leq k-1} H^{k-j-\frac{1}{2}}(\Gamma) \quad$ onto $\quad T_1 = \prod_{j \in I} H^{k-j-\frac{1}{2}}(\Gamma)$

and such that

(2-16) $\quad \sigma_2' = 1 - \sigma_1'$ is the projector from T' onto $T_2' = \prod_{j \in J} H^{-k+j-\frac{1}{2}}(\Gamma)$.

Let us consider the following closed subspace W of H of $H^k(\Omega)$:

(2-17) $\quad W = \ker \sigma_1 \gamma = \{u \in H^k(\Omega) \quad$ such that $\quad \gamma_j u = 0 \text{ for } j \in I\}$.

We deduce Corollary 2-3 from Theorem 6.2-2.

COROLLARY 2-3 Let $a(u, v)$ be defined by (2-1) and Λ by (2-3). Let $f \in L^2(\Omega)$ and let $t_j \in H^{-k+j+\frac{1}{2}}(\Gamma)$ for $j \in J$. Then the two following problems are equivalent:

1. Boundary-value problem for Λ
Look for u satisfying

$$(2\text{-}18) \quad \begin{cases} \text{(i) } u \in H^k(\Omega, \Lambda), \\ \text{(ii) } \Lambda u = f, \\ \text{(iii) } \gamma_j u = 0 \quad \text{for } j \in I, \, \delta_{2k-1-j}u = t_j \quad \text{for } j \in J. \end{cases}$$

2. Variational problem for Λ
Look for u satisfying

$$(2\text{-}19) \quad \begin{cases} \text{(i) } u \in W, \\ \text{(ii) } a(u, v) = (f, v) + \sum_{j \in J} \langle t_j, \gamma_j v \rangle \quad \text{for any } v \in W. \end{cases}$$

Furthermore, if $a(u, v)$ is W-elliptic, there exists a unique solution of these problems and the operator $(\Lambda, \ldots, \gamma_j, \ldots, \delta_{2k-1-l}, \ldots)$ (when $j \in I$, $l \in J$) is an isomorphism from $H^k(\Omega, \Lambda)$ onto $L^2(\Omega) \times \prod_{j \in I} H^{k-j-\frac{1}{2}}(\Gamma) \times \prod_{l \in J} H^{-k+l+\frac{1}{2}}(\Gamma)$. ▲

2-6. Boundary Value Problems for $\Delta^2 + \lambda$

We view $\Delta^2 + \lambda$ as the formal operator associated with the bilinear form

$$(2\text{-}20) \quad a(u, v) = \int_{\Omega} \Delta u \, \Delta v \, dx + \lambda \int_{\Omega} uv \, dx.$$

We have to choose a Hilbert space V such that

$$(2\text{-}21) \quad \begin{cases} \text{(i) } a(u, v) \text{ is continuous and } V\text{-elliptic}, \\ \text{(ii) there exist spaces } H \text{ and } T \text{ and } \gamma \in L(V, T) \quad \text{such that} \\ \quad \text{the assumption (2-1) of Section 6.2-1 holds.} \end{cases}$$

Such a choice of a Hilbert space V depends on the boundary conditions.

Remark 2-1 *(Neumann problem)* If we choose $V = H(\Omega, \Delta)$, it is obvious that $a(u, v)$ is V-elliptic (when $\lambda > 0$). ■

Remark 2-2 The form $a(u, v)$ is continuous on $H^2(\Omega)$, but it is not $H^2(\Omega)$-elliptic. ■

On the other hand, we have seen that the operators γ_0 and γ_1 map $H(\Omega, \Delta)$ onto $H^{-\frac{1}{2}}(\Gamma)$ and $H^{-\frac{3}{2}}(\Gamma)$, respectively (Theorem 1-2 with $\Lambda = \Delta$) and that the kernel of $\gamma = (\gamma_0, \gamma_1)$ is $H_0^2(\Omega)$. But γ is not surjective.

Let T be the image of $H(\Omega, \Delta)$ by γ. We only know that

(2-22) $H^{3/2}(\Gamma) \times H^{1/2}(\Gamma) \subset T \subset H^{-1/2}(\Gamma) \times H^{-3/2}(\Gamma)$.

(We do not characterize here the space T.)

Then the formal operator associated with $a(u, v)$ is

(2-23) $\Lambda u = \Delta^2 u + \lambda u,$

and its domain $V(\Lambda)$ is the space $H(\Omega, \Delta, \Delta^2)$ defined by

(2-24) $H(\Omega, \Delta, \Delta^2) = \{u \in L^2(\Omega) \text{ such that } \Delta u \text{ and } \Delta^2 u \in L^2(\Omega)\}.$

Theorem 6.2-1 told us that there exists an operator $\delta = (\delta_0, \delta_1)$ mapping $H(\Omega, \Delta, \Delta^2)$ into T' such that the Green formula holds

(2-25) $\begin{cases} a(u, v) = (\Lambda u, v) + \langle \delta_0 u, \gamma_0 v \rangle - \langle \delta_1 u, \gamma_1 v \rangle \\ \qquad \text{for any} \quad u \in H(\Omega, \Delta, \Delta^2) \quad \text{and} \quad v \in H(\Omega, \Delta). \end{cases}$

By comparing (2-25) with the usual Green formula, we deduce that δ_0 and δ_1 are the extensions to $H(\Omega, \Delta, \Delta^2)$ of the trace operators $\delta_0 = \delta_1 \Delta$ and $\delta_1 = \gamma_0 \Delta$. We thus deduce from Proposition 6.2-1 the following theorem:

THEOREM 2-2   Let us assume that Ω is smooth and that $\lambda > 0$ or that $\lambda \leq 0$ is not an eigenvalue. Let $f \in L^2(\Omega)$, $t_0 \in H^{1/2}(\Gamma)$, and $t_1 \in H^{3/2}(\Gamma)$. Then there exists a unique solution of one of the two following equivalent problems:

1. The Neumann problem for $\Delta^2 + \lambda$
Look for $u \in H(\Omega, \Delta, \Delta^2)$ satisfying

(2-26) $\Delta^2 u + \lambda u = f, \gamma_0 \Delta u = t_0, \text{ and } \gamma_1 \Delta u = t_1.$

2. The variational Neumann problem for $\Delta^2 + \lambda$
Look for $u \in H(\Omega, \Delta)$ satisfying

(2-27) $a(u, v) = (f, v) + \langle t_1, \gamma_0 v \rangle - \langle t_0, \gamma_1 v \rangle \qquad \text{for any} \quad v \in H(\Omega, \Delta).$ ▲

Remark 2-3   Actually, we deduce from Theorem 6.2-6 that $((\Delta^2 + \lambda), \gamma_0 \Delta, \gamma_1 \Delta)$ is an isomorphism from $H(\Omega, \Delta, \Delta^2)$ onto $L^2(\Omega) \times T'$. ■

We now see that the form $a(u, v)$ is $H^2(\Omega)$-elliptic on convenient closed subspaces of $H^2(\Omega)$ and deduce other theorems of existence. To accomplish this, we choose the following spaces V, H, T, and operator γ:

(2-28) $\begin{cases} V = \{u \in H^2(\Omega) \text{ such that } \gamma_1 u = 0\}, H = L^2(\Omega), \\ T = H^{3/2}(\Gamma) \quad \text{and} \quad \gamma = \gamma_0. \end{cases}$

Assumption 2-1 of Section 6.2-1 is satisfied in this case: by Theorem 6.3-1, γ_0 is surjective from V onto T and its kernel is $H_0^2(\Omega)$, which is dense in H. Moreover, there exists a right inverse μ_1 of γ_1 mapping $H^{1/2}(\Gamma)$ into V.

On the other hand, $a(u, v)$ is a continuous V-elliptic form (when $\lambda > 0$). Indeed, since $\Delta^2 + \lambda$ is an isomorphism from V onto $L^2(\Omega)$ (homogeneous Neumann problem), the norms $\|u\|_{H^2(\Omega)}$ and $\|u\|_{H(\Omega,\Delta)}$ are equivalent on V, and again the formal operator associated with $a(u, v)$ is $\Delta^2 + \lambda$.

Let us consider the space $H^2(\Omega, \Delta^2)$ defined by

$$(2\text{-}29) \qquad H^2(\Omega, \Delta^2) = \{u \in H^2(\Omega) \quad \text{such that} \quad \Delta^2 u \in L^2(\Omega)\}.$$

THEOREM 2-3 Assuming that Ω is smooth and that $\lambda > 0$, we let $f \in L^2(\Omega)$, $t_0 \in H^{1/2}(\Gamma)$, and $t_1 \in H^{-3/2}(\Gamma)$. Then there exists a unique solution $u \in H^2(\Omega, \Delta^2)$ of the following problem:

$$(2\text{-}30) \qquad \Delta^2 u + \lambda u = f, \gamma_1 u = t_0, \quad \text{and} \quad \gamma_1 \Delta u = t_1.$$

If u is the solution of (2-30), then $w = u - \mu_1 t_0$ is the solution of the variational equation in which we look for $w \in V$ satisfying

$$(2\text{-}31) \quad a(w, v) = (f, v) + \langle t_1, \gamma_0 v \rangle - a(\mu_1 t_0, v) \qquad \text{for any} \quad v \in V.$$

Conversely, if w is the solution of (2-31), then $u = w + \mu_1 t_0$ satisfies (2-30).

Proof The proof of Theorem 2-3 is left as an exercise. ▲

Remark 2-4 In other words, the operator $((\Delta^2 + \lambda), \gamma_1, \gamma_1 \Delta)$ is an isomorphism from $H^2(\Omega, \Delta^2)$ onto $L^2(\Omega) \times H^{1/2}(\Gamma) \times H^{-3/2}(\Gamma)$. By permuting the roles played by γ_0 and γ_1, we can prove in the same way that the operator $((\Delta^2 + \lambda), \gamma_0, \gamma_0 \Delta)$ is an isomorphism from $H^2(\Omega, \Delta^2)$ onto $L^2(\Omega) \times H^{3/2}(\Gamma) \times H^{-1/2}(\Gamma)$.

Replacing V by $H_0^2(\Omega)$, we would obtain the variational formulation of the Dirichlet problem: we can prove that the operator $((\Delta^2 + \lambda), \gamma_0, \gamma_1)$ is an isomorphism from $H^2(\Omega, \Delta^2)$ onto $L^2(\Omega) \times H^{3/2}(\Gamma) \times H^{1/2}(\Gamma)$.

We can obtain other theorems of isomorphism by studying the regularity of the solutions of the homogeneous boundary-value problems and by using the results of Section 6.2-7 (see also Section 2-4 of this chapter). ■

Remark 2-5 The operator $\Delta^2 + \Delta + 1$ can be viewed as the formal operator associated with the bilinear form

$$(2\text{-}32) \qquad a(u, v) = \int_\Omega \Delta u \, \Delta v \, dx + \int_\Omega \text{grad } u \cdot \text{grad } v \, dx + \int_\Omega uv \, dx.$$

It is clear that this form is continuous and V-elliptic when we choose $V = H^1(\Omega, \Delta)$. We know by Theorem 1-2 that the operators γ_0 and γ_1 map

$H^1(\Omega, \Delta)$ onto $H^{1/2}(\Gamma)$ and $H^{-1/2}(\Gamma)$, respectively, and that the kernel of $\gamma = (\gamma_0, \gamma_1)$ is $H_0^2(\Omega)$, which is dense in $L^2(\Omega)$.

The image S of $H^1(\Omega, \Delta)$ by γ satisfies

$$(2\text{-}33) \qquad H^{3/2}(\Gamma) \times H^{1/2}(\Gamma) \subset S \subset H^{1/2}(\Gamma) \times H^{-1/2}(\Gamma).$$

Then the formal operator associated with $a(u, v)$ is $\Lambda = \Delta^2 + \Delta + 1$ and its domain $V(\Lambda)$ is the space $H^1(\Omega, \Delta, \Delta^2)$ defined by

$$(2\text{-}34) \quad H^1(\Omega, \Delta, \Delta^2) = \{u \in H^1(\Omega) \text{ such that } \Delta u \text{ and } \Delta^2 u \in L^2(\Omega)\}.$$

We thus deduce that the operator $(\Delta^2 + \Delta + 1, \gamma_0\Delta, \gamma_1\Delta + \gamma_1)$ is an isomorphism from the space $H^1(\Omega, \Delta, \Delta^2)$ onto $L^2(\Omega) \times S'$.

In particular, if $f \in L^2(\Omega)$, $t_0 \in H^{1/2}(\Gamma)$, and $t_1 \in H^{-1/2}(\Gamma)$, there exists a unique solution of the Neumann problem

$$(2\text{-}35) \qquad \begin{cases} \text{(i)} \quad u \in H^1(\Omega, \Delta, \Delta^2), \\ \text{(ii)} \quad \Delta^2 u + \Delta u + u = f, \\ \text{(iii)} \quad \gamma_1\Delta u + \gamma_1 u = t_1, \\ \text{(iv)} \quad \gamma_0\Delta u = t_0. \end{cases}$$

■

CHAPTER 8

Approximation of Neumann-Type Problems

In this chapter we continue the study of internal approximations of variational equations made in Section 3.1-7, where these variational equations are treated as equivalent to Neumann-type problems.

We apply the results thus obtained for constructing approximate problems to Neumann problems for elliptic differential operators of order $2k$ and to several other examples of Neumann-type problems.

1. THEOREMS OF CONVERGENCE AND ERROR ESTIMATES

The solution $u \in V(\Lambda)$ of the Neumann problem

$$\begin{cases} \text{(i)} & u \in V(\Lambda), \\ \text{(ii)} & \Lambda u = f, \\ \text{(iii)} & \delta u = t, \end{cases}$$

where f and t are given in H and T, respectively, is a solution of the variational equation

$$\begin{cases} \text{(i)} & u \in V, \\ \text{(ii)} & a(u, v) = (f, v) + \langle t, \gamma v \rangle \quad \text{for any} \quad v \in V. \end{cases}$$

Therefore we approximate it by the solution $u_h \in V_h$ of the variational equation

$$a(p_h u_h, p_h v_h) = (f, p_h v_h) + \langle t, \gamma p_h v_h \rangle \quad \text{for any} \quad v_h \in V_h.$$

We already know that if $a(u, v)$ is (V, H)-coercive and 0 is not an eigenvalue of the Neumann problem, then there exists a constant M such that

$$\| u - p_h u_h \|_V \leq M \| u - p_h \hat{r}_h u \|_V \leq M e_U^V(p_h) \| u \|_U \quad \text{when} \quad u \in U \subset V.$$

Furthermore, if $U_0 = V_0 = \ker \gamma$, we obtain the following error estimate:

$$\begin{aligned} \| \Lambda(u - p_h u_h) \|_{U_0'} &\leq M e_{U_0}^V(p_h) \| u - p_h \hat{r}_h u \|_V \\ &\leq M e_{U_0}^V(p_h) e_U^V(p_h) \| u \|_U. \end{aligned}$$

If we assume certain properties of regularity—namely, (Λ, δ) is an isomorphism from U into $H \times R'$ and from $H(\Lambda)$ onto $H \times S'$ where $U = V(\Lambda)$, $S \subset T \subset R$, and $H(\Lambda) = \{u \in H$ such that $\Lambda u \in H\}$—we deduce the following error estimates:

$$\begin{cases} \text{(i)} \ \|u - p_h u_h\|_H \leq M e_U{}^V(p_h) \ \|u - p_h \hat{r}_h u\|_V \leq M e_U{}^V(p_h)^2 \ \|u\|_U, \\ \text{(ii)} \ \|\gamma(u - p_h u_h)\|_R \leq M e_U{}^V(p_h) \|u - p_h \hat{r}_h u\|_V \leq M e_U{}^V(p_h)^2 \ \|u\|_U. \end{cases}$$

Finally, by using convergent quasi-optimal approximations (V_h, p_h, r_h) of U, which are quasi-optimal for the injection from U into V, we can prove the regularity of the convergence

$$\begin{cases} \text{(i)} \ \lim \|u - p_h u_h\|_U = 0 & \text{if} \ t \in R', \\ \text{(ii)} \ \lim \|u - p_h u_h\|_V = 0 & \text{if} \ t \in T', \\ \text{(iii)} \ \lim \|u - p_h u_h\|_H = 0 & \text{if} \ t \in S'. \end{cases}$$

1-1.   Internal Approximation of a Neumann-Type Problem

In Section 3.1-7 we studied the internal approximation of variational equations. Here we continue this investigation when the variational equation is equivalent to a Neumann-type problem (see Section 6.2-4).

Let us consider a bilinear form $a(u, v)$ continuous on a Hilbert space V, where we assume that there exist a pivot space H, a Hilbert space T, and an operator $\gamma \in L(V, T)$ such that

(1-1)
$$\begin{cases} \text{(i)} \ V \text{ is contained in } H \text{ with a stronger topology,} \\ \text{(ii)} \ \text{the kernel } V_0 = \ker \gamma \text{ of } \gamma \text{ is dense in } H, \\ \text{(iii)} \ \gamma \text{ maps } V \text{ onto } T. \end{cases}$$

Let $\Lambda \in L(V, V_0') \cap L(V(\Lambda), H)$ be the formal operator and $\delta \in L(V(\Lambda), T')$ be the Neumann operator associated with $a(u, v)$.

When $f \in H$ and $t \in T'$ are given, a solution $u \in V(\Lambda)$ of the Neumann problems

(1-2)
$$\Lambda u = f \quad \text{and} \quad \delta u = t$$

is a solution $u \in V$ of the variational equation

(1-3)
$$a(u, v) = (f, v) + \langle t, \gamma v \rangle \quad \text{for any} \ v \in V,$$

and conversely.

Considering convergent approximations $(V_h, p_h, \hat{r}_h)$ of the Hilbert space V, we approximate a solution u of the Neumann-type problem (1-2) by the solution $u_h \in V_h$ of the discrete variational equation

(1-4)
$$a(p_h u_h, p_h v_h) = (f, p_h v_h) + \langle t, \gamma p_h v_h \rangle \quad \text{for any} \ v_h \in V_h.$$

We assume either that

(1-5) $\qquad\qquad\qquad a(u, v)$ is V-elliptic

or that

(1-6) $\begin{cases} \text{(i) } a(u, v) \text{ is } (V, H)\text{-coercive,} \\ \text{(ii) the injection from } V \text{ into } H \text{ is compact,} \\ \text{(iii) } A \in L(V, V') \text{ defined by } a(u, v) \text{ is an isomorphism,} \\ \text{(iv) } V_h \text{ are finite-dimensional spaces.} \end{cases}$

Then the conclusions of Theorem 3.1-6 hold. In particular, we obtain the following estimates of error.

THEOREM 1-1 If we assume (1-1) and either (1-5) or (1-6), then there exist unique solutions $u \in V(\Lambda)$ of the Neumann problem (1-2) and $u_h \in V_h$ of the approximate problem (1-4) for h small enough, and

(1-7) $\quad \|u - p_h u_h\|_V \leq M \|u - p_h \hat{r}_h u\|_V \leq M e_U^V(p_h) \|u\|_U$ if $u \in U \subset V$.

Let $U_0 \subset V_0$ be a Hilbert space, and assume that the solution u of (1-2) belongs actually to a Hilbert space U. Then

(1-8) $\qquad \|\Lambda(u - p_h u_h)\|_{U_0'} \leq M e_{U_0}^V(p_h) e_U^V(p_h) \|u\|_U .$ ▲

Proof The first part of Theorem 1-1 is a restatement of Theorem 3.1-6; in fact, we can see that estimate (1-8) follows from estimate (1-40) of Theorem 3.1-6 by taking $X = U_0$. Indeed, since $U_0 \subset V_0$, we can write

$$a(u - p_h u_h, v) = (\Lambda(u - p_h u_h), v) \qquad \text{for any } v \in U_0. \quad ■$$

Beginning the study of the regularity of the convergence, we assume that the solution u belongs to a Hilbert space U, and we introduce a Hilbert space K such that

(1-9) $\qquad U \subset K \subset V$, the injections being continuous.

Let us consider a prolongation q_h mapping V_h into K, related to p_h by the stability function

(1-10) $\qquad\qquad s_K^V(q_h, p_h) = \sup_{v_h \in V_h} \dfrac{\|q_h v_h\|_K}{\|p_h v_h\|_V}$

(see Section 2.2-1).

THEOREM 1-2 Let us take the assumptions of Theorem 1-1. Then, if u belongs to U, the error $u - q_h u_h$ in K is estimated by

(1-11) $\quad \|u - q_h u_h\| \leq M(e_U^K(q_h r_h) + s_K^V(q_h, p_h) e_U^V(p_h r_h)) \|u\|_U$

for any restriction r_h. Therefore $q_h u_h$ converges to u in K—strongly if the right-hand side of (1-11) converges to 0 and weakly if it is bounded. ▲

Proof The proof is trivial: we deduce from the stability that

$$(1\text{-}12) \qquad \|p_h(u_h - r_h u)\|_V \leq M e_U^V(p_h r_h) \|u\|_U .$$

Therefore, by writing $u - q_h u_h = u - q_h r_h u + q_h(r_h u - u_h)$ we deduce (1-11) from (1-12), (1-10), and the definition of truncation errors. ∎

1-2. Convergence and Estimates of Error in Larger Spaces

We assume in this section the same hypotheses of regularity used in Sections 6.2-7 and 6.2-8: there exist Hilbert spaces U, R, and S such that

$$(1\text{-}13) \qquad \begin{cases} \text{(i) } U \subset V(\Lambda) \cap V(\Lambda^*), \ S \subset T \subset R, \\ \text{(ii) } D_0, D_0^*, D, \text{ and } D^* \text{ are contained in } U, \\ \text{(iii) } (\gamma, \delta) \text{ and } (\gamma, \delta^*) \text{ map } U \text{ onto } S \times R', \\ \text{(iv) } U_0 = D \cap D_0 = D^* \cap D_0^* \text{ is dense in } H. \end{cases}$$

(See assumption 6.2-43).)

Let us begin by proving some estimates of the error.

THEOREM 1-3 Assuming (1-1), (1-13), and either (1-5) or (1-6) let u and u_h be the solutions of (1-2) and (1-4), respectively. Then the following estimates of error hold:

$$(1\text{-}14) \qquad \begin{cases} \text{(i) } \quad \|u - p_h u_h\|_H \leq M e_U^V(p_h) \|u - p_h \hat{r}_h u\|_V, \\ \text{(ii) } \|\gamma(u - p_h u_h)\|_R \leq M e_U^V(p_h) \|u - p_h \hat{r}_h u\|_V. \end{cases}$$ ▲

Proof The estimates of Theorem 1-3 follow from estimate (1-40) of Theorem 3.1-6 for different choices of space X.

First, let us choose $X = D^*$. Then we can write

$$(1\text{-}15) \qquad a(u - p_h u_h, v) = a_*(v, u - p_h u_h) = (\Lambda^* v, u - p_h u_h).$$

Since Λ^* is an isomorphism from D^* onto H, we obtain

$$(1\text{-}16) \qquad \sup_{v \in D^*} \frac{|a(u - p_h u_h, v)|}{\|v\|_{D^*}} = \|u - p_h u_h\|_H.$$

Since D^* is a closed subspace of U, we have the inequality

$$e_{D^*}^V(p_h) \leq e_U^V(p_h).$$

Finally, let us take

$$X = \{u \in U \quad \text{such that} \quad \Lambda^* u = 0\}.$$

Then δ^* is an isomorphism from X onto R', since (Λ_0^*, δ^*) is an isomorphism from U onto $H \times R'$. However, when v ranges over X, we can write

$$(1\text{-}17) \quad a(u - p_h u_h, v) = a_*(v, u - p_h u_h) = \langle \delta^* v, \gamma(u - p_h u_h) \rangle \quad \text{for any } v \in X.$$

Therefore

$$\sup_{v \in X} \frac{|a(u - p_h u_h, v)|}{\|v\|_X} = \|\gamma(u - p_h u_h)\|_R,$$

which implies   [(1-14)(ii)], since $e_X^V(p_h) \le e_U^V(p_h)$.   ∎

By Theorem 6.2-7, we know that the solution u of the Neumann problem (1-2) belongs to U when $t \in R'$, to $V(\Lambda) \subset V$ when $t \in T'$ and to $H(\Lambda)$ when $t \in S'$.

We have already proved the convergence in V and in U (Theorems 1-1 and 1-2). Now we study the convergence of $p_h u_h$ to u in H when $t \in S'$.

Let us assume that there exist a prolongation q_h mapping V_h into U and a restriction r_h such that

$$(1\text{-}18) \quad \begin{cases} \text{(i) } s_U^V(q_h, p_h) e_U^V(p_h r_h) \le M, \\[2mm] \text{(ii) } \lim_{h \to 0} \|u - q_h r_h u\|_U = 0, \\[2mm] \text{(iii) } e_U^V(p_h r_h) \quad \text{and} \quad e_U^V(q_h r_h) \text{ converge to 0 with } h, \\[2mm] \text{(iv) } s_V^V(q_h, p_h) \le M. \end{cases}$$

If t is given in S', we approximate the solution $u \in H(\Lambda)$ of the Neumann problem (1-2) by the solution u_h of the discrete variational equation

$$(1\text{-}19) \quad a(p_h u_h, p_h v_h) = (f, q_h v_h) + \langle t, \gamma q_h v_h \rangle \quad \text{for any } v_h \in V_h.$$

Equation 1-19 has a meaning because $\gamma q_h v_h \in S$ and $t \in S'$. We should note that the approximate operator A_h is the same in (1-4) and (1-18): the irregularity of the data t is involved only on the right-hand side of the approximate equation.

THEOREM 1-4   Let us assume (1-1), (1-13), (1-18), and either (1-5) or (1-6). In addition, let $f \in H$ and $t \in S'$ be given, $u \in H$ the solution of (1-2), and $u_h \in V_h$ the solution of the approximate problem (1-19). Then

$$(1\text{-}20) \quad \begin{cases} \text{(i) } \lim_{h \to 0} \|u - p_h u_h\|_H = 0, \\[2mm] \text{(ii) } \lim_{h \to 0} \|\gamma(u - p_h u_h)\|_R = 0. \end{cases}$$

▲

Proof Let us recall that the restriction A^* of Λ^* to $D^* = \ker \delta^*$ is an isomorphism from D^* onto H; then its transpose $(\Lambda^*)'$ is an isomorphism from H onto $(D^*)'$.

Let π be the canonical injection from D^* into U, and consider the operator $B_h \in L(D^*, U)$ defined by

$$(1\text{-}21) \qquad B_h u = (\pi - q_h A_h^{*-1} p_h' \Lambda^*) u = u - q_h u_h,$$

where u_h is the solution of the approximate equation

$$(1\text{-}22) \quad \begin{cases} (A_h^* u_h, v_h)_h = a_*(p_h u_h, p_h v_h) = a_*(u, p_h v_h) = (\Lambda^* u, p_h v_h) \\ \qquad\qquad\qquad\qquad\qquad\qquad\qquad \text{for any} \quad v_h \in V_h. \end{cases}$$

Then, by applying Theorem 1-2 with $K = U$ and $K = V$ and by using assumptions (1-18), we deduce that

$$(1\text{-}23) \quad \begin{cases} \text{(i)} \ \ \|B_h\|_{L(D^*, U)} = \|B_h'\|_{L(U', (D^*)')} \leq M, \\ \text{(ii)} \ \ \|B_h\|_{L(D^*, V)} = \|B_h'\|_{L(V', (D^*)')} \leq \varepsilon(h), \end{cases}$$

where $\varepsilon(h) \leq M(e_U^V(q_h r_h) + s_V^V(q_h, p_h) e_U^V(p_h))$ converges to 0 with h.
This implies that

$$(1\text{-}24) \qquad B_h' l \text{ converges to 0 in } (D^*)' \qquad \text{for any} \ \ l \in U'.$$

Indeed, since the operators B_h' are bounded in $L(U', (D^*)')$, Theorem 2.1-9 implies (1-24), if we can prove that $B_h' l$ converges to 0 for any l belonging to the dense subspace V' of U'. But this last property follows from [(1-23)(ii)].

Now let us consider linear forms $l = f + \gamma' t$ where $f \in H$ and $t \in S'$. Then

$$(1\text{-}25) \qquad B_h' l = \pi' l - (\Lambda^*)' p_h(A_h^{-1}) q_h' l = (\Lambda^*)'(u - p_h u_h),$$

where u and u_h are the solutions of (1-2) and (1-19), respectively.

It is clear that $p_h(A_h)^{-1} q_h' l = p_h u_h$. On the other hand, $\pi' l = (\Lambda^*)' u$; since $u \in H(\Lambda)$, we deduce from the Green formula (2-44) of Section 6.2-7, that for any $v \in D^* = \ker \delta^*$,

$$(1\text{-}26) \quad \begin{cases} ((\Lambda^*)'u, v) = (u, \Lambda^* v) = (\Lambda u, v) + \langle \delta u, \gamma v \rangle \\ \qquad\qquad = (f, v) + \langle t, \gamma v \rangle = (f, \pi v) + \langle t, \gamma \pi v \rangle = (\pi' l, v). \end{cases}$$

Then we deduce from (1-24) and (1-25) that $(\Lambda^*)'(u - p_h u_h)$ converges to 0 in $(D^*)'$. Since $(\Lambda^*)'$ is an isomorphism from H onto $(D^*)'$, we have proved [(1-20)(i)].

Now let us prove [(1-20)(ii)]. For that purpose, we introduce the operators $C_h \in L(R', U)$ defined by

$$(1-27) \qquad C_h t = \pi(\Lambda^*)^{-1}\gamma' t - q_h A_h^* p_h' \gamma' t = u - q_h u_h,$$

where u is the solution of the Neumann problem $\Lambda^* u = 0$ and $\delta^* u = t$, and u_h its approximation, solution of the discrete equation

$$(1-28) \qquad (A_h^* u_h, v_h)_h = a_*(p_h u_h, p_h v_h) = \langle t, \gamma p_h v_h \rangle.$$

By applying Theorem 1-2 with $K = U$ and $K = V$, we deduce that

$$(1-29) \qquad \begin{cases} \text{(i)} \ \ \|C_h\|_{L(R',U)} = \|C_h'\|_{L(U',R)} \leq M, \\ \text{(ii)} \ \ \|C_h\|_{L(R',V)} = \|C_h'\|_{L(V',R)} \leq \varepsilon(h). \end{cases}$$

Then, since V' is dense in U', we deduce from (1-29) and Theorem 2.1-9 that

$$(1-30) \qquad C_h' l \text{ converges to 0 in } R \qquad \text{for any} \quad l \in U'.$$

Let us take $l = f + \gamma' t$ where $f \in H$ and $t \in S'$. Then

$$(1-31) \qquad C_h' l = \gamma(\Lambda^*)'^{-1}\pi' l - \gamma p_h(A_h)^{-1} q_h' l = \gamma(u - p_h u_h),$$

where u is the solution of the Neumann problem 1-2 and u_h the solution of the approximate equation (1-19). $\blacksquare$

In particular, let us assume that p_h maps V_h into U and that

$$(1-32) \qquad \begin{cases} \text{(i)} \ \ \sup_h s_U{}^V(p_h) e_U{}^V(p_h r_h) \leq M, \\ \text{(ii)} \ \ \lim_{h \to 0} \|u - p_h r_h u\|_U = 0 \qquad \text{for any} \quad u \in U. \end{cases}$$

THEOREM 1-5 Let us assume (1-1), (1-13), (1-32), and either (1-5) or (1-6). Assume that $f \in H$ is given, u is the solution of (1-2), and u_h is the solution of (1-4).

If $t \in R'$, then $u \in U$, $\gamma u \in S$ and

$$(1-33) \qquad \lim_{h \to 0} \|u - p_h u_h\|_U = 0, \qquad \lim_{h \to 0} \|\gamma(u - p_h u_h)\|_S = 0.$$

If $t \in T'$, then $u \in V$, $\gamma u \in T$, and

$$(1-34) \qquad \lim_{h \to 0} \|u - p_h u_h\|_V = 0, \qquad \lim_{h \to 0} \|\gamma(u - p_h u_h)\|_T = 0.$$

If $t \in s'$, then $u \in H$, $\gamma u \in R$, and

$$(1-35) \qquad \lim_{h \to 0} \|u - p_h u_h\|_H = 0, \qquad \lim_{h \to 0} \|\gamma(u - p_h u_h)\|_R = 0.$$

Furthermore, if $t \in R'$ and if the dimension of V_h is equal to $n(h)$, then there exists a constant M such that

$$(1\text{-}36) \qquad \|u - p_h u_h\|_V \leq M E_U^{\ V}(n(h) - 1) \|u\|_U,$$

where $E_U^{\ V}(n)$ is the n-width of the injection from U into V. ▲

2. APPROXIMATION OF NEUMANN PROBLEMS FOR ELLIPTIC OPERATORS OF ORDER $2k$

We apply the results of Section 1 for approximating solutions of Neumann problems to elliptic differential operators of order $2k$. After studying the properties of the matrix A_h of the approximate Neumann problem, we deduce theorems of convergence, error estimates, and the regularity of the convergence. Also, we point out the special properties obtained when piecewise-polynomial approximations are employed.

Finally, we prove that convergent finite-element approximations of the Sobolev spaces $H^{2k}(\Omega)$ are convergent approximations of the spaces $H^k(\Omega, \Lambda)$ and $H(\Omega, \Lambda)$.

2-1. Approximation of Neumann Problems for Elliptic Differential Operators

Let Ω be a smooth bounded open subset of R^n, Γ its boundary and Λ the formal operator defined by

$$(2\text{-}1) \qquad \Lambda u = \sum_{|p|,|q| \leq k} (-1)^{|q|} D^q(a_{pq}(x) D^p u)$$

associated with the bilinear form

$$(2\text{-}2) \qquad a(u, v) = \sum_{|p|,|q| \leq k} \int_\Omega a_{pq}(x) D^p u D^q v \, dx$$

(see Section 7.2-3). By corollary 7.2-3, a solution $u \in H^k(\Omega, \Lambda)$ of the Neumann problem

$$(2\text{-}3) \qquad \begin{cases} \text{(i) } \Lambda u = f; \quad f \text{ is given in } L^2(\Omega), \\ \text{(ii) } \delta_j u = t_j \quad \text{for} \quad k \leq j \leq 2k - 1; \quad t_j \text{ is given in } H^{k-j-\frac{1}{2}}(\Gamma), \end{cases}$$

is a solution u of the variational equation

$$(2\text{-}4) \qquad \begin{cases} \text{(i) } u \in H^k(\Omega), \\ \text{(ii) } a(u, v) = (f, v) + \sum_{0 \leq j \leq k-1} \langle t_{2k-j-1}, \gamma_j v \rangle \quad \text{for any} \quad v \in H^k(\Omega). \end{cases}$$

We assume either that

$$(2\text{-}5) \quad \sum_{|p|,|q|\leq k} a_{pq}(x)z^p z^q \geq c \sum_{|p|\leq k} |z^p|^2; \quad c > 0; \quad x \in \Omega \text{ almost everywhere,}$$

or that

$$(2\text{-}6) \quad \begin{cases} \text{(i)} \ a(v, v) + \lambda |v|^2 \geq c \|v\|^2; \quad c > 0; \quad \lambda \geq 0, \\ \text{(ii) } 0 \text{ is not an eigenvalue of the Neumann problem.} \end{cases}$$

Let us consider finite-element approximations $(H_h{}^\mu(\Omega), p_h, r_{h,\Omega})$ of the space $H^k(\Omega)$ (see Section 5.2-2) where p_h is associated with a function $\mu \in H^k(R^n)$ with compact support by

$$(2\text{-}7) \quad p_h u_h = \sum_{j \in \mathscr{R}_h{}^\mu(\Omega)} u_h{}^j \mu\left(\frac{x}{h} - j\right).$$

We approximate the solution u of the Neumann problem (2-3) by the solution $u_h \in H_h{}^\mu(\Omega)$ of the discrete variational equation

$$(2\text{-}8) \quad \begin{cases} \text{(i) } u_h \in H_h{}^\mu(\Omega), \\ \text{(ii) } a(p_h u_h, p_h v_h) = (f, p_h v_h) + \sum_{0 \leq j \leq k-1} \langle t_{2k-1-j}, \gamma_j p_h v_h \rangle \\ \qquad\qquad\qquad\qquad\qquad\qquad\qquad\qquad \text{for any } v_h \in H_h{}^\mu(\Omega). \end{cases}$$

If we supply the discrete space $H_h{}^\mu(\Omega)$ with the duality pairing

$$(2\text{-}9) \quad (u_h, v_h)_h = h \sum_{j \in \mathscr{R}_h{}^\mu(\Omega)} u_h{}^j v_h{}^j,$$

the discrete variational equation (2-8) is equivalent to the discrete operational equation

$$(2\text{-}10) \quad A_h u_h = l_h,$$

where $l_h \in H_h{}^\mu(\Omega)$ is the vector of components

$$(2\text{-}11) \quad \begin{cases} l_h{}^i = h^{-1} \int_\Omega f(x)\mu\left(\frac{x}{h} - i\right) dx \\ \qquad + \sum_{0 \leq j \leq k-1} h^{-1} \int_\Gamma t_{2k-j-1}(x)\gamma_j \mu\left(\frac{x}{h} - i\right) d\sigma(x), \end{cases}$$

and A_h is the matrix whose entries $d_h(i, j)$ are defined by

$$(2\text{-}12) \quad \begin{cases} d_h(i, j) = h^{-1} a\left(\mu\left(\frac{x}{h} - j\right), \mu\left(\frac{x}{h} - i\right)\right) \\ \qquad = h^{-1} \sum_{|p|,|q|\leq k} h^{-p-q} \int_\Omega a_{pq}(x)(D^p\mu)\left(\frac{x}{h} - j\right)(D^q\mu)\left(\frac{x}{h} - i\right) dx. \end{cases}$$

Definition 2-1   The largest number of nonzero entries in each row and each column of a matrix A_h is the "number of levels" of this matrix.   ▲

PROPOSITION 2-1   The number of levels of the discrete matrix A_h of the Neumann problem (2-3) is at most equal to the number of levels of the function μ.

The Neumann conditions are involved in the entries $d_h(i, j)$ when both the multi-integers i and j belong to $\mathscr{R}_h{}^\mu(\Gamma)$ and in the components $l_h{}^i$ when $i \in \mathscr{R}_h{}^\mu(\Gamma)$.

If the form $a(u, v)$ is $H^k(\Omega)$-elliptic, the matrix A_h is positive-definite and the entries $d_h(i, i)$ are positive.

The matrix A_h is symmetric whenever

$$a_{pq}(x) = a_{qp}(x) \qquad \text{for} \quad |p|, |q| \leq k. \qquad ▲$$

Proof   The proof of Proposition 2-1 is left as an exercise.   ■

PROPOSITION 2-2   If the coefficients $a_{pq}(x)$ are constant and if one of the multi-integers i, j belongs to $\mathscr{R}_{0,h}^\mu(\Omega)$, then

$$(2\text{-}13) \qquad d_h(i, j) = \sum_{|p|, |q| \leq k} (-1)^{|q|} h^{-p-q} a_{pq} D^{p+q} \mu * \tilde{\mu}(i - j),$$

where $\tilde{\mu}(x) = \mu(-x)$. If the data $f(x)$ are constant and if $i \in \mathscr{R}_{0,h}^\mu(\Omega)$, then

$$(2\text{-}14) \qquad\qquad l_h{}^i = f. \qquad ▲$$

Proof   Since the coefficients a_{pq} are constant, and since $i \in \mathscr{R}_{0,h}^\mu(\Omega)$, we obtain

$$d_h(i, j) = h^{-1} \sum_{|p|, |q| \leq k} h^{-p-q} a_{pq} \int_{R^n} (D^p \mu)\left(\frac{x}{h} - j\right)(D^q \mu)\left(\frac{x}{h} - i\right) dx$$

$$= \sum_{|p|, |q| \leq k} h^{-p-q} a_{pq} \int_{R^n} D^p \mu(y) D^q \mu(y + j - i) dy$$

$$= \sum_{|p|, |q| \leq k} (-1)^{|q|} h^{-p-q} a_{pq} D^{p+q} \mu * \tilde{\mu}(i - j).$$

If f is constant and $i \in \mathscr{R}_{0,h}^\mu(\Omega)$, however, we obtain

$$l_h{}^i = h^{-1} f \int_{R^n} \mu\left(\frac{x}{h} - i\right) dx = f. \qquad ■$$

Now let us assume that μ satisfies the criterion of m-convergence

$$(2\text{-}15) \qquad \sum_{k \in Z^n} \frac{k^j}{j!}(x - k) = \sum_{0 \leq p \leq j} b^{j-p} \frac{x^p}{p!} \qquad \text{for} \quad 0 \leq |j| \leq m.$$

This criterion of m-convergence implies the following properties of A_h:

PROPOSITION 2-3 If we assume that the function μ satisfies the criterion of m-convergence, then the entries $d_h(i,j)$ satisfy the following relations:

$$(2\text{-}16) \quad \begin{cases} \sum_{j \in \mathscr{R}_h^\mu(\Omega)} \frac{j^r}{r!} d_h(i,j) = \sum_{|q| \le k, 0 \le s \le r} b^s h^{-(q+r-s)} \\ \qquad \times \int_\Omega \left(\sum_{0 \le p \le r-s} a_{pq}(x) \frac{x^{r-s-p}}{(r-s-p)!} \right) (D^q \mu) \left(\frac{x}{h} - i \right) dx \end{cases}$$

and

$$(2\text{-}17) \quad \begin{cases} \sum_{i,j \in \mathscr{R}_h^\mu(\Omega)} \frac{i^t}{t!} \frac{j^r}{r!} d_h(i,j) = \sum_{\substack{0 \le s \le r \\ 0 \le l \le t}} h^{-(r+t-s-l)} b^s b^l \\ \qquad \times \int_\Omega \sum_{\substack{0 \le p \le r-s \\ 0 \le q \le t-l}} a_{pq}(x) \frac{x^{r+t-s-l-p-q}}{(r-p-s)!(t-q-l)!} dx. \end{cases} \quad \blacktriangle$$

Proof Formula 2-16 follows from

$$(2\text{-}18) \quad \sum_j \frac{j^r}{r!} d_h(i,j) = h^{-1} a \left[\sum_j \frac{j^r}{r!} \mu\left(\frac{x}{h} - j\right), \mu\left(\frac{x}{h} - i\right) \right]$$

and from

$$(2\text{-}19) \quad D^p \left[\sum_j \frac{j^r}{r!} \mu\left(\frac{x}{h} - j\right) \right] = \sum_{0 \le s \le r-p} b^s \frac{x^{r-s-p}}{(r-s-p)!} h^{s-r}.$$

The proof of (2-17) is analogous.   ∎

In particular, for $r = t = 0$, we obtain the following two relations:

$$(2\text{-}20) \quad \begin{cases} (i) \quad \sum_{j \in \mathscr{R}_h^\mu(\Omega)} d_h(i,j) = \sum_{|q| \le k} h^{-q} \int_\Omega a_{0q}(x)(D^q\mu)\left(\frac{x}{h} - i\right) dx \\ (ii) \quad \sum_{i,j \in \mathscr{R}_h^\mu(\Omega)} d_h(i,j) = \int_\Omega a_{00}(x)\, dx. \end{cases} \quad \blacksquare$$

Now let us study the condition number of the matrix A_h.

PROPOSITION 2-4 Let us assume that μ satisfies the stability property and that Ω satisfies the property of μ-stability for a subsequence of h.

If we assume either (2-5) or (2-6), the condition number of the matrix h satisfies

$$(2\text{-}21) \quad \chi(A_h) \le M |h|^{-2k} \quad \text{when} \quad h \text{ ranges over this subsequence.} \quad \blacktriangle$$

Proof Proposition 2-4 follows from Theorem 3.1-7 and S.2-3.   ∎

2-2. Convergence Properties of finite Element Approximations of Neumann Problems

We must assume the following regularity properties (see Section 7.2-4):

$$(2\text{-}22) \begin{cases} \text{(i) the coefficients are } k\text{-times continuously differentiable,} \\ \text{(ii) the subspaces } D, D^*, D_0, \text{ and } D_0^* \text{ are contained in } H^{2k}(\Omega), \\ \text{(iii) the operators } (\gamma, \delta) \text{ and } (\gamma, \delta^*) \text{ map } H^{2k}(\Omega) \text{ onto} \\ \qquad \prod_{0 \le j \le 2k-1} H^{2k-j-\frac{1}{2}}(\Gamma), \\ \text{(iv) } H_0^{2k}(\Omega) = D \cap D_0 = D^* \cap D_0^*. \end{cases}$$

THEOREM 2-1 Let us assume that Ω is a smooth bounded open subset of R^n and that μ satisfies the criterion of m-convergence for $m \ge k$. We also have $f \in L^2(\Omega)$ and $t_j \in H^{k-j-\frac{1}{2}}(\Gamma)$ given $(k \le j \le 2k - 1)$, $u \in H^k(\Omega, \Lambda)$ the solution of the Neumann problem (2-3), and $u_h \in H_h^\mu(\Omega)$ the solution of its internal approximation (2-8).

Let us assume further that either (2-5) or (2-6) holds. Then

$$(2\text{-}23) \qquad \lim_{h \to 0} \| u - p_h u_h \|_{k,\Omega} = 0.$$

Moreover, if $u \in H^s(\Omega)$ for $k \le s \le m + 1$, there exists a constant M such that

$$(2\text{-}24) \begin{cases} \text{(i) } \| u - p_h u_h \|_{k,\Omega} \le M \, |h|^{s-k} \, \| u \|_{s,\Omega} \\ \qquad\qquad\qquad\qquad \text{for } k \le s \le m + 1, \\ \text{(ii) } \| \Lambda(u - p_h u_h) \|_{-t,\Omega} \le M \, |h|^{s+t-2k} \, \| u \|_{s,\Omega} \\ \qquad\qquad\qquad\qquad \text{for } k \le s, t \le m + 1. \end{cases}$$

Finally, if the regularity properties (2-22) hold, there exists a constant M such that

$$(2\text{-}25) \begin{cases} \text{(i) } \| u - p_h u_h \|_{0,\Omega} \le M \, |h|^{s-k+\min(k,m+1-k)} \, \| u \|_{s,\Omega} \\ \qquad\qquad\qquad\qquad \text{for } k \le s \le m + 1, \\ \text{(ii) } \| \gamma_j(u - p_h u_h) \|_{-j-\frac{1}{4},\Gamma} \le M \, |h|^{s-k+\min(k,m+1-k)} \, \| u \|_{s,\Omega} \\ \qquad\qquad\qquad\qquad \text{for } k \le s \le m + 1. \quad \blacktriangle \end{cases}$$

Proof The finite-element approximations are convergent, and therefore (2-23) and (2-24) follow from Theorems 1-1 and 5.2-1 with $V = H^k(\Omega)$, $U = H^s(\Omega)$, and $U_0 = H_0^t(\Omega)$.

Since assumptions (2-22) imply assumptions (1-13) of Theorem 1-3 with $U = H^{2k}(\Omega)$ and $R = \prod_{0 \le j \le k-1} H^{-j-\frac{1}{2}}(\Omega)$, then (2-25) follows from Theorem 1-3: indeed, in this case, Theorem 5.2-1 implies that

$$e_U^V(p_h) \le \begin{cases} c\,|h|^{m+1-k} & \text{if } H^{m+1}(\Omega) \text{ is contained in } H^{2k}(\Omega), \\ c\,|h|^k & \text{if } H^{2k}(\Omega) \text{ is contained in } H^{m+1}(\Omega). \end{cases} \qquad \blacksquare$$

We now study the regularity of convergence.

THEOREM 2-2 Let us assume that μ satisfies the criterion of m-convergence for $m \ge 2k$ and the stability property. Let Ω be a smooth bounded open subset satisfying the property of μ-stability for a subsequence of h. Finally, let us assume that (2-22) and either (2-5) or (2-6) hold. Take $f \in L^2(\Omega)$ as given, u as the solution of the Neumann problem (2-3), and $u_h \in H_h^\mu(\Omega)$ as the solution of its internal approximation (2-8).

If $t_j \in H^{2k-j-\frac{1}{2}}(\Gamma)$, then $u \in H^{2k}(\Omega)$, $\gamma_j u \in H^{2k-j-\frac{1}{2}}(\Gamma)$, and

$$(2\text{-}26) \quad \lim_{h \to 0} \|u - p_h u_h\|_{2k,\Omega} = 0, \qquad \lim_{h \to 0} \|\gamma_j(u - p_h u_h)\|_{2k-j-\frac{1}{2},\Gamma} = 0.$$

If $t_j \in H^{k-j-\frac{1}{2}}(\Gamma)$, then $u \in H^k(\Omega)$, $\gamma_j u \in H^{k-j-\frac{1}{2}}(\Gamma)$, and

$$(2\text{-}27) \quad \lim_{h \to 0} \|u - p_h u_h\|_{k,\Omega} = 0, \qquad \lim_{h \to 0} \|\gamma_j(u - p_h u_h)\|_{k-j-\frac{1}{2},\Gamma} = 0.$$

If $t_j \in H^{-j-\frac{1}{2}}(\Gamma)$, then $u \in L^2(\Gamma)$, $\gamma_j u \in H^{-j-\frac{1}{2}}(\Gamma)$, and

$$(2\text{-}28) \quad \lim_{h \to 0} \|u - p_h u_h\|_{0,\Omega} = 0, \qquad \lim_{h \to 0} \|\gamma_j(u - p_h u_h)\|_{-j-\frac{1}{2},\Gamma} = 0.$$

Furthermore, if $t_j \in H^{2k-j-\frac{1}{2}}(\Gamma)$ and if $n(h)$ is the dimension of $H_h^\mu(\Omega)$, then there exists a constant M such that

$$(2\text{-}29) \qquad \|u - p_h u_h\|_{k,\Omega} \le M E_{2k}^k(n(h) - 1)\,\|u\|_{2k,\Omega},$$

where $E_{2k}^k(n)$ is the n-width of the injection from $H^{2k}(\Omega)$ into $H^k(\Omega)$. $\blacktriangle$

Proof Theorem 2-2 follows obviously from Theorem 1-5 and Theorems 5.2-1 and 5.2-3. $\blacksquare$

Now let us consider the case of $m = 2k - 1$, where we use either the $(2m + 1)^n$-level or the $[2(2m)^n - (2m - 1)^n]$-level piecewise-polynomial approximations of $H^k(\Omega)$. For simplicity, we state the following result only in the case of the $(2m + 1)^n$-level piecewise-polynomial approximations.

THEOREM 2-3 Let us assume that Ω is a smooth bounded open subset satisfying the property (5.2-29) of Section 5.2-4 for $m = 2k$. Let us assume (2-22) and either (2-5) or (2-6). Let $f \in L^2(\Omega)$ and $t_j \in H^{-j-\frac{1}{2}}(\Gamma)$ be given, with $u \in L^2(\Omega)$ the solution of the Neumann problem (2-3) and $u_h \in H_h^{(2k-1)}(\Omega)$ the solution of the discrete variational equation

$$(2\text{-}30) \quad \begin{cases} a(p_h^{(2k-1)}u_h, p_h^{(2k-1)}v_h) = (f, p_h^{(2k)}v_h) + \sum_{0 \leq j \leq k-1} \langle t_{2k-j-1}, \gamma_j p_h^{(2k)}v_h \rangle \\ \qquad\qquad\qquad\qquad\qquad\qquad\qquad \text{for any} \quad v_h \in H_h^{(2k-1)}(\Omega). \end{cases}$$

Then

$$(2\text{-}31) \quad \begin{cases} \text{(i)} \ \lim_{h \to 0} \|u - p_h^{(2k-1)}u_h\|_{0,\Omega} = 0, \\ \text{(ii)} \ \lim \|\gamma_j(u - p_h^{(2k-1)}u_h)\|_{-j-\frac{1}{2},\Gamma} = 0. \end{cases} \quad \blacktriangle$$

Proof Theorem 2-3 follows from Theorem 1-4 and from Corollary 5.2-1, with $U = H^{2k}(\Omega)$, $V = H^k(\Omega)$, $V_h = H_h^{(2k-1)}(\Omega)$, $p_h = p_h^{(2k-1)}$, and $q_h = p_h^{(2k)}$. Then Corollary 5.2-1 implies assumptions (1-18) of Theorem 1-4. ∎

2-3. The $(2m + 1)^n$-Level Approximations of the Neumann Problem

We study the special properties we obtain by using the $(2m + 1)^n$-level piecewise-polynomial approximations of the Sobolev space $H^k(\Omega)$.

Definition 2-2 The discrete variational equation

$$(2\text{-}32) \quad a(p_h^{(m)}u_h, p_h^{(m)}v_h) = l(p_h^{(m)}v_h) \quad \text{for any} \quad v_h \in H_h^{(m)}(\Omega),$$

where $l(v) = (f, v) + \sum_{0 \leq j \leq k-1} \langle t_{2k-j-1}, \gamma_j v \rangle$ is the $(2m + 1)^n$-level approximation of the Neumann problem (2-3). We denote the operational formulation of (2-32) by

$$(2\text{-}33) \quad A_h^{(m)}u_h = l_h^{(m)}. \quad \blacktriangle$$

Let us recall that

$$(2\text{-}34) \quad \pi_{(m+1)}(x) = \sum_{k \leq (m)} \alpha_{(m)}^k(x - k)\theta(x - k),$$

where

$$(2\text{-}35) \quad \begin{cases} \text{(i)} \quad \alpha_{(m)}^k(x) = \sum_{r \leq (m)} a_{(m)}(k, r)\dfrac{x^r}{r!}, \\[2mm] \text{(ii)} \ a_{(m)}(k, r) = a_m(k_1, r_1) \cdots a_m(k_n, r_n) \quad \text{if} \ k, r \in Z^n, \\[2mm] \text{(iii)} \ a_m(k, r) = \displaystyle\sum_{0 \leq i \leq k} (-1)^i \binom{m + 1}{i}\dfrac{(k - i)^{m-r}}{(m - r)!} \quad \text{if} \ k, r \in Z. \end{cases}$$

PROPOSITION 2-5 The matrix $A_h^{(m)}$ has at most $(2m + 1)^n$-levels. The entries $d_h^{(m)}(i, j)$ are equal to

(2-36)
$$
\begin{cases}
d_h^{(m)}(i, j) = \sum_{\substack{|p|,|q| \le k \\ t, l \le (m)}} \sideset{}{'}\sum_{\substack{p \le r \le (m) \\ q \le s \le (m)}} h^{-p-q} a_{(m)}(l, r) a_{(m)}(l + j - i, s) \\
\qquad\qquad \times \int_{\omega_h^{i+j} \cap \Omega} a_{pq}(x) \frac{((x/h) - t - j)^{r+s-p-q}}{(r - p)!(s - q)!} dx,
\end{cases}
$$

where $\omega_h^k = (kh, (k + (1))h)$. The components $l_h^{(m),i}$ of $l_h^{(m)}$ are equal to

(2-37)
$$
\begin{cases}
l_h^{(m),i} = \sum_{k \le (m)} \sum_{r \le (m)} a_{(m)}(k, r) h^{-1} \left[\int_{\Omega \cap \omega_h^{k+j}} f(x) \frac{(x/h - i - k)^r}{r!} dx \right. \\
\qquad\qquad \left. + \sum_{0 \le j \le k-1} \int_{\Gamma \cap \omega_h^{k+j}} t_{2k-j-1}(x) \gamma_j \frac{(x/h - i - k)^r}{r!} d\sigma(x) \right].
\end{cases}
$$

If the coefficients $a_{pq}(x)$ are constant and if one of the multi-integers i, j belongs to $\mathscr{R}_{0,h}^{(m)}(\Omega)$, we obtain

(2-38) $\quad d_h^{(m)}(i, j) = \sum_{|p|,|q| \le k} (-1)^{|q|} h^{-p-q} a_{(2m+1)}((m + 1) + i - j, p + q).$ ▲

Proof Indeed, by using (2-34) and (2-35), we deduce that

(2-39)
$$
\begin{cases}
D^p \pi_{(m+1)} \left(\frac{x}{h} - j \right) = h^{-p} \sum_{k \le (m)} \sum_{p \le r \le (m)} a_{(m)}(k, r) \\
\qquad\qquad \times \frac{(x/h - k - j)^{r-p}}{(r - p)!} \theta_{(k+j)h}(x).
\end{cases}
$$

Therefore,

(2-40)
$$
\begin{cases}
D^p \pi_{(m+1)} \left(\frac{x}{h} - j \right) D^q \pi_{(m+1)} \left(\frac{x}{h} - i \right) \\
\quad = h^{-p-q} \sum_{t \le (m)} \sum_{\substack{p \le r \le (m) \\ q \le s \le (m)}} a_{(m)}(t, r) a_{(m)}(t + j - i, s) \\
\qquad\qquad \times \frac{(x/h - t - j)^{r+s-p-q}}{(r - p)!(s - q)!} \theta_{(t+j)h}(x)
\end{cases}
$$

By multiplying by $a_{pq}(x)$ and integrating on Ω, we deduce (2-36).

Equation 2-37 follows from (2-39) with $p = 0$. Finally, we deduce (2-38) from (2-13), since

$$(2\text{-}41) \quad \begin{cases} D^{p+q}\pi_{m+1} * \tilde{\pi}_{(m+1)}(i-j) = D^{p+q}\pi_{(2m+2)}((m+1)+i-j) \\ \qquad\qquad = a_{(2m+1)}((m+1)+i-j, p+q) \end{cases}$$

(see Corollary 4.2-1). ∎

EXAMPLE 2-1 Let us assume that $n = 2$, $k = 1$, and $m = 1$. Let Λ be a second-order linear differential operator and $(H_h^{(1)}(\Omega), p_h^{(1)}, r_{h,\Omega})$ be the 9-level piecewise-polynomial approximations of multi-degree 1 of $H^1(\Omega)$.

The matrix $A_h^{(1)}$ has 9 levels: if i is a fixed multi-integer of $\mathcal{R}_{0,h}^{(1)}(\Omega)$, the multi-integers j such that $d_h^{(1)}(i,j)$ does not necessarily vanish are as follows:

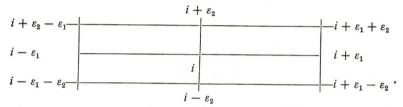

Table 1 presents the values of the coefficients $a_{(3)}(p,j)$ when $n = 2$.

Table 1. Values of $a_{(3)}(p,j)$ when $n = 2$

p \ j	$(0,0)$	$(1,0)$	$(0,1)$	$(-1,0)$	$(0,-1)$	$(1,1)$	$(-1,1)$	$(-1,-1)$	$(1,-1)$
$(0,0)$	$\frac{16}{36}$	$\frac{4}{36}$	$\frac{4}{36}$	$\frac{4}{36}$	$\frac{4}{36}$	$\frac{1}{36}$	$\frac{1}{36}$	$\frac{1}{36}$	$\frac{1}{36}$
$(1,0)$	0	$-\frac{4}{12}$	0	$\frac{4}{12}$	0	$-\frac{1}{12}$	$\frac{1}{12}$	$\frac{1}{12}$	$-\frac{1}{12}$
$(0,1)$	0	0	$-\frac{4}{12}$	0	$\frac{4}{12}$	$-\frac{1}{12}$	$-\frac{1}{12}$	$\frac{1}{12}$	$\frac{1}{12}$
$(2,0)$	$-\frac{8}{6}$	$\frac{4}{6}$	$-\frac{2}{6}$	$\frac{4}{6}$	$-\frac{2}{6}$	$\frac{1}{6}$	$\frac{1}{6}$	$\frac{1}{6}$	$\frac{1}{6}$
$(0,2)$	$-\frac{8}{6}$	$-\frac{2}{6}$	$\frac{4}{6}$	$-\frac{2}{6}$	$\frac{4}{6}$	$\frac{1}{6}$	$\frac{1}{6}$	$\frac{1}{6}$	$\frac{1}{6}$
$(1,1)$	0	0	0	0	0	$\frac{1}{4}$	$-\frac{1}{4}$	$-\frac{1}{4}$	$\frac{1}{4}$

For instance, let us consider the Neumann problem for the operator $A = -\Delta + \lambda$.

We deduce that when $n = 2$ and $i \in \mathcal{R}_{0,h}^{(1)}(\Omega)$, the ith component of the vector $A_h^{(1)}u_h$ is equal to

$$(2\text{-}42) \quad \begin{cases} (A_h^{(1)}u_h)^i = \dfrac{2}{3h^2}\left(8u_h^{\,i} - \sum_{\substack{k \neq 0 \\ k \leq (1)}} u_h^{\,i-k}\right) \\ \qquad + \dfrac{\lambda}{36}\left(16u_h^{\,i} + 4\sum_{\substack{k \neq 0 \\ |k| \leq 1}} u_h^{\,i-k} + \sum_{\substack{k \leq (1) \\ |k| > 1}} u_h^{\,i-k}\right). \end{cases}$$

2-4. The $[2(2m)^n-(2m-1)^n]$-Level Approximations of the Neumann Problem

Now let us use the $[2(2m)^n - (2m - 1)^n]$-level piecewise-polynomial approximations of the Sobolev space $H^k(\Omega)$.

Definition 2-3 The discrete variational equation

$$(2\text{-}43) \qquad a(\bar{p}_h^{(m)}u_h, \bar{p}_h^{(m)}v_h) = (\bar{p}_h^{(m)}v_h) \qquad \text{for any} \quad v_h \in \bar{H}_h^{(m)}(\Omega)$$

is the $[2(2m)^n - (2m - 1)^n]$-level approximation of the Neumann problem (2-3). We denote the operational formulation of (2-43) by

$$(2\text{-}44) \qquad \bar{A}_h^{(m)}u_h = \bar{\rho}_h^{(m)}. \qquad \blacktriangle$$

By writing $\mu_{(m)} = \psi * \pi_{(m)}$ in the form

$$(2\text{-}45) \qquad \mu_{(m)}(x) = \sum_{\sigma} \sum_{k \leq (m)} \alpha_{(m)}^{k,\sigma}(x - k)\theta_k^{\sigma}(x),$$

where σ ranges over the permutations of $(1, \ldots, n)$, θ_k^{σ} denotes the characteristic function of the simplex ω_{σ}^k (see Section 5.1-4), and $\alpha_{(m)}^{k,\sigma}$ are the polynomials

$$(2\text{-}46) \qquad \alpha_{(m)}^{k,\sigma}(x) = \sum_{r \leq (m)} a_{(m)}^{\sigma}(k, r) \frac{x^r}{r!}$$

defined by (2-45).

PROPOSITION 2-6 The matrix $\bar{A}_h^{(m)}$ has at most $2(2m)^n - (2m - 1)^n$ levels. The entries $d_h^{(m)}(i, j)$ are equal to

$$(2\text{-}47) \quad
\begin{cases}
d_h^{(m)}(i, j) = \displaystyle\sum_{\substack{|p| \\ t, l \leq (m) \\ \sigma}} \sum_{\substack{|q| \leq k \ p \leq r \leq (m) \\ q \leq s \leq (m)}} h^{-p-q}a_{(m)}^{\sigma}(l, r)a_{(m)}^{\sigma}(l + j - i, s) \\
\qquad\qquad \times \displaystyle\int_{\omega_{\sigma,h}^{t+j} \cap \Omega} a_{pq}(x)\frac{(x/h - t - j)^{r+s-p-q}}{(r - p)!\,(s - q)!}\,dx.
\end{cases}$$

$\blacktriangle$

Proof The proof is analogous to the proof of Proposition (2-5) and is left as an exercise. $\blacksquare$

EXAMPLE 2-2 Let us assume that $n = 2$, $k = 1$, $m = 1$. Let Λ be a second-order linear differential operator and $(\bar{H}_h^{(1)}, \bar{p}_h^{(1)}, r_{h,\Omega})$ be the 7-level piecewise-polynomial approximations of multidegree 1 of $H^1(\Omega)$.

The matrix $\bar{A}_h^{(1)}$ has 7 levels: if i is a fixed multi-integer of $\bar{\mathcal{R}}_{0,h}^{(1)}(\Omega)$, the multi-integers j such that $d_h^{(1)}(i,j)$ does not necessarily vanish are the following:

For instance, let us consider the case of the Neumann problem for the operator $\Lambda = -\Delta + \lambda$. We deduce that, when $n = 2$ and $i \in \bar{\mathcal{R}}_{0,h}^{(1)}(\Omega)$, the ith component of the vector $\bar{A}_h^{(1)}u_h$ is equal to

$$(2\text{-}48) \quad \bar{A}_h^{(1)}u_h = h^{-2}\left(4u_h{}^i - \sum_{\substack{|k|\leq 1 \\ k\neq 0}} u_h^{i-k}\right) + \frac{\lambda}{12}\left(6u_h{}^i + \sum_{\substack{|k|\leq 1 \\ k\neq 0}} u_h^{i-k} + u_h^{i+(1)} + u_h^{i-(1)}\right)$$

We give in Table 2 the values of the coefficients $D^p(\mu_{(1)}*\tilde{\mu}_{(1)})(j)$ when $n = 2$

Table 2.   Values of $D^p(\mu_{(1)} * \tilde{\mu}_{(1)})(j)$ when $n = 2$

p \ j	$(0,0)$	$(1,0)$	$(0,1)$	$(-1,0)$	$(0,-1)$	$(1,1)$	$(-1,-1)$
$(0,0)$	$\frac{1}{12}$	$\frac{1}{12}$	$\frac{1}{12}$	$\frac{1}{12}$	$\frac{1}{12}$	$\frac{1}{12}$	$\frac{1}{12}$
$(1,0)$	0	$\frac{1}{3}$	$-\frac{1}{6}$	$-\frac{1}{3}$	$\frac{1}{6}$	$\frac{1}{6}$	$-\frac{1}{6}$
$(0,1)$	0	$-\frac{1}{6}$	$\frac{1}{3}$	$\frac{1}{6}$	$-\frac{1}{3}$	$\frac{1}{6}$	$-\frac{1}{6}$
$(2,0)$	-2	1	0	1	0	0	0
$(0,2)$	-2	0	1	0	1	0	0
$(1,1)$	1	$-\frac{1}{2}$	$-\frac{1}{2}$	$-\frac{1}{2}$	$-\frac{1}{2}$	$\frac{1}{2}$	$\frac{1}{2}$

2-5.   Approximations of the Spaces $H^k(\Omega, \Lambda)$ and $H(\Omega, \Lambda)$

Knowing that finite-element approximations $(H_h{}^\mu(\Omega), p_h, r_{h,\Omega})$ are convergent approximations of the Sobolev space $H^{2k}(\Omega)$ when μ satisfies the criterion of $2k$-convergence, we deduce that these approximations are convergent approximations of the spaces $H^k(\Omega, \Lambda) = \{u \in H^k(\Omega)$ such that $\Lambda u \in L^2(\Omega)\}$ and $H(\Omega, \Lambda) = \{u \in L^2(\Omega)$ such that $\Lambda u \in L^2(\Omega)$.

THEOREM 2-4   Let Ω be a smooth bounded open subset of R^n, and assume (2-6) and (2-22). Let $\mu \in H^m(\Omega)$ be a function with compact support satisfying $\int \mu(x)\,dx = 1$ and the criterion of m-convergence for $m \leq 2k$.

Then the finite-element approximations $(H_h^{\mu}(\Omega), p_h, \hat{r}_h)$ [where $\hat{r}_h$ is the optimal restriction associated with p_h in either $H^k(\Omega, \Lambda)$ or $H(\Omega, \Lambda)$] are convergent approximations of the spaces $H^k(\Omega, \Lambda)$ and $H(\Omega, \Lambda)$. Furthermore, there exists a constant c such that

$$(2\text{-}49) \quad \begin{cases} e_U{}^V(p_h) \leq c \, |h|^{s-2k} \text{ when } U = H^s(\Omega) \text{ and } V \text{ is either} \\ \qquad\qquad\qquad\qquad\qquad\qquad\qquad H^k(\Omega, \Lambda) \text{ or } H(\Omega, \Lambda) \end{cases}$$

for $2k \leq s \leq m + 1$. ▲

Proof We prove Theorem 2-4 only when $V = H^k(\Omega, \Lambda)$. Let $\hat{r}_h$ be the optimal restriction associated with p_h in $H^k(\Omega, \Lambda)$. Since $\|p_h \hat{r}_h\|_{L(V,V)} = 1$, the approximations will be convergent if we prove that $\|u - p_h \hat{r}_h u\|_{H^k(\Omega,\Lambda)}$ converges to 0 when u ranges over a dense subspace U (see Theorem 2.1-9). But $U = H^{2k}(\Omega)$ is dense in $H^k(\Omega, \Lambda)$ (by Theorem 6.2-8) and, because of the convergence of the approximations $(H_h^{\mu}(\Omega), p_h, r_{h,\Omega})$ in $H^{2k}(\Omega)$, we deduce from the inequalities

$$\|u - p_h \hat{r}_h u\|_{H^k(\Omega,\Lambda)} \leq \|u - p_h r_{h,\Omega} u\|_{H^k(\Omega,\Lambda)} \leq c \, \|u - p_h r_{h,\Omega} u\|_{2k,\Omega}$$

that $u - p_h \hat{r}_h u$ converges to 0 in $H^k(\Omega, \Lambda)$ when u ranges over $H^{2k}(\Omega)$.

On the other hand, if u belongs to $U = H^s(\Omega)$ with $2k \leq s \leq m + 1$, we obtain

$$\|u - p_h \hat{r}_h u\|_{H^k(\Omega,\Lambda)} \leq c \, \|u - p_h r_{h,\Omega} u\|_{2k,\Omega} \leq c \, |h|^{s-2k} \, \|u\|_{s,\Omega},$$

from which we deduce estimates (2-49). ■

3. APPROXIMATION OF OTHER NEUMANN-TYPE PROBLEMS

3-1. Approximation of the Value of the Solution at a Point of the Boundary

Here we approximate the problem set forth in Section 7.1-6. Let $\Omega \subset R^n$, $n \leq 3$, be a smooth bounded open subset and $a \in \Gamma$ a point of the boundary Γ of Ω. We can now approximate the value $u(a)$ of the solution $u \in H^2(\Omega)$ of the Neumann problem

$$(3\text{-}1) \qquad\qquad \Lambda^* u = f; \qquad \frac{\partial u}{\partial n_{\Lambda^*}} = t$$

when f ranges over $L^2(\Omega)$ and t ranges over $H^{1/2}(\Gamma)$.

Here Λ is the differential operator defined by

$$(3\text{-}2) \qquad\qquad \Lambda u = \sum_{|p|,|q| \leq 1} (-1)^{|q|} D^q(a_{pq}(x) D^p u),$$

where the coefficients $a_{pq}(x)$ are continuously differentiable and satisfy

$$(3\text{-}3) \qquad a(u, v) = \sum_{|p|,|q|\leq 1} \int_\Omega a_{pq}(x)D^p u D^q v \, dx$$

is $H^1(\Omega)$-elliptic.

By Corollary 7.1-6, we know that

$$(3\text{-}4) \qquad u(a) = \int_\Omega k(x)f(x) \, dx + \int_\Gamma k(x)t(x) \, d\sigma(x),$$

where $k(x)$ is the solution of the Neumann problem

$$(3\text{-}5) \qquad \Lambda k = 0; \qquad \frac{\partial k}{\partial n_\Lambda} = \delta(a).$$

Since the Dirac measure $\delta(a)$ at the point $a \in \Gamma$ belongs to $H^{-3/4}(\Gamma)$ when $n \leq 3$, (see Theorem 6.3-8), we can approximate the solution $k \in L^2(\Omega)$ of (3-5) by using Theorem 2-3 (or Theorem 2-2). Here we need only state the case in which k is approximated by the 3^n-level approximate Neumann problem. (We choose $m = 1$.)

THEOREM 3-1 Let a belong to the boundary Γ of a smooth bounded open subset Ω of R^n, with $n \leq 3$. Let Λ be a second-order differential operator satisfying the assumptions of Theorem 7.1-1, and let f belong to $L^2(\Omega)$ and t to $H^{1/2}(\Gamma)$, with u the solution of the Neumann problem (3-1). Finally, let $k_h = (k_h{}^j)_j \in H_h^{(1)}(\Omega)$ be the solution of the 3^n-level approximate Neumann problem

$$(3\text{-}6) \qquad A_h^{(1)}k_h = l_h^{(2)},$$

where the components $l_h^{(2),j}$ of $l_h^{(2)}$ are defined by

$$(3\text{-}7) \qquad l_h^{(2),j} = h^{-1}\pi_{(3)}(ah^{-1} - j); j \in \mathscr{R}_h^{(1)}(\Omega).$$

Then $\int_\Omega f(x)p_h^{(1)}k_h \, dx + \int_\Gamma t(x)p_h^{(1)}k_h \, d\sigma(x)$ converges to $u(a)$:

$$(3\text{-}8) \qquad \begin{cases} \lim_{h\to 0} \left| u(a) - \sum_{j\in\mathscr{R}_h^{(1)}(\Omega)} k_h{}^j \right. \\ \left. \times \left[\int_\Omega f(x)\pi_{(2)}(xh^{-1} - j) \, dx + \int_\Gamma t(x)\pi_{(2)}(xh^{-1} - j) \, d\sigma(x) \right] \right| = 0. \end{cases}$$

▲

Proof By Theorem 2-3, we know that $p_h^{(1)}k_h$ converges to k in $L^2(\Omega)$ and that, for $m = k = 1$ and $j = 0$, $\gamma_0 p_h^{(1)}k_h$ converges to $\gamma_0 k$ in $H^{-1/2}(\Gamma)$. Since f belongs to $L^2(\Omega)$ and t belongs to $H^{1/2}(\Gamma)$, $\int_\Omega f(x)p_h^{(1)}k_h \, dx + \int_\Gamma t(x)p_h^{(1)}k_h \, d\sigma(x)$ converges to $\int_\Omega f(x)k(x) \, dx + \int_\Gamma t(x)k(x) \, d\sigma(x) = u(a)$. ∎

3-2. Approximation of Oblique Boundary-Value Problems

Let us consider an example of an oblique boundary-value problem for a differential operator Λ of order $2k$ defined by $\Lambda u = \sum_{|p|,|q|\leq k} (-1)^{|q|} D^q(a_{pq}(x)D^p u)$ (if $k = 1$, see Section 7.1-6). Let $\alpha_j(x)$ be functions of $L^\infty(\Gamma)$ for $k \leq j \leq 2k - 1$; we approximate the solution u of the following oblique boundary-value problem:

$$(3\text{-}9) \quad \begin{cases} \text{(i)} \quad \Lambda u = \sum_{|p|,|q|\leq k} (-1)^{|q|} D^q(a_{pq}(x)D^p u) = f(x) \quad \text{in} \quad \Omega, \\ \text{(ii)} \quad \delta_{2k-j-1}u + \alpha_{2k-j-1}(x)\gamma_j u = \iota_j(x) \quad \text{in} \quad \Gamma \\ \qquad\qquad\qquad\qquad\qquad\qquad \text{for} \quad k \leq j \leq 2k - 1. \end{cases}$$

It is possible to prove that this boundary-value problem is equivalent to the following variational equation on $H^k(\Omega)$: look for $u \in H^k(\Omega)$ satisfying

$$(3\text{-}10) \quad \begin{cases} a(u, v) + m(\gamma u, \gamma v) = \int_\Omega fv \, dx + \sum_{j=0}^{k-1} \int_\Gamma \iota_{2k-j-1}(x)\gamma_j v \, d\sigma(x) \\ \qquad\qquad\qquad\qquad\qquad\qquad\qquad \text{for any} \quad v \in H^k(\Omega), \end{cases}$$

where

$$(3\text{-}11) \quad \begin{cases} a(u, v) = \sum_{|q|,|q|\leq k} \int_\Omega a_{pq}(x)D^p u D^q v \, dx \\ \qquad m(\gamma u, \gamma v) = \sum_{j=0}^{k-1} \int_\Omega \alpha_{2k-1-j}(x)\gamma_j u \gamma_j v \, d\sigma(x). \end{cases}$$

Let us consider finite-element approximations $(H_h{}^\mu(\Omega), p_h, r_{h,\Omega})$ of the Sobolev space $H^k(\Omega)$ associated with a function μ of $H^m(R^n)$ for $m \geq k$. We approximate the solution u of the oblique problem (3-9) by the internal approximation of the variational equation (3-10): this approximate variational equation on $H_h{}^\mu(\Omega)$ is:

$$(3\text{-}12) \quad \begin{cases} a(p_h u_h, p_h v_h) + m(\gamma p_h u_h, \gamma p_h v_h) = (f, p_h v_h) + \sum_{j=0}^{k-1} \langle \iota_{2k-j-1}, \gamma_j p_h v_h \rangle \\ \qquad\qquad\qquad\qquad\qquad\qquad \text{for any} \quad v_h \in H_h{}^\mu(\Omega), \end{cases}$$

which is equivalent to the operational equation

$$(3\text{-}13) \qquad\qquad\qquad A_h u_h + M_h u_h = l_h.$$

We have already studied the entries of the matrix A_h and the components of the vector l_h (see Propositions 2-1-2-4), so let us now investigate the entries $C_h(i, j)$ of the matrix M_h; they are defined by

$$(3\text{-}14) \quad C_h(i, j) = \sum_{r=0}^{k-1} \int_\Gamma \alpha_{2k-r-1}(x)\gamma_r \mu(xh^{-1} - i)\gamma_r \mu(xh^{-1} - j) \, d\sigma(x).$$

Therefore, these entries vanish when one of the multi-integers i, j belongs to $\mathscr{R}^{\mu}_{0,h}(\Omega)$. On the other hand, the number of levels of matrix M_h is at most equal to the number of levels of μ. Then the properties of the matrix $A_h + M_h$ are analogous to the properties of the matrix A_h (see Propositions 2-1-2-4).

The convergence properties of the internal approximation of an oblique boundary-value problem are analogous to the convergence properties of the internal approximation of the Neumann problem stated in Section 2-2.

THEOREM 3-2 We assume that the coefficients $a_{pq}(x)$ of the operator Λ and the functions $\alpha_j(x)$ $(k \in j \leq 2k - 1)$ belong to $L^{\infty}(\Omega)$ and $L^{\infty}(\Gamma)$, respectively. We assume further that the form $a(u, v) + m(\gamma u, \gamma v)$ is $H^k(\Omega)$-elliptic and that μ satisfies the criterion of m-convergence for $m \geq k$.

Let f belong to $L^2(\Omega)$ and t_j to $H^{k-j-\frac{1}{2}}(\Gamma)$ for $k \leq j \leq 2k - 1$. If u is the solution of the oblique boundary-value problem (3-9) and if u_h is the solution of its internal approximation (3-12), then

$$(3\text{-}15) \qquad \lim_{h \to 0} \|u - p_h u_h\|_{k,\Omega} = 0.$$

Moreover, if u belongs to $H^s(\Omega)$, then

$$(3\text{-}16) \qquad \|u - p_h u_h\|_{k,\Omega} \leq M|h|^{s-k}\|u\|_{s,\Omega} \qquad \text{for} \quad k \leq s \leq m + 1. \qquad \blacktriangle$$

The other theorems of convergence (Theorems 2-2 and 2-3) regarding the Neumann problem can be extended to the oblique problem when we replace assumptions (2-22) by the following assumptions:

$$(3\text{-}17) \quad \begin{cases} \text{(i) the coefficients } a_{pq}(x) \text{ are } k\text{-times continuously differentiable,} \\ \text{(ii) the subspaces } \bar{D}, \bar{D}^*, D_0, D_0^* \text{ are contained in } H^{2k}(\Omega), \\ \text{(iii) the operators } (\gamma, \delta) \text{ and } (\gamma, \delta^*) \text{ map } H^{2k}(\Omega) \text{ onto} \\ \qquad\qquad\qquad\qquad\qquad \prod_{0 \leq j \leq 2k-1} H^{2k-j-\frac{1}{2}}(\Gamma), \\ \text{(vi) } H_0^{2k}(\Omega) = \bar{D} \cap D_0 = \bar{D}^* \cap D_0^*, \end{cases}$$

where we set $\delta = (\delta_k, \ldots, \delta_j, \ldots, \delta_{2k-1})$ and $\delta^* = (\delta_k^*, \ldots, \delta_{2k-1}^*)$ with $\delta_j = \delta_j + \alpha_j(x)\gamma_{2k-1-j}$ and $\delta_j^* = \delta_j^* + \alpha_j(x)\gamma_{2k-j-1}$, $\bar{D} = \ker \delta$ and $\bar{D}^* = \ker \delta^*$.

3-3. Approximation of a Problem with Elliptic Boundary Conditions

For the sake of simplicity in constructing an internal approximation of a boundary-value problem with an elliptic differential boundary condition (see Section 7.1-2), let us take $\Omega = R^2_+$ and $\Gamma = R$. We denote by (x, y)

a point of R_+^2 and by x a point of $\Gamma = R$. By Corollary 7.1-7, we know that the solution u of the boundary-value problem

$$(3\text{-}18) \quad \begin{cases} \text{(i)} & -\Delta u + \lambda u = f(x, y) \quad \text{in} \quad \Omega = R_+^2, \\ \text{(ii)} & -D_y u(x, 0) - \mu D_x^2 u(x, 0) + \nu u(x, 0) = t(x) \quad \text{in} \quad \Gamma = R, \end{cases}$$

is equivalent to the V-elliptic variational equation

$$(3\text{-}19) \quad \begin{cases} \text{(i)} & u \in V, \\ \text{(ii)} & a(u, v) + m(\gamma_0 u, \gamma_0 v) = (f, v) + \langle t, \gamma_0 v \rangle \quad \text{for any} \quad v \in V, \end{cases}$$

where

$$(3\text{-}20) \quad \begin{cases} \text{(i)} \ V = \{u(x, y) \in H^1(R_+^2) \ \text{such that} \ \gamma_0 u(x) = u(x, 0) \in H^1(R)\}, \\[1mm] \text{(ii)} \ a(u, v) = \displaystyle\int_{R_+^2} D_x u D_x v \, dx \, dy + \int_{R_+^2} D_y u D_y v \, dx \, dy \\[2mm] \qquad\qquad\qquad\qquad\qquad\qquad + \lambda \displaystyle\int_{R_+^2} uv \, dx \, dy, \\[2mm] \text{(iii)} \ m(\gamma_0 v, \gamma_0 v) = \mu \displaystyle\int_{R} D_x u(x, 0) D_x v(x, 0) \, dx \\[2mm] \qquad\qquad\qquad\qquad\qquad\quad + \nu \displaystyle\int_{R} u(x, 0) v(x, 0) \, dx. \end{cases}$$

In order to approximate such a variational equation, we need approximations of the space V defined by [(3-20)(i)]. Let us consider the piecewise-multilinear approximations $(H_h^{(1)}(\Omega), p_h^{(1)}, r_{h,\Omega})$ of $H^1(\Omega)$.

PROPOSITION 3-1 The approximations $(H_h^{(1)}(\Omega), p_h^{(1)}, r_{h,\Omega})$ are convergent approximations of the space V defined by [(3-20)(i)]. Furthermore, if u belongs to the space U of functions u of $H^2(\Omega)$ such that $\gamma_0 u$ belongs to $H^2(\Gamma)$ and $D_x D_y u$ belongs to $L^2(\Omega)$, then there exists a restriction $r_{h,\Omega}$ such that

$$(3\text{-}21) \qquad\qquad \| u - p_h^{(1)} r_{h,\Omega} u \|_V \leq M \, |h| \, \| u \|_U \, . \qquad\qquad \blacktriangle$$

We prove Proposition 3-1 at the end of the section.   ■

Since the approximations $(H_h^{(1)}(\Omega), p_h^{(1)} r_{h,\Omega})$ are convergent approximations of the space V, the internal approximation of variational equation (3-19) will be a 9-level approximation of the boundary-value problem (3-18) described by the following variational equation on $H_h^{(1)}(\Omega)$: look for $u_h \in H_h^{(1)}(\Omega)$ satisfying

$$(3\text{-}22) \quad \begin{cases} a(p_h^{(1)} u_h, p_h^{(1)} v_h) + m(\gamma_0 p_h^{(1)} u_h, \gamma_0 p_h^{(1)} v_h) = (f, p_h^{(1)} v_h) + \langle t, \gamma_0 p_h^{(1)} v_h \rangle \\ \qquad\qquad\qquad\qquad\qquad\qquad\qquad\qquad \text{for any} \quad v_h \in H_h^{(1)}(\Omega). \end{cases}$$

Therefore, Theorem 1-1 and Proposition 3-1 imply the Theorem 3-3.

THEOREM 3-3 If we let u and u_h be the solutions of the problems (3-18) and (3-22), then

$$(3\text{-}23) \quad \lim_{h \to 0} \|u - p_h^{(1)}u_h\|_{1,\Omega} = 0 \quad \text{and} \quad \lim_{h \to 0} \|\gamma_0(u - p_h^{(1)}u_h)\|_{1,\Gamma} = 0.$$

Moreover, if u belongs to the space U defined in Proposition 3-1, then

$$(3\text{-}24) \quad \|u - p_{h,\Omega}^{(1)}u_h\|_{1,\Omega}^2 + \|\gamma_0(u - p_{h,\Omega}^{(1)}u_h)\|_{1,\Gamma}^2 \leq M \, |h|^2 \, \|u\|_U^2. \quad \blacktriangle$$

Let us study now the operational equation $B_h^{(1)}u_h = l_h^{(1)}$ equivalent to the variational approximate equation (3-22).

Since $\Omega = R_+^2$, $\mathcal{R}_h^{(1)}(\Omega) = \{(i,j) \in Z^2 \text{ such that } j \geq -1\}$ and $\mathcal{R}_{0,h}^{(1)}(\Omega) = \{(i,j) \in Z^2 \text{ such that } j \geq 0\}$.

Therefore, if $j \geq 0$, the entries $C_h^{(1)}((i,j);(k,p))$ of the matrix $B_h^{(1)}$ defined by $a(p_h^{(1)}u_h, p_h^{(1)}v_h) + m(\gamma_0 p_h^{(1)}u_h, \gamma_0 p_h^{(1)}v_h)$ do not depend on the boundary conditions.

Actually, by Proposition 2-5 and by (2-42), the (i,j)th component of $B_h^{(1)}u_h$ is equal to

$$(3\text{-}25) \quad \begin{cases} (B_h^{(1)}u_h)^{(i,j)} = \dfrac{2}{3h^2} [8u_h^{i,j} - u_h^{i-1,j} - u_h^{i+1,j} - u_h^{i,j-1} - u_h^{i,j+1} \\ \qquad\qquad - u_h^{i+1,j+1} - u_h^{i-1,j-1} - u_h^{i+1,j-1} - u_h^{i-1,j+1}] \\ \qquad + \dfrac{\lambda}{36} [16u_h^{i,j} + 4u_h^{i,j} + 4u_h^{i+1,j} + 4u_h^{i,j-1} + 4u_h^{i,j+1} \\ \qquad\qquad + u_h^{i+1,j+1} + u_h^{i-1,j-1} + u_h^{i+1,j-1} + u_h^{i-1,j+1}] \qquad j \geq 0. \end{cases}$$

The boundary conditions [(3-18)(ii)] are involved in the components $(B_h^{(1)}u_h)^{(i,-1)}$ of the vector $B_h^{(1)}u_h$. Such a component is obtained by choosing in (3-22) the sequence v_h such that $v_h^{(i,-1)} = 1$ and $v_h^{(k,p)} = 0$ for $(k,p) \neq (i,-1)$. We thus deduce that

$$(B_h^{(1)}u_h)^{(i,-1)} = h^{-2} \sum_{\substack{k \in Z, \\ p > -1}} u_h^{k,p}$$

$$\cdot \left\{ \left[\int_R D\pi_2(xh^{-1} - k)D\pi_2(xh^{-1} - i)\, dx \int_R \pi_2(yh^{-1} - p)\pi_2(yh^{-1} + 1)\, dy \right.\right.$$

$$+ \int_R \pi_2(xh^{-1} - k)\pi_2(xh^{-1} - i)\, dx \int_R D\pi_2(yh^{-1} - p)D\pi_2(yh^{-1} + 1)\, dy$$

$$(3\text{-}26) \quad + \lambda \int_R \pi_2(xh^{-1} - k)\pi_2(xh^{-1} - i)\, dx \int_R \pi_2(yh^{-1} - p)\pi_2(yh^{-1} + 1)\, dy$$

$$+ \mu \int_R D\pi_2(xh^{-1} - k)D\pi_2(xh^{-1} - i)\, dx \, \pi_2(-p)\pi_2(1)$$

$$\left.+ \nu \int_R \pi_2(xh^{-1} - k)\pi_2(xh^{-1} - i)\, dx \, \pi_2(-p)\pi_2(1) \right\}.$$

Now we notice that

$$\int_R \pi_2(yh^{-1} - p)\pi_2(yh^{-1} + 1)\, dy = \begin{cases} \dfrac{h^2}{3} & \text{if } p = -1, \\ \dfrac{h^2}{6} & \text{if } p = 0, \\ 0 & \text{if } p > 1, \end{cases}$$

$$\int_R D\pi_2(yh^{-1} - p)D\pi_2(yh^{-1} + 1)\, dy = \begin{cases} 1 & \text{if } p = -1, \\ -1 & \text{if } p = 0, \\ 0 & \text{if } p > 1, \end{cases}$$

and that

$$\pi_2(k) = \begin{cases} 1 & \text{if } k = 1, \\ 0 & \text{if } k \neq 1. \end{cases}$$

Therefore, we deduce that the component $(B_h^{(1)}u_h)^{(i,-1)}$ of $B_h^{(1)}u_h$ is equal to

$$(B_h^{(1)}u_h)^{(i,-1)} = \left(\frac{1}{6h^2}\right)(8u_h^{i,-1} - u_h^{i-1,-1} - u_h^{i+1,-1} - 2u_h^{i+1,0} - 2u_h^{i,0} - 2u_h^{i-1,0})$$

$$(3\text{-}27) \quad + \frac{\lambda}{36}(8u_h^{i,-1} + 2u_h^{i-1,-1} + 2u_h^{i+1,-1} + u_h^{i+1,0} + 4u_h^{i,0} + u_h^{i-1,0})$$

$$+ \frac{\mu}{h^3}(-u_h^{i-1,-1} + 2u_h^{i,-1} - u_h^{i+1,-1}) + \frac{\nu}{6h}(4u_h^{i,-1} + u_h^{i-1,-1} + u_h^{i+1,-1}).$$

On the other hand, the component $l_h^{(i,-1)}$ of the approximate data is equal to

$$(3\text{-}28) \quad \begin{cases} l_h^{(1,-1)} = \dfrac{1}{h^2}\int_{-\infty}^{+\infty} dx \int_0^h f(x, y)\pi_2(xh^{-1} - i)(1 - yh^{-1})\, dy \\ \qquad\qquad + \dfrac{1}{h^2}\int t(x)\pi_2(xh^{-1} - i)\, dx. \end{cases}$$

In other words, the elliptic differential boundary-value conditions are involved in the equations $(B_h^{(1)}u_h)^{(j,-1)} = l_h^{(i,-1)}$ for all the values $i \in Z$.

Proof of Proposition 3-1 To begin with, we must prove that $\gamma_0 p_h^{(1)}u_h$ belongs to $H^1(\Gamma) = H^1(R)$. Since $(\gamma_0 u)(x) = u(x, 0)$, we see that

$$(3\text{-}29) \quad \gamma_0 p_h^{(1)}u_h = \sum_{Z \in i} u_h^{i,-1}\pi_2(xh^{-1} - i) = p_{h,\Gamma}^1\gamma_0 u_h,$$

where we set

(3-30)
$$\begin{cases} \text{(i)} \ (\gamma_0 u_h)^i = u_h^{i,-1}, \\ \text{(ii)} \ p_{h,\Gamma}^1 u_h = \sum_{i \in Z} u_h^i \pi_2(xh^{-1} - i). \end{cases}$$

Therefore, since $p_{h,\Gamma}^1$ is the prolongation of the piecewise-linear approximation (of degree 1) of $H^1(\Gamma) = H^1(R)$, $\gamma_0 p_h^{(1)} u_h = p_{h,\Gamma}^1 \gamma_0 u_h$ belongs to $H^1(\Gamma)$.

Now, we prove that there exists a restriction $r_{h,\Omega}$ such that $\gamma_0 p_h^{(1)} r_{h,\Omega} u$ converges to $\gamma_0 u$ in $H^1(\Gamma)$. Let λ be a function with compact support such that

(3-31)
$$\int \lambda(t)\,dt = 1 \qquad \text{and} \qquad \int t\lambda(t)\,dt = 0,$$

and let $\omega \in L(H^2(R_+^2), H^2(R^2))$ be the operator of extension defined by

(3-32)
$$\omega u(x, y) = \begin{cases} u(x, y) & \text{if } y \geq 0, \\ 3u(x, -y) - 2u(x, 2y) & \text{if } y \leq 0. \end{cases}$$

(See Section 6.3-5.)

Let us define $r_{h,\Omega}$ by

(3-33)
$$(r_{h,\Omega}u)^{i,j} = \iint \omega u(x, y)\lambda(xh^{-1} - i)\lambda(yh^{-1} - j)\,dx\,dy.$$

Then we have

(3-34)
$$\gamma_0 r_{h,\Omega} u = r_{h,\Gamma}\gamma_{0h} u,$$

where $r_{h,\Gamma}$ is a restriction of a piecewise-polynomial approximation; that is, $r_{h,\Gamma}$ is defined by

(3-35)
$$(r_{h,\Gamma}u)^i = \int u(x)\lambda(xh^{-1} - i)\,dx,$$

and where $\gamma_{0h} \in L(H^2(R_+^2), H^1(\Gamma))$ is defined by

(3-36)
$$(\gamma_{0h}u)(x) = h^{-1}\int \omega a(x, y)\lambda(yh^{-1} + 1)\,dy.$$

It is clear that the norms of γ_{0h} in $L(H^1(R_+^2), H^1(\Gamma))$ are bounded and that $\gamma_{0h}u$ converges to $\gamma_0 u$ in $H^1(\Gamma)$. Therefore $\gamma_0 p_h^{(1)} r_{h,\Omega} u = p_{h,\Gamma}^1 r_{h,\Gamma}\gamma_{0h} u$ converges to $\gamma_0 u$ in $H^1(\Gamma)$ since, by Theorem 4.2-2, $p_{h,\Gamma}^1 r_{h,\Gamma} v$ converges to v in $H^1(\Gamma)$. Since $p_h^{(1)} r_{h,\Omega} u$ converges to u in $H^1(\Omega)$, the approximations of V are convergent.

On the other hand, we can write

(3-37)
$$\gamma_0 p_h^{(1)} r_{h,\Omega} u - \gamma_0 u = p_{h,\Gamma}^1 r_{h,\Gamma}\gamma_0 u - \gamma_0 u + p_{h,\Gamma}^1 r_{h,\Gamma}(\gamma_{0h} u - \gamma_0 u),$$

By (3-31) and Theorem 4.3-3, we know that

$$(3\text{-}38) \qquad \| p_{h,\Gamma}^1 r_{h,\Gamma} \gamma_0 u - \gamma_0 u \|_{1,\Gamma} \le ch \, \| \gamma_0 u \|_{2,\Gamma}.$$

Let us prove that

$$(3\text{-}39) \qquad |D_x(\gamma_0 u - \gamma_{0h} u)|_{L^2(\Gamma)} \le ch \, |D_x D_y^2 u|_{L^2(R_+^2)}.$$

Indeed, expanding $\omega u(x, y)$ by the Taylor formula with respect to y and using (3-31), we deduce that

$$(3\text{-}40) \quad D_x(\gamma_0 u - \gamma_{0h} u) = \frac{1}{h} \int dy \int_0^y (y - t) D_y^2 D_x \omega u(x, t) \lambda(y h^{-1} + 1) \, dt.$$

Inequality (3-39) follows from (3-40) if we apply the Cauchy-Schwarz inequality several times. In the same way, we can prove that

$$(3\text{-}41) \qquad |\gamma_0 u - \gamma_{0h} u|_{L^2(\Gamma)} \le ch \, |D_y^2 u|_{L^2(R_+^2)}$$

Finally, we know that

$$(3\text{-}42) \qquad \| u - p_h^{(1)} r_{h,\Omega} u \|_{H^1(R_+^2)} \le ch \, \| u \|_{H^2(R_+^2)}.$$

Therefore, since U is the space of functions u of $H^2(R_+^2)$ such that $D_y^2 D_x u$ belongs to $L^2(R_+^2)$ and $\gamma_0 u$ belongs to $H^2(\Gamma)$, we deduce estimate (3-31) from the foregoing inequalities. ∎

3-4. Approximation of Interface Problems

In considering an interface problem, recall Section 7.1-7 and assume that Ω is halved in two open subsets Ω_1 and Ω_2 by a manifold Σ. We look for $u^1 \in H^1(\Omega_1)$ and $u^2 \in H^1(\Omega_2)$ satisfying

$$(3\text{-}43) \quad \begin{cases} \text{(i)} \quad \Lambda^k \mu^k = f^k & \text{on} \quad \Omega_k \ (k = 1, 2), \\[2mm] \text{(ii)} \quad \dfrac{\partial u^k}{\partial n_{\Lambda^k}} = 0 & \text{on} \quad \Gamma_k - \Sigma \ (k = 1, 2), \\[2mm] \text{(iii)} \quad u^1 = u^2 & \text{on} \quad \Sigma \ \text{ and } \ \dfrac{\partial u^1}{\partial n_{\Lambda^1}} + \dfrac{\partial u^2}{\partial n_{\Lambda^2}} = 0 \quad \text{on} \quad \Sigma, \end{cases}$$

where Λ^k is a second-order linear differential operator

$$(3\text{-}44) \qquad \Lambda^k u^k = \sum_{|p|, |q| \le 1} (-1)^{|q|} D^q(a_{pq}^{\ k}(x) D^p u).$$

By Corollary 1-5, we know that this problem is equivalent to the following variational problem, in which we view the Sobolev space $H^1(\Omega)$ as the closed subspace W of the space $V = H^1(\Omega_1) \times H^1(\Omega_2)$ consisting of the pairs

$u = (u^1, u^2)$ such that $u^1 = u^2$ on Σ. If we set

$$(3\text{-}45) \qquad a^k(u^k, v^k) = \sum_{|p|,|q| \leq 1} \int_{\Omega_k} a_{pq}{}^k(x) D^p u D^q v \, dx,$$

then the solution $u = (u^1, u^2)$ of (3-43) is equivalent to the solution u of

$$(3\text{-}46) \quad \begin{cases} \text{(i)} \qquad u = (u^1, u^2) \in W, \\[2mm] \text{(ii)} \;\; a(u, v) = \sum_{k=1}^{2} a^k(u^k, v^k) = \sum_{k=1}^{2} (f^k, v^k) = l(v) \qquad \text{for any} \;\; v \in W. \end{cases}$$

Now let us assume that the bilinear forms $a^k(u^k, v^k)$ are $H^1(\Omega_k)$-elliptic ($k = 1, 2$). Then $a(u, v)$ is V-elliptic and there exists a unique solution $u = (u^1, u^2)$ of the equivalent problems (3-43) and (3-46).

In order to approximate the solution u of these problems, we use finite-element approximations $(H_h(\Omega), p_h, r_{h,\Omega})$ (associated with $\mu \in H^m(R^n)$, $m \geq 1$) of the space $W = H^1(\Omega)$ and construct the internal approximation of the variational equation (3-46).

Let us set

$$p_{h,\Omega_k} u_h = p_h{}^\mu u_h |_{\Omega_k}.$$

Then the approximate variational equation is defined by

$$(3\text{-}47) \quad \begin{cases} \text{(i)} \;\; u_h \in H_h{}^\mu(\Omega), \\[2mm] \text{(ii)} \;\; \sum_{k=1}^{2} a^k(p_{h,\Omega_k} u_h, p_{h,\Omega_k} v_h) = \sum_{k=1}^{2} (f^k, p_{h,\Omega_k} v_h) \qquad \text{for any} \;\; v_h \in H_h{}^\mu(\Omega). \end{cases}$$

Theorem 3-4 follows from Theorem 1-1 (where V is replaced by W) and Theorem 5.2-1 of Section 5.2-2.

THEOREM 3-4 Let us assume that the forms $a^k(u^k, v^k)$ are $H^1(\Omega_k)$-elliptic and that μ satisfies the criterion of m-convergence for $m \geq 1$. Then, if $u = (u^1, u^2)$ is the solution of the interface problem (3-43) and u_h is the solution of (3-47), we obtain

$$(3\text{-}48) \qquad \lim_{h \to 0} \| u^k - p_{h,\Omega_k} u_h \|_{1,\Omega_k} = 0 \qquad \text{for} \;\; k = 1, 2.$$

Moreover, if u belongs to $H^{m+1}(\Omega)$, the following estimate holds:

$$(3\text{-}49) \qquad \| u^k - p_{h,\Omega_k} u_h \|_{1,\Omega_k} \leq C |h|^m \|u\|_{m+1,\Omega} \qquad \text{for} \;\; k = 1, 2. \qquad \blacktriangle$$

Now let us study the operational equation equivalent to the approximate variational equation (3-46); we denote it by

$$(3\text{-}50) \qquad A_h u_h = l_h,$$

and here the entries $d_h(i, j)$ of the matrix A_h are defined by

$$(3\text{-}51) \quad d_h(i, j) = \sum_{k=1}^{2} \sum_{|p|, |q| \leq 1} \int_{\Omega_k} a_{pq}^{\ k}(x) D^p \mu(xh^{-1} - j) D^q \mu(xh^{-1} - i) \, dx.$$

Then the number of levels of matrix A_h is at most equal to the number of levels of μ. When one of the indices i, j belongs to $\mathscr{R}_{0,h}^{\mu}(\Omega_k)$, then the entry $d_h(i, j)$ does not involve the boundary conditions or the interface conditions.

Let us denote by $\mathscr{R}_h^{\mu}(\Sigma)$ the set of multi-integers j such that

$$(3\text{-}52) \qquad \text{support } \mu(xh^{-1} - j)) \cap \Omega_k \neq \varnothing \text{ for } k = 1 \text{ and } 2.$$

Then the interface conditions are involved in the entries $d_h(i, j)$ if both i and j belong to $\mathscr{R}_h^{\mu}(\Sigma)$.

The matrix A_h satisfies the other properties stated in Section 2-1.

3-5. Approximation of the Neumann Problem for $\Delta^2 + \gamma$

By Theorem 7.2-2 we know that the solution $u \in H(\Omega, \Delta, \Delta^2)$ of the Neumann problem

$$(3\text{-}53) \quad \begin{cases} \text{(i) } \Delta^2 u + \lambda u = f, \\[2mm] \text{(ii) } \gamma_0 \Delta u = \Delta u|_\Gamma = t_0, \\[2mm] \text{(iii) } \gamma_1 \Delta u = \dfrac{\partial \Delta u}{\partial n} = t_1 \end{cases}$$

is the solution of the variational equation

$$(3\text{-}54) \quad \begin{cases} \text{(i) } u \in H(\Omega, \Delta), \\[2mm] \text{(ii) } a(u, v) = \displaystyle\int_\Omega \Delta u \, \Delta v \, dx + \lambda \int_\Omega uv \, dx = l(v) = (f, v) + \langle t_1, \gamma_0 v \rangle \\[2mm] \qquad\qquad\qquad\qquad\qquad - \langle t_0, \gamma_1 v \rangle \quad \text{for any } v \in H(\Omega, \Delta), \end{cases}$$

and conversely, whenever $f \in L^2(\Omega)$, $t_0 \in H^{3/2}(\Gamma)$, and $t_1 \in H^{1/2}(\Gamma)$.

Let us introduce finite-element approximations $(H_h^{\mu}(\Omega), p_h, r_h)$ of $H(\Omega, \Lambda)$ associated with a function $\mu \in H^m(\Omega)$ for $m \geq 2k$ (see Section 2-5), so that the internal approximation of the variational formulation of the Neumann problem (3-53) is the following discrete variational equation.

$$(3\text{-}55) \quad \begin{cases} \text{(i) } u_h \in H_h^{\mu}(\Omega), \\[2mm] \text{(ii) } a(p_h u_h, p_h v_h) = l(p_h v_h) \quad \text{for any } v_h \in H_h^{\mu}(\Omega). \end{cases}$$

We deduce Theorem 3-5 from Theorems 1-1 and 2-4.

THEOREM 3-5 Let us assume that $\Omega \subset R^n$ is a smooth bounded subset and that μ satisfies the criterion of m-convergence for $m \geq 2k$. If λ is not an eigenvalue of the Neumann problem for $\Delta^2 + \lambda$, there exist unique solutions $u \in H(\Omega, \Delta, \Delta^2)$ and $u_h \in H_h^{(m)}(\Omega)$ of (3-53) and (3-55) (for h small enough), and

$$(3\text{-}56) \quad \begin{cases} \text{(i)} \ \lim_{h \to 0} \|u - p_h u_h\|_{H(\Omega, \Delta)} = 0, \\ \text{(ii)} \ \|(\Delta^2 + \lambda)(u - p_h u_h)\|_{H^{-m-1}(\Omega)} \leq c \, |h|^{m-1} \, \|u\|_{H(\Omega, \Delta)} . \end{cases}$$

Furthermore, if the solution u belongs to $H^{m+1}(\Omega)$, we deduce that

$$(3\text{-}57) \quad \begin{cases} \text{(i)} \ \|u - p_h u_h\|_{H(\Omega, \Delta)} \leq c \, |h|^{m-1} \, \|u\|_{H^{m+1}(\Omega)}, \\ \text{(ii)} \ \|(\Delta^2 + \lambda)(u - p_h u_h)\|_{H^{-m-1}(\Omega)} \leq c \, |h|^{2(m-1)} \, \|u\|_{H^{m+1}(\Omega)} . \end{cases} \quad \blacktriangle$$

Proof The proof of Theorem 3-5 is left as an exercise. ∎

CHAPTER 9

Perturbed Approximations and Least-Squares Approximations

This chapter is devoted to the study of methods of approximations of boundary-value problems which are no longer of the Neumann type: the usual variational formulation of such boundary-value problems leads to variational equations on closed subspaces of Sobolev spaces defined by homogeneous boundary conditions.

We avoid the use of approximations of these closed subspaces in two ways. First, we use a perturbation method for approximating such a boundary-value problem by a Neumann-type boundary-value problem and we take the internal approximation of this Neumann-type problem. Second, we use a least-squares method.

1. PERTURBED APPROXIMATIONS

Let us consider the solution $u \in V(\Lambda)$ of the Dirichlet problem

> (i) $\Lambda u = f$ where f is given in H,
>
> (ii) $\gamma u = t$ where t is given in T.

Let ε be a parameter going to 0, so that u becomes the limit of the solutions u_ε of the variational equation on V:

$$\begin{cases} \text{(i) } u_\varepsilon \in V, \\ \text{(ii) } a(u_\varepsilon, v) + \varepsilon^{-1}\langle \gamma u_\varepsilon, \gamma v \rangle = (f, v) + \varepsilon^{-1}\langle t, \gamma v \rangle \quad \text{for any } v \in V. \end{cases}$$

Then, if (V_h, p_h, r_h) are approximations of V, we approximate u by the solution u_h of the discrete variational equation

$$a(p_h u_h, p_h v_h) + \varepsilon(h)^{-1} \langle \gamma p_h u_h, \gamma p_h v_h \rangle = (f, p_h v_h) + \varepsilon(h)^{-1} \langle t, \gamma p_h v_h \rangle$$
$$\text{for any } v_h \in V_h$$

and prove convergence theorems and error estimates from a convenient choice of $\varepsilon(h)$.

1-1. Internal Approximation of a Variational Boundary-Value Problem

In this chapter we approximate a (σ_1, m)-boundary-value problem, when σ_1 is a nonzero projector (see Section 6.2-5). In Chapter 8 we studied the particular case where $\sigma_1 = 0$ (Neumann-type problems), and it is now appropriate to recall the definition of a (σ, m) problem.

We introduce the following items:

1. Spaces V, H, and T and $\gamma \in L(V, T)$ satisfying

$$
(1\text{-}1) \quad
\begin{cases}
\text{(i) } \gamma \text{ is a linear continuous operator from } V \text{ onto } T, \\
\text{(ii) } V \text{ is contained in } H \text{ with a stronger topology}, \\
\text{(iii) the kernel } V_0 = \ker \gamma \text{ of } \gamma \text{ is dense in } H.
\end{cases}
$$

2. A continuous bilinear form $a(u, v)$ on $V \times V$, its associated formal operator Λ, and its Neumann operator δ.

3. A continuous bilinear form $m(t, s)$ on $T \times T$, its associated operator $M \in L(T, T')$.

4. A continuous projector σ_1 of T.

We set $\sigma_2 = 1 - \sigma_1$ and, for $j = 1, 2$,

$$
T_j = \sigma_j T; \; T'_j = \sigma'_j T', \; \gamma_j = \sigma_j \gamma; \; \delta_j = \sigma'_j \delta; \; \mu_j = \sigma_j \mu,
$$

where μ is a right inverse of γ.

With the following data—$f \in H$, $t_1 \in T_1$, and $t_2 \in T'_2$—and using Theorem 6.2-2, we know that the (σ_1, m)-boundary-value problem in which we look for $u \in V(\Lambda)$ satisfying

$$
(1\text{-}2) \quad
\begin{cases}
\text{(i) } \Lambda u = f, \\
\text{(ii) } \gamma_1 u = t_1, \\
\text{(iii) } \delta_2 u + \sigma'_2 M \gamma u = t_2
\end{cases}
$$

is equivalent to the variational equation in which we look for $w \in W = \ker \gamma_1$ satisfying

$$
(1\text{-}3) \quad
\begin{cases}
a(w, v) + m(\gamma w, \gamma v) = (f, v) + \langle t_2, \gamma_2 v \rangle \\
\qquad\qquad - a(\mu_1 t_1, v) - m(\gamma \mu_1 t_1, \gamma v) \quad \text{for any } v \in W,
\end{cases}
$$

where u and w are related by $u = w + \mu_1 t_1$.

Therefore, we are able to approximate the solution w of (1-3) by the solution of its internal approximate equation (see Section 3.1-7) if we can solve the two following problems:

- construct an element $\mu_1 t_1$ of V such that $\gamma_1 \mu_1 t_1 = t_1$
- construct convergent approximations (W_h, p_h, r_h) of the closed subspace $W = \ker \gamma_1$ of V.

The internal approximation is the following discrete variational equation on W_h: look for $w_h \in W_h$ satisfying

$$(1\text{-}4) \quad \begin{cases} a(p_h w_h, p_h v_h) + m(\gamma p_h w_h, \gamma p_h v_h) = (f, p_h v_h) + \langle t_2, \gamma_2 p_h v_h \rangle \\ \qquad\qquad -a(\mu_1 t_1, p_h v_h) - m(\gamma \mu_1 t_1, \gamma p_h v_h) \quad \text{for any} \ v_h \in W_h. \end{cases}$$

In this case, we can use Theorem 3.1-6. The answer to the first question is obvious only when $t_1 = 0$ or when t_1 is defined as the trace $\gamma_1 u$ of a given element u (see Section 10.2-1 in the case of the Dirichlet problem). On the other hand, we cannot hope to construct convergent approximations of any closed subspace W of V containing V_0. Finally, in order to solve the problem of the regularity of the convergence, we have to construct quasi-optimal approximations of W.

In order to avoid these difficulties, we first replace the (σ_1, m)-problem by a $(0, m_\varepsilon)$-problem (i.e., by an oblique problem depending on a parameter ε). We know how to approximate such oblique problems (see Chapter 8, Section 3-2).

This method is useful when we only need to know approximations of the space V (and knowledge of approximations of all the closed subspaces W of V is unnecessary) and where we can approximate nonhomogeneous problems without constructing a solution u_1 of $\gamma_1 u_1 = t_1$.

1-2. Perturbed Approximation of a Variational Boundary-Value Problem

It is possible to assume that the form $m(t, s)$ is equal to 0: we replace $a(u, v)$ by $\tilde{a}(u, v) = a(u, v) + m(\gamma u, \gamma v)$; the associated formal operator Λ remains the same, and the Neumann operator $\tilde{\delta}$ is equal to $\delta + \gamma' M$.

Therefore, in approximating the solution u of the σ_1-problem, we look for $u \in V(\Lambda)$ such that

$$(1\text{-}5) \quad \begin{cases} \text{(i)} \ \Lambda u = f \in H, \\ \text{(ii)} \ \gamma_1 u = t_1 \in T_1, \\ \text{(iii)} \ \delta_2 u = t_2 \in T_2' \end{cases}$$

Although we assume that $a(u, v)$ is continuous and V-elliptic:

$$(1\text{-}6) \quad \begin{cases} |a(u, v)| \leq M \, \|u\| \, \|v\| \quad \text{and} \quad a(v, v) \geq c \, \|v\|^2 \\ \qquad\qquad\qquad\qquad\qquad\qquad\qquad\qquad \text{for any} \ u, v \in V, \end{cases}$$

we also assume that there exist a space W and a pivot space Q (i.e., identified with its dual) such that

$$(1\text{-}7) \quad \begin{cases} \text{(i)} \; T \subset Q; \text{ the injection is continuous and dense} \\ \text{(ii)} \; V \subset W; \text{ the injection is continuous and dense} \\ \text{(iii)} \; \gamma_1 \text{ is a linear continuous operator from } W \text{ into } Q. \end{cases}$$

(We do not require that γ_1 map W onto Q.)

Finally, we introduce convergent approximations (V_h, p_h, r_h) of the space V. Let $\varepsilon(h)$ be a function converging to 0 with h. We approximate the solution u of (1-5) by the solution u_h of the following discrete variational equation:

$$(1\text{-}8) \quad \begin{cases} a(p_h u_h, p_h v_h) + \varepsilon(h)^{-1} \langle \gamma_1 p_h u_h - t_1, \gamma_1 p_h v_h \rangle \\ \qquad\qquad = (f, p_h v_h) + \langle t_2, \gamma_2 p_h v_h \rangle \qquad \text{for all} \quad v_h \in V_h. \end{cases}$$

Definition 1-1   The discrete variational equation (1-8) is the perturbed (internal) approximation of the boundary-value problem (1-5.)   ▲

1-3.   Convergence in the Initial Space

THEOREM 1-1   Let us assume (1-1), (1-6), and (1-7). Let us assume moreover

$$(1\text{-}9) \quad \begin{cases} \text{(i)} \; \lim_{h \to 0} \| u - p_h r_h u \|_V = 0, \\[2mm] \text{(ii)} \; \lim_{h \to 0} e_V^{\,W}(p_h r_h) = 0, \\[2mm] \text{(iii)} \; \lim_{h \to 0} \dfrac{e_V^{\,W}(p_h r_h)}{\sqrt{\varepsilon(h)}} = 0. \end{cases}$$

Let u and u_h be the solutions of (1-5) and (1-8), respectively. Then

$$(1\text{-}10) \quad \begin{cases} \text{(i)} \; \lim_{h \to 0} \| u - p_h u_h \|_V = 0, \\[1mm] \text{(ii)} \; \| \gamma_1 (u - p_h u_h) \|_{T_1'} \le M \varepsilon(h), \\[1mm] \text{(iii)} \; \varepsilon(h)^{-1} \gamma_1 (u - p_h u_h) \text{ converges weakly to } \delta_1 u \text{ in } T_1'. \end{cases}$$

Finally, let $U_0 \subset V_0$ be a dense subspace of V_0. Then the following error estimate holds:

$$(1\text{-}11) \quad \begin{cases} \| \Lambda(u - p_h u_h) \|_{U_0'} \le M e_{U_0}^{\,V}(p_h r_h)(\| u - p_h r_h u \|_V \\ \qquad\qquad\qquad + \| \delta_1 u + \varepsilon(h)^{-1} \gamma_1 (u - p_h r_h u) \|_{T'}). \end{cases}$$

▲

Proof We subtract $\varepsilon(h)^{-1}\langle\gamma_1 p_h r_h u, \gamma_1 p_h v_h\rangle + a(p_h r_h u, p_h v_h)$ from the two sides of (1-8). Since $t_1 = \gamma_1 u$, we obtain

$$(1\text{-}12)\quad\begin{cases} a(p_h u_h - p_h r_h u, p_h v_h) + \varepsilon(h)^{-1}\langle\gamma_1 p_h(u_h - r_h u), \gamma_1 p_h v_h\rangle = (f, p_h v_h) \\ \quad - a(p_h r_h u, p_h v_h) + \varepsilon(h)^{-1}\langle\gamma_1(u - p_h r_h u), \gamma_1 p_h v_h\rangle + \langle t_2, \gamma_2 p_h v_h\rangle. \end{cases}$$

Now we use the Green formula (see Theorem 6.2-1)

$$(f, p_h v_h) = a(u, p_h v_h) - \langle\delta_1 u, \gamma_1 p_h v_h\rangle - \langle\delta_2 u, \gamma_2 p_h v_h\rangle.$$

Since $\delta_2 u = t_2$, we deduce the relation

$$(1\text{-}13)\quad\begin{cases} a(p_h u_h - p_h r_h u, p_h v_h) + \varepsilon(h)^{-1}\langle\gamma_1 p_h(u_h - r_h u), \gamma_1 p_h v_h\rangle \\ = a(u - p_h r_h u, p_h v_h) + \varepsilon(h)^{-1}\langle\gamma_1(u - p_h r_h u), \gamma_1 p_h v_h\rangle - \langle\delta_1 u, \gamma_1 p_h v_h\rangle. \end{cases}$$

Let us choose $v_h = \varphi_h = u_h - r_h u$. Thus we deduce from (1-13) the V-ellipticity and the continuity of $a(u, v)$ the following inequality

$$(1\text{-}14)\quad\begin{cases} c\,\|p_h\varphi_h\|^2 + \varepsilon(h)^{-1}\,\|\gamma_1 p_h\varphi_h\|_Q^2 \leq M[\|u - p_h r_h u\|\,\|p_h\varphi_h\| \\ \quad + \varepsilon(h)^{-1/4}\,\|u - p_h r_h u\|_W\,\varepsilon(h)^{-1/4}\,\|\gamma_1 p_h\varphi_h\|_T + \|\delta_1 u\|_{T'}\,\|p_h\varphi_h\|_V]. \end{cases}$$

Now we apply the inequality $ab \leq \epsilon a^2 + (1/4\epsilon)b^2$ to the terms of the right-hand side of (1-14). This implies that

$$(1\text{-}15)\quad\begin{cases} \|p_h\varphi_h\|^2 + \varepsilon(h)^{-1}\,\|\gamma_1 p_h\varphi_h\|_Q^2 < M(\|u - p_h r_h u\|_V^2 \\ \quad + \varepsilon(h)^{-1}e_V{}^W(p_h r_h)^2\,\|u\|_V^2 + \|\delta_1 u\|_{T'}{}^2). \end{cases}$$

Therefore, assumptions (1-9) imply that the right-hand side of (1-15) is bounded.

Thus we can extract from $p_h\varphi_h$ a subsequence (again denoted $p_h\varphi_h$) such that $p_h\varphi_h$ converges weakly in V to an element φ. Furthermore, (1-15) implies that $\varphi \in V_1 = \ker\gamma_1$, since $\|\gamma_1 p_h\varphi_h\|_Q^2 \leq M\varepsilon(h)$ converges to 0.

What we actually prove is that $p_h\varphi_h$ converges strongly to $\varphi = 0$ in V. For that purpose, let us take $v_h = \varphi_h$ in (1-13). The right-hand side of (1-13) converges to 0 since, by (1-15) and [(1-9)(iii)],

$$(1\text{-}16)\quad\begin{cases} \varepsilon(h)^{-1}\langle\gamma_1(u - p_h r_h u), \gamma_1 p_h\varphi_h\rangle \\ \qquad\qquad \leq M\varepsilon(h)^{-1/2}\,\|\gamma_1 p_h\varphi_h\|_Q \cdot \varepsilon(h)^{-1/2}e_V{}^W(p_h r_h)\,\|u\|_V. \end{cases}$$

We thus have proved that

$$(1\text{-}17)\quad\begin{cases} \text{(i) } \|u - p_h u_h\|_V \leq \|u - p_h r_h u\|_V + \|p_h\varphi_h\|_V \text{ converges to 0,} \\ \text{(ii) } \|\gamma_1(u - p_h u_h)\|_Q \leq \|\gamma_1(u - p_h r_h u)\|_Q + \|\gamma_1 p_h\varphi_h\|_Q \\ \qquad\qquad \leq M(e_V{}^W(p_h) + \varepsilon(h)^{1/4}) \leq M\varepsilon(h)^{1/4}. \end{cases}$$

Now let us prove [(1-10)(iii)]. When v ranges over V, we can write

$$(1\text{-}18) \quad \begin{cases} \varepsilon(h)^{-1}\langle \gamma_1(p_h u_h - u), \gamma_1 v \rangle = (f, p_h r_h v) - a(p_h u_h, p_h r_h v) \\ \qquad + \langle t_2, \gamma_2 p_h r_h v \rangle + \varepsilon(h)^{-1}\langle \gamma_1(p_h u_h - u), \gamma_1(v - p_h r_h v) \rangle. \end{cases}$$

The right-hand side of this relation converges to 0 because

$$|\varepsilon(h)^{-1}\langle \gamma_1(u - p_h u_h), \gamma_1(v - p_h r_h v) \rangle| \leq \varepsilon(h)^{-\frac{1}{2}} \|\gamma_1(u - p_h u_h)\|_Q$$
$$\cdot \varepsilon(h)^{-\frac{1}{2}} e_V{}^W(p_h r_h) \|v\|_V \text{ converges to 0}$$

by [(1-17)(ii)] and [(1-9)(ii)]. Then $(\varepsilon(h)^{-1}\gamma_1'\gamma_1(p_h u_h - u), v)$ converges to $(f, v) - a(u, v) + \langle t_2, \gamma_2 v \rangle = -\langle \delta_1 u, \gamma_1 v \rangle = -(\gamma_1' \delta_1 u, v)$ for any $v \in V$.

Since γ_1 maps V onto T_1, we deduce that $\varepsilon(h)^{-1}\gamma_1(u - p_h u_h)$ converges weakly to $\delta_1 u$ in T_1'. On the other hand, [(1-10)(ii)] is an obvious consequence of [(1-10)(iii)].

Finally, we have to prove (1-11). For that purpose, we deduce from the Green formula the relation

$$(1\text{-}19) \quad \begin{aligned} a(u - p_h u_h, v) &= a(u - p_h u_h, v - p_h r_h v) \\ &\quad - \langle \delta_1 u - \varepsilon(h)^{-1}\gamma_1(u - p_h u_h), \gamma_1 p_h r_h v \rangle. \end{aligned}$$

When v ranges over $U_0 \subset V_0$, we can write $a(u - p_h u_h, v) = (\Lambda(u - p_h u_h), v)$ and estimate the right-hand side of (1-19) by

$$|a(u - p_h u_h, v - p_h r_h v) - \langle \delta_1 u - \varepsilon(h)^{-1}\gamma_1(u - p_h u_h), \gamma_1(v - p_h r_h v) \rangle|$$
$$\leq M e_{U_0}{}^V(p_h r_h) \|u\|_{U_0}(\|u - p_h u_h\|_V + \|\delta_1 u - \varepsilon(h)^{-1}\gamma_1(u - p_h u_h)\|_{T_1'}).$$

We thus have proved inequality (1-11), but we should not neglect to notice that this inequality implies

$$(1\text{-}20) \qquad \|\Lambda(u - p_h u_h)\|_{U_0'} \leq M e_{U_0}{}^V(p_h r_h),$$

since $u - p_h u_h$ is bounded in V and $\delta_1 u - \gamma_1(u - p_h u_h)$ is bounded in T_1'.

Next, by assuming hypotheses of regularity of the solution u, we estimate the second term of the right-hand side of (1-11).

1-4. Estimates of Error

In order to obtain estimates of error, we have to assume that u belongs to a space U contained in V. Actually, we assume that U satisfies the following properties:

$$(1\text{-}21) \qquad\qquad \delta_1 \text{ maps } \quad U \quad \text{into} \quad Q.$$

THEOREM 1-2 Let us assume again (1-1), (1-6), and (1-7). Let us assume moreover (1-21), and let us choose $\varepsilon(h)$ such that

$$(1\text{-}22) \qquad\qquad M^{-1} \leq \lim_{h \to 0} \varepsilon(h)^{-1} e_U{}^W(p_h r_h) \leq M.$$

Then the error $u - p_h u_h$ obeys the following estimate:

$$(1\text{-}23) \qquad \|u - p_h u_h\|_V \leq M(\sqrt{e_U^W(p_h r_h)} + e_U^V(p_h r_h)) \|u\|_U),$$

Furthermore, if $e_U^V(p_h)^2 \leq e_U^W(p_h r_h)$,

$$(1\text{-}24) \qquad \begin{cases} \text{(i)} \quad \varepsilon(h)^{-1}\gamma_1(u - p_h u_h) \text{ converges weakly to } \delta_1 u \text{ in } Q, \\ \text{(ii)} \quad \|\Lambda(u - p_h u_h)\|_{U_0'} = M e_U^W(p_h) \|u\|_U. \end{cases} \qquad \blacktriangle$$

Proof Let us set

$$(1\text{-}25) \qquad \varphi_h = u_h - r_h u \qquad \text{and} \qquad \tau_h = \delta_1 u - \varepsilon(h)^{-1}\gamma_1(u - p_h u_h).$$

We estimate the expression X_h defined by

$$(1\text{-}26) \qquad X_h = a(p_h\varphi_h, p_h\varphi_h) + \varepsilon(h) \|\tau_h\|_Q^2.$$

Since we can write the approximate equation in the form

$$a(p_h u_h, p_h v_h) = (f, p_h u_h) + \langle t_2, \gamma_2 p_h v_h \rangle - \varepsilon(h)^{-1}.$$

$$\langle \gamma_1(p_h u_h - u), \gamma_1 p_h v_h \rangle = a(u, p_h v_h) - \langle \tau_h, \gamma_1 p_h v_h \rangle,$$

we deduce that

$$a(p_h u_h - p_h r_h u, p_h v_h) = a(u - p_h r_h u, p_h v_h) - \langle \tau_h, \gamma_1 p_h v_h \rangle.$$

Therefore

$$\begin{aligned} X_h &= a(p_h\varphi_h, p_h\varphi_h) + \varepsilon(h) \|\tau_h\|_Q^2 \\ &= a(u - p_h r_h u, p_h\varphi_h) + \langle \tau_h, \varepsilon(h)\tau_h + \gamma_1 p_h u_h - \gamma_1 p_h r_h u \rangle \\ &= a(u - p_h r_h u, p_h\varphi_h) + \langle \tau_h, \gamma_1(u - p_h r_h u) \rangle - \varepsilon(h)\langle \tau_h, \delta_1 u \rangle. \end{aligned}$$

Thus we deduce the following inequality:

$$(1\text{-}27) \quad \begin{cases} X_h \leq M \|u - p_h r_h u\| \|p_h\varphi_h\| + \|\tau_h\|_Q \|u - p_h r_h u\|_W \\ \quad + \varepsilon(h) \|\tau_h\|_Q \|\delta_1 u\|_Q \leq M e_U^V(p_h r_h) \|u\|_U \|p_h\varphi_h\| \\ \quad + \varepsilon(h)^{1/2} \|\tau_h\|_Q \varepsilon(h)^{-1/2} e_U^W(p_h r_h) \|u\|_U + \varepsilon(h)^{1/2} \|\tau_h\|_Q \varepsilon(h)^{1/2} \|\delta_1 u\|_Q \end{cases}$$

Using the inequalities $ab \leq \varepsilon a^2 + (1/4\varepsilon)b^2$, we deduce that

$$\|p_h\varphi_h\|^2 + \varepsilon(h) \|\tau_h\|_Q^2 \leq M(e_U^V(p_h r_h)^2 + \varepsilon(h)^{-1}e_U^W(p_h r_h)^2 + \varepsilon(h)).$$

By assumption (1-22), we know that there exists a positive constant M such that $M^{-1} \leq \varepsilon(h)^{-1} e_U^W(p_h r_h) \leq M$. We deduce from this property and from assumption [(1-21)(ii)] the estimate

$$\|p_h\varphi_h\|^2 + \varepsilon(h) \|\tau_h\|_Q^2 \leq M(e_U^W(p_h r_h) + e_U^V(p_h r_h)^2).$$

This implies the estimate (1-23) and that τ_h is bounded in Q when $e_U{}^V(p_h r_h)^2 \leq M e_U{}^W(p_h r_h)$. Therefore τ_h converges weakly to τ in Q. Since τ_h converges to 0 in T', we deduce that τ_h converges weakly to 0 in Q, and thus [(1-24)(i)] is proved. Finally, we deduce from (3-19) that when v ranges over $D_0 = \{v \in U$ such that $\gamma v = 0\}$, we obtain

$$
\begin{aligned}
(\Lambda(u - p_h u_h), v) &= a(u - p_h u_h, v - p_h r_h v) \\
&\quad - \langle \delta_1 u - \varepsilon(h)^{-1}\gamma_1(u - p_h u_h), \gamma_1(v - p_h r_h v)\rangle \leq M(e_U{}^V(p_h r_h) \|u - p_h u_h\|_V \\
&\quad + \|\delta_1 u - \varepsilon(h)^{-1}\gamma_1(u - p_h u_h)\|_Q e_U{}^W(p_h r_h) \leq M e_U{}^W(p_h r_h),
\end{aligned}
$$

by using (1-23) and the assumption that $e_U{}^V(p_h r_h)^2 \leq M e_U{}^W(p_h r_h)$.

1-5. Convergence in Smaller Spaces

Let us introduce a space L satisfying

$$(1\text{-}28) \qquad\qquad U \subset L \subset V,$$

the injections being continuous and dense.

Let q_h be a prolongation mapping V_h into L related to p_h by the stability function

$$(1\text{-}29) \qquad s_L{}^V(q_h, p_h) = \sup_{v_h \in V_h} \frac{\|q_h v_h\|_L}{\|p_h v_h\|_V}.$$

THEOREM 1-3   We assume the hypotheses of Theorem 1-2, and we set

$$(1\text{-}30) \quad \mu_U{}^L(h) = s_L{}^V(q_h, p_h)(\sqrt{e_U{}^W(p_h r_h)} + e_U{}^V(p_h r_h)) + e_U{}^L(q_h r_h).$$

Then, if u belongs to the space U, the norm in L of the error $u - q_h u_h$ is estimated by

$$(1\text{-}31) \qquad\qquad \|u - q_h u_h\|_L \leq M \mu_U{}^L(h) \|u\|_U,$$

and thus $q_h u_h$ converges to u in L if $\mu_U{}^L(h)$ converges to 0.   ▲

Proof   We deduce from (1-23) that

$$
\|q_h(u_h - r_h u)\|_L \leq M s_L{}^V(q_h, p_h)(\sqrt{e_U{}^W(p_h r_h)} + e_U{}^V(p_h r_h))\|u\|_U
$$
$$
\text{Therefore} \qquad \|u - q_h u_h\|_L \leq M \mu_U{}^L(h) \|u\|_U. \qquad ■
$$

1-6. Convergence in Larger Spaces

For the sake of simplicity, we restrict ourselves to the case of the Dirichlet problem (where the projector σ_1 is the identity). Let us recall the results stated in Sections 2-7 and 2-8 of Chapter 6. We introduce Hilbert spaces U, R, and S such that

(1-32)

$$\begin{cases} \text{(i)} \ \ U \subset V(\Lambda) \cap V(\Lambda^*); \ S \subset T \subset R; \ \text{the injections being continuous and dense,} \\ \text{(ii)} \ \ D = \ker \delta, \ D^* = \ker \delta^*, \ D_0 = \ker \gamma, \text{ and } D_0^* = \ker \gamma \text{ are contained in } U, \\ \text{(iii)} \ \ (\gamma, \delta) \text{ and } (\gamma, \delta^*) \text{ map } U \text{ onto } S \times R', \\ \text{(iv)} \ \ U_0 = \ker (\gamma, \delta) = \ker (\gamma, \delta^*) \text{ is dense in } H. \end{cases}$$

Then, by Theorem 6.2-7, there exists a unique solution of the nonhomogeneous Dirichlet problem

$$(1\text{-}33) \qquad\qquad \Lambda u = f \ ; \ \gamma u = t$$

which belongs to $H(\Lambda) \subset H$ when f belongs to H and t belongs to R.

On the other hand, by the Green formula (2-44) of Section 6.2-7, we deduce that problem (1-33) is equivalent to

$$(1\text{-}34) \quad (u, \Lambda^*v) = (f, v) - \langle t, \delta^*u \rangle \quad \text{for any} \quad v \in V \quad \text{such that} \quad \gamma v = 0.$$

Now let us introduce a Hilbert space L such that

(1-35)

$$\begin{cases} \text{(i)} \ \ U \subset L \subset V, \\ \text{(ii)} \ \ \delta^* \text{ maps } L \text{ onto } X' \text{ where } T \subset X \subset R, \end{cases}$$

the injections being continuous and dense in both cases.

Thus, if $t \in X \subset R$, there exists a unique solution u of (1-33).

Let q_h be a prolongation mapping V_h into L such that

(1-36)

$$\begin{cases} \mu_U{}^L(h) = s_L{}^V(q_h, p_h)(\sqrt{e_U{}^W(p_h r_h)} + e_U{}^V(p_h r_h)) + e_U{}^L(q_h r_h) \\ \text{converges to 0.} \end{cases}$$

We shall approximate the solution u of the nonhomogeneous Dirichlet problem (1-33) by the solution u_h of the following variational equation on V_h:

(1-37)

$$\begin{cases} \text{if } f \in H \text{ and } t \in X, \text{ look for } u_h \in V_h \text{ satisfying } a(p_h u_h, p_h v_h) \\ + \ \varepsilon(h)^{-1} \langle \gamma p_h u_h, \gamma p_h v_h \rangle = (f, q_h v_h) - \langle t, \delta^* q_h v_h \rangle \quad \text{for any} \quad v_h \in V_h. \end{cases}$$

This equation has a meaning because q_h maps V_h into $L \subset V$ and since $\delta^* q_h u_h$ belongs to X'.

We now prove the convergence of $p_h u_h$ to u in H.

THEOREM 1-4 Let us assume the hypotheses of Theorem 1-2. Furthermore, let us suppose (1-32), (1-35), and (1-36).

Let f belong to H and t belong to X, with u the solution of the non-homogeneous Dirichlet problem (1-33) and u_h the solution of the approximate equation (1-37). Then $p_h u_h$ converges to u in H and the error is estimated by

$$(1\text{-}38) \qquad \|u - p_h u_h\| \leq M\mu_U{}^L(h)(\|f\|_H{}^2 + \|t\|_X{}^2)^{\frac{1}{2}}. \qquad \blacktriangle$$

Proof The proof of Theorem 1-4 is analogous to the proof of Theorem 8.1-4 (case of the Neumann problem): we deduce Theorem 1-4 from Theorem 1-3 by transposition.

Let us introduce the operators B_h and C_h defined by

$$(1\text{-}39) \qquad \begin{cases} \text{(i)} \ (C_h u_h, v_h)_h = a(p_h u_h, p_h v_h) + \varepsilon(h)^{-1}\langle \gamma p_h u_h, \gamma p_h v_h \rangle, \\ \text{(ii)} \ B_h v = v - q_h C_h'^{-1} p_h' \Lambda_0^* v, \end{cases}$$

where Λ_0^* is the restriction of Λ^* to the closed subspace $D_0 = \ker \gamma$. When v ranges over $D_0 = \ker \gamma$, $\Lambda_0^* v$ ranges over H and $B_h v$ ranges over L. Moreover, by Theorem 1-3, where $a(u, v)$ is replaced by $a_*(u, v) = a(v, u)$, with σ_1 the identity and $t = 0$, we deduce the inequality

$$(1\text{-}40) \qquad \|B_h\|_{L(D_0, L)} \leq M\mu_U{}^L(h).$$

Therefore, since the norm of an operator is equal to the norm of its transpose, we deduce the inequality

$$(1\text{-}41) \qquad \|B_h' l\|_{D_0'} \leq M\mu_U{}^L(h)\, \|l\|_{L'} \qquad \text{for any} \quad l \in L'.$$

To compute $B_h' l$ when $l(v) = (f, v) - \langle t, \delta^* v \rangle$, we obtain

$$(1\text{-}42) \qquad B_h' l = l - (\Lambda_0^*)' p_h C_h^{-1} q_h' l = (\Lambda_0^*)'(u - p_h u_h),$$

where u and u_h are the solutions of (1-33) and (1-37), respectively. Indeed, the solution u of $(\Lambda_0^*)' u = l = f - (\delta^*)' t$ is the solution u of (1-34), and thus the solution u of (1-33). On the other hand, if u_h is the solution of (1-37), we can write $p_h u_h = p_h C_h^{-1} q_h' l$.

Since Λ_0^* is an isomorphism from D_0 onto E', $(\Lambda_0^*)'$ is an isomorphism from E onto D_0' and we obtain the inequality

$$(1\text{-}43) \qquad \|u - p_h u_h\|_E \leq M\mu_U{}^L(h)\, \|l\|_{L'},$$

which implies Theorem 1-4. $\qquad \blacksquare$

2. PERTURBED APPROXIMATIONS OF BOUNDARY-VALUE PROBLEMS

2-1. Perturbed Approximations by Finite-Element Approximations

Let Λ be the differential operator defined by

$$(2\text{-}1) \qquad \Lambda u = \sum_{|p|, |q| \leq k} (-1)^{|q|} D^q(a_{pq}(x) D^p u), \qquad a_{pq}(x) \in L^\infty(\Omega),$$

which is the formal operator associated with the bilinear form

$$(2\text{-}2) \qquad a(u, v) = \sum_{|p|,|q| \leq k} \int_\Omega a_{pq}(x) D^p u D^q v \, dx$$

continuous on $H^k(\Omega)$, where $\Omega \subset R^n$ is a smooth bounded open subset.

Let us recall the Green formula (see Section 7.2-1)

$$(2\text{-}3) \quad \begin{cases} a(u, v) = (\Lambda u, v) + \sum_{0 \leq j \leq k-1} \langle \delta_{2k-1-j} u, \gamma_j v \rangle \\ \qquad\qquad\qquad \text{for any} \quad u \in H^k(\Omega, \Lambda), v \in H^k(\Omega). \end{cases}$$

Let us consider the following boundary-value problem (see Section 7.2-5),

$$(2\text{-}4) \quad \begin{cases} \text{(i)} \ \Lambda u = f \quad \text{where } f \text{ is given in } L^2(\Omega), \\ \text{(ii)} \ \gamma_j u = g_j \quad \text{for } 0 \leq j \leq p-1, 1 \leq p \leq k \\ \qquad\qquad\qquad\qquad \text{where } g_j \in H^{k-j-\frac{1}{2}}(\Gamma), \\ \text{(iii)} \ \delta_j u = h_j \quad \text{for } k \leq j \leq 2k-1-p \\ \qquad\qquad\qquad\qquad \text{where } h_j \in H^{k-j-\frac{1}{2}}(\Gamma). \end{cases}$$

We can approximate the solution $u \in H^k(\Omega, \Lambda)$ of (2-4), if such exists, by its perturbed internal approximation. We set

$$(2\text{-}5) \quad \begin{cases} \text{(i)} \ V = H^k(\Omega), H = L^2(\Omega), \\ \text{(ii)} \ T_1 = \prod_{0 \leq j \leq p-1} H^{k-j-\frac{1}{2}}(\Gamma), T_2 = \prod_{p \leq j \leq k-1} H^{k-j-\frac{1}{2}}(\Gamma), \\ \text{(iii)} \ \gamma_1 = (\gamma_0, \ldots, \gamma_{p-1}, 0, \ldots, 0), \gamma_2 = (0, \ldots, 0, \gamma_p, \ldots, \gamma_{k-1}) \\ \text{(iv)} \ \delta_1 = (\delta_{2k-1}, \ldots, \delta_{2k-p}, 0, \ldots, 0), \\ \qquad\qquad\qquad\qquad \delta_2 = (0, \ldots, 0, \delta_{2k-p-1}, \ldots, \delta_k). \end{cases}$$

So assumption (1-7) of Theorem 1-1 is satisfied if we choose

$$(2\text{-}6) \quad \begin{cases} \text{(i)} \ Q = L^2(\Gamma)^k, \\ \text{(ii)} \ W = H^{p-\alpha-\frac{1}{2}}(\Omega) \quad \text{where } \alpha > 0. \end{cases}$$

Indeed, by Theorem 6.3-9 we know that γ_1 maps W into

$$\prod_{0 \leq j \leq p-1} H^{p+\alpha-j-1}(\Gamma) \times \prod_{p \leq j \leq k-1} L^2(\Gamma) \subset Q.$$

In order to approximate the boundary-value problem (2-4), we need approximations of the space $H^k(\Omega)$. Let us use, for instance, the finite-element approximations $(H_h^\mu(\Omega), p_h, r_{h,\Omega})$ associated with a function μ of $H^m(R^n)$ with $m \geq k$.

Since W is a space of order $\theta = (k + \frac{1}{2} - p - \alpha)/k$ between $V = H^k(\Omega)$ and $H = L^2(\Omega)$ (see Theorem 6.3-7), we deduce from Proposition 2.4-1 and Theorem 5.2-1 that

$$(2\text{-}7) \qquad e_V{}^W(p_h r_{h,\Omega}) \leq c \, |h|^{k-p-\alpha+\frac{1}{2}}.$$

Therefore, assumption (1-9) of Theorem 1-1 is satisfied if we choose, for instance,

$$\varepsilon(h) = M \, |h|^{2(k-p-2\alpha+1)} \qquad \text{where} \quad M \text{ is a given constant.}$$

Then we approximate the solution $u \in H^k(\Omega, \Lambda)$ of boundary-value problem (2-4) by the solution $u_h \in H_h{}^\mu(\Omega)$ of the discrete variational equation

$$(2\text{-}8) \quad \begin{cases} a(p_h u_h, p_h v_h) + \varepsilon(h)^{-1} \displaystyle\sum_{0 \leq j \leq p-1} \int_\Gamma \gamma_j p_h u_h \cdot \gamma_j p_h v_h \, d\sigma(x) \\[2mm] \quad = \displaystyle\int_\Omega f(x) p_h v_h \, dx + \varepsilon(h)^{-1} \sum_{0 \leq j \leq p-1} \int_\Gamma g_j(x) \gamma_j p_h v_h \, d\sigma(x) \\[2mm] \qquad + \displaystyle\sum_{p \leq j \leq k-1} \int_\Gamma h_{2k-1-j}(x) \gamma_j p_h v_h \, d\sigma(x) \qquad \text{for any} \quad v_h \in H_h{}^\mu(\Omega) \end{cases}$$

where $a(u, v)$ is defined by (2-2).

This discrete variational equation is equivalent to an equation of the form

$$(2\text{-}9) \qquad A_{\varepsilon(h)} u_h = l_{\varepsilon(h)}.$$

The properties of the matrix $A_{\varepsilon(h)}$ and of the vector $l_{\varepsilon(h)} \in H_h{}^\mu(\Omega)$ were examined in Sections 8.2-1 and 8.3-1. In particular, note that the number of levels of matrix $A_{\varepsilon(h)}$ is at most equal to the number of levels of μ.

We deduce from Theorem 1-1 the following consequence.

THEOREM 2-1 Let us assume that Ω is a smooth bounded subset of R^n and that μ satisfies the criterion of m-convergence for $m \geq k$. We also assume that the form $a(u, v)$ defined by (2-2) is $H^k(\Omega)$-elliptic, and we let u be the solution of the boundary-value problem (2-4) and $u_h \in H_h{}^\mu(\Omega)$ the solution of its perturbed approximation, with $\varepsilon(h) = M \, |h|^{2(k-p+1-2\alpha)}$. Then

$$(2\text{-}10) \quad \begin{cases} \text{(i)} \quad \displaystyle\lim_{h \to 0} \|u - p_h u_h\|_{k,\Omega} = 0, \\[2mm] \text{(ii)} \quad M^{-1} \, |h|^{-2(k-p+1-2\alpha)} \gamma_j(u - p_h u_h) \text{ converges weakly to} \\[2mm] \qquad \delta_{2k-1-j} u \text{ in } H^{-k+j+\frac{1}{2}}(\Gamma) \qquad \text{for} \quad 0 \leq j \leq p - 1. \end{cases}$$

Furthermore, we obtain the following estimates of the error

$$(2\text{-}11) \quad \begin{cases} \text{(i)} \quad \|\Lambda(u - p_h u_h)\|_{H^{-m-1}(\Omega)} \leq M \, |h|^{m+1-k} \|u\|_{H^k(\Omega)}, \\[2mm] \text{(ii)} \quad \|\gamma_j(u - p_h u_h)\|_{H^{-j+k+\frac{1}{2}}(\Gamma)} \leq M \, |h|^{2(k-p+1-2\alpha)} \|u\|_{H^k(\Omega)}. \end{cases} \qquad \blacktriangle$$

2-2.   Error Estimates and Regularity of the Convergence

Assumption 1-21 of Theorem 1-2 is satisfied when the solution u of the boundary-value problem (2-4) belongs to the space

$$(2\text{-}12) \qquad\qquad U = H^{2k}(\Omega) \cap H^{m+1}(\Omega).$$

Indeed, we deduce from the trace Theorem 6.3-1 that $\delta_1 = (\delta_{2k-1}, \dots, \delta_{2k-p}, 0, \dots, 0)$ maps $H^{2k}(\Omega)$ into $Q = (L^2(\Gamma))^k$. For this choice of space U, we obtain from Theorem 5.3-1 the estimates

$$(2\text{-}13) \qquad\qquad \begin{cases} \text{(i)} \;\; e_U{}^W(p_h r_{h,\Omega}) \le c\, |h|^{m+1-p-\alpha+\frac{1}{2}}, \\[2mm] \text{(ii)} \;\; e_U{}^V(p_h r_{h,\Omega}) \le c\, |h|^{m+1-k}. \end{cases}$$

Therefore, assumption (1-22) of Theorem 1-2 is satisfied if we choose, for instance,

$$(2\text{-}14) \qquad\qquad \varepsilon(h) = M\, |h|^{m+1-p-\alpha+\frac{1}{2}}.$$

Theorem 1-2 shows us that the error in $H^k(\Omega)$ may be estimated in the following way:

$$(2\text{-}15) \qquad \|u - p_h u_h\|_{H^k(\Omega)} \le c(|h|^{m+1-k} + |h|^{(m+1-p-\alpha+\frac{1}{2})/2})\, \|u\|_U.$$

In other words, we have proved Theorem 2-2.

THEOREM 2-2   Let us suppose the assumptions of Theorem 2-1, letting u be the solution of the boundary-value problem (2-4) and u_h the solution of its perturbed approximation, with $\varepsilon(h)$ defined by (2-14).
If $m \le 2k - p - \alpha - \frac{1}{2}$, the error satisfies

$$(2\text{-}16) \qquad\qquad \|u - p_h u_h\|_{H^k(\Omega)} \le c\, |h|^{m+1-k}\, \|u\|_{H^{2k}(\Omega)}.$$

If $m \ge 2k - p - \alpha - \frac{1}{2}$, the error is estimated by

$$(2\text{-}17) \qquad \begin{cases} \text{(i)} \;\; \|u - p_h u_h\|_{H^k(\Omega)} \le c\, |h|^{(m+1-p-\alpha+\frac{1}{2})/2}\, \|u\|_U, \\[2mm] \text{(ii)} \;\; \|\Lambda(u - p_h u_h)\|_{H^{-m-1}(\Omega)} \le c\, |h|^{m+1-p-\alpha+\frac{1}{2}}\, \|u\|_{H^{2k}(\Omega)\cap H^{m+1}(\Omega)}. \end{cases}$$

▲

Now let us study the convergence in the spaces $L = H^s(\Omega)$ for $k \le s \le m + 1$ in the case, for instance, of $(2m + 1)^n$-level piecewise-polynomial approximations. We choose

$$(2\text{-}18) \qquad\qquad q_h = \begin{cases} p_h^{(m)} & \text{if } \; k \le s \le m, \\[2mm] p_h^{(m+1)} & \text{if } \; m < s \le m + 1. \end{cases}$$

The following estimate of the stability function comes logically from Theorem 5.2-5 and from Proposition 2.4-2:

$$(2\text{-}19) \qquad s_L{}^V(q_h, p_h^{(m)}) \leq c \, |h|^{k-s}.$$

Then Theorems 1-3 and 2-2 imply the results of Theorem 2-3.

THEOREM 2-3   Let us suppose the assumptions of Theorem 2-2, assuming that Ω satisfies the property of $\pi_{(m+1)}$-stability and defining q_h by (2-18). Then, if $m \leq 2k - p - \alpha - \frac{1}{2}$,

$$(2\text{-}20) \qquad \|u - q_h u_h\|_{H^s(\Omega)} \leq c \, |h|^{m+1-s} \, \|u\|_{H^{2k}(\Omega)}.$$

If $m \geq 2k - p - \alpha - \frac{1}{2}$ and if $s \leq (m + 2k - p + \frac{3}{2} - \alpha)/2$, then

$$(2\text{-}21) \qquad \|u - q_h u_h\|_{H^s(\Omega)} \leq c \, |h|^{(m+2k-p-\alpha+3/2)/2-s} \, \|u\|_U. \qquad \blacktriangle$$

Remark 2-1   In the case of the Dirichlet problem, we must take $p = k$, which means that estimates (2-16) and (2-20) cannot hold, since $m \geq k$. Only estimates (2-17) and (2-21) can be used. ∎

In particular, the space U coincides with $H^{2k}(\Omega)$ when we choose $m = 2k - 1$.

COROLLARY 2-1   Let us assume that Ω is a smooth bounded open subset of R^n satisfying the property of $\pi_{(2k)}$-stability. Let us assume also that the operator $(\Lambda, \gamma_0, \ldots, \gamma_{k-1})$ is an isomorphism from $H^{2k}(\Omega)$ onto $L^2(\Omega) \times \prod_{0 \leq j \leq k-1} H^{2k-j-\frac{1}{2}}(\Gamma)$ and that $a(u, v)$ is $H^k(\Omega)$-elliptic.

Let u_h be the solution of the solution of the perturbed approximation (2-8) with $\varepsilon(h) = M \, |h|^{k-\alpha+\frac{1}{2}}$. Then, for $s \leq (3k + \frac{1}{2} - \alpha)/2$, the following estimate holds:

$$(2\text{-}22) \qquad \|u - p_h^{(2k)} u_h\|_{H^s(\Omega)} \leq c \, |h|^{(3k-2s-\alpha+\frac{1}{2})/2} \, \|u\|_{H^{2k}(\Omega)}. \qquad \blacktriangle$$

In the next section, we investigate in particular the case of the Dirichlet problem for second-order differential operators.

2-3.   The 3^n-level Perturbed Approximation of the Dirichlet Problem

Let Λ be a second-order differential operator and assume that the assumptions of Theorem 7.1-2 are satisfied. Then, if f is given in $L^2(\Omega)$ and t in $H^{-\frac{1}{2}}(\Gamma)$, there exists a unique solution $u \in H(\Omega, \Lambda) \subset L^2(\Omega)$ of the Dirichlet problem

$$(2\text{-}23) \qquad \Lambda u = f \quad \text{and} \quad \gamma_0 u = t.$$

THEOREM 2-4 Supposing the assumptions of Theorem 7.1-2 and the assumptions of Theorem 2-2, let u be the solution of the Dirichlet problem (2-23) and u_h the solution of the 3^n-level perturbed approximate equation

$$(2\text{-}24) \quad \begin{cases} a(p_h^{(1)}u_h, p_h^{(1)}v_h) + M^{-1}|h|^{\alpha-3/4}\int_\Gamma p_h^{(1)}u_h p_h^{(1)}v_h\,d\sigma(x) \\ = \int_\Omega f(x)p_h^{(2)}v_h\,dx - \int_\Gamma t(x)\frac{\partial}{\partial n_\Lambda}p_h^{(2)}v_h\,d\sigma(x) \quad \text{for any} \quad v_h \in H_h^{(1)}(\Omega). \end{cases}$$

Then if $\mu \leq \frac{1}{4} - 2\alpha$ and $t \in H^{\mu+\alpha-1/4}(\Gamma)$, the solution u_h of (2-24) converges to the solution u of (2-23) and

$$(2\text{-}25) \quad \|u - p_h^{(1)}u_h\|_{L^2(\Omega)} \leq c\,|h|^\mu(\|f\|_{L^2(\Omega)}^2 + \|t\|_{H^{\mu+\alpha-1/4}(\Gamma)}^2)^{1/2}. \quad \blacktriangle$$

Proof Theorem 2-4 follows from Theorem 1-4 and from Corollary 2-1, with $k = 1$ and $\mu = \frac{7}{4} - s - \alpha$. $\blacksquare$

In particular, Theorem 2-4 can be used for the approximation of

$$(2\text{-}26) \quad \int_\Gamma j(x)\,\partial u/\partial n_{\Lambda^*}\,d\sigma(x) \quad \text{where} \quad u \text{ satisfies } \Lambda^* u = f \text{ and } \gamma_0 u = 0,$$

when j is a given function of $L^2(\Gamma)$ and f ranges over $L^2(\Omega)$. By a proof analogous to the proof of Corollary 7.1-6, we know that

$$(2\text{-}27) \quad \int_\Gamma j(x)\frac{\partial u}{\partial n_{\Lambda^*}}\,d\sigma(x) = \int_\Omega f(x)k(x)\,dx,$$

where $k(x) \in H(\Omega, \Lambda) \subset L^2(\Omega)$ is the solution of

$$(2\text{-}28) \quad \Lambda k = 0 \quad \text{and} \quad \gamma_0 k = -j.$$

Indeed, we deduce from the Green formula that

$$-\left\langle j, \frac{\partial u}{\partial n_{\Lambda^*}} \right\rangle = \left\langle \gamma_0 k, \frac{\partial u}{\partial n_{\Lambda^*}} \right\rangle$$
$$= (\Lambda k, u) - (k, \Lambda^* u) + \left\langle \frac{\partial k}{\partial n_\Lambda}, \gamma_0 u \right\rangle = -(k, f),$$

since $u \in H^2(\Omega)$ and $k \in H(\Omega, \Lambda)$.

Then, in order to approximate the value $\int_\Gamma j(x)\,\partial u/\partial n_{\Lambda^*}\,d\sigma(x)$, we approximate once and for all the solution k of (2-28) instead of approximating the solution u for any $f \in L^2(\Omega)$. (See Sections 7.1-10 and 8.3-1.) Since $L^2(\Gamma) = H^{\mu+\alpha-1/4}(\Gamma)$ for $\mu = \frac{1}{4} - \alpha$, we deduce Corollary 2-2 from 2-4.

COROLLARY 2-2 The assumptions are those of Theorem 2-4. If k_h is the solution of the 3^n-level approximate equation (2-24) with $f = 0$ and $t = -j \in L^2(\Gamma)$, then

$$(2\text{-}29) \quad \left| \int_\Gamma j(x) \frac{\partial u}{\partial n_{\Lambda^*}} \, d\sigma(x) - \int_\Omega f(x) p_h^{(1)} k_h \, dx \right| \le c \, |h|^{\frac{1}{4}-\alpha} \|f\|_{L^2(\Omega)} \|j\|_{L^2(\Gamma)}$$

▲

3. LEAST-SQUARES APPROXIMATIONS

3-1. Least-Squares Approximation Schemes

Let V, H, and T be Hilbert spaces, Λ an operator from V into H, and γ an operator from V into T. Let us assume further that

$$(3\text{-}1) \qquad (\Lambda, \gamma) \text{ is an isomorphism from } V \text{ onto } H \times T$$

and let us consider the solution $u \in V$ of

$$(3\text{-}2) \qquad \begin{cases} \text{(i) } \Lambda u = f & \text{where } f \text{ is given in } H, \\ \text{(ii) } \gamma u = t & \text{where } t \text{ is given in } T. \end{cases}$$

We also assume that

$$(3\text{-}3) \qquad \begin{cases} \text{(i) } H \text{ is a pivot space [for the duality pairing } (f, v)], \\ \text{(ii) } Q \text{ is a pivot space (for the duality pairing } \langle t, s \rangle), \\ \text{(iii) } T \subset Q, \text{ the injection being continuous and dense.} \end{cases}$$

Let $(V_h, p_h, \hat{r}_h)$ be convergent approximations of the Hilbert space V and $\beta(h)$ a positive function of the parameter h. Now we can approximate the solution u of (3-2) by the solution $u_h \in V_h$ of the discrete variational equation

$$(3\text{-}4) \qquad \begin{cases} (\Lambda p_h u_h, \Lambda p_h v_h) + \beta(h)^{-2} \langle \gamma p_h u_h, \gamma p_h v_h \rangle \\ \qquad = (f, \Lambda p_h v_h) + \beta(h)^{-2} \langle t, \gamma p_h v_h \rangle \qquad \text{for any } v_h \in V_h. \end{cases}$$

PROPOSITION 3-1 Let us assume (3-1) and (3-3). The solution u of (3-2) minimizes on V the functional $J(v)$ defined by

$$(3\text{-}5) \qquad J(v) = \|f - \Lambda v\|_H^2 + \beta(h)^{-2} \|t - \gamma v\|_Q^2,$$

and the solution u_h of (3-4) minimizes the functional $J(p_h v_h)$ on V_h:

$$(3\text{-}6) \qquad \begin{cases} J(p_h u_h) = \|\Lambda(u - p_h u_h)\|_H^2 + \beta(h)^{-2} \|\gamma(u - p_h u_h)\|_Q^2 \\ \qquad \le \|\Lambda(u - p_h v_h)\|_H^2 + \beta(h)^{-2} \|\gamma(u - p_h v_h)\|_Q^2 = J(p_h v_h). \end{cases}$$

▲

Proof The first statement is obvious. Let us prove that the solution u_h of (3-4) minimizes $J(p_h v_h)$ on V_h. Since $(\Lambda p_h u_h - f, \Lambda p_h v_h) + \beta(h)^{-2} \langle \gamma p_h u_h - t, \gamma p_h v_h \rangle = 0$ for any $v_h \in V_h$, we deduce from the Cauchy-Schwarz inequality that

$$J(p_h u_h) = (f - \Lambda p_h u_h, f - \Lambda p_h u_h) + \beta(h)^{-2} \langle t - \gamma p_h u_h, t - \gamma p_h u_h \rangle$$
$$= (f - \Lambda p_h u_h, f - p_h v_h) + \beta(h)^{-2} \langle t - \gamma p_h u_h, t - \gamma p_h v_h \rangle$$
$$\leq J(p_h u_h)^{1/2} J(p_h v_h)^{1/2}.$$

Therefore, $J(p_h u_h) \leq J(p_h v_h)$ for any $v_h \in V_h$.

Conversely, let us assume that u_h minimizes $J(p_h v_h)$ on V_h. Then for any $\theta > 0$ and for any $v_h \in V_h$, $J(p_h u_h + \theta p_h v_h) - J(p_h u_h) \geq 0$. This implies that

$$(f - \Lambda p_h u_h, \Lambda p_h v_h) + \beta(h)^{-2} \langle t - \gamma p_h u_h, \gamma p_h v_h \rangle \geq -\frac{\theta}{2} J(p_h v_h).$$

Therefore, when θ converges to 0, we deduce that u_h is a solution of (3-4). ∎

Definition 3-1 The discrete variational equation (3-4) is the least-squares approximation of (3-2). ▲

Now the problem is to choose $\beta(h)$ in such way that $J(p_h u_h)$ converges to 0, that is, such that $\|\Lambda(u - p_h u_h)\|_H$ and $\|\gamma(u - p_h u_h)\|_Q$ converge to 0 and to obtain error estimates.

For instance, if we choose $\beta(h) = \beta$, we obtain the inequalities

$$(3-7) \quad \begin{cases} \text{(i)} \quad \|\Lambda(u - p_h u_h)\|_H \leq M \|u - p_h \hat{r}_h u\|_V \leq M e_U^V(p_h) \|u\|_U, \\ \text{(ii)} \quad \|\gamma(u - p_h u_h)\|_Q \leq M \|u - p_h \hat{r}_h u\|_V \leq M e_U^V(p_h) \|u\|_U, \end{cases}$$

when the solution $u \in U \subset V$. [We take $v_h = \hat{r}_h u$ in inequality (3-6).]

We can improve these estimates by choosing a function $\beta(h)$ that does not converge to 0 too fast: indeed, it is clear that the inequality

$$(3-8) \quad \begin{cases} J(p_h u_h) = \|\Lambda(u - p_h u_h)\|_H^2 \\ \qquad\qquad + \beta(h)^{-2} \|\gamma(u - p_h u_h)\|_Q^2 \leq M^2 e_U^V(p_h)^2 \|u\|_U^2 \end{cases}$$

implies inequalities

$$(3-9) \quad \begin{cases} \text{(i)} \quad \|\Lambda(u - p_h u_h)\|_H \leq M e_U^V(p_h) \|u\|_U, \\ \text{(ii)} \quad \|\gamma(u - p_h u_h)\|_Q \leq M \beta(h) e_U^V(p_h) \|u\|_U. \end{cases}$$

3-2. Error Estimates (I)

In this section, we assume that there exists a space W such that

(3-10) $\begin{cases} \text{(i) } V \subset W, \\ \text{(ii) } \|\gamma v\|_Q \leq c_W \|v\|_W \quad \text{for any } v \in W. \end{cases}$

PROPOSITION 3-2 Let us assume (3-1), (3-3), and (3-10), choosing

$$(3\text{-}11) \qquad \beta(h) = M^{-1} c_W \frac{e_U{}^W(p_h r_h)}{e_U{}^V(p_h r_h)},$$

where (V_h, p_h, r_h) are approximations of V. If u is the solution of (3-2) and u_h is the solution of (3-4) with $\beta(h)$ defined by (3-11), there exists a constant M such that

$$(3\text{-}12) \qquad \begin{cases} \text{(i) } \|\Lambda(u - p_h u_h)\|_H \leq \sqrt{1 + M^2} e_U{}^V(p_h r_h) \|u\|_U, \\ \text{(ii) } \|\gamma(u - p_h u_h)\|_Q \leq \sqrt{1 + M^2} c_W e_U{}^V(p_h r_h) \|u\|_U. \end{cases} \qquad \blacktriangle$$

Proof The proof is straightforward. We take $v_h = r_h u$ in (3-6) and estimate $J(p_h r_h u)$ in the following way:

$$\|\Lambda(u - p_h r_h u)\|_H{}^2 + \beta(h)^{-2} \|\gamma(u - p_h r_h u)\|_Q{}^2$$
$$\leq (e_U{}^V(p_h r_h)^2 + c_W{}^2 \beta(h)^{-2} e_U{}^W(p_h r_h)^2) \|u\|_U$$
$$\leq (M^2 + 1) e_U{}^V(p_h r_h)^2 \|u\|_U{}^2. \quad \blacksquare$$

EXAMPLE 3-1 Let us choose the following spaces:

$$(3\text{-}13) \qquad V = H^2(\Omega), \ H = L^2(\Omega), \ T = H^{3/4}(\Gamma), \ Q = L^2(\Gamma),$$

where Ω is a smooth bounded open subset of R^n and Γ is its boundary. Then assumptions (3-3) are satisfied.

Let Λ be an elliptic differential operator satisfying the assumptions of Theorem 7.1-1. Assumption 3-1 is satisfied by Theorem 7.1-1 if we take $\gamma = \dot{\gamma}_0$ to be the trace operator. If we take

$$(3\text{-}14) \qquad W = H^{1/2 + \alpha}(\Omega) \quad \text{for } \alpha > 0, \gamma u = \gamma_0 u = u|_\Gamma,$$

then assumption (3-10) is satisfied (see Theorem 6.3-9).

Let us introduce finite-element approximations associated with a function μ satisfying the criterion of m-convergence for $m \geq 2$, and let us take

$$(3\text{-}15) \qquad U = H^{m+1}(\Omega) \quad \text{and} \quad \beta(h) = M |h|^{(3-\alpha)/2}.$$

Let $f \in L^2(\Omega)$ and $t \in H^{3/4}(\Gamma)$ be given, and assume that $u \in H^2(\Omega)$ is the solution of the Dirichlet problem

$$(3\text{-}16) \qquad\qquad \Lambda u = f \quad \text{and} \quad \gamma_0 u = t.$$

Also, let $u_h \in H_h^{\mu}(\Omega)$ be the solution of the discrete variational equation

$$(3\text{-}17) \quad \begin{cases} (\Lambda p_h u_h, \Lambda p_h v_h) + M^{-2} |h|^{-3+\alpha} \langle \gamma_0 p_h u_h, \gamma_0 p_h v_h \rangle \\ \quad = (f, \Lambda p_h v_h) + M^{-2} |h|^{-3+\alpha} \langle t, \gamma_0 p_h v_h \rangle \qquad \text{for any} \quad v_h \in H_h^{\mu}(\Omega). \end{cases}$$

COROLLARY 3-1   Let Ω be a smooth bounded open subset of R^n, Λ be an elliptic second-order differential operator satisfying the assumptions of Theorem 7.1-1 and $\mu \in H^m(\Omega)$ a function with compact support satisfying the criterion of m-convergence for $m \geq 2$. Let u be the solution of the Dirichlet problem (3-16) and u_h the solution of its least-squares approximation (3-17). Then there exists a constant M_α such that

$$(3\text{-}18) \quad \begin{cases} \text{(i)} \quad \|\Lambda(u - p_h u_h)\|_{0,\Omega} \leq M_\alpha |h|^{m-1} \|u\|_{m+1,\Omega}, \\ \text{(ii)} \quad \|\gamma_0(u - p_h u_h)\|_{0,\Gamma} \leq M_\alpha |h|^{m+1/2-\alpha} \|u\|_{m+1,\Omega}. \end{cases} \qquad \blacktriangle$$

Later we prove another theorem which permits us to take $\alpha = 0$ in (3-15), (3-17), and (3-18).

3-3.   Error Estimates (II)

Here we take an assumption of regularity:

$$(3\text{-}19) \quad \begin{cases} \text{(i)} \ (\Lambda, \gamma) \text{ is an isomorphism from } V \text{ onto } H \times T, \\ \text{(ii)} \ (\Lambda, \gamma) \text{ is an isomorphism from } U \text{ onto } E \times S, \end{cases}$$

where

$$(3\text{-}20) \quad \begin{cases} \text{(i)} \ H \text{ and } Q \text{ are pivot spaces for the duality pairings } (f, v) \text{ and} \\ \qquad \langle t, s \rangle, \\ \text{(ii)} \ U \subset V, E \subset H, S \subset T \subset Q, \text{ the injections being continuous} \\ \qquad \text{and dense,} \\ \text{(iii)} \ T' \text{ is a space of order } \theta \text{ between } Q' \text{ and } S'. \\ \qquad \text{(see Definition (2.4-6).)} \end{cases}$$

Let (V_h, p_h, r_h) be approximations of the space V and call $e_U^V(p_h)$ their error function. We approximate the solution u of

$$(3\text{-}21) \qquad\qquad \Lambda u = f, \ \gamma u = t$$

by the solution u_h of its least-squares approximation

$$(3\text{-}22) \quad \begin{cases} (\Lambda p_h u_h, \Lambda p_h v_h) + \beta(h)^{-2} \langle \gamma p_h u_h, \gamma p_h v_h \rangle \\ \qquad\qquad = (f, \Lambda p_h v_h) + \beta(h)^{-2} \langle t, \gamma p_h v_h \rangle \qquad \text{for any} \quad v_h \in V_h, \end{cases}$$

where $\beta(h)$ satisfies

$$(3\text{-}23) \qquad M^{-1}\beta(h) \leq e_U{}^V(p_h)^{\theta/(1-\theta)} \leq M\beta(h).$$

THEOREM 3-1 Let us assume (3-19), (3-20), and (3-23). If $u \in U$ is the solution of (3-21) and if $u_h \in V_h$ is the solution of (3-22), then there exists a constant M such that

$$(3\text{-}24) \quad \begin{cases} \text{(i)} \quad \|\Lambda(u - p_h u_h)\|_H \leq M e_U{}^V(p_h) \|u\|_U, \\ \text{(ii)} \quad \|\gamma(u - p_h u_h)\|_Q \leq M e_U{}^V(p_h)^{1/(1-\theta)} \|u\|_U. \end{cases} \qquad \blacktriangle$$

EXAMPLE 3-2 To continue the study of Example 3-1, we assume that (Λ, γ_0) is an isomorphism from $H^{m+1}(\Omega)$ onto $H^{m-1}(\Omega) \times H^{m+\frac{1}{2}}(\Gamma)$. In other words, we take $U = H^{m+1}(\Omega)$, $E = H^{m-1}(\Omega)$, and $S = H^{m+\frac{1}{2}}(\Gamma)$. By Theorem 6.3-7, we know that $T' = H^{-\frac{3}{2}}(\Gamma)$ is a space of order $\theta = 3/(2m + 1)$ between $L^2(\Gamma)$ and $H^{-m-\frac{3}{2}}(\Gamma)$. Since $e_U{}^V(p_h) \leq M |h|^{m-1}$ when μ satisfies the criterion of m-convergence, we deduce that

$$(3\text{-}25) \qquad \beta(h) = M^{\frac{1}{2}} |h|^{\frac{3}{2}}$$

satisfies inequality (3-23). $\blacksquare$

We continue the study of least-squares approximations in Section 3-4.

Proof of Theorem 3-1 We prove Theorem 3-1 by proving the following lemmas. First of all, let us notice that

$$(3\text{-}26) \quad \begin{cases} B_h(f, t) = J(p_h u_h) \\ \qquad = \inf_{v_h \in V_h} J(p_h v_h) = \inf_{v_h \in V_h} (\|f - \Lambda p_h v_h\|_H{}^2 + \beta(h)^{-2} \|t - \gamma p_h v_h\|_Q{}^2 \end{cases}$$

can be viewed as the norm of the factor space X/N_h of the space $X = H \times Q$ supplied with the norm

$$(3\text{-}27) \qquad A_h(f, t) = (\|f\|_H{}^2 + \beta(h)^{-2} \|t\|_Q{}^2)^{\frac{1}{2}}$$

by the space $N_h = \Lambda p_h V_h \times \gamma p_h V_h$

For simplicity, we shall set $e(h) = e_U{}^V(p_h)$ during the proof of the theorem. $\blacksquare$

LEMMA 3-1 There exists a constant C independent of f, t, and h such that

$$(3\text{-}28) \qquad B_h(f, t)^2 \leq 2 \|f\|_H{}^2 + C^2 e(h)^{2(1-\delta)} \|t\|_S{}^2$$

where $\delta = \theta/(1 - \theta)$. $\blacktriangle$

Proof Since $B_h(f, t)$ is a norm, we obtain

$$(3\text{-}29) \qquad B_h(f, t) \le B_h(f, 0) + B_h(0, t) \le \|f\|_H + B_h(0, t).$$

Since the injection from T into Q is continuous, we can estimate $B_h(0, t)$ in the following way:

$$
\begin{aligned}
B_h(0, t)^2 &= \inf \left(\|\Lambda p_h v_h\|_H^2 + \beta(h)^{-2} \|t - \gamma p_h v_h\|_Q^2 \right) \\
&\le \inf \left(\|\Lambda p_h v_h\|_H^2 + \beta(h)^{-2} c^2 \|t - \gamma p_h v_h\|_T^2 \right) \\
&\le c^2 \beta(h)^{-2} \inf \left(\|\Lambda p_h v_h\|_H^2 + \|t - \gamma p_h v_h\|_T^2 \right) \\
&= c^2 \beta(h)^{-2} \inf \left(\|\Lambda(u_0 - p_h v_h)\|_H^2 + \|\gamma(u_0 - p_h v_h)\|_T^2 \right) \\
&\le c_1^2 \beta(h)^{-2} \inf \|u_0 - p_h v_h\|_V^2 \le c_2^2 \beta(h)^{-2} e(h)^2 \|u_0\|_U^2 \\
&\le c_3^2 \beta(h)^{-2} e(h)^2 \|t\|_S^2 \le c_4^2 e(h)^{2(1-\delta)} \|t\|_S^2,
\end{aligned}
$$

where u_0 is the solution of $\Lambda u_0 = 0$ and $\gamma u_0 = t$.

Therefore, there exists a constant C such that

$$(3\text{-}30) \quad \begin{cases} B_h(f, t)^2 \le \left(\|f\|_H + \dfrac{C}{\sqrt{2}} e(h)^{1-\delta} \|t\|_S \right)^2 \\ \qquad\qquad\qquad \le 2\|f\|_H^2 + C^2 e(h)^{2(1-\delta)} \|t\|_S^2. \quad \blacksquare \end{cases}$$

LEMMA 3-2 If $B_h(f, t)$ satisfies inequality (3-28), then there exists a constant M such that

$$(3\text{-}31) \qquad B_h(f, t) \le \|f\|_H + C^\theta M^{(1-\theta)} e(h)^{-\theta\delta} \|t\|_T. \qquad \blacktriangle$$

Proof Here again, we write $B_h(f, t) \le B_h(f, 0) + B_h(0, t) \le \|f\|_H + B_h(0, t)$. In order to estimate $B_h(0, t)$, we observe

$$(3\text{-}32) \quad \begin{cases} \text{(i)} \quad B_h(0, t) \le C e(h)^{1-\delta} \|t\|_S \qquad \text{by} \quad (3\text{-}28), \\ \text{(ii)} \quad B_h(0, t) \le \beta(h)^{-1} \|t\|_Q \le M e(h)^{-\theta/(1-\theta)} \|t\|_Q \\ \qquad\qquad\qquad\qquad\qquad \text{by } (3\text{-}26) \quad \text{with} \quad v_h = 0, \end{cases}$$

and since T' is a space of order θ between $Q = Q'$ and S', we deduce from Theorem 2.4-8 that

$$(3\text{-}33) \quad B_h(0, t) \le C^\theta e(h)^{\theta(1-\delta)} M^{1-\theta} e(h)^{-\theta} \|t\|_T \le C^\theta M^{1-\theta} e(h)^{-\theta\delta} \|t\|_T.$$

(we took $Y = X/N_h$, $V = S$, $H = Q$, $W = T$.) Therefore

$$(3\text{-}34) \quad \begin{cases} B_h(f, t)^2 \le (\|f\|_H + C^\theta M^{1-\theta} e(h)^{-\theta\delta} \|t\|_T)^2 \\ \qquad\qquad \le 2(\|f\|_H^2 + C^{2\theta} M^{2(1-\theta)} e(h)^{-2\theta\delta} \|t\|_T^2). \quad \blacksquare \end{cases}$$

LEMMA 3-3 Let us assume that $B_h(f, t)$ satisfies inequality (3-31). Then there exists a constant m independent of h, θ, f, and t such that

$$(3\text{-}35) \qquad B_h(f, t)^2 \leq 2 \, \|f\|_H^2 + m^2 M^{2(1-\theta)} C^{2\theta} e(h)^{2(1-\theta\delta)} \, \|t\|_S^2. \qquad \blacktriangle$$

Proof Let us consider the space $X_\theta = H \times T$ supplied with the norm

$$(3\text{-}36) \qquad A_h^{\,\theta}(f, t) = (\|f\|_H + C^\theta M^{(1-\theta)} e(h)^{-\theta\delta} \|t\|_T)$$

and the factor space X_θ/N_h supplied with the factor norm

$$(3\text{-}37) \qquad B_h^{\,\theta}(f, t) = \inf_{v_h \in V_h} (\|f - \Lambda p_h v_h\|_H + C^\theta M^{(1-\theta)} e(h)^{-\theta\delta} \|t - \gamma p_h v_h\|_T).$$

Inequality 3-31 amounts to writing

$$(3\text{-}38) \qquad B_h(f, t) \leq A_h^{\,\theta}(f, t) \qquad \text{for any} \quad (f, t) \in X_\theta = H \times T.$$

Then we apply Theorem 2.1-7 with $U = X_\theta$, $V = X$, and $P = N_h$: inequality (3-38) implies the inequality

$$(3\text{-}39) \qquad B_h(f, t) \leq B_h^{\,\theta}(f, t) \qquad \text{for any} \quad (f, t) \in X_\theta = H \times T.$$

It remains to estimate $B_h^{\,\theta}(f, t)$ in the same way we estimated $B_h(f, t)$ in Lemma 3-1. First, we write $B_h^{\,\theta}(f, t) \leq B_h^{\,\theta}(f, 0) + B_h^{\,\theta}(0, t) \leq \|f\|_H + B_h^{\,\theta}(0, t)$. Next, we deduce the following inequalities:

$$
\begin{aligned}
B_h^{\,\theta}(0, t) &= \inf (\|\Lambda p_h v_h\|_H + C^\theta M^{(1-\theta)} e(h)^{-\theta\delta} \|t - \gamma p_h v_h\|_T) \\
&\leq C^\theta M^{(1-\theta)} e(h)^{-\theta\delta} \inf (\|\Lambda p_h v_h\|_H + \|t - \gamma p_h v_h\|_T) \\
&= C^\theta M^{(1-\theta)} e(h)^{-\theta\delta} \inf (\|\Lambda(u_0 - p_h v_h)\|_H + \|\gamma(u_0 - p_h v_h)\|_T) \\
&\leq m_1 C^\theta M^{(1-\theta)} e(h)^{-\theta\delta} \inf \|u_0 - p_h v_h\|_V \leq m_1 C^\theta M^{(1-\theta)} e(h)^{(1-\theta\delta)} \|u_0\|_U \\
&\leq m_2 C^\theta M^{(1-\theta)} e(h)^{(1-\theta\delta)} \|t\|_S,
\end{aligned}
$$

where u_0 is the solution of $\Lambda u_0 = 0$ and $\gamma u_0 = t$.

Therefore

$$(3\text{-}40) \qquad B_h^{\,\theta}(f, t) \leq \|f\|_H + m_2 C^\theta M^{1-\theta} e(h)^{1-\theta\delta} \|t\|_S.$$

Then (3-39) and (3-40) imply

$$(3\text{-}41) \qquad \begin{cases} B_h(f, t)^2 \leq \left(\|f\|_H + \dfrac{m}{\sqrt{2}} C^\theta M^{1-\theta} e(h)^{1-\theta\delta} \|t\|_S \right)^2 \\[2mm] \qquad\qquad \leq 2 \, \|f\|_H^2 + m^2 C^{2\theta} M^{2(1-\theta)} e(h)^{2(1-\theta\delta)} \|t\|_S^2. \qquad \blacksquare \end{cases}$$

LEMMA 3-4 Let us assume (3-19), (3-20), and (3-23). Then there exists a constant M such that

$$(3\text{-}42) \qquad B_h(f, t) \leq (2 \, \|f\|_H^2 + M^2 e(h)^2 \|t\|_S^2)^{1/4}. \qquad \blacktriangle$$

Proof Lemmas 3-2 and 3-3 state that inequality (3-28) implies inequality (3-35). Then, by using these lemmas s times, we deduce the inequality

$$(3\text{-}43) \quad B_h(f, t)^2 \leq 2 \|f\|_H^2 + [m^{\theta^{s-1}} C^{\theta^s} M^{(1-\theta)(1+\cdots+\theta^s)} e(h)^{(1-\theta^s \delta)}]^2 \|t\|_S^2.$$

Therefore, when s goes to ∞, inequalities (3-43) imply (3-42).   ■

LEMMA 3-5   Let S_λ be a space such that

$$(3\text{-}44) \qquad S_\lambda' \text{ is a space of order } \lambda \text{ between } Q = Q' \text{ and } S'.$$

Then there exists a constant M such that

$$(3\text{-}45) \qquad B_h(f, t)^2 \leq \|f\|_H^2 + M^2 e(h)^{2(\lambda-\theta)/(1-\theta)} \|t\|_S^2.$$ ▲

Proof The proof of Lemma 3-5 is analogous to the proof of Lemma 3-2. First we write that $B_h(f, t) \leq \|f\|_H + B_h(0, t)$ and then we estimate $B_h(0, t)$ in the following way:

$$(3\text{-}46) \quad \begin{cases} \text{(i)} \quad B_h(0, t) \leq Me(h) \|t\|_S & \text{by Lemma 3-4,} \\ \text{(ii)} \quad B_h(0, t) \leq Me(h)^{-\theta/(1-\theta)} \|t\|_Q & \text{by (3-26),} \quad \text{with} \quad v_h = 0. \end{cases}$$

Since S_λ' is a space of order λ between Q' and S', Theorem 2.4-8 with $Y = X/N_h$, $V = S$, $H = Q$, and $W = S$ implies

$$(3\text{-}47) \quad B_h(0, t) \leq Me(h)^\lambda e(h)^{-\theta(1-\lambda)/(1-\theta)} \|t\|_S \leq Me(h)^{(\lambda-\theta)/(1-\theta)} \|t\|_S.$$ ■

In particular, since T' is a space of order θ between Q' and S', we deduce the inequality

$$(3\text{-}48) \qquad J(p_h u_h) = B_h(f, t) \leq M(\|f\|_H^2 + \|t\|_T^2)^{1/2}.$$

LEMMA 3-6   There exists a constant M such that

$$(3\text{-}49) \qquad B_h(f, t) \leq Me(h)(\|f\|_E^2 + \|t\|_S^2)^{1/2} \leq Me(h) \|u\|_U.$$ ▲

Proof The proof of Lemma 3-6 is analogous to the proof of Lemma 3-3. If we supply $X_0 = H \times T$ with the norm $A_h^0(f, t) = \|f\|_H + \|t\|_T$ and the factor space X_0/N_h with the factor norm $B_h^0(f, t) = \inf (\|f - \Lambda p_h v_h\|_H + \|t - \gamma p_h v_h\|_T)$, we deduce from (3-48) that

$$(3\text{-}50) \qquad B_h(f, t) \leq MA_h^0(f, t) \quad \text{for any} \quad (f, t) \in X_0 = H \times T.$$

Then we apply Theorem 2.1-7 with $U = X_0$, $V = X$, and $P = N_h$: inequality (3-50) implies

$$(3\text{-}51) \qquad B_h(f, t) \leq MB_h^0(f, t) \quad \text{for any} \quad (f, t) \in H \times T.$$

It remains to estimate $B_h^0(f, t)$:

$$B_h^0(f, t) = \inf \left(\|f - \Lambda p_h v_h\|_H + \|t - \gamma p_h v_h\|_T \leq c \inf \|u - p_h v_h\|_V \right.$$

$$(3\text{-}52) \qquad\qquad \leq ce(h) \|u\|_U \leq c_1 e(h)(\|f\|_E^2 + \|t\|_S^2)^{1/2}. \quad \blacksquare$$

It is clear that Lemma 3-6 implies Theorem 3-1. Actually, we are going to prove a more precise result.

THEOREM 3-2 Let us assume (3-19), (3-20), and (3-23). If we let S_λ and E_μ be spaces satisfying

$$(3\text{-}53) \qquad \begin{cases} \text{(i) } S'_\lambda \text{ is a space of order } \lambda \text{ between } Q = Q' \text{ and } S', \\ \text{(ii) } E'_\mu \text{ is a space of order } \mu \text{ between } H = H' \text{ and } E', \end{cases}$$

then there exists a constant M such that

$$(3\text{-}54) \quad J(p_h u_h) = B_h(f, t) \leq M(e_U^V(p_h)^{2\mu} \|f\|_E^2 + e_U^V(p_h)^{2(\lambda-\theta)/(1-\delta)} \|t\|_S^2)$$

$$\blacktriangle$$

Proof Indeed, $B_h(f, t) \leq B_h(f, 0) + B_h(0, t)$, and we have already estimated $B_h(0, t)$ (see Lemma 3-5). On the other hand, $B_h(f, 0)$ obeys the following estimates:

$$(3\text{-}55) \qquad \begin{cases} \text{(i) } B_h(f, 0) \leq \|f\|_H, \\ \text{(ii) } B_h(f, 0) \leq Me(h) \|f\|_E \qquad \text{by Lemma 3-6.} \end{cases}$$

Since E'_μ is a space of order μ between H' and E', we deduce that

$$(3\text{-}56) \qquad\qquad B_h(f, 0) \leq Me(h)^\mu \|f\|_E$$

by applying Theorem 2.4-8. $\qquad\qquad \blacksquare$

3-4. Least-Squares Approximations of Dirichlet Problems

Let Ω be a smooth bounded open subset of R^n, Γ its boundary, Λ a second-order differential operator, and γ_0 the trace operator defined by $\gamma_0 u = u|_\Gamma$. Let us assume that

$$(3\text{-}57) \quad \begin{cases} \text{(i) } (\Lambda, \gamma_0) \text{ is an isomorphism from } H^2(\Omega) \text{ onto } L^2(\Omega) \times H^{3/2}(\Gamma), \\ \text{(ii) } (\Lambda, \gamma_0) \text{ is an isomorphism from } H^{m+1}(\Omega) \text{ onto} \\ \qquad H^{m-1}(\Omega) \times H^{m+1/2}(\Gamma). \end{cases}$$

We approximate the solution u of the Dirichlet problem

$$(3\text{-}58) \quad \begin{cases} \text{(i) } \Lambda u = f \qquad \text{where } f \text{ is given in } L^2(\Omega), \\ \text{(ii) } \gamma_0 u = t \qquad \text{where } t \text{ is given in } H^{3/2}(\Gamma). \end{cases}$$

For that purpose let us introduce finite-element approximations $(H_h{}^\mu(\Omega),$ $p_h, r_{h,\Omega})$ associated with a function $\mu \in H^m(R^n)$ with compact support satisfying the m-criterion of convergence for $m \geq 2$ and the least-squares approximation of the Dirichlet problem (3-58):

$$(3\text{-}59) \quad \begin{cases} (\Lambda p_h u_h, \Lambda p_h v_h) + M^{-1} |h|^{-3} \langle \gamma_0 p_h u_h, \gamma_0 p_h v_h \rangle \\ \qquad = (f, \Lambda p_h v_h) + M^{-1} |h|^{-3} \langle t, \gamma_0 p_h v_h \rangle \qquad \text{for any } v_h \in H_h{}^\mu(\Omega). \end{cases}$$

Let us use $B_h(f, t)$ to denote the following measure of the error

$$(3\text{-}60) \quad B_h(f, t) = (\|f - \Lambda p_h u_h\|_{0,\Omega}^2 + M^{-1} h^{-3} \|t - \gamma_0 p_h u_h\|_{0,\Gamma}^2)^{\frac{1}{2}}.$$

THEOREM 3-3 Let us assume that (3-57) holds and that μ satisfies the criterion of m-convergence for $m \geq 2$. If $f \in L^2(\Omega)$ and $t \in H^{\frac{3}{2}}(\Gamma)$ are given, $u \in H^2(\Omega)$ is the solution of the Dirichlet problem (3-58), and $u_h \in H_h{}^\mu(\Omega)$ is the solution of its least-squares approximation (3-59), then

$$(3\text{-}61) \quad \lim_{h \to 0} B_h(f, t) = 0.$$

Moreover, if $f \in H^s(\Omega)$ for $0 \leq s \leq m - 1$ and $t \in H^r(\Omega)$ for $0 \leq r \leq m + \frac{1}{2}$, then there exists a constant M such that

$$(3\text{-}62) \quad B_h(f, t) \leq M(|h|^s \|f\|_{s,\Omega} + |h|^{r-\frac{3}{2}} \|t\|_{r,\Gamma}). \qquad \blacktriangle$$

Remark 3-1 Inequality (3-62) implies

$$(3\text{-}63) \quad \begin{cases} \text{(i) } \|\Lambda(u - p_h u_h)\|_{0,\Omega} \leq M(|h|^s \|f\|_{s,\Omega} + |h|^{r-\frac{3}{2}} \|t\|_{r,\Gamma}), \\ \text{(ii) } \|\gamma_0(u - p_h u_h)\|_{0,\Gamma} \leq M(|h|^{s+\frac{3}{2}} \|f\|_{s,\Omega} + |h|^r \|t\|_{r,\Gamma}). \end{cases}$$

In particular, if we take $s = m - 1$ and $r = m + \frac{1}{2}$, we obtain

$$(3\text{-}64) \quad \begin{cases} \text{(i) } \|\Lambda(u - p_h u_h)\|_{0,\Omega} \leq M |h|^{m-1} \|u\|_{m+1,\Omega}, \\ \text{(ii) } \|\gamma_0(u - p_h u_h)\|_{0,\Gamma} \leq M |h|^{m+\frac{1}{2}} \|h\|_{m+1,\Omega}. \end{cases} \qquad \blacksquare$$

Proof Theorem 3-3 follows from Theorem 3-2 with $V = H^2(\Omega)$, $H = L^2(\Omega)$, $T = H^{\frac{3}{2}}(\Gamma)$, $U = H^{m+1}(\Omega)$, $E = H^{m-1}(\Omega)$, $S = H^{m+\frac{1}{2}}(\Gamma)$.

By Theorem 6.3-7 we deduce that $H^{-\frac{3}{2}}(\Gamma)$ is a space of order $\theta = 3/(2m + 1)$ between $L^2(\Gamma)$ and $H^{-(m+\frac{1}{2})}(\Gamma)$ and that

$$(3\text{-}65) \quad \begin{cases} \text{(i) } H^s(\Omega)' \text{ is a space of order } \mu = \dfrac{s}{m + 1} \text{ between } L^2(\Omega) \text{ and } \\ \quad H^{m-1}(\Omega), \\ \text{(ii) } H^{-r}(\Gamma) \text{ is a space of order } \lambda = \dfrac{2r}{2m + 1} \text{ between } L^2(\Gamma) \text{ and } \\ \quad H^{-(m+\frac{1}{2})}(\Gamma). \end{cases}$$

Then assumptions (3-20) and (3-53) are satisfied.

Since by Theorem 5.2-1 $e_{m+1}^2(p_h) \leq c\,|h|^{m-1}$, assumption (3-23) will be satisfied if we choose

$$(3\text{-}66) \qquad \beta(h) = M^{\frac{1}{4}}\,|h|^{\frac{3}{4}}.$$

Then estimate (3-62) follows from estimate (3-54) when we choose

$$S_\lambda = H^r(\Gamma) \qquad \text{and} \qquad E_\mu = H^s(\Omega). \qquad \blacksquare$$

Now we estimate $\|u - p_h u_h\|_{0,\Omega}$.

LEMMA 3-7 Let us suppose the assumptions of Theorem 3-3. Then the inequality

$$(3\text{-}67) \qquad \left\{ \begin{aligned} &|(\Lambda(u - p_h u_h), \Lambda v) + M^{-1}\,|h|^{-3}\langle\gamma_0(u - p_h u_h), \gamma_0 v\rangle| \\ &\qquad \leq M B_h(f, t)(|h|^s\,\|\Lambda v\|_{s,\Omega} + |h|^{r-\frac{3}{4}}\,\|\gamma_0 v\|_{r,\Gamma}) \\ &\qquad\qquad\qquad \text{for}\quad 0 \leq s \leq m-1,\, 0 \leq r \leq m + \tfrac{1}{2}. \end{aligned} \right.$$

holds when v ranges over $H^{m+1}(\Omega)$. $\qquad \blacktriangle$

Proof Indeed, if u is the solution of the Dirichlet problem (3-58) and if u_h is the solution of (3-59), we can write

$$\begin{aligned} &|(\Lambda(u - p_h u_h), \Lambda v) + M^{-1}\,|h|^{-3}\langle\gamma_0(u - p_h u_h), \gamma_0 v\rangle| \\ &= |(\Lambda(u - p_h u_h), \Lambda(v - p_h v_h) + M^{-1}\,|h|^{-3}\langle\gamma_0(u - p_h u_h), \gamma_0(v - p_h v_h)\rangle| \\ &\leq B_h(f, t) \inf_{v_h \in V_h} (\|\Lambda(v - p_h v_h)\|_{0,\Omega}^2 + M^{-1}\,|h|^{-3}\,\|\gamma_0(v - p_h v_h)\|_{0,\Gamma}^2)^{\frac{1}{2}} \\ &\leq B_h(f, t) \cdot B_h(\Lambda v, \gamma_0 v) \leq M \cdot B_h(f, t)(|h|^s\,\|f\|_{s,\Omega} + |h|^{r-\frac{3}{4}}\,\|t\|_{r,\Gamma}), \end{aligned}$$

by using succesively the Cauchy-Schwarz inequality and Theorem 3-3. $\qquad \blacksquare$

Also, we need regularity assumptions. Since (γ_0, γ_1) maps $H^2(\Omega)$ onto $H^{\frac{3}{2}}(\Gamma) \times H^{\frac{1}{2}}(\Gamma)$ and $H_0^2(\Omega)$ is the kernel of (γ_0, γ_1) (see Theorem 6.3-1), we deduce from Theorem 6.2-6 the following Green formula: there exists a unique operator (δ_0, δ_1) mapping $H(\Omega, \Lambda^*)$ into $H^{\frac{1}{2}}(\Omega) \times H^{-\frac{3}{2}}(\Omega)$ such that

$$(3\text{-}68) \qquad \left\{ \begin{aligned} &(\Lambda u, v) = (u, \Lambda^* v) + \langle\gamma_0 u, \delta_1 v\rangle - \langle\gamma_1 u, \delta_0 v\rangle \\ &\qquad\qquad \text{for any}\quad u \in H^2(\Omega),\, v \in H(\Omega, \Lambda^*), \end{aligned} \right.$$

where Λ^* is the formal adjoint of Λ and $H(\Omega, \Lambda^*)$ is its domain.

Since (Λ, γ_0) is an isomorphism from $H^2(\Omega)$ onto $L^2(\Omega) \times H^{\frac{3}{2}}(\Gamma)$, we deduce that (Λ^*, δ_0) is an isomorphism from $H(\Omega, \Lambda^*)$ onto $L^2(\Omega) \times H^{-\frac{1}{2}}(\Gamma)$.

We further assume that

$$(3\text{-}69) \quad \begin{cases} \text{(i) } (\delta_0, \delta_1) \text{ maps } H^2(\Omega) \text{ onto } H^{3/2}(\Gamma) \times H^{1/2}(\Gamma), \\ \text{(ii) } H_0^2(\Omega) \text{ is the kernel of } (\delta_0, \delta_1), \\ \text{(iii) } (\Lambda^*, \delta_0) \text{ is an isomorphism from } H^2(\Omega) \text{ onto } L^2(\Omega) \times H^{3/2}(\Gamma). \end{cases}$$

These assumptions imply in particular that

$(3\text{-}70)$ (Λ, γ_0) is an isomorphism from $H(\Omega, \Lambda)$ onto $L^2(\Omega) \times H^{-1/2}(\Gamma)$.

LEMMA 3-8 Let us suppose the assumptions of Theorem 3-3 and (3-69). Then, for any r such that $0 \le r \le m + \frac{1}{2}$, there exists a constant M such that

$$(3\text{-}71) \quad \|\gamma_0(u - p_h u_h)\|_{-r, \Gamma} \le M \, |h|^{r + 3/2} \, B_h(f, t). \qquad \blacktriangle$$

Proof By (3-70), the Dirichlet problem $\Lambda v = 0$ and $\gamma_0 v = t$ has a unique solution $v \in H(\Omega, \Lambda)$ when t ranges over $H^r(\Omega)$. Then Lemma 3-7 implies that

$$(3\text{-}72) \quad \begin{cases} |\langle \gamma_0(u - p_h u_h), t \rangle| \le M \, |h|^{3 + r - 3/2} B_h(f, t) \, \|t\|_{r, \Gamma} \quad \text{for any} \\ \hspace{8cm} t \in H^r(\Gamma). \end{cases}$$

Inequalities 3-72 imply (3-71). $\qquad\qquad\qquad\qquad\qquad\qquad\qquad\qquad\qquad$ ■

THEOREM 3-4 If the assumptions of Theorem 3-3 and assumptions (3-69) hold, there exists a constant M such that

$$(3\text{-}73) \quad \|u - p_h u_h\|_{0, \Omega} \le M \, |h|^{\min(2, m-1)} \, B_h(f, t). \qquad \blacktriangle$$

Remark 3-2 Theorems 3-3 and 3-4 imply that if $u \in H^{m+1}(\Omega)$, there exists a constant M such that

$$(3\text{-}74) \quad \|u - p_h u_h\|_{0, \Omega} \le M \, |h|^{m - 1 + \min(2, m-1)} \, \|u\|_{m+1, \Omega} \quad \text{for } m \ge 2. \qquad \blacksquare$$

Proof Let us consider the unique solution v of the Dirichlet problem $\Lambda v = w$ and $\gamma_0 v = 0$ when w ranges over $H^2(\Omega)$. Then Lemma 3-7 implies the following inequalities:

$$(3\text{-}75) \quad \begin{cases} (\Lambda(u - p_h u_h), w) \le M \, |h|^s \, B_h(f, t) \, \|w\|_{2, \Omega} \\ \hspace{5cm} \text{where} \quad s = \min(2, m - 1). \end{cases}$$

By applying the Green formula to the left-hand side of (3-75), we obtain

$$(3\text{-}76) \quad \begin{cases} |(u - p_h u_h, \Lambda^* w)| \le M \, |h|^s \, B_h(f, t) \, \|w\|_{2, \Omega} + |\langle \gamma_0(u - p_h u_h), \delta_1 w \rangle| \\ \hspace{3cm} + |\langle \gamma_1(u - p_h u_h), \delta_0 w \rangle| \quad \text{for any} \quad w \in H^2(\Omega). \end{cases}$$

Since (Λ^*, δ_0) is an isomorphism from $H^2(\Omega)$ onto $L^2(\Omega) \times H^{3/2}(\Gamma)$, we deduce that for any $v \in L^2(\Omega)$, there exists a unique solution $w \in H^2(\Omega)$ satisfying $\Lambda w = v$, $\gamma_0 w = 0$, and $\|\delta_1 w\|_{1/2,\Gamma} \le c \|v\|_{0,\Omega}$. Therefore (3-76) implies that

$$(3\text{-}77) \quad \begin{cases} |(u - p_h u_h, v)| \le M |h|^s B_h(f, t) \|v\|_{0,\Omega} + c \|\gamma_0(u - p_h u_h)\|_{-1/2,\Gamma} \\ \qquad \times \|v\|_{0,\Omega} \le M |h|^s B_h(f, t) \|v\|_{0,\Omega} + M |h|^2 B_h(f, t) \|v\|_{0,\Omega} \end{cases}$$

by Lemma 3-8 with $r = \frac{1}{2}$.

Finally, it is clear that (3-74) follows from (3-77). ∎

CHAPTER 10

Conjugate Problems and A Posteriori Error Estimates

We begin by giving two examples of conjugate problems of a boundary-value problem and their applications to obtaining a posteriori error estimates. Then we study a general method of construction of conjugate problems.

In Section 2 we approximate the solution of a Dirichlet problem by the internal approximation of its usual variational formulation (Section 2-1) and by the internal approximation of the first example of conjugate problem (Section 2-2).

For approximating the second example of conjugate problems, we need approximations of the domain $\mathbf{H}^k(\Omega, D^*)$ of the operator D^* mapping a vector function $\mathbf{u} = (u^q)_{|q| \le k}$ such that each component u^q belongs to $L^2(\Omega)$ onto $D^*\mathbf{u} = \sum_{|q| \le k} (-1)^{|q|} D^q u^q$. Such approximations are constructed in Section 3, and the internal approximations of the second example of conjugate problems are examined in Section 4.

1. CONJUGATE PROBLEMS OF BOUNDARY-VALUE PROBLEMS

1-1. First Example of a Conjugate Problem

Let Ω be a smooth bounded open subset of R^n, Γ its boundary, $a_{pq}(x)$ functions belonging to $L^\infty(\Omega)$, and Λ the differential operator defined by

$$(1-1) \qquad \Lambda u = \sum_{|p|, |q| \le k} (-1)^{|q|} D^q(a_{pq}(x) D^p u),$$

which is the formal operator associated with the bilinear form

$$(1-2) \qquad a(u, v) = \sum_{|p|, |q| \le k} \int_\Omega a_{pq}(x) D^p u D^q v \, dx$$

by the Green formula

$$a(u, v) = (\Lambda u, v) + \sum_{0 \le j \le k-1} \langle \delta_{2k-j-1} u, \gamma_j v \rangle$$

(1-3) for any $u \in H^k(\Omega, \Lambda), v \in H^k(\Omega)$.

(See Section 7.2-1.)

Let us introduce the following data:

(1-4) $\begin{cases} \text{(i) } f \in H = L^2(\Omega), \\ \text{(ii) } g_j \in H^{k-j-\frac{1}{2}}(\Gamma) \quad \text{for } 0 \le j \le p-1 \quad \text{where } 1 \le p \le k, \\ \text{(iii) } h_j \in H^{k-j-\frac{1}{2}}(\Gamma) \quad \text{for } k \le j \le 2k-p-1, \end{cases}$

and let us consider a solution $u \in H^k(\Omega, \Lambda)$ (if any) of the following boundary-value problem:

(1-5) $\begin{cases} \text{(i) } \Lambda u + \lambda u = f, \\ \text{(ii) } \gamma_j u = g_j \quad \text{for } 0 \le j \le p-1, \\ \text{(iii) } \delta_j u = h_j \quad \text{for } k \le j \le 2k-p-1. \end{cases}$

We already know that any solution $u \in H^k(\Omega, \Lambda)$ of (1-5) solves the variational equation

(1-6) $\begin{cases} \text{(i) } u \in H^k(\Omega), \\ \text{(ii) } \gamma_j u = g_j \quad \text{for } 0 \le j \le p-1, \\ \text{(iii) } a(u, v) = (f, v) + \sum_{p \le j \le k-1} \langle h_{2k-j-1}, \gamma_j v \rangle \quad \text{for any } v \in V_1, \end{cases}$

where

(1-7) $V_1 = \{v \in H^k(\Omega) \quad \text{such that } \gamma_j v = 0 \quad \text{for } 0 \le j \le p-1\}.$

In this variational formulation, the boundary conditions [(1-5)(ii)] are forced and the boundary conditions [(1-5)(iii)] are natural. However, we can construct another variational formulation of boundary-value problem (1-5) *for which the boundary conditions [(1-5)(ii)] are natural and the boundary conditions [(1-5)(iii)] are forced.*

THEOREM 1-1 Any solution $u \in H^k(\Omega, \Lambda)$ of boundary-value problem (1-5) is a solution of the variational equation

(1-8) $\begin{cases} \text{(i) } u \in H^k(\Omega, \Lambda), \\ \text{(ii) } \delta_j u = h_j \quad \text{for } k \le j \le 2k-p-1, \\ \text{(iii) } (\Lambda u, \Lambda v) + \lambda a_*(u, v) = (f, \Lambda v) + \lambda \sum_{0 \le j \le p-1} \langle g_j, \delta_{2k-j-1} v \rangle \\ \qquad\qquad\qquad\qquad \text{when } v \text{ ranges over } H_2^k(\Omega, \Lambda) \end{cases}$

where

$$H_2{}^k(\Omega, \Lambda) = \{v \in H^k(\Omega, \Lambda) \quad \text{such that} \quad \delta_j v = 0 \quad \text{for}$$
$$(1\text{-}9) \qquad\qquad\qquad\qquad\qquad\qquad\qquad k \leq j \leq 2k - p - 1\}.$$

Conversely, if $a(u, v)$ is $H^k(\Omega)$-coercive and if 0 is not an eigenvalue of the Neumann problem for Λ, any solution u of the variational equation (1-8) is solution of the boundary-value problem (1-5). ▲

Proof We deduce Theorem 1-1 from Corollary 1-1 of Theorem 1-5. ■

In particular, in the case of Dirichlet problems, we have obtained a variational formulation for which the Dirichlet conditions are natural. Furthermore, we prove the following a posteriori error estimates. (see Theorem 1-6).

THEOREM 1-2 Let u be the solution of the boundary-value problem (1-5), $v \in H^k(\Omega)$ any functions satisfying the boundary conditions $\gamma_j v = g_j$ for $0 \leq j \leq p - 1$ and $\hat{v} \in H^k(\Omega, \Lambda)$ be any function satisfying the boundary conditions $\delta_j \hat{v} = h_j$ for $k \leq j \leq 2k - p - 1$. We assume that

$$(1\text{-}10) \quad \begin{cases} \text{(i)} \quad |a(u, v)| \leq M \, \|u\|_{k,\Omega} \|v\|_{k,\Omega} & \text{for any} \quad u, v \in H^k(\Omega), \\ \text{(ii)} \quad a(v, v) \geq c \, \|v\|_{k,\Omega}^2; c > 0 & \text{for any} \quad v \in H^k(\Omega). \end{cases}$$

Then, if $\lambda > 0$, the following a posteriori error estimates hold:

$$(1\text{-}11) \quad \begin{cases} \text{(i)} \quad c \, \|u - v\|_{k,\Omega}^2 + \lambda \, \|u - v\|_{0,\Omega}^2 \\ \qquad\qquad \leq \lambda^{-1} \|f - \lambda v - \Lambda \hat{v}\|_{0,\Omega}^2 + M^2 c^{-1} \, \|v - \hat{v}\|_{k,\Omega}^2, \\ \text{(ii)} \quad c \, \|u - \hat{v}\|_{k,\Omega}^2 + \lambda^{-1} \|\Lambda(u - \hat{v})\|_{0,\Omega}^2 \\ \qquad\qquad \leq \lambda^{-1} \|f - \lambda v - \Lambda \hat{v}\|_{0,\Omega}^2 + M^2 c^{-1} \, \|v - \hat{v}\|_{k,\Omega}^2. \end{cases}$$
▲

In other words, knowing approximate solutions $v \in H^k(\Omega)$ satisfying [(1-15)(ii)] and approximate solutions $\hat{v} \in H^k(\Omega, \Lambda)$ satisfying [(1-5)(iii)], we can estimate the actual errors $u - v$ and $u - \hat{v}$ by the right-hand side of inequalities (1-11). *First of all, this term does not depend on the unknown solution u of (1-5) and second, it converges to 0 whenever both v and $\hat{v}$ converge to u.*

1-2. Second Example of a Conjugate Problem

Another variational formulation of the boundary-value problem (1-5) exchanges the forced boundary conditions for natural boundary conditions, and vice versa. For that purpose, we need the space

(1-12) $\quad \mathbf{H}(\Omega) = (L^2(\Omega))^N = \{\mathbf{u} = (u^q)_{|q| \leq k} \quad$ such that $\quad u^q \in L^2(\Omega)\}$

supplied with the following inner product and norm:

(1-13) $\qquad\qquad [\mathbf{u}, \mathbf{v}] = \sum_{|q| \leq k} \int_\Omega u^q v^q \, dx, \quad [\mathbf{u}] = ([\mathbf{u}, \mathbf{u}])^{\frac{1}{2}}$

and the following operators

(1-14) $\begin{cases} \text{(i)} \ D \in L(H^k(\Omega), \mathbf{H}(\Omega)) \quad \text{defined by} \quad Du = (D^q u)_{|q| \leq k}, \\[2mm] \text{(ii)} \ \mathscr{A} \in L(\mathbf{H}(\Omega)), \mathbf{H}(\Omega)) \quad \text{defined by} \quad (\mathscr{A}\mathbf{u})^q = \sum_{|p| \leq k} a_{pq}(x) u^p, \\[2mm] \text{(iii) the operator } D^* \quad \text{defined by} \quad D^*\mathbf{u} = \sum_{|p| \leq k} (-1)^{|p|} D^p u^p \\[2mm] \qquad \text{mapping} \quad \mathbf{H}^k(\Omega, D^*) = \{\mathbf{u} \in \mathbf{H}(\Omega) \quad \text{such that} \\[2mm] \qquad\qquad\qquad\qquad\qquad\qquad D^*\mathbf{u} \in L^2(\Omega)\} \text{ into } L^2(\Omega). \end{cases}$

The operator $D^* = (D|_{H_0^k(\Omega)})' \in L(\mathbf{H}(\Omega), H^{-k}(\Omega))$ can be viewed as the formal adjoint of the operator D. Thus, by the Green formula proved in Theorem 6.2-6 and the trace Theorem 6.3-1, there exist unique operators $\beta_j \in L(\mathbf{H}^k(\Omega, D^*), H^{k-j-1/2}(\Gamma))$ for $k \leq j \leq 2k - 1$ such that

(1-15) $\begin{cases} [\mathbf{u}, Dv] = (D^*\mathbf{u}, v) + \sum_{0 \leq j \leq k-1} \langle \beta_{2k-j-1}\mathbf{u}, \gamma_j v \rangle, \\[2mm] \qquad \text{when } \mathbf{u} \text{ ranges over } \mathbf{H}^k(\Omega, D^*) \text{ and } v \text{ over } H^k(\Omega). \end{cases}$

Then it is clear that

(1-16) $\begin{cases} \text{(i)} \ \Lambda u = D^* \ \mathscr{A} \ Du \quad \text{for any} \quad u \in H^k(\Omega), \\[2mm] \text{(ii)} \ \delta_j u = \beta_j \ \mathscr{A} \ Du \quad \text{for any} \quad u \in H^k(\Omega, \Lambda). \end{cases}$

We can deduce from Corollary 1-2 of Theorem 1-5 the following result.

THEOREM 1-3 Let us assume that

(1-17) $\qquad\qquad \mathscr{A}$ is an isomorphism from $\mathbf{H}(\Omega)$ onto itself.

If $u \in H^k(\Omega, \Lambda)$ is a solution of boundary-value problem (1-5), then $\mathbf{u} = \mathscr{A} Du$ is a solution of the variational equation

(1-18) $\begin{cases} \text{(i)} \ \mathbf{u} \in \mathbf{H}^k(\Omega, D^*), \\[2mm] \text{(ii)} \ \beta_j \mathbf{u} = h_j \quad \text{for} \quad k \leq j \leq 2k - p - 1, \\[2mm] \text{(iii)} \ (D^*\mathbf{u}, D^*\mathbf{v}) + \lambda[\mathscr{A}^{-1}\mathbf{u}, \mathbf{v}] = (f, D^*\mathbf{v}) + \lambda \sum_{0 \leq j \leq p-1} \langle g_j, \beta_{2k-j-1}\mathbf{v} \rangle \\[2mm] \qquad\qquad\qquad \text{when } \mathbf{v} \text{ ranges over } \mathbf{H}_2^k(\Omega, D^*), \end{cases}$

where

$$(1\text{-}19) \quad \mathbf{H}_2^k(\Omega, D^*) = \{\mathbf{u} \in \mathbf{H}^k(\Omega, D^*) \quad \text{such that} \quad \beta_j \mathbf{u} = 0$$
$$\text{for} \quad k \le j \le 2k - p - 1\}.$$

Conversely, if $\mathbf{u} \in \mathbf{H}^k(\Omega, D^*)$ is a solution of the variational equation (1-18), then the function $u = \lambda^{-1}(f - D^*\mathbf{u})$ is a solution of boundary-value problem (1-5).   ▲

Remark 1-1 We can prove that a solution u of (1-18) is a solution of the system of differential equations

$$(1\text{-}20) \quad \begin{cases} \text{(i)} \quad \sum_{|p| \le k} [(-1)^{|p|} D^{p+q} u^p + b_{pq}(x) u^p] = D^q f \quad \text{for} \quad |q| \le k, \\[2mm] \text{(ii)} \quad \beta_j \mathbf{u} = h_j \quad \text{for} \quad k \le j \le 2k - p - 1, \\[2mm] \text{(iii)} \quad \gamma_j \left(f - \sum_{|p| \le k} (-1)^{|p|} D^p u^p \right) = \lambda g_j \quad \text{for} \quad 0 \le j \le p - 1 \end{cases}$$

where the functions $b_{pq}(x)$ are the entries of the matrix $\mathscr{A}^{-1}$.   ■

We also prove the following error estimates (see Theorem 1-6).

THEOREM 1-4 Let us assume that

$$(1\text{-}21) \quad \begin{cases} \text{(i)} \quad |[\mathscr{A}\mathbf{u}, \mathbf{v}]| \le M[\mathbf{u}][\mathbf{v}] \quad \text{for any} \quad \mathbf{u}, \mathbf{v} \in \mathbf{H}^k(\Omega, D^*), \\[2mm] \text{(ii)} \quad [\mathscr{A}\mathbf{v}, \mathbf{v}] \ge c[\mathbf{v}]^2 \quad \text{for any} \quad \mathbf{v} \in \mathbf{H}^k(\Omega, D^*); \ c > 0. \end{cases}$$

Let $u \in H^k(\Omega, \Lambda)$ be the solution of the boundary-value problem (1-5), $v \in H^k(\Omega)$ any function satisfying $\gamma_j v = g_j$ for $0 \le j \le p - 1$, and $\mathbf{v} \in \mathbf{H}^k(\Omega, D^*)$ any vector function satisfying $\beta_j \mathbf{v} = h_j$ for $k \le j \le 2k - p - 1$. Then if $\lambda > 0$, the following a posteriori estimates hold:

$$(1\text{-}22) \quad \begin{cases} \text{(i)} \quad c^2 \|u - v\|_{k,\Omega}^2 + \lambda \|u - v\|_{0,\Omega}^2 \le \lambda^{-1} \|f - v - D^*\mathbf{v}\|_{0,\Omega}^2 \\[1mm] \hspace{4cm} + M^2 c^{-1} [\mathbf{v} - Dv]^2, \\[2mm] \text{(ii)} \quad c[\mathbf{u} - \mathbf{v}]^2 + \lambda^{-1} \|D^*(\mathbf{u} - \mathbf{v})\|_{0,\Omega}^2 \le \lambda^{-1} \|f - v - D^*\mathbf{v}\|_{0,\Omega}^2 \\[1mm] \hspace{4cm} + M^2 c^{-1} [\mathbf{v} - Dv]^2. \end{cases}$$

The right-hand side of these a posteriori estimates does not depend on the unknown solution u of (1-5) and it converges to 0 when v converges to the solution u of (1-5) and $\mathbf{v}$ converges to the solution $\mathbf{u}$ of (1-18).   ▲

1-3. Construction of Conjugate Problems

The two foregoing examples are particular cases of the following general construction of variational formulations and conjugate variational formulations of a boundary-value problem. As usual, we assume that there exists a pivot space H, Hilbert spaces V and T, and an operator $\gamma \in L(V, T)$ satisfying

$$(1\text{-}23) \quad \begin{cases} \text{(i) } \gamma \text{ maps } V \text{ onto } T, \\ \text{(ii) } V \text{ is contained in } H \text{ with a stronger topology,} \\ \text{(iii) the kernel } V_0 \text{ of } \gamma \text{ is dense in } H. \end{cases}$$

Let E be another Hilbert space, and introduce the operators

$$(1\text{-}24) \quad P \in L(V, E) \quad Q \in L(E, E') \quad G = QP \in L(V, E')$$

and the formal adjoint $G^* = (G|_{V_0})' \in L(E, V_0')$ of G defined by $(G^*\hat{u}, v) = [\hat{u}, Gv]$ for any $\hat{u} \in E$, $v \in V_0$ (where $[\hat{u}, \hat{v}]$ denotes the duality pairing on $E' \times E$),

We consider the class of operators Λ defined by

$$(1\text{-}25) \quad \Lambda = G^*P \in L(V, V_0').$$

Let $E(G^*)$ and $V(\Lambda)$ be the domains of G^* and Λ defined by

$$(1\text{-}26) \quad \begin{cases} \text{(i) } E(G^*) = \{\hat{u} \in E \quad \text{such that} \quad G^*\hat{u} \in H\}, \\ \text{(ii) } V(\Lambda) = \{u \in V \quad \text{such that} \quad \Lambda u \in H\} = \{u \in V \quad \text{such that} \\ \qquad\qquad\qquad\qquad\qquad\qquad\qquad\qquad Pu \in E(G^*)\}. \end{cases}$$

Then Theorem 6.2-6 implies the existence of an operator $\beta \in L(E(G^*), T')$ such that the Green formula

$$(1\text{-}27) \quad (u, G^*\hat{v}) = [Gu, \hat{v}] - \langle \gamma u, \beta \hat{v} \rangle \quad \text{for any} \quad u \in V, \hat{v} \in E(G^*)$$

holds. When we consider the continuous bilinear form

$$a(u, v) = [Pu, Gv] = [Pu, QPv]$$

defined on V, it is clear that Λ is its formal operator and that

$$(1\text{-}28) \quad \begin{cases} \text{(i) } a(u, v) = (\Lambda u, v) + \langle \delta u, \gamma v \rangle \quad \text{for any} \quad u \in V(\Lambda), v \in V, \\ \text{(ii)} \qquad \delta = \beta P \in L(V(\Lambda), T') \end{cases}$$

(see Theorem 6.2-1).

We now study variational formulations of the following σ_1-boundary-value problem:

$$(1\text{-}29) \quad \begin{cases} \text{(i) } u \in V(\Lambda), \\ \text{(ii) } \Lambda u + \lambda u = f \quad \text{where} \quad f \text{ is given in } H, \\ \text{(iii) } \gamma_1 u = t_1 \quad \text{where} \quad t_1 \text{ is given in } T_1 = \sigma_1 T, \ \gamma_1 = \sigma_1 \gamma, \\ \text{(iv) } \delta_2 u = t_2 \quad \text{where} \quad t_2 \text{ is given in } T_2' = \sigma_2' T, \ \sigma_2 = 1 - \sigma_1, \\ \hspace{11cm} \delta_2 = \sigma_2' \delta \end{cases}$$

(see Section 6.2-5). For $i = 1, 2$, let us set

$$(1\text{-}30) \quad \begin{cases} V_i = \{u \in V \quad \text{such that} \quad \gamma_i u = 0\}, \\ \hspace{2cm} E_i(G^*) = \{u \in E(G^*) \quad \text{such that} \quad \delta_i u = 0\}. \end{cases}$$

Theorem 6.2-2 implies that this boundary-value problem is equal to the variational equation

$$(1\text{-}31) \quad \begin{cases} \text{(i) } u \in V, \\ \text{(ii) } \gamma_1 u = t_1, \\ \text{(iii) } a(u, v) + \lambda(u, v) = (f, v) + \lambda\langle t_2, \gamma_2 v \rangle \quad \text{for any} \quad v \in V_1. \end{cases}$$

To construct the conjugate problem associated with the splitting $\Lambda = G^*P$, we assume that

$$(1\text{-}32) \quad E_0(G^*) = \langle \hat{v} \in E(G^*) \quad \text{such that} \quad \beta\hat{v} = 0\} \text{ is dense in } E.$$

Then,

$(1\text{-}33)$ G can be extended to a continuous operator from H into $E_0(G^*)'$.

Indeed, the adjoint $G^{**} = (G^*|_{E_0(G^*)})' \in L(H, E_0(G^*)')$ of G^* is an extension of $G \in L(V, E')$, since V is dense in H, E' is dense in $E_0(G^*)'$ and, furthermore, $[G^{**}u, \hat{v}] = (u, G^*\hat{v}) = [Gu, \hat{v}]$ when $u \in V$ and $\hat{v} \in E_0(G^*)$. Then $G^{**}u = Gu$ for any $u \in V$ and we can state $(1\text{-}33)$ by setting $G = G^{**}$. Also, we assume that

$$(1\text{-}34) \quad \begin{cases} \text{(i) } V = H(G) = \{v \in H \quad \text{such that} \quad Gv \in E'\}, \\ \text{(ii) } Q \in L(E, E') \text{ is one to one.} \end{cases}$$

THEOREM 1-5 Let us assume $(1\text{-}23)$ through $(1\text{-}25)$. If u is a solution of the boundary-value problem $(1\text{-}29)$, then

$$(1\text{-}35) \hspace{4cm} \hat{u} = Pu$$

is a solution of the variational equation

$$(1\text{-}36) \quad \begin{cases} \text{(i)} \ \hat{u} \in E(G^*), \\ \text{(ii)} \ \beta_2 \hat{u} = t_2, \\ \text{(iii)} \ (G^*\hat{u}, G^*\hat{v}) + \lambda[Q\hat{u}, \hat{v}] = (f, G^*\hat{v}) + \lambda\langle t_1, \beta_1\hat{v}\rangle \\ \qquad\qquad\qquad\qquad\qquad\qquad \text{for any} \quad \hat{v} \in E_2(G^*). \end{cases}$$

Conversely, if the assumptions (1-32) and (1-34) are satisfied and $\hat{u}$ is a solution of the variational equation (1-36), then

$$(1\text{-}37) \qquad\qquad\qquad u = \lambda^{-1}(f - G^*\hat{u})$$

is a solution of boundary-value problem (1-29). $\quad\blacktriangle$

Proof When $\hat{v}$ ranges over $E_2(G^*)$, $G^*\hat{v}$ belongs to E and thus, by applying the Green formula to

$$(1\text{-}38) \qquad\qquad (\Lambda u, G^*\hat{v}) + \lambda(u, G^*\hat{v}) = (f, G^*\hat{v}),$$

we obtain

$$(1\text{-}39) \quad (\Lambda u, G^*\hat{v}) + \lambda[Gu, \hat{v}] = (f, G^*\hat{v}) + \lambda\langle\gamma_1 u, \beta_1\hat{v}\rangle \text{ for any } \hat{v} \in E_2(G^*).$$

Then if we set $\hat{u} = Pu \in E(G^*)$, we deduce (1-36) from (1-39), because $\Lambda u = G^*Pu = G^*\hat{u}$, $\quad Gu = QPu = Q\hat{u}$, $\quad \gamma_1 u_1 = t_1$, and $\quad \delta_2 u = \beta_2 Pu = \beta_2\hat{u} = t_2$.

Conversely, let $\hat{u}$ be a solution of (1-36) and set $u = \lambda^{-1}(f - G^*\hat{u})$. When $\hat{v}$ ranges over $E_0(G^*)$, [(1-36)(iii)] implies that $GG^*\hat{u} + \lambda Q\hat{u} = Gf$. Therefore, $Gu = QPu = G(\lambda^{-1}(f - G^*u)) = Q\hat{u}$. This implies that $u \in V$—by [(1-34)(i)] —and that $Pu = \hat{u}$—by [(1-34)(ii)]. Furthermore, we obtain $u \in V(\Lambda) = \{u \in V \text{ such that } Pu \in E(G^*)\}$, $\delta_2 u = \beta_2 Pu = t_2$, and $\Lambda u + \lambda u = G^*\hat{u} + \lambda u = f$. It remains to prove that $\gamma_1 u = t_1$.

By applying the Green formula to [(1-36)(iii)] written in the following way

$$(\Lambda u, G^*\hat{v}) + \lambda[Gu, \hat{v}] = (f, G^*\hat{v}) + \lambda\langle t, \beta_1\hat{v}\rangle \qquad \text{for any} \quad \hat{v} \in E_2(G^*)$$

we deduce that

$$(1\text{-}40) \qquad\qquad \langle t_1 - \gamma_1 u, \beta_1\hat{v}\rangle = 0 \qquad \text{for any} \quad \hat{v} \in E_2(G^*). \quad\blacksquare$$

We deduce that $\gamma_1 u = t_1$ from Lemma 1-1.

LEMMA 1-1 Under the assumptions of Theorem 1-5, β_1 maps $E_2(G^*)$ onto T_1'. $\quad\blacktriangle$

Proof First, the space $V = H(G)$ is a Hilbert space for the graph norm $(\|G\hat{v}\|_{E'}{}^2 + \|\hat{v}\|_H{}^2)^{1/2}$. Indeed, if u_n is a Cauchy sequence for this graph norm, then u_n is a Cauchy sequence in H and Gu_n is a Cauchy sequence in E'. Therefore

(1-41) u_n converges to u in H and Gu_n converges to f in E'.

We have to show that $f = Gu$—which implies that $u \in V$ by [(1-34)(i)]. Indeed, when $\hat{w} \in E_0(G^*)$, $(u_n, G^*\hat{w}) = [Gu_n, \hat{w}]$ converges to $(u, G^*\hat{w}) = [f, \hat{w}] = [Gu, \hat{w}]$. Since $E_0(G^*)$ is dense in E, this implies that $f = Gu$. Now let us prove that β_1 maps $E_2(G^*)$ onto T_1'. For that purpose, consider the solution $u \in V$ of the variational equation

(1-42) $b(u, v) = [KGu, Gv] + (u, v) = \langle t_1, \gamma v \rangle$ for any $v \in V$,

where $t_1 \in T_1'$ and $K \in L(E', E)$ is the canonical isometry from E' onto E.

The bilinear form $b(u, v)$ is V-elliptic, since $b(v, v)^{1/4}$ is the graph norm of V. Then there exists a unique solution $u \in V$ of (1-42), which is the solution of

(1-43) $G^*KGu = -u$ $\beta_1 KGu = t_1$ and $\beta_2 KGu = 0$.

Then $\hat{u} = KGu$ belongs to $E_2(G^*)$ and is a solution of $\beta_1 \hat{u} = t_1$. ∎

Definition 1-1 The variational equation (1-36) is the conjugate problem of the boundary-value problem (1-29) associated with the splitting $\Lambda = G^*P$ of Λ. ▲

Remark 1-2 A solution $\hat{u}$ of the variational equation (1-29) is a solution of the boundary-value problem

(1-44)
$$\begin{cases} \text{(i) } GG^*\hat{u} + Q\hat{u} = Gf, \\ \text{(ii) } \beta_2\hat{u} = t_2, \\ \text{(iii) } \gamma_1(f - G^*\hat{u}) = \lambda t_1. \end{cases}$$

Let us notice that solutions of the boundary-value problems

(1-45)
$$\begin{cases} \text{(i) } \hat{u} \in E(GG^*) = \{\hat{u} \in E(G^*) \quad \text{such that} \quad G\hat{u} \in E'\}, \\ \text{(ii) } GG^*\hat{u} + Q\hat{u} = f \quad \text{where } f \text{ is given in } E', \\ \text{(iii) } \gamma_1 G^*\hat{u} = t_1 \quad \text{where } t_1 \text{ is given in } T_1', \\ \text{(iv) } \beta_2\hat{u} = t_2 \quad \text{where } t_2 \text{ is given in } T_2' \end{cases}$$

are solutions of the variational equation

(1-46)
$$\begin{cases} \text{(i) } \hat{u} \in E(G^*), \\ \text{(ii) } \beta_2\hat{u} = t_2, \\ \text{(iii) } (G^*\hat{u}, G^*\hat{v}) + \lambda[Q\hat{u}, \hat{v}] = [f, \hat{v}] + \langle t_1, \beta_1\hat{v} \rangle \\ \qquad\qquad\qquad\qquad\qquad \text{for any } \hat{v} \in E_2(G^*), \end{cases}$$

and conversely. Such solutions exist and are unique if, for instance, either Q is E-elliptic and $\lambda > 0$, or, more generally, if Q is compact from $E(G^*)$ into E and λ is not an eigenvalue of (1-44). ∎

THEOREM 1-6 Let us assume that

$$(1\text{-}47) \quad \begin{cases} \text{(i)} \;\; |[Q\hat{u}, \hat{v}]| \leq M \, \|u\|_E \, \|v\|_E & \text{for any} \;\; u, v \in E, \\ \text{(ii)} \;\; [Q\hat{v}, \hat{v}] \geq c \, \|v\|_E^2 & \text{for any} \;\; v \in E; \, c > 0. \end{cases}$$

Let u be a solution of the boundary-value problem (1-29), $v \in V$ satisfying $\gamma_1 v = t_1$ and $\hat{v} \in E(G^*)$ satisfying $\beta_2 \hat{v} = t_2$. Then the following a posteriori error estimates hold:

$$(1\text{-}48) \quad \begin{cases} \text{(i)} \;\; c \, \|P(u-v)\|_E^2 + \lambda \, \|u-v\|_H^2 \leq \lambda^{-1} \, \|f - \lambda v - G^*v\|_H^2 \\ \qquad\qquad\qquad\qquad\qquad\qquad\qquad + M^2 c^{-1} \, \|Pv - \hat{v}\|_E^2, \\ \text{(ii)} \;\; c \, \|\hat{u} - \hat{v}\|_E^2 + \lambda^{-1} \, \|G^*(\hat{u} - \hat{v})\|_H^2 \leq \lambda^{-1} \, \|f - \lambda v - G^*\hat{v}\|_H^2 \\ \qquad\qquad\qquad\qquad\qquad\qquad\qquad + M^2 c^{-1} \, \|Pv - \hat{v}\|_E^2. \;\; \blacktriangle \end{cases}$$

Proof Indeed, $w = u - v$ belongs to V_1 because $\gamma_1(u - v) = 0$. Then

$$(1\text{-}49) \quad a(u-v, w) + \lambda(u-v, w) = (f - \lambda v, w) - [Pv, QPw] + \langle t_2, \gamma_2 w \rangle.$$

Now by the Green formula we deduce that

$$(1\text{-}50) \quad \langle t_2, \gamma_2 w \rangle = \langle \beta_2 \hat{v}, \gamma_2 w \rangle = [\hat{v}, Gw] - (G^*\hat{v}, w).$$

Then (1-49) and (1-50) imply that

$$(1\text{-}51) \quad a(u-v, w) + \lambda(u-v, w) = (f - \lambda v - G^*\hat{v}, w) + [\hat{v} - Pv, QPw].$$

Since $w = u - v$, we deduce from the Cauchy-Schwarz inequality that

$$(1\text{-}52) \quad \begin{cases} c \, \|P(u-v)\|_E^2 + \lambda \, \|u-v\|_H^2 \\ \quad \leq \lambda^{-\frac{1}{2}} \, \|f - \lambda v - G^*\hat{v}\|_H \, \lambda^{\frac{1}{2}} \, \|u-v\|_H \\ \quad + Mc^{-\frac{1}{2}} \, \|\hat{v} - Pv\|_E \, c^{\frac{1}{2}} \, \|P(u-v)\|_E \leq (c \, \|P(u-v)\|_E^2 \\ \quad + \lambda \, \|u-v\|_H^2)^{\frac{1}{2}} (\lambda^{-1} \, \|f - \lambda v - G^*\hat{v}\|_H^2 + M^2 c^{-1} \, \|\hat{v} - Pv\|_E^2)^{\frac{1}{2}}. \end{cases}$$

We can prove [(1-48)(ii)] in the same way: $\hat{w} = \hat{u} - \hat{v}$ belongs to $E_2(G^*)$, and thus

$$(1\text{-}53) \quad \begin{cases} \text{(i)} \;\; (G^*(\hat{u} - \hat{v}), G^*\hat{w}) + \lambda[Q(\hat{u} - \hat{v}), \hat{w}] = (f - G^*\hat{v}, G^*\hat{w}) \\ \qquad\qquad\qquad\qquad\qquad\qquad\qquad\qquad\qquad - [Q\hat{v}, \hat{w}] + \langle t_1, \beta_1 \hat{w} \rangle, \\ \text{(ii)} \;\; \langle t_1, \beta_1 \hat{w} \rangle = \langle \gamma_1 v, \beta_1 \hat{w} \rangle = [\hat{w}, Gv] - (G^*\hat{w}, v). \end{cases}$$

Therefore we obtain

$$(G^*(\hat{u} - \hat{v}), G^*\hat{w}) + \lambda[Q(\hat{u} - \hat{v}), \hat{w}] = (f - \lambda v - G^*\hat{v}, G^*\hat{w})$$

$$+ [Q(Pv - \hat{v}), w].$$

By taking $\hat{w} = \hat{u} - \hat{v}$ and by using the Cauchy-Schwarz inequality, we deduce [(1-48)(ii)]. ∎

EXAMPLE 1-1 Let us assume (1-23), letting $a(u, v)$ be a continuous bilinear form on V; and $\Lambda \in L(V(\Lambda), H)$ and $\delta \in L(V(\Lambda), T')$ are, respectively, its formal operator and its Neumann operator.

COROLLARY 1-1 Any solution u of (1-29) is a solution of the variational equation

$$(1\text{-}54) \quad \begin{cases} \text{(i) } u \in V(\Lambda), \\ \text{(ii) } \delta_2 u = t_2, \\ \text{(iii) } (\Lambda u, \Lambda v) + \lambda a_*(u, v) = (f, \Lambda v) + \langle t_1, \delta_1 v \rangle \\ \qquad\qquad\qquad\qquad\qquad\qquad \text{for any } \quad v \in V_2(\Lambda). \end{cases}$$

Conversely, if $a(u, v)$ is (V, H) coercive and if 0 is not an eigenvalue of the Neumann problem, any solution u of (1-54) is a solution of the boundary-value problem (1-29). ▲

Proof We take $E = V, P = 1, Q = G = A^* \in L(V, V')$ where A^* is the operator associated with the bilinear form $a_*(u, v) = a(v, u)$. Then

$$(1\text{-}55) \qquad\qquad G^* = (A^*|_{V_0})' = \Lambda,$$

since $(G^*u, v) = (u, A^*v) = a_*(v, u) = a(u, v) = (\Lambda u, v)$ when $u \in V, v \in V_0$.

Therefore, $E(G^*) = V(\Lambda)$, $\beta = \delta$, and, by Theorem 1-5, any solution u of the boundary-value problem (1-29) is a solution of the conjugate variational equation (1-54). The converse is true, since assumptions (1-32) and (1-34) of theorem 1-5 are satisfied. Indeed,

$$(1\text{-}56) \qquad E_0(G^*) = \{u \in V(\Lambda) \quad \text{such that} \quad \delta u = 0\} = D$$

is dense in $E = V$ by Theorem 6.2-5. On the other hand, since $A^* + \lambda$ is an isomorphism from V onto V' for λ large enough, V is the domain of $G = A^*$. Since 0 is not an eigenvalue of the Neumann problem, $Q = A^*$ is invertible, and thus one to one.

EXAMPLE 1-2 Let us assume that

$$(1\text{-}57) \qquad\qquad \Lambda = D^*\mathscr{A}D,$$

where

$$(1\text{-}58) \quad \begin{cases} \text{(i)} \ \ D \text{ maps } V \text{ into } E = E' \quad \text{ where } \ E \text{ is supposed to be} \\ \qquad\qquad\qquad\qquad\qquad\qquad\qquad\qquad \text{a pivot space} \\ \text{(ii)} \ \ \mathscr{A} \in L(E,E), \\ \text{(iii)} \ \ D^* = (D|_{V_0})' \in L(E,V_0') \cap L(E(D^*),H) \text{ is the} \\ \qquad\qquad\qquad\qquad\qquad\qquad\qquad \text{formal adjoint of } D \end{cases}$$

and D^* is related to D by the Green formula

$$(1\text{-}59) \quad (D^*\hat{u}, v) = [\hat{u}, Dv] - \langle \beta\hat{u}, \gamma v \rangle \quad \text{ for any } \ \hat{u} \in E(D^*), v \in V.$$

COROLLARY 1-2 Let us assume (1-57), (1-58), and

$$(1\text{-}60) \qquad\qquad \mathscr{A} \in L(E, E) \text{ is an isomorphism.}$$

Then, if u is a solution of the boundary-value problem (1-29), $\hat{u} = \mathscr{A}Du$ is a solution of the conjugate variational equation

$$(1\text{-}61) \quad \begin{cases} \text{(i)} \ \ \hat{u} \in E(D^*), \\ \text{(ii)} \ \ \beta_2\hat{u} = t_2, \\ \text{(iii)} \ \ (D^*\hat{u}, D^*\hat{v}) + \lambda[\mathscr{A}^{-1}\hat{u}, \hat{v}] = (f, D^*\hat{v}) + \lambda\langle t_1, \beta_1\hat{v}\rangle \\ \qquad\qquad\qquad\qquad\qquad\qquad\qquad\qquad \text{for any } \ \hat{v} \in E_2(D^*). \end{cases}$$

Conversely, if we assume that

$$(1\text{-}62) \quad \begin{cases} \text{(i)} \ \ E_0(D^*) = \{\hat{u} \in E(D^*) \quad \text{ such that } \quad \beta\hat{u} = 0\} \text{ is dense in } E, \\ \text{(ii)} \ \ V = \{u \in H \quad \text{ such that } \quad Du \in E\}, \end{cases}$$

and if $\hat{u}$ is a solution of the conjugate variational equation (1-61), then $u = \lambda^{-1}(f - D^*\hat{u})$ is a solution of the boundary-value problem (1-29). ▲

Proof We take here $P = \mathscr{A}D$, $Q = \mathscr{A}^{-1}$, $G = D$, and $G^* = D^*$. Then we have the splitting $\Lambda = D^*\mathscr{A}D = G^*P$ and $G = QP$.

Now [(1-62)(i)] implies assumption (1-32) of Theorem 1-5 and [(1-62)(ii)] implies assumption [(1-34)(i)]. Finally, [(1-34)(ii)] holds, since $Q = \mathscr{A}^{-1}$ is invertible. Then Corollary 1-2 follows from Theorem 1-5. ∎

Remark 1-3 We can replace assumption (1-60) by the weaker assumption: there exists an operator $\mathscr{B} \in L(E, E)$ such that $\mathscr{B}\mathscr{A}Du = Du$ for any $u \in V$. Then we can replace $\mathscr{A}^{-1}$ by $\mathscr{B}$ in Corollary 1-2. ∎

Remark 1-4 Theorem 1-3 can be shown to follow from Corollary 1-2 if we set

(1-63) $E = \mathbf{H}(\Omega), \; Du = (u^q)_{|q| \le k}, \; (\mathscr{A}\mathbf{u})^q = \sum_{|p| \le k} a_{pq}(x)u^p.$

Assumption 1-62 follows from the very definition of the Sobolev space $H^k(\Omega)$. We prove later that (1-62) holds (see Theorem 3-1). ∎

There are several ways to split a given differential operator Λ in the form (1-57). For instance, if

(1-64) $\Lambda u = -\sum_{i,j=1,\dots,n} D_j(a_{ij}(x)D_i u),$

we can write $\Lambda = D^*\mathscr{A}D$ where we take $E = (L^2(\Omega))^n \quad Du = (D_i u)_{i=1,\dots,n}$, and $\mathscr{A}$ as the matrix

(1-65)
$$\begin{pmatrix} a_{11}(x) & \cdots & a_{n1}(x) \\ \cdot & & \cdot \\ \cdot & & \cdot \\ \cdot & & \cdot \\ a_{1n}(x) & \cdots & a_{nn}(x) \end{pmatrix}.$$

Also, we can use the splitting given in Section 1-2 by replacing $\Lambda = \Lambda - \lambda + \lambda$ and writing $\Lambda - \lambda = D^*\mathscr{A}D$, with $E = (L^2(\Omega))^{n+1}$, $Du = (u, D_1 u, \dots, D_n u)$, and $\mathscr{A}$ as the matrix

(1-66)
$$\begin{pmatrix} -\lambda & 0 & \cdots & 0 \\ 0 & a_{11}(x) & \cdots & a_{n1}(x) \\ \cdot & \cdot & & \cdot \\ \cdot & \cdot & & \cdot \\ \cdot & \cdot & & \cdot \\ 0 & a_{1n}(x) & \cdots & a_{nn}(x) \end{pmatrix}$$

Also, let us consider the case of $\Lambda = \Delta^2 + 1$. We can choose the splitting given in Section 1-2 with $k = 2$, but we can choose the following splittings as well:

(1-67) $\begin{cases} \text{(i) } V = H(\Omega, \Delta), \; E = (L^2(\Omega))^2, \; D = (1, \Delta) \quad \mathscr{A} = 1 \\ \text{(ii) } V = H(\Omega, \Delta), \; E = (L^2(\Omega))^{n+1}, \; D = (1, D_1^2, \dots, D_n^2), \mathscr{A} = 1. \end{cases}$

EXAMPLE 1-3 Let us suppose assumptions (1-57) and (1-58). We choose

(1-68) $P = D, \; Q = \mathscr{A}', \; G = QP = \mathscr{A}'D, \; G^* = D^*\mathscr{A},$

and we obtain Corollary 1-3.

COROLLARY 1-3 Let us assume (1-57) and (1-58). If u is a solution of (1-29), then $\hat{u} = Du$ is a solution of the following conjugate variational equation

$$(1\text{-}69) \quad \begin{cases} \text{(i)} \ \hat{u} \in E(D^* \mathscr{A}), \\ \text{(ii)} \ \beta_2 \mathscr{A} \hat{u} = t_2, \\ \text{(iii)} \ (D^* \mathscr{A} \hat{u}, D^* \mathscr{A} \hat{v}) + \lambda[\mathscr{A}' \hat{u}, \hat{v}] = (f, D^* \mathscr{A} \hat{v}) + \lambda \langle t_1, \beta_1 \mathscr{A} \hat{v} \rangle \\ \qquad\qquad\qquad\qquad\qquad\qquad \text{for any } \ \hat{v} \in E_2(D^* \mathscr{A}). \quad \blacktriangle \end{cases}$$

2. APPLICATIONS TO THE APPROXIMATION OF DIRICHLET PROBLEMS

2-1. Approximation of the Dirichlet Problem (I)

Let us consider a solution $u \in H^k(\Omega, \Lambda)$ of the Dirichlet problem

$$(2\text{-}1) \quad \begin{cases} \text{(i)} \ \Lambda u + \lambda u = \displaystyle\sum_{|p|, |q| \le k} (-1)^{|q|} D^q(a_{pq}(x) D^p u) + \lambda u = f \\ \qquad\qquad\qquad\qquad\qquad\qquad\qquad\qquad f \text{ given in } L^2(\Omega), \\ \text{(ii)} \ \gamma_j u = t_j \quad \text{ for } \ 0 \le j \le k - 1, \ \ t_j \text{ given in } H^{k-j-\frac{1}{2}}(\Gamma). \end{cases}$$

Let us assume that we know a solution $u_1 \in H^k(\Omega)$ of

$$(2\text{-}2) \qquad\qquad \gamma_j u_1 = t_j \quad \text{ for } \ 0 \le j \le k - 1.$$

Let us set

$$(2\text{-}3) \qquad\qquad w = u - u_1 \in H_0^k(\Omega).$$

Then w is a solution of the variational equation

$$(2\text{-}4) \quad \begin{cases} \text{(i)} \ w \in H_0^k(\Omega), \\ \text{(ii)} \ a(w, v) + \lambda(w, v) = l(v) \quad \text{ for any } \ v \in H_0^k(\Omega), \end{cases}$$

where

$$(2\text{-}5) \qquad\qquad l(v) = (f, v) - a(u_1, v) - \lambda(u_1, v).$$

Conversely, if w is solution of (2-4), $u = w + u_1$ is a solution of the Dirichlet problem (2-2) (see Theorem 6.2-2).

We thus can approximate a solution u of the variational equation (2-4) on $H_0^k(\Omega)$ by introducing finite-element approximations $(H_{0,h}^v(\Omega), q_h, r_{0,h})$ of the space $H_0^k(\Omega)$ associated with a function $v \in H^m(R^n)$ with compact support satisfying the criterion of m-convergence for $m \ge k$ (see Section 5.3-2). Let w_h be the solution of the discrete variational equation

$$(2\text{-}6) \quad a(q_h w_h, q_h v_h) + \lambda(q_h w_h, q_h v_h) = l(q_h v_h) \quad \text{ for any } \ v_h \in H_{0,h}^v(\Omega),$$

equivalent to an equation

(2-7) $$A_{0,h}w_h = l_{0,h},$$

where the entries

(2-8) $$d_{0,h}(i,j) = h^{-1}a\left[v\left(\frac{x}{h}-j\right), v\left(\frac{x}{h}-i\right)\right) + \lambda\left(v\left(\frac{x}{h}-j\right), v\left(\frac{x}{h}-i\right)\right)\right]$$
$$i,j \in \mathscr{R}^v_{0,h}(\Omega)$$

of the matrix $A_{0,h}$ are as set forth in Section 8.2-1.

In particular, the number of levels of the matrix $A_{0,h}$ is at most equal to the number of levels of the function v, $A_{0,h}$ is positive definite (respectively symmetric) whenever $a(u, v)$ is $H_0^k(\Omega)$-elliptic (respectively symmetric).

If v satisfies the stability property, then there exists a constant M such that

(2-9) $\chi(A_{0,h}) \leq M\,|h|^{-2k}$ where $\chi(A_{0,h})$ is the condition number of $A_{0,h}$.

Remark 2-1 Proposition 8.2-3 does not hold in this case.   ∎

THEOREM 2-1 Let us assume that v satisfies the criterion of m-convergence for $m \geq k$, that Ω satisfies the property of v-convergence (see Section 5.3-2), that $a(u, v)$ is coercive, and that λ is not an eigenvalue of the Dirichlet problem.

Let $u \in H^k(\Omega, \Lambda)$ be the solution of the Dirichlet problem (2-1) and $w_h \in H_{0,h}^v(\Omega)$ the solution of the discrete variational equation (2-6). Then

(2-10) $$\lim_{h \to 0} \|u - u_1 - q_h w_h\|_{k,\Omega} = 0$$

and there exists a constant M such that

(2-11) $\|\Lambda(u - u_1 - q_h w_h)\|_{-t,\Omega} \leq M\,|h|^{t-k}\,\|u\|_{k,\Omega}$ for $k \leq t \leq m+1$

Furthermore, if $u - u_1 \in H_0^s(\Omega)$ for $k \leq s \leq m+1$, then there exists a constant M such that

(2-12) $$\begin{cases} \text{(i) } \|u - u_1 - q_h w_h\|_{k,\Omega} \leq M\,|h|^{s-k}\,\|u\|_{s,\Omega} \text{ for } k \leq s \leq m+1, \\ \text{(ii) } \|\Lambda(u - u_1 - q_h w_h)\|_{-t,\Omega} \leq M\,|h|^{t+s-2k}\,\|u\|_{s,\Omega} \\ \qquad\qquad\qquad\qquad\qquad\qquad \text{for } k \leq s, t \leq m+1, \end{cases}$$

and

(2-13) $$\lim_{h \to 0} \|u - u_1 - q_h w_h\|_{s,\Omega} = 0 \quad \text{for } k \leq s \leq m.$$ ▲

Proof Theorem 2-1 follows from Theorem 3.1-7 with $V = H^k(\Omega)$, $X = H_0^t(\Omega)$, and $U = H_0^s(\Omega)$ and from Theorems 5.3-1 and 5.3-2.   ∎

2-2. Approximation of the Dirichlet Problem (II)

To approximate the solution u of the Dirichlet problem (2-1) by the internal approximation of its conjugate variational equation

$$(2\text{-}14) \quad \begin{cases} \text{(i)} \ u \in H^k(\Omega, \Lambda), \\ \text{(ii)} \ (\Lambda u, \Lambda v) + \lambda a_*(u, v) = (f, \Lambda v) + \lambda \sum_{0 \le j \le k-1} \langle t_j, \delta_{k-j-1} v \rangle \\ \qquad\qquad\qquad\qquad\qquad\qquad\qquad\qquad \text{for any} \ \ v \in H^k(\Omega, \Lambda), \end{cases}$$

we introduce finite-element approximations $(H_h{}^\mu(\Omega),\ p_h,\ \hat{r}_h)$ of $H^k(\Omega, \Lambda)$ associated with a function $\mu \in H^m(R^n)$ satisfying the m-criterion of convergence for $m \ge 2k$ (see Section 8.2-5).

Let $u_h \in H_h{}^\mu(\Omega)$ be the solution of the discrete variational equation

$$(2\text{-}15) \quad \begin{aligned} (\Lambda p_h u_h, \Lambda p_h v_h) &+ \lambda a_*(p_h u_h, p_h v_h) = (f, \Lambda p_h v_h) \\ &+ \lambda \sum_{0 \le j \le k-1} \langle t_j, \delta_{2k-j-1} p_h v_h \rangle \quad \text{for any} \ \ v_h \in H_h{}^\mu(\Omega), \end{aligned}$$

which is equivalent to an equation $A_h u_h = l_h$ having the same properties as the internal approximations of a Neumann problem (see Section 8.2-1).

THEOREM 2-2 Let us assume that μ satisfies the criterion of m-convergence for $m \ge 2k$, that $a(u, v)$ is coercive, that λ is not an eigenvalue of the Dirichlet problem, and that 0 is not an eigenvalue of the Neumann problem.

If u is the solution of the Dirichlet problem (2-1) and if u_h is the solution of the discrete variational equation (2-15), then

$$(2\text{-}16) \quad \lim_{h \to 0} \|u - p_h u_h\|_{k,\Omega} = 0 \qquad \text{and} \qquad \lim_{h \to 0} \|\Lambda(u - p_h u_h)\|_{0,\Omega} = 0.$$

Furthermore, if $u \in H^s(\Omega)$ for $2k \le s \le m + 1$, there exists a constant M such that

$$(2\text{-}17) \quad \|\Lambda(u - p_h u_h)\|_{0,\Omega} \le M \, |h|^{s-2k} \, \|u\|_{s,\Omega} \qquad \text{for} \ \ 2k \le s \le m + 1. \quad \blacktriangle$$

Proof The proof follows from Theorem 3.1-7 with $V = H^k(\Omega, \Lambda)$ and $U = H^s(\Omega)$ and from Theorem 8.2-4 of Section 8.2-5. ∎

THEOREM 2-3 Let us suppose that $a(u, v)$ is $H^k(\Omega)$-elliptic, that $\lambda > 0$ and that the assumptions of Theorems 2-1 and 2-2 are satisfied. If c is the constant of $H^k(\Omega)$-ellipticity and M the norm of $a(u, v)$, then

$$(2\text{-}18) \quad \begin{cases} \text{(i)} \ c \, \|u - u_1 - q_h w_h\|_{k,\Omega}^2 + \lambda \, \|u - u_1 - q_h w_h\|_{0,\Omega}^2 \le X_h(u_h, w_h), \\ \text{(ii)} \ c \, \|u - p_h u_h\|_{k,\Omega}^2 + \lambda^{-1} \, \|\Lambda(u - p_h u_h)\|_{0,\Omega}^2 \le X_h(u_h, w_h), \end{cases}$$

where u_h is the solution of the discrete equation (2-15), w_h the solution of the discrete equation (2-6), and $X_h(u_h, w_h)$ is defined by

$$(2-19) \quad \begin{cases} X_h(u_h, w_h) = \lambda^{-1} \| f - \lambda u_1 - \lambda q_h w_h - \Lambda p_h u_h \|_{k,\Omega}^2 \\ \qquad\qquad\qquad + M^2 c^{-1} \| u_1 + q_h w_h - p_h u_h \|_{0,\Omega}^2. \quad \blacktriangle \end{cases}$$

Proof   Theorem 2-3 follows from Theorems 1-2, 2-1, and 2-2.   ∎

2-3. The Case of Second-Order Differential Operators

For simplicity, we estimate the error $\| u - p_h u_h \|_{0,\Omega}$ in the case of $k = 1$ (i.e., Λ is a second-order differential operator). Let us assume that

$$(2-20) \quad \begin{cases} \text{(i) } (\Lambda, \delta) \text{ is an isomorphism from } H^4(\Omega) \text{ onto } L^2(\Omega) \times H^{3/2}(\Gamma), \\ \text{(ii) } (\Lambda + \lambda)^*\Lambda \text{ is an isomorphism from } D^4(\Omega) \text{ onto } L^2(\Omega), \end{cases}$$

where

$$(2-21) \quad D^4(\Omega) = \{ v \in H^4(\Omega) \quad \text{such that} \quad \gamma \Lambda v = \delta \Lambda v = 0 \}.$$

THEOREM 2-4   Let us suppose the assumptions of Theorem 2-2 and (2-20). Then there exists a constant M such that

$$(2-22) \quad \| u - p_h u_h \|_{0,\Omega} \leq M \, |h|^{\min(2, m-1)} (\| u - p_h u_h \|_{k,\Omega} + \| \Lambda (u - p_h u_h) \|_{0,\Omega}).$$
$$\blacktriangle$$

Proof   Since u is a solution of (2-14) and u_h a solution of (2-15), we deduce from the Green formula the inequality

$$(2-23) \quad \begin{cases} [(\Lambda + \lambda)(u - p_h u_h), \Lambda v] + \lambda \langle \gamma(u - p_h u_h), \delta v \rangle \\ = [\Lambda(u - p_h u_h), \Lambda v] + \lambda a_*(u - p_h u_h, v) \\ = [\Lambda(u - p_h u_h), \Lambda(v - p_h \hat{r}_h v)] + \lambda a_*(u - p_h u_h, v - p_h \hat{r}_h v) \\ \qquad\qquad\qquad \leq M \| u - p_h u_h \|_{H^1(\Omega, \Lambda)} \| v - p_h \hat{r}_h v \|_{H^1(\Omega, \Lambda)}. \end{cases}$$

When g ranges over $H^{3/2}(\Gamma)$, there exists a unique solution $v \in H^4(\Omega)$ such that $\Lambda v = 0$, $\delta v = g$. Therefore, we deduce from (2-23)

$$(2-24) \quad \begin{cases} |\langle \gamma(u - p_h u_h), g \rangle| \leq M \| u - p_h u_h \|_{H^1(\Omega, \Lambda)} \| v - p_h \hat{r}_h v \|_{H^1(\Omega, \Lambda)} \\ \qquad\qquad \leq M \, |h|^{\min(2, m-1)} \| u - p_h u_h \|_{H^1(\Omega, \Lambda)} \| v \|_{4,\Omega} \\ \qquad\qquad \leq M \, |h|^{\min(2, m-1)} \| u - p_h u_h \|_{H^1(\Omega, \Lambda)} \| g \|_{3/2, \Gamma} \end{cases}$$

by Theorem 8.2-4.

This implies that

$$(2-25) \quad \| \gamma(u - p_h u_h) \|_{-3/2, \Gamma} \leq M \, |h|^{\min(2, m-1)} \| u - p_h u_h \|_{H^1(\Omega, \Gamma)}.$$

Again using the Green formula, we obtain, when $v \in D^4(\Omega)$

$$(2\text{-}26) \quad \begin{cases} |(u - p_h u_h, (\Lambda + \lambda)^* \Lambda v)| \leq \lambda |\langle \gamma(u - p_h u_h), \delta v \rangle| \\ \qquad + M \|u - p_h u_h\|_{H^1(\Omega, \Lambda)} \|v - p_h \hat{r}_h v\|_{H^1(\Omega, \Lambda)} . \end{cases}$$

When w ranges over $L^2(\Omega)$, there exists a unique solution $v \in D^4(\Omega)$ such that $(\Lambda + \lambda)^* \Lambda v = w$. Then (2-26) implies that

$$(2\text{-}27) \quad \begin{cases} |(u - p_h u_h, w)| \leq M \|u - p_h u_h\|_{H^1(\Omega, \Lambda)} \|v - p_h r_h v\|_{H^1(\Omega, \Lambda)} \\ + \|(u - p_h u_h)\|_{-\frac{1}{2}, \Gamma} \|v\|_{\frac{1}{2}, \Gamma} \leq M |h|^{\min(2, m-1)} \|u - p_h u_h\|_{H^1(\Omega, \Lambda)} \|w\|_{0, \Omega} \end{cases}$$

by (2-25) and Theorem 8.2-4. Then it is clear that (2-27) implies (2-22). ∎

3. FINITE-ELEMENT APPROXIMATIONS OF THE SPACES $H^k(\Omega, D^*)$

3-1. Spaces $H^k(\Omega, D^*)$

Let Ω be a smooth bounded open subset of R^n and Γ its boundary. Recalling that

$$(3\text{-}1) \quad H^k(\Omega, D^*) = \{\mathbf{u} \in H(\Omega) \text{ such that } D^* \mathbf{u} = \sum_{|p| \leq k} (-1)^{|p|} D^p u^p \in L^2(\Omega)\}$$

let us set

$$(3\text{-}2) \quad \begin{cases} \text{(i)} \quad [\mathbf{u}, \mathbf{v}] = \sum_{|q| \leq k} \int_\Omega u^q v^q \, dx, \\ \text{(ii)} \quad \|\mathbf{u}\|_{0,\Omega} = \left(\sum_{|q| \leq k} \|u^q\|_{0,\Omega} \right)^{\frac{1}{2}}, \\ \text{(iii)} \quad \|\mathbf{u}\|_{k,\Omega} = (\|\mathbf{u}\|_{0,\Omega}^2 + \|D^* \mathbf{u}\|_{0,\Omega}^2)^{\frac{1}{2}}. \end{cases}$$

In addition, we need the following theorems of density.

THEOREM 3-1 Let $s \geq k$. Then $\mathbf{H}^s(\Omega) = (H^s(\Omega))^N$ is dense in $\mathbf{H}^k(\Omega, D^*)$. The kernel $\mathbf{H}_0^k(\Omega, D^*)$ of the operator $\beta = (\beta_k, \ldots, \beta_{2k-1})$ is the closure in $\mathbf{H}^k(\Omega, D^*)$ of the space $(\Phi_0(\Omega))^N$. ▲

Proof Let $l(\mathbf{v})$ be a continuous linear form on $\mathbf{H}^k(\Omega, D^*)$ vanishing on $\mathbf{H}^s(\Omega)$. According to Theorem 2.1-2 we have to prove that $l = 0$. Since $\mathbf{H}^k(\Omega, D^*)$ is the domain of D^*, we can write, by Theorem 2.1-8

$$(3\text{-}3) \quad l(\mathbf{v}) = [\mathbf{f}, \mathbf{v}] - (u, D^* \mathbf{v}) \quad \text{where } \mathbf{f} \in \mathbf{H}(\Omega) \quad \text{and} \quad u \in L^2(\Omega).$$

Since $l(\mathbf{v})$ vanishes on $\mathbf{H}_0^s(\Omega) = (H_0^s(\Omega))^N$, we deduce that

$$(3\text{-}4) \quad l(\mathbf{v}) = [\mathbf{f} - Du, \mathbf{v}] \quad \text{for any } \mathbf{v} \in \mathbf{H}_0^s(\Omega).$$

This implies that $\mathbf{f} = Du$ and that $u \in H^k(\Omega)$. Therefore, we can apply the Green formula and obtain

$$(3\text{-}5) \qquad l(\mathbf{v}) = \langle \gamma u, \beta \mathbf{v} \rangle \qquad \text{where} \quad u \in H^k(\Omega).$$

On the other hand, the Neumann operator $\delta = \beta D^*$ associated with the continuous bilinear form $a(u, v) = [Du, Dv]$ maps $H^{s+k}(\Omega)$ onto $R' = \prod_{k \leq j \leq k-1} H^{s+k-j-\frac{1}{2}}(\Gamma)$ and β maps $H^k(\Omega, D^*)$ onto $T' = \prod_{k \leq j \leq 2k-1} H^{k-j-\frac{1}{2}}(\Gamma)$. Furthermore, for every $t \in R'$, there exists $v \in H^{s+k}(\Omega)$ satisfying $\delta v = \beta Dv = t$. In other words, there exists $\hat{v} = Dv \in H^s(\Omega)$ such that $\beta \mathbf{v} = t$.

Since $l(\mathbf{v})$ vanishes on $\mathbf{H}^s(\Omega)$, we deduce from (3-5) that $\langle \gamma u, t \rangle = 0$ for any $t \in R'$. Since R' is dense in T', this implies that $\gamma u = 0$, and thus that $l = 0$.

To prove the second statement of the theorem, we must recall that $(\Phi_0(\Omega))^N$ is contained in $\mathbf{H}_0{}^k(\Omega, D^*)$. Hence it suffices to prove that the orthogonal of $(\Phi_0(\Omega))^N$ is contained in the orthogonal of $\mathbf{H}_0{}^k(\Omega, D^*)$, which is the closed range of the transpose β' of β, since $\mathbf{H}_0{}^k(\Omega, D^*)$ is the kernel of the surjective operator β mapping $\mathbf{H}^k(\Omega, D^*)$ onto T.

We have seen that a continuous linear form $l(\mathbf{v})$ on $\mathbf{H}^k(\Omega, D^*)$ vanishing on $(\Phi_0(\Omega))^N$ can be written in the form

$$(3\text{-}6) \qquad l(\mathbf{v}) = \langle \gamma u, \beta \mathbf{v} \rangle = [\beta' \gamma u, \mathbf{v}] \qquad \text{where} \quad u \in H^k(\Omega)$$

Then $l(\mathbf{v})$ belongs to the range of β' and the theorem is proved. ∎

3-2. Approximations of the Space $\mathbf{H}^k(\Omega, D^*)$

Let us begin by the case of $\Omega = R^n$, calling $\mu \in H^m(R^n)$ a function with compact support satisfying $\int \mu(x)\, dx = 1$ and the criterion of m-convergence. We say also that $V_h = L_h{}^2$, and p_h is the prolongation associated with μ by

$$(3\text{-}7) \qquad p_h u_h = \sum_{j \in Z^n} u_h{}^j \mu\!\left(\frac{x}{h} - j\right).$$

Let us set

$$(3\text{-}8) \qquad
\begin{cases}
\text{(i)} \quad p_h{}^q = \pi_{q,h} * p_h u_h \quad \text{associated with} \quad \mu_q = \pi_q * \mu, \\[4pt]
\text{(ii)} \quad \mathbf{u}_h = (u_h{}^q)_{|q| \leq k} \in \mathbf{H}_h{}^k(R^n, D^*) = (L_h{}^2)^N, \\[4pt]
\text{(iii)} \quad \mathbf{p}_h \mathbf{u}_h = (p_h{}^q u_h{}^q)_{|q| \leq k}, \\[4pt]
\text{(iv)} \quad D_h^* = \sum_{|q| \leq k} (-1)^{|q|} \nabla_h{}^q u_h.
\end{cases}$$

Therefore $\mathbf{p}_h$ maps $\mathbf{H}_h{}^k(R^n, D^*)$ into $\mathbf{H}^k(R^n, D^*)$ and satisfies the commutation formula

$$(3\text{-}9) \qquad D^* \mathbf{p}_h \mathbf{u}_h = p_h D_h^* \mathbf{u}_h.$$

Let $\lambda \in L^\infty(\Omega)$ be a function with compact support satisfying $\int \lambda(x)\, dx = 1$ and set

$$(3\text{-}10) \qquad \lambda_q = \dot{\lambda} * \tilde{\pi}_{(k)-q} \qquad \text{for any} \quad |q| \leq k.$$

We associate with λ the following restrictions:

$$(3\text{-}11) \quad \begin{cases} \text{(i)} \quad r_h u = \left\{ \int \lambda^j_{0,h}(x) u(x)\, dx \right\}_{j \in Z^n} \qquad \text{where} \quad \lambda_0 = \lambda * \tilde{\pi}_{(k)}, \\[2mm] \text{(ii)} \quad r_h^q u = \left\{ \int \lambda^j_{q,h}(x) u(x)\, dx \right\}_{j \in Z^n}, \\[2mm] \text{(iii)} \quad \mathbf{r}_h \mathbf{u} = (r_h^q u^q)_{|q| \leq k}. \end{cases}$$

Therefore $\mathbf{r}_h$ maps $\mathbf{H}^k(R^n, D^*)$ onto $\mathbf{H}_h^k(R^n, D^*)$ and satisfies the commutation formula

$$(3\text{-}12) \qquad D_h^* \mathbf{r}_h \mathbf{u} = r_h D^* \mathbf{u}.$$

Indeed

$$(3\text{-}13) \quad \begin{cases} \nabla_h^q r_h^q u^q = r_h^q \nabla_h^q u^q = r_h^q(\pi_{q,h} * D^q u^q) \\[2mm] \qquad = \left\{ \int \lambda^j_{q,h}(x) \pi_{q,h} * D^q u^q(x)\, dx \right\}_j \\[2mm] \qquad = \left\{ \int (\lambda^j_{q,h} * \tilde{\pi}_{q,h})(x) D^q u^q(x)\, dx \right\}_j \\[2mm] \qquad = \left\{ \int \lambda_{0,h}(x) D^q u^q(x)\, dx \right\}_j = r_h(D^q u^q). \end{cases}$$

Definition 3-1 Approximations $(\mathbf{H}_h^k(R^n, D^*), \mathbf{p}_h, \mathbf{r}_h)$ are finite-element approximations of $\mathbf{H}^k(R^n, D^*)$ associated with the functions μ and λ. ▲

THEOREM 3-2 Let us assume that

$$(3\text{-}14) \quad \begin{cases} \text{(i)} \quad \mu \text{ satisfies the criterion of } m\text{-convergence for } m \geq 0, \\[2mm] \text{(ii)} \quad \iint \mu(x) \lambda_0(y)(x - y)^k\, dx\, dy = 0 \qquad \text{for} \quad 0 = |k| \leq m. \end{cases}$$

Then the approximations $(\mathbf{H}_h^k(R^n, D^*), \mathbf{p}_h, \mathbf{r}_h)$ associated with μ and λ are convergent in $\mathbf{H}^k(R^n, D^*)$ and there exists a constant M such that

$$(3\text{-}15) \quad \|\mathbf{u} - \mathbf{p}_h \mathbf{r}_h \mathbf{u}\|_k \leq M\, |h|^s \left(\sum_{|q| \leq k} \|u^q\|_s + \|D^* \mathbf{u}\|_s \right) \quad \text{for } 0 \leq s \leq m + 1.$$
▲

Proof Commutation formulas (3-9) and (3-12) imply that

$$(3\text{-}16) \qquad p_h r_h D^* \mathbf{u} = D^* \mathbf{p}_h \mathbf{r}_h \mathbf{u} \qquad \text{for any} \quad \mathbf{u} \in \mathbf{H}^k(R^n, D^*).$$

Then we must prove that

$$(3\text{-}17) \quad \begin{cases} \text{(i)} \ p_h{}^q r_h{}^q u^q \quad \text{converges to} \quad u^q \text{ in } L^2(\Omega), \\ \text{(ii)} \ p_h r_h D^* \mathbf{u} = D^* \mathbf{p}_h \mathbf{r}_h \mathbf{u} \quad \text{converges to} \quad D^* \mathbf{u} \text{ in } L^2(\Omega). \end{cases}$$

By Theorem 5.1-1, we also have to prove that

$$(3\text{-}18) \quad \begin{cases} \text{(i)} \ \displaystyle\iint \mu_q(x) \lambda_q(y)(x - y)^k \, dx \, dy = 0 \quad \text{for} \ \ 0 \leq |k| \leq m, \\ \text{(ii)} \ \displaystyle\iint \mu(x) \lambda_0(y)(x - y)^k \, dx \, dy = 0 \quad \text{for} \ \ 0 \leq |k| \leq m. \end{cases}$$

Condition [(3-18)(ii)] holds by assumption, let us now prove [(3-18)(i)]:

$$(3\text{-}19) \quad \begin{cases} \displaystyle\iint \mu_q(x) \lambda_q(y)(x - y)^k \, dx \, dy = \int (\mu_q * \tilde{\lambda}_q)(z) z^k \, dz \\ \qquad\qquad = \displaystyle\int (\mu * \pi_{(k)} * \tilde{\lambda})(z) z^k \, dz \\ \qquad\qquad = \displaystyle\int \mu * \tilde{\lambda}_0(z) z^k \, dz \\ \qquad\qquad = \displaystyle\iint \mu(x) \lambda_0(y)(x - y)^k \, dx \, dy \\ \qquad\qquad = 0 \quad \text{for} \ \ 0 \leq |k| \leq m. \end{cases}$$

Thus we have proved the convergence of the finite-element approximations in $\mathbf{H}^k(R^n, D^*)$. Since $\mu_q = \mu * \pi_q$ satisfies the criterion of m-convergence whenever μ does, we deduce from Theorem 5.1-1 that

$$(3\text{-}20) \quad \begin{cases} \text{(i)} \ \|u^q - p_h{}^q r_h{}^q u^q\|_0 \leq c \, |h|^s \, \|u^q\|_s \quad \text{for} \ \ 0 \leq s \leq m + 1, \\ \text{(ii)} \ \|D^*(\mathbf{u} - \mathbf{p}_h \mathbf{r}_h \mathbf{u})\|_0 = \|(1 - p_h r_h) D^* \mathbf{u}\|_0 \leq c \, |h|^s \, \|D^* \mathbf{u}\|_0 \\ \qquad\qquad\qquad\qquad\qquad\qquad\qquad\qquad\qquad \text{for} \ \ 0 \leq s \leq m + 1. \end{cases}$$

Then (3-15) follows from (3-20). ∎

When Ω is a smooth bounded subset of R^n, we associate with the functions μ and $\mu_q = \mu * \pi_q$

$$(3\text{-}21) \quad \begin{cases} \text{(i)} \ \mathbf{H}_h{}^k(\Omega, D^*) = \displaystyle\prod_{|q| \leq k} H_h{}^{\mu_q}(\Omega), \\ \text{(ii)} \ p_h{}^q u_h{}^q = \displaystyle\sum_{j \in \mathcal{R}_{h\mu_q}(\Omega)} u_h{}^j \mu_q \left(\frac{x}{h} - j \right) \quad \text{when} \ \ x \in \Omega, \\ \text{(iii)} \ \mathbf{p}_h \mathbf{u}_h = (p_h{}^q u_h{}^q)_{|q| \leq k}. \end{cases}$$

We choose the restriction $\hat{\mathbf{r}}_h$ to be the optimal restriction associated with $\mathbf{p}_h$ in $\mathbf{H}^k(\Omega, D^*)$.

Definition 3-2   The approximations $(H_h{}^k(\Omega, D^*), \mathbf{p}_h, \hat{\mathbf{r}}_h)$ associated with a function $\mu \in H^m(R^n)$ with compact support satisfying $\int \mu(x)\, dx = 1$ are finite-element approximations of the space $\mathbf{H}^k(\Omega, D^*)$.   ▲

THEOREM 3-3   Let assume that μ satisfies the criterion of m-convergence for $m \geq 0$. Then the finite-element approximations associated with μ are convergent approximations of $\mathbf{H}^k(\Omega, D^*)$. Furthermore, there exists a constant M such that

$$(3\text{-}22) \quad \|\mathbf{u} - \mathbf{p}_h \mathbf{r}_h \mathbf{u}\|_{k,\Omega} \leq M\, |h|^s \left(\sum_{|q| \leq k} \|u^q\|_{s,\Omega} + \|D^*\mathbf{u}\|_{s,\Omega} \right)$$
$$\text{for } \quad 0 \leq s \leq m + 1. \quad ▲$$

Proof   Since the norms $\|1 - \mathbf{p}_h \hat{\mathbf{r}}_h\|_{L(\mathbf{H}^k(\Omega, D^*), \mathbf{H}^k(\Omega, D^*))} \leq 1$, it suffices to prove that

$$(3\text{-}23) \qquad \|\mathbf{u} - \mathbf{p}_h \hat{\mathbf{r}}_h \mathbf{u}\|_{k,\Omega} \text{ converges to 0 for any } \mathbf{u} \in \mathbf{H}^k(\Omega)$$

since $\mathbf{H}^k(\Omega)$ is dense in $\mathbf{H}^k(\Omega, D^*)$ (see Theorems 3-1 and 2.1-9).

Let $\omega \in L(H^{m+1+k}(\Omega), H^{m+1+k}(R^n))$ be a right inverse of the operator ρ of restriction to Ω. Then we have inequality

$$(3\text{-}24) \quad \|\mathbf{u} - \mathbf{p}_h \hat{\mathbf{r}}_h \mathbf{u}\|_{k,\Omega} \leq \|\mathbf{u} - \mathbf{p}_h \mathbf{r}_h \omega \mathbf{u}\|_{k,\Omega} \leq c\, \|\mathbf{u} - \mathbf{p}_h \mathbf{r}_h \omega \mathbf{u}\|_{\mathbf{H}^k(R^n)}$$
$$\text{if } \quad \mathbf{u} \in \mathbf{H}^k(\Omega)$$

where $\omega \mathbf{u} = (\omega u^q)_{|q| \leq k}$ and $\mathbf{r}_h$ is a restriction associated with a function λ satisfying $[(3\text{-}14)(\text{ii})]$. Equation 3-24 implies (3-23) (by Theorem 5.2-1), and thus the convergence of the approximations in $\mathbf{H}^k(\Omega, D^*)$.

Furthermore, if $\mathbf{u} \in \mathbf{H}^s(\Omega)$ and $D^*\mathbf{u} \in H^s(\Omega)$ for $0 \leq s \leq m + 1$, we deduce from (3-24) and from Theorem 3-2 that

$$(3\text{-}25) \quad \begin{cases} \|\mathbf{u} - \mathbf{p}_h \hat{\mathbf{r}}_h \mathbf{u}\|_{k,\Omega} \leq c\, \|\omega \mathbf{u} - \mathbf{p}_h \mathbf{r}_h \omega \mathbf{u}\|_{k,(R^n)} \\[2mm] \qquad \leq c\, |h|^s \left(\sum_{|q| \leq k} \|\omega u^q\|_s + \|D^*\omega \mathbf{u}\|_s \right) \\[2mm] \qquad \leq c\, |h|^s \left(\sum_{|q| \leq k} \|u^q\|_{s,\Omega} + \|D^*\mathbf{u}\|_{s,\Omega} \right). \end{cases} \quad ■$$

Finally, we can construct finite-element approximations of the space $H_0^k(\Omega, D^*)$ by associating with μ

$$(3\text{-}26) \quad \begin{cases} \text{(i) } \mathbf{H}_{0,h}(\Omega, D^*) = \prod_{|q| \leq k} H_{0,h}^{\mu_q}(\Omega) \\ \text{(ii) } \mathbf{p}_h \mathbf{u}_h = (p_h^q u_h^q)_{|q| \leq k} \quad \text{defined by} \quad [(3\text{-}21) \text{ (ii)}], \\ \text{(iii) the optimal restriction } \mathbf{r}_h \text{ associated with } \mathbf{p}_h \text{ in } \mathbf{H}_0^k(\Omega, D^*). \end{cases}$$

Definition 3-3   The approximations $(\mathbf{H}_{0,h}^k(\Omega, D^*), \mathbf{p}_h, \hat{\mathbf{r}}_h)$ associated with a function $\mu \in H^m(R^n)$ with compact support satisfying $\int \mu(x)\, dx = 1$ are the finite-element approximations of the space $\mathbf{H}_0^k(\Omega, D^*)$.   ▲

THEOREM 3-4   Let us assume that μ satisfies the criterion of m-convergence and that Ω satisfies the property of $\mu * \pi_{(k)}$-convergence. Then the finite-element approximations of $\mathbf{H}_0^k(\Omega, D^*)$ are convergent.   ▲

Proof   The proof is analogous to the proof of Theorems 3-3 and 5.3-2 and is left as an exercise.   ■

4. APPROXIMATION OF THE SECOND EXAMPLE OF A CONJUGATE PROBLEM

4-1.   Approximation of the Conjugate Dirichlet Problem

Let $u \in H^k(\Omega, \Lambda)$ be the solution of the Dirichlet problem

$$(4\text{-}1) \quad \begin{cases} \text{(i) } \Lambda u + \lambda u = f \quad f \text{ is given in } L^2(\Omega), \\ \text{(ii) } \gamma_j u = t_j \quad \text{for} \quad 0 \leq j \leq k - 1, \quad t_j \text{ is given in } H^{k-j-\frac{1}{2}}(\Gamma), \end{cases}$$

where

$$(4\text{-}2) \qquad\qquad \Lambda = D^* \mathscr{A} D$$

with

$$(4\text{-}3) \quad \begin{cases} \text{(i) } Du = (D^q u)_{|q| \leq k}; \ D \in L[H^k(\Omega), \mathbf{H}(\Omega)], \\ \text{(ii) } (\mathscr{A}\mathbf{u})^q = \sum_{|p| \leq k} a_{pq}(x) u^p, \ \mathscr{A} \in L[\mathbf{H}(\Omega), \mathbf{H}(\Omega)], \\ \text{(iii) } D^*\mathbf{u} = \sum_{|p| \leq k} (-1)^{|p|} D^p u^p, \ D^* \in L[\mathbf{H}^k(\Omega), D^*), L^2(\Omega)]. \end{cases}$$

Let us assume that

$$(4\text{-}4) \qquad\qquad \mathscr{A} \text{ is } \mathbf{H}(\Omega)\text{-elliptic}$$

and denote its inverse by

(4-5) $$\mathscr{B} = \mathscr{A}^{-1}, \quad (\mathscr{B}\mathbf{u})^q = \sum_{|p| \le k} b_{pq}(x)u^p.$$

Let us consider the solution $\mathbf{u} = \mathscr{A}\,Du$ of the conjugate problem

(4-6) $$\begin{cases} \text{(i)} \;\; \mathbf{u} \in \mathbf{H}^k(\Omega, D^*) \\ \text{(ii)} \;\; (D^*\mathbf{u}, D^*\mathbf{v}) + \lambda[\mathscr{B}\mathbf{u}, \mathbf{v}] = (f, D^*\mathbf{v}) + \sum_{0 \le j \le k-1} \langle t_j, \beta_{2k-j-1}\mathbf{v} \rangle \\ \qquad\qquad\qquad\qquad\qquad\qquad\qquad \text{for any} \;\; \mathbf{v} \in \mathbf{H}^k(\Omega, D^*) \end{cases}$$

(see Section 1-2). By using finite-element approximations $[\mathbf{H}_h^k(\Omega, D^*), \mathbf{p}_h, \mathbf{r}_h)$ of $\mathbf{H}^k(\Omega, D^*)]$, we approximate the solution $\mathbf{u} = \mathscr{A}\,Du$ of the conjugate problem (4-6) by the solution $\mathbf{u}_h \in \mathbf{H}_h^k(\Omega, D^*)$ of its internal approximation

(4-7) $$\begin{cases} (p_h D_h^*\mathbf{u}_h, p_h D_h^*\mathbf{v}_h) + \lambda \sum_{|p|,|q| \le k} \int_\Omega b_{pq}(x) p_h^p u_h^p p_h^q u_h^q \, dx \\ = (f, p_h D_h^*\mathbf{v}_h) + \sum_{0 \le j \le k-1} \langle t_j, \beta_{2k-1-j} p_h \mathbf{v}_h \rangle \quad \text{for any} \;\; \mathbf{v}_h \in \mathbf{H}_h^k(\Omega, D^*). \end{cases}$$

THEOREM 4-1 Let us assume that Ω is a smooth bounded open subset of R^n, that μ satisfies the criterion of $(m - k)$-convergence where $m \ge k$, that $\mathscr{A}$ is $\mathbf{H}(\Omega)$-elliptic, and that $\lambda > 0$.

Let $u \in H^k(\Omega, \Lambda)$ be a solution of the Dirichlet problem (4-1), with $\mathbf{u} \in \mathbf{H}^k(\Omega, D^*)$ the solution of its conjugate problem (4-6) and $\mathbf{u}_h \in \mathbf{H}_h^k(\Omega, D^*)$ the solution of the discrete variational equation (4-7). Then

(4-8) $$\lim_{h \to 0} \| \mathbf{u} - \mathbf{p}_h \mathbf{u}_h \|_{k,\Omega} = 0.$$

Furthermore, if the coefficients $a_{pq}(x)$ are smooth, if $u \in H^s(\Omega)$ and $\Lambda u = D^*\mathbf{u} \in H^{s-k}(\Omega)$ for $k \le s \le m + 1$, there exists a constant M such that

(4-9) $$\begin{cases} \text{(i)} \;\; \| p_h^q u_h^q - \sum_{|p| \le k} a_{pq}(x) D^p u \|_{0,\Omega} \le c \, |h|^{s-k} (\|u\|_{s,\Omega} + \|\Lambda u\|_{s-k,\Omega}), \\ \text{(ii)} \;\; \| \Lambda u - p_h D_h^* \mathbf{u}_h \|_{0,\Omega} = \| f - \lambda u - p_h D_h^* \mathbf{u}_h \|_{0,\Omega} \\ \qquad\qquad\qquad\qquad \le c \, |h|^{s-k} (\|u\|_{s,\Omega} + \|\Lambda u\|_{s-k,\Omega}) \\ \qquad\qquad\qquad\qquad\qquad\qquad \text{for} \;\; k \le s \le m + 1. \quad \blacktriangle \end{cases}$$

Proof The first statement follows from Theorem 3.1-6 and from Theorem 3-3. Actually, these theorems imply that

(4-10) $$\| \mathbf{u} - \mathbf{p}_h \mathbf{u}_h \|_{k,\Omega} \le M \, |h|^r \left(\sum_{|q| \le k} \|u^q\|_{r,\Omega} + \|D^*\mathbf{u}\|_{r,\Omega} \right)$$
$$\text{for} \;\; 0 \le r \le m - k + 1.$$

If the solution $u \in H^s(\Omega)$, and if the coefficients $a_{pq}(x)$ are smooth enough, then $u_q = \sum_{|p| \le k} a_{pq}(x) D^p u$ belongs to $H^{s-k}(\Omega)$ for $|q| \le k$. Then if $\Lambda u = D^* \mathbf{u} = f - \lambda u$ belongs to $H^{s-k}(\Omega)$, inequality (4-10) implies (with $r = s - k$)

$$(4\text{-}11) \qquad \|u - \mathbf{p}_h \mathbf{u}_h\|_{k,\Omega} \le c \, |h|^{s-k} (\|u\|_{s,\Omega} + \|\Lambda u\|_{s-k,\Omega})$$

Then (4-9) follows from (4-11), since

$$(4\text{-}12) \quad \begin{cases} \|\mathbf{u} - \mathbf{p}_h \mathbf{u}_h\|_{k,\Omega}^2 = \sum_{|q| \le k} \left\| \sum_{|p| \le k} a_{pq}(x) D^p u - p_h{}^q u_h{}^q \right\|_{0,\Omega}^2 \\ \qquad\qquad\qquad\qquad\qquad + \|\Lambda u - p_h D_h^* \mathbf{u}_h\|_{0,\Omega}^2. \quad \blacksquare \end{cases}$$

Remark 4-1 By using the approximations scheme (4-7), we approximate each derivative $D^q u$ of the solution u by $\sum_{|p| \le k} b_{pq}(x) p_h{}^p u_h{}^p$ and $\Lambda u = f - \lambda u$ by $p_h D_h^* \mathbf{u}_h$. $\blacksquare$

THEOREM 4-2 Let us suppose the assumptions of Theorem 4-1. If we further assume that

$$(4\text{-}13) \quad \begin{cases} (\Lambda^* + \lambda, \gamma) \text{ is an isomorphism from } H^{2k}(\Omega) \text{ onto} \\ \qquad\qquad\qquad\qquad L^2(\Omega) \times \prod_{0 \le j \le k-1} H^{2k-j-\frac{1}{2}}(\Gamma), \end{cases}$$

then there exists a constant M such that

$$(4\text{-}14) \quad \begin{cases} \text{(i)} \ \|\Lambda u - p_h D_h^* \mathbf{u}_h\|_{-k,\Omega} \le c \, |h|^{\min(k, m+1-k)} \|\mathbf{u} - \mathbf{p}_h \mathbf{u}_h\|_{k,\Omega}, \\ \text{(ii)} \ \|\beta_j(\mathbf{u} - \mathbf{p}_h \mathbf{u}_h)\|_{-j-\frac{1}{2},\Gamma} \le c \, |h|^{\min(k, m+1-k)} \|\mathbf{u} - \mathbf{p}_h \mathbf{u}_h\|_{k,\Omega} \\ \qquad\qquad\qquad\qquad\qquad\qquad\qquad\qquad \text{for} \quad k \le j \le 2k - 1. \end{cases}$$

Furthermore, if we assume that

$$(4\text{-}15) \quad \begin{cases} DD^* + \lambda \mathscr{B}' \text{ is an isomorphism from } \mathbf{H}^{k,k}(\Omega, D^*) \text{ onto} \\ \qquad\qquad\qquad\qquad \mathbf{H}(\Omega) \times \prod_{0 \le j \le k-1} H^{2k-j-\frac{1}{2}}(\Gamma), \end{cases}$$

where

$$(4\text{-}16) \quad \mathbf{H}^{k,k}(\Omega, D^*) = \{\mathbf{u} \in \mathbf{H}^k(\Omega) \qquad \text{such that} \qquad D^* \mathbf{u} \in H^k(\Omega)\},$$

then there exists a constant M such that

$$(4\text{-}17) \quad \|u^q - p_h{}^q u_h{}^q\|_{0,\Omega} \le c \, |h|^{\min(k, m+1-k)} \|\mathbf{u} - \mathbf{p}_h \mathbf{u}_h\|_{k,\Omega} \qquad \text{for} \quad |q| \le k. \quad \blacktriangle$$

Proof We begin by using the inequality

$$(4\text{-}18) \quad \begin{cases} (D^*(\mathbf{u} - \mathbf{p}_h \mathbf{u}_h), D^* \mathbf{v}) + \lambda[\mathscr{B}(\mathbf{u} - \mathbf{p}_h \mathbf{u}_h), \mathbf{v}] \\ = (D^*(\mathbf{u} - \mathbf{p}_h \mathbf{u}_h), D^*(\mathbf{v} - \mathbf{p}_h \mathbf{r}_h \mathbf{v})) + \lambda[\mathscr{B}(\mathbf{u} - \mathbf{p}_h \mathbf{u}_h), \mathbf{v} - \mathbf{p}_h \mathbf{r}_h \mathbf{v}] \\ \qquad\qquad\qquad\qquad\qquad \le M \, \|\mathbf{u} - \mathbf{p}_h \mathbf{u}_h\|_{k,\Omega} \, \|\mathbf{v} - \mathbf{p}_h \mathbf{r}_h \mathbf{v}\|_{k,\Omega}. \end{cases}$$

When w ranges over $L^2(\Omega)$ and g_j ranges over $H^{2k-j-\frac{1}{2}}(\Gamma)$, there exists a unique solution $v \in H^{2k}(\Omega)$ of the problem

$$(4\text{-}19) \qquad D^*\mathscr{A}'Dv + \lambda v = w, \quad \gamma_j v = -g_j \quad \text{for} \quad 0 \leq j \leq k - 1.$$

However, if $w \in H_0^k(\Omega)$, the solution $\mathbf{v} = \mathscr{A}'Dv$ of the conjugate problem of (4-19) satisfies

$$(4\text{-}20) \quad \begin{cases} \text{i)} \ DD^*\mathbf{v} + \lambda\mathscr{B}'\mathbf{v} = Dw; \ \gamma_j(D^*\mathbf{v}) = g_j \quad \text{for} \quad 0 \leq j \leq k - 1, \\ \text{(ii)} \ \mathbf{v} \in \mathbf{H}^k(\Omega), \ D^*\mathbf{v} = \Lambda v = w - \lambda v \in H^k(\Omega) \end{cases}$$

(see Remark 1-2).

Moreover, when $w \in H_0^k(\Omega)$ and $g_j \in H^{2k-j-\frac{1}{2}}(\Gamma)$, we can write

$$(4\text{-}21) \quad \begin{cases} (D^*(\mathbf{u} - \mathbf{p}_h\mathbf{u}_h), D^*\mathbf{v}) + \lambda[\mathscr{B}(\mathbf{u} - \mathbf{p}_h\mathbf{u}_h), \mathbf{v}] \\ \quad = [\mathbf{u} - \mathbf{p}_h\mathbf{u}_h, (DD^* + \lambda\mathscr{B}')\mathbf{v}] + \langle \beta(\mathbf{u} - \mathbf{p}_h\mathbf{u}_h), \gamma D^*\mathbf{v} \rangle \\ \quad = [\mathbf{u} - \mathbf{p}_h\mathbf{u}_h, Dw] + \langle \beta(\mathbf{u} - \mathbf{p}_h\mathbf{u}_h), g \rangle \\ \quad = (D^*(\mathbf{u} - \mathbf{p}_h\mathbf{u}_h), w) + \langle \beta(\mathbf{u} - \mathbf{p}_h\mathbf{u}_h), g \rangle. \end{cases}$$

Then we deduce from (4-18), (4-21), and Theorem 3-3 that

$$(4\text{-}22) \quad \begin{cases} |(D^*(\mathbf{u} - \mathbf{p}_h\mathbf{u}_h), w)| + |\langle \beta(\mathbf{u} - \mathbf{p}_h\mathbf{u}_h), g \rangle| \\ \quad \leq M |h|^s \|\mathbf{u} - \mathbf{p}_h\mathbf{u}_h\|_{k,\Omega}(\|w\|_{k,\Omega} + \|g\|_{\underset{0 \leq j \leq k-1}{\Pi} H^{2k-j-\frac{1}{2}}(\Gamma)}), \end{cases}$$

where $s = \min(k, m + 1 - k)$, from which we obtain inequalities (4-14). Now let us assume (4-19) and let $\mathbf{v} \in \mathbf{H}^{k,k}(\Omega)$ be the solution of

$$(4\text{-}23) \qquad DD^*\mathbf{v} + \lambda\mathscr{B}'\mathbf{v} = \mathbf{w}, \quad \gamma D^*\mathbf{v} = 0,$$

when $\mathbf{w}$ ranges over $\mathbf{H}(\Omega)$. Then we deduce from (4-21) and (4-18) that

$$(4\text{-}24) \quad \begin{cases} [\mathbf{u} - \mathbf{p}_h\mathbf{u}_h, (DD^* + \lambda\mathscr{B}')\mathbf{v}] = [\mathbf{u} - \mathbf{p}_h\mathbf{u}_h, \mathbf{w}] \leq M \|\mathbf{u} - \mathbf{p}_h\mathbf{u}_h\|_{\Gamma,\Omega} \\ \quad \times \|\mathbf{v} - \mathbf{p}_h\mathbf{r}_h\mathbf{v}\|_{k,\Omega} \leq M |h|^s \|\mathbf{u} - \mathbf{p}_h\mathbf{u}_h\|_{k,\Omega} \|\mathbf{v}\|_{\mathbf{H}^{k,k}(\Omega,D^*)} \\ \qquad\qquad\qquad\qquad\qquad \leq M |h|^s \|\mathbf{u} - \mathbf{p}_h\mathbf{u}_h\|_{k,\Omega} \|\mathbf{w}\|_{\mathbf{H}(\Omega)}. \end{cases}$$

Then (4-24) implies inequalities (4-17). ∎

Now we state a posteriori error estimates (see Theorem 1-4).

THEOREM 4-3 Let us suppose the assumptions of Theorems 2-1 and 4-1. Let c be the constant of $\mathbf{H}(\Omega)$-ellipticity of $\mathscr{A}$ and M the norm of $\mathscr{A}$. Let u be the solution of the Dirichlet problem (4-1), w_h the solution of the discrete variational equation (2-6), $\mathbf{u}_h$ the solution of (4-7), and

$$(4\text{-}25) \quad \begin{aligned} X_h(\mathbf{u}_h, w_h) &= \lambda^{-1}\| f - \lambda u_1 - \lambda q_h w_h - p_h D_h^*\mathbf{u}_h\|_{0,\Omega}^2 \\ &\quad + M^2c^{-1} \|\mathbf{p}_h\mathbf{u}_h - \mathscr{A}D(u_1 - q_h w_h)\|_{0,\Omega}^2. \end{aligned}$$

Then the following a posteriori error estimates hold;

$$(4\text{-}26) \begin{cases} \text{(i)} \quad c^2 \|u - u_1 - q_h w_h\|_{k,\Omega}^2 + \lambda \|u - u_1 - q_h w_h\|_{0,\Omega}^2 \leq X_h(\mathbf{u}_h, w_h), \\ \text{(ii)} \quad c \|\mathbf{u} - p_h \mathbf{u}_h\|_{0,\Omega}^2 + \lambda^{-1} \|\Lambda u - p_h D_h^* \mathbf{u}_h\|_{0,\Omega}^2 \leq X_h(\mathbf{u}_h, w_h). \quad \blacktriangle \end{cases}$$

4-2. Properties of the Discrete Conjugate Problem

The solution $\mathbf{u}_h$ of the discrete variational equation (4-7) is a solution of an equation

$$(4\text{-}27) \qquad\qquad A_h \mathbf{u}_h = l_h.$$

For the purpose of studying (4-27) we introduce the matrices J_h and $\mathscr{B}_h^{p,q}$ defined by

$$(4\text{-}28) \begin{cases} \text{(i)} \quad (J_h u_h, v_h)_h = \displaystyle\int_\Omega p_h u_h p_h v_h \, dx, \\ \text{(ii)} \quad (\mathscr{B}_h^{p,q} u_h, v_h)_h = \displaystyle\int_\Omega b_{pq}(x) p_h^p u_h^p p_h^q v_h^q \, dx, \end{cases}$$

whose entries, respectively, are given by

$$(4\text{-}29) \begin{cases} \text{(i)} \quad h^{-1} \displaystyle\int_\Omega \mu\left(\frac{x}{h} - j\right)\mu\left(\frac{x}{h} - i\right) dx, \\ \text{(ii)} \quad h^{-1} \displaystyle\int_\Omega b_{pq}(x)\mu_p\left(\frac{x}{h} - j\right)\mu_q\left(\frac{x}{h} - i\right) dx, \qquad \mu_q = \mu * \pi_q. \end{cases}$$

Let us introduce the vectors l_h^q, whose components $l_h^{q,i}$ are defined by

$$(4\text{-}30) \begin{cases} l_h^{q,i} = h^{-1} \displaystyle\int_\Omega f(x)(-1)^{|q|} \nabla_h^q \mu\left(\frac{x}{h} - i\right) dx \\ \qquad\qquad + \lambda h^{-1} \displaystyle\sum_{0 \leq j \leq k-1} \int_\Gamma t_j(x)\beta_{2k-j-1}^q \mu_q\left(\frac{x}{h} - i\right) d\sigma(x), \end{cases}$$

where we set

$$(4\text{-}31) \qquad \beta_j \mathbf{v} = \sum_{|q| \leq k} \beta_j^q v^q \qquad \text{for} \quad k \leq j \leq 2k - 1.$$

Finally, let us set

$$(4\text{-}32) \qquad\qquad \tilde{\nabla}_h^q = (-1)^{|q|} \nabla_h^q.$$

Then the discrete conjugate problem (4-7) is equivalent to the following system of equations: find $\mathbf{u}_h = (u_h^q)_{|q| \leq k} \in \prod_{|q| \leq k} H_h^{\mu_q}(\Omega)$ satisfying for $|q| \leq k$

$$(4\text{-}33) \qquad \tilde{\nabla}_h^q J_h\left(\sum_{|p| \leq k} (-1)^{|p|} \nabla_h^p u_h^p\right) + \sum_{|p| \leq k} \mathscr{B}_h^{p,q} u_h^p = l_h^q.$$

CHAPTER 11

External and Partial Approximations

In Section 1 we define external and partial approximations, and we extend several results of Chapters 2 and 3.

Section 2 constitutes an investigation of external and partial approximations of boundary-value problems, wherein we generalize some of the results of Sections 8.2 and 9.1.

In view of approximating boundary-value problems for elliptic differential operators, we construct partial approximations of the Sobolev spaces in Section 11.3.

Finally, Section 4 deals with the construction of partial approximations of several boundary-value problems.

1. EXTERNAL APPROXIMATIONS; STABILITY, CONVERGENCE, AND ERROR ESTIMATES

Let $\bar{V}$ and $\bar{F}$ be two Hilbert spaces and $\bar{A} \in L(\bar{V}, \bar{F})$ a continuous linear operator from $\bar{V}$ into $\bar{F}$. We associate with these spaces and this operator the following items:

$$(1\text{-}1) \quad \begin{cases} \text{(i)} \ \ V \text{ is a closed subspace of } \bar{V}, \\ \text{(ii)} \ \ F = \bar{F}/F_0 \text{ is a factor space of } \bar{F} \text{ by a closed subspace } F_0 \text{ of } \bar{F}, \\ \text{(iii)} \ \ A = \mu\bar{A}\pi \in L(V, F), \end{cases}$$

where

$$\begin{cases} \text{(i)} \ \ \pi \text{ is the canonical injection from } V \text{ into } \bar{V}, \\ \text{(ii)} \ \ \mu \text{ is the canonical surjection from } \bar{F} \text{ onto } F = \bar{F}/F_0. \end{cases}$$

Let $g \in \bar{F}$ and $f = \mu g \in F$ be given. We would like to study the approximation of the solution u of

$$(1\text{-}2) \qquad \qquad Au = f \quad \text{ or } \quad \mu\bar{A}\pi u = \mu g$$

by using approximations of $\bar{V}$ and $\bar{F}$ instead of approximations of V and F.

For that purpose, we need the following notions of approximations of $\bar{V}$ which are convergent in V and of approximations of $\bar{V}$ and $\bar{F}$ which are "external approximations" of V and F.

Definition 1-1 Approximations (V_h, p_h, r_h) of $\bar{V}$ *are convergent in V if*

(1-3) $$\|u - p_h r_h u\|_{\bar{V}} \text{ converges to 0 for any } u \text{ in } V.$$

We say that (V_h, p_h, r_h) are *external approximations of the closed subspace V of $\bar{V}$* if the approximations (V_h, p_h, r_h) of $\bar{V}$ satisfy the following property:

(1-4) if $p_h u_h$ converges weakly to u in $\bar{V}$, then u belongs actually to V.

We say that (F_h, q_h, s_h) are *external approximations of the factor space $F = \bar{F}/F_0$* if the approximations (F_h, q_h, s_h) of $\bar{F}$ satisfy the property

(1-5) $$\|s_h g\|_{F_h} \text{ converges to 0 for any } g \in F_0.$$

Furthermore, the operator $A_h = s_h \bar{A} p_h$ is the *external approximation* of $A = \mu \bar{A} \pi$, and the discrete equation

(1-6) $$s_h \bar{A} p_h u_h = s_h g \qquad \text{where } f = \mu g$$

is the *external approximation of* (1-2) associated with approximations (V_h, p_h, r_h) and (F_h, q_h, s_h) of $\bar{V}$ and $\bar{F}$. ▲

Remark 1-1 Since we use only the prolongation p_h of the approximation of $\bar{V}$ and the restriction s_h of the approximation of $\bar{F}$ in the actual construction of the external approximate equation (1-6), we can assume that

(1-7) $$\begin{cases} \text{(i) } r_h = \hat{r}_h \text{ is the optimal restriction associated with } p_h \text{ in } V, \\ \text{(ii) } q_h = \hat{q}_h \text{ is the optimal prolongation associated with } s_h \text{ in } F \end{cases}$$

(see Theorem 2.4-2).

Usually we supply V_h with the discrete norm $\|v_h\|_{V_h} = \|p_h v_h\|_{\bar{V}}$ associated with p_h and the space F_h with the discrete norm $\|f_h\|_{F_h} = \|\hat{q}_h f_h\|_{\bar{F}}$ associated with s_h (see Section 2.4-1). With the latter choice, condition (1-5) is equivalent to

(1-8) $$\hat{q}_h s_h g \text{ converges to 0 for any } g \in F_0.$$

Otherwise, if the norms $\|f_h\|_{F_h}$ satisfy the property

(1-9) $$\|s_h f\|_{F_h} \leq c \|f\|_F \qquad \text{where } c \text{ is independent of } h,$$

then (1-8) implies (1-5). ■

Remark 1-2 (*Duality properties*) Let V be a closed subspace of $\bar{V}$ and choose

(1-10) $$\bar{F} = \bar{V}', F = V', \text{ and } F_0 = V^\perp.$$

Then, by Theorem 2.1-6, we can identify V' to the space $\bar{F}/F_0 = \bar{V}'/V^\perp$. ∎

Are the dual approximations of the external approximations of V external approximations of the dual V'?

Let (V_h, p_h, r_h) be approximations of $\bar{V}$ and (V_h', q_h, s_h) the dual approximations of $\bar{F} = \bar{V}'$, where $s_h = p_h'$ and $q_h = r_h'$ (see Section 2.4-2).

THEOREM 1-1 The following statements are equivalent:

1. The approximations $(V_h, p_h, \hat{r}_h)$ are external approximations of the closed subspace V of $\bar{V}$.

2. The dual approximations $(V_h', \hat{q}_h, s_h)$ are external approximations of $V' = \bar{V}'/V^\perp$. ▲

Proof Let us begin by proving that statement 1 implies 2. Let $g \in F_0 = V^\perp$. Then

(1-11) $$\|s_h g\|_{V_h'} = \sup_{v_h \in V_h} \frac{|(s_h g, v_h)_h|}{\|p_h v_h\|} = \sup_{v_h \in V_h} \frac{|(g, p_h v_h)|}{\|p_h v_h\|}.$$

The supremum in (1-11) is achieved at a point u_h of the unit ball of V_h. Then $p_h u_h$ is bounded in $\bar{V}$ and there exists a subsequence of h (again denoted by h) such that $p_h u_h$ converges weakly to u in $\bar{V}$.

By assumption, u belongs to V. Then

$$\|s_h g\|_{V_h'} = |(g, p_h u_h)| \text{ converges to } |(g, u)| = 0,$$

since $g \in V^\perp$ and $u \in V$.

Conversely, let us assume statement 2, with $p_h u_h$ converging weakly to u in $\bar{V}$. Then for any $g \in \bar{V}'$,

$$(s_h g, u_h)_h = (g, p_h u_h) \text{ converges to } (g, u).$$

If $g \in V^\perp = F_0$, then $\|s_h g\|_{V_h'}$ converges to 0 and $|(s_h g, u_h)_h| \leq \|s_h g\|_{V_h'} \|p_h u_h\| \leq M\|s_h g\|_{V_h'}$ converges to 0.

We deduce that $(g, u) = 0$ for any $g \in V^\perp$. This amounts to saying that u belongs to $V^{\perp\perp} = V$. ∎

Remark 1-3 By using the optimal restriction $\hat{r}_h$ associated with p_h, we can characterize approximations of $\bar{V}$ that are both convergent in V and external approximations of V. ∎

COROLLARY 1-1   Approximations $(V_h, p_h, \hat{r}_h)$ of $\bar{V}$ are convergent in V and are external approximations of V if and only if $p_h \hat{r}_h$ converges pointwise to the orthogonal projector τ on V:

$$(1-12) \qquad \|\tau u - p_h \hat{r}_h u\| \text{ converges to } 0 \qquad \text{for any} \quad u \in \bar{V}. \qquad \blacktriangle$$

Proof   Let $u = u_0 + u_1$ where $u_0 \in V$ and $u_1 \in V^\oplus$. Since the approximations are convergent in V, $p_h r_h u_0$ converges to u_0, and conversely. Let us prove that $p_h r_h u_1$ converges to 0, denoting by J the isomorphism from $\bar{V}$ into $\bar{V}'$. First, $J u_1 = g$ belongs to $V^\perp$. Second, we deduce from Theorem 2.4-2 that

$$p_h \hat{r}_h u_1 = p_h \hat{r}_h J^{-1} g = J^{-1} \hat{q}_h s_h g.$$

We thus deduce from Theorem 1-1 that $p_h \hat{r}_h u_1$ converges to 0 for any $u_1 \in V^\oplus$ if and only if the approximations of $\bar{V}$ are external approximations of V. $\blacksquare$

1-2.   Example: Partial Approximations of a Finite Intersection of Spaces

In describing an important example of external approximations of a finite intersection $V = \bigcap V^i$ of Hilbert spaces V^i, we consider the product $\bar{V} = \Pi V^i$ of the spaces V^i, consisting of sequences $\bar{u} = (u^i)_i$, where the components $u^i \in V^i$.

We identify V with the diagonal of $\bar{V} = \Pi V^i$ (which is a closed subspace of $\bar{V}$) by saying that $\bar{u} = (u^i)_i$ belongs to V if and only if all the components u^i of $\bar{u}$ are equal to a same element u (which belongs necessarily to $V = \bigcap V^i$).

If V_h is a discrete space, any prolongation p_h from V_h into $\bar{V} = \Pi V^i$ can be written in the form

$$(1-13) \qquad p_h u_h = (p_h{}^i u_h)_i \qquad \text{where} \quad p_h{}^i \in L(V_h, V^i)$$

and any restriction r_h from V onto V_h can be written in the form

$$(1-14) \quad r_h \bar{u} = \sum_i r_h{}^i u^i \qquad \text{where} \quad \bar{u} = (u^i)_i \qquad \text{and} \qquad r_h{}^i \in L(V^i, V_h).$$

Definition 1-2   Approximations $(V_h, p_h{}^i, r_h{}^i)$ of the spaces V^i are partial approximations of the intersection $V = \bigcap V^i$ if the approximations (V_h, p_h, r_h) of $\bar{V} = \Pi V^i$ defined by (1-13) and (1-14) are external approximations of V identified to the diagonal of $\bar{V}$. This amounts to saying that

$$(1-15) \quad \begin{cases} \text{(i) } p_h \text{ is an isomorphism from } V_h \text{ onto its closed range in } \bar{V} = \Pi V^i, \\ \text{(ii) if } p_h{}^i u_h \text{ converges weakly to } u^i \text{ in } V^i \text{ for any } i, \text{ then all the } u^i \text{ are} \\ \qquad \text{equal to a same element } u \text{ (which belongs to } V). \end{cases} \qquad \blacktriangle$$

1-3. Stability and Convergence of External Approximations of Operators

Let $A_h = s_h \bar{A} p_h$ be the external approximations of $A = \mu \bar{A} \pi$ defined by the approximations $(V_h, p_h, \hat{r}_h)$ and $(F_h, \hat{q}_h, s_h)$ of the spaces $\bar{V}$ and $\bar{F}$. Since p_h and s_h are given explicitly, we supply the discrete spaces V_h and F_h with the norms associated with p_h and s_h

$$(1\text{-}16) \quad \|v_h\|_{V_h} = \|p_h v_h\|_{\bar{V}} \quad \text{and} \quad \|f_h\|_{F_h} = \sup_{g_h \in F_h'} \frac{|(g_h, f_h)_h|}{\|s_h' g_h\|} = \|\hat{q}_h f_h\|_{\bar{F}}.$$

Definition 1-3 The external approximations $A_h = s_h \bar{A} p_h$ of A are stable (for the norms of $\bar{V}$ and $\bar{F}$) if there exists a constant S independent of h such that

$$(1\text{-}17) \quad \|p_h u_h\|_{\bar{V}} \leq S \|s_h A_h u_h\|_{F_h} = S \|\hat{q}_h s_h A_h u_h\|_{\bar{F}} \quad \text{for all} \quad u_h \in V_h. \quad \blacktriangle$$

Then stable external approximations A_h of A are isomorphisms from V_h into F_h.

Let us consider the external approximation

$$(1\text{-}18) \quad s_h \bar{A} p_h u_h = s_h g \quad \text{where} \quad f = \mu g$$

of the equation

$$(1\text{-}19) \quad Au = f \quad \text{or} \quad \mu \bar{A} \pi u = f = \mu g,$$

and let us assume that

$$(1\text{-}20) \quad \begin{cases} \text{(i) the approximations } (F_h, \hat{q}_h, s_h) \text{ of } F \text{ are external approxima-} \\ \quad \text{tions of } F = \bar{F}/F_0, \\ \text{(ii) the approximations } (V_h, p_h, \hat{r}_h) \text{ of } \bar{V} \text{ are convergent in } V, \\ \text{(iii) the operators } A_h = s_h \bar{A} p_h \text{ are stable.} \end{cases}$$

THEOREM 1-2 Let us assume that there exists a solution $u \in V$ of (1-19). Then the assumptions (1-20) imply that

$$(1\text{-}21) \quad \begin{cases} \text{(i) there exists a unique solution } u_h \text{ of (1-18) for any } f \in F, \\ \text{(ii) } \|u - p_h u_h\|_{\bar{V}} \text{ converges to 0} \quad \text{for any} \quad f \in F. \end{cases} \quad \blacktriangle$$

Proof By the stability of A_h we deduce [(1-21)(i)] and the inequality

$$(1\text{-}22) \quad \|p_h(u_h - \hat{r}_h u)\|_{\bar{V}} \leq S \|\hat{q}_h s_h \bar{A}(p_h u_h - p_h \hat{r}_h u)\|_{\bar{F}}.$$

On the other hand, we can write

$$(1\text{-}23) \quad \hat{q}_h s_h \bar{A} p_h (u_h - \hat{r}_h u) = \hat{q}_h s_h (g - \bar{A} u) + \hat{q}_h s_h \bar{A}(u - p_h \hat{r}_h u).$$

But $g - \bar{A}u$ belongs to F_0 since, by (1-19), $\mu(g - \bar{A}u) = 0$. Then $\hat{q}_h s_h(g - \bar{A}u)$ converges to 0, since the approximations of $\bar{F}$ are external approximations of $F = \bar{F}/F_0$. However, since u belongs to V and since the approximations of $\bar{V}$ are convergent in V, $u - p_h \hat{r}_h u$ converges to 0 and $\|\hat{q}_h s_h A(u - p_h \hat{r}_h u)\|_{\bar{F}} \leq M \|u - p_h \hat{r}_h u\|_{\bar{V}}$ converges to 0.

Therefore, $\|p_h u_h - p_h \hat{r}_h u\|_{\bar{V}}$ converges to 0 and, since the approximations of $\bar{V}$ are convergent in V, $u - p_h u_h = u - p_h \hat{r}_h u + p_h(\hat{r}_h u - u_h)$ converges to 0. ∎

Theorem 1-2 shows that conditions (1-20) are sufficient in order to imply the convergence of $p_h u_h$ to a solution u of (1-19). They are necessary when A is an isomorphism from V onto F.

THEOREM 1-3 Let us assume that A is an isomorphism from V onto F. Then the conditions (1-20) and (1-21) are equivalent. ▲

Proof We have to prove that (1-21) implies (1-20). First, the approximations of $\bar{V}$ are convergent in V, since

(1-24) $\|u - p_h \hat{r}_h u\| \leq \|u - p_h u_h\|$ converges to 0 for any $u \in V$.

Second, the operators $A_h = s_h \bar{A} p_h$ are stable. Indeed, since

(1-25) $p_h u_h = p_h A_h^{-1} s_h g$ converges to $u = A^{-1} \mu g$ for any $g \in \tilde{F}$,

then there exists a constant S independent of h such that

(1-26) $$\|p_h A_h^{-1} s_h g\|_{\bar{V}} \leq S \|g\|_F$$

Replacing g by $\hat{q}_h g_h$ in (1-26), we deduce the inequality

(1-27) $$\|p_h A_h^{-1} g_h\|_{\bar{V}} \leq S \|\hat{q}_h g_h\|_{\tilde{F}},$$

which is equivalent to (1-17) (when we set $u_h = A_h^{-1} g_h$).

Finally, let us prove that the approximations of $\bar{F}$ are external approximations of F. Let $g \in F_0$. Then $A^{-1} \mu g = 0$ and if u_h is the solution of $A_h u_h = s_h A p_h u_h = s_h g$, $p_h u_h$ converges to 0. Then $\|\hat{q}_h s_h g\|_{\bar{F}} = \|\hat{q}_h s_h A p_h u_h\|_{\bar{F}} \leq M \|p_h u_h\|_{\bar{V}}$ converges to 0. ∎

1-4. Estimates of Error and Regularity of the Convergence

We give estimates of error $u - p_h u_h$ in terms of the error and external error function in which (V_h, p_h, r_h) are approximations of $\bar{V}$ and $U \subset V$ is a Hilbert space contained in V with a stronger topology.

Definition 1-4   We say that

$$(1\text{-}28) \qquad e_U^{\,V}(p_h r_h) = \sup_{u \in U} \frac{\|u - p_h r_h u\|_V}{\|u\|_U}$$

is the *truncation error* of the approximation. If we take $r_h = \hat{r}_h$, we say that

$$(1\text{-}29) \qquad e_U^{\,V}(p_h) = e_U^{\,V}(p_h \hat{r}_h)$$

is the *error function*.   ▲

It is clear that $e_U^{\,V}(p_h) \leq e_U^{\,V}(p_h r_h)$ for any restriction r_h. Now let $\bar{E}$ and $\bar{F}$ be two Hilbert spaces such that

$$(1\text{-}30) \quad \begin{cases} \text{(i)} & \bar{E} \subset \bar{F} \text{ with a stronger topology,} \\ \text{(ii)} & F_0 \text{ is a closed subspace of } \bar{F} \text{ and } E_0 = \bar{E} \cap F_0, \\ \text{(iii)} & F = \bar{F}/F_0 \quad \text{and} \quad E = \bar{E}/E_0 . \end{cases}$$

Let (F_h, q_h, s_h) be approximations of $\bar{F}$.

Definition 1-5   We say that

$$(1\text{-}31) \qquad t_{E_0}^{\,\bar{F}}(q_h s_h) = \sup_{f \in E_0} \frac{\|q_h s_h f\|_{\bar{F}}}{\|f\|_{\bar{E}}}$$

is the *external truncation error* of the approximation.

If we take $q_h = \hat{q}_h$, we say that

$$(1\text{-}32) \qquad t_{E_0}^{\,\bar{F}}(s_h) = t_{E_0}^{\,\bar{F}}(\hat{q}_h s_h)$$

is the *external error function* of the approximation.   ▲

THEOREM 1-4   Let $A_h = s_h \bar{A} p_h$ be stable external approximations of the operator $A = \mu \bar{A} \pi$ for the approximations (V_h, p_h, r_h) and (F_h, q_h, s_h) of $\bar{V}$ and $\bar{F}$.

Let us assume that the solution u of (1-19) belongs to $U \subset V$ and that g and $\bar{A}u$ belongs to $\bar{E} \subset \bar{F}$. Then, for any restriction r_h, the following estimate of the error holds:

$$(1\text{-}33) \quad \|u - p_h u_h\|_V \leq M(e_U^{\,V}(p_h r_h)\, \|u\|_U + t_{E_0}^{\,\bar{F}}(s_h)\, \|g - \bar{A}u\|_{\bar{E}}). \quad ▲$$

Proof   We deduce from (1-22) and (1-23) (with $\hat{r}_h$ replaced by any r_h) the inequality

$$(1\text{-}34) \quad \|p_h(u_h - r_h u)\|_V \leq S(\|\hat{q}_h s_h (\bar{A}u - g)\|_F + \|u - p_h r_h u\|_V).$$

Then (1-33) follows from (1-34) and the definitions (1-28) and (1-32) of the truncation error and the external error function. ∎

To study the regularity of the convergence, we introduce a prolongation p_h^1 mapping V_h into U, as well as the stability function linking the prolongations p_h and p_h^1:

$$(1\text{-}35) \qquad s_U^{\,V}(p_h^{\,1}, p_h) = \sup_{v_h \in V_h} \frac{\|p_h^{\,1} v_h\|_U}{\|p_h v_h\|_V}.$$

THEOREM 1-5 Let us suppose the assumptions of Theorem 1-4. Furthermore, let us assume that there exists a restriction r_h such that

$$(1\text{-}36) \quad \begin{cases} \text{(i)} & \|u - p_h^{\,1} r_h u\|_U \quad \text{converges to } 0 \quad \text{for any } u \in U, \\ \text{(ii)} & s_U^{\,V}(p_h^{\,1}, p_h)\, \|\hat{q}_h s_h g\| \quad \text{converges to } 0 \quad \text{for any } \; g \in E_0, \\ \text{(iii)} & s_U^{\,V}(p_h^{\,1}, p_h)\, \|u - p_h r_h u\|_V \quad \text{converges to } 0 \quad \text{for any } \; u \in U. \end{cases}$$

Then $p_h^1 u_h$ converges to u in U. ▲

Proof Indeed, $\|p_h^{\,1}(u_h - r_h u)\|_U \leq s_U^{\,V}(p_h^{\,1}, p_h)\, \|p_h(u_h - r_h u)\|_V$ converges to 0 by (1-34) and [(1-36)(ii) and (iii)]. Therefore, [(1-36)(i)] implies that $- p_h^1 u_h$ converges to 0 in U.

1-5. Properties of the External Error Functions

LEMMA 1-1 The external truncation errors and the external error functions are related by the following inequalities:

$$(1\text{-}37) \quad \begin{cases} \text{(i)} & t_{E_0}^{\,F}(s_h) \leq t_{E_0}^{\,F}(q_h s_h) \quad \text{if } s_h q_h = 1, \\ \text{(ii)} & t_{E_0}^{\,F}(q_h s_h) \leq \|q_h s_h\|_{L(F,F)} t_{E_0}^{\,F}(s_h). \end{cases}$$
▲

Proof By the very definition of optimal prolongation associated with s_h (see (4-5), Section 2.4-1),

$$\|\hat{q}_h s_h f\|_F \leq \|q_h s_h f\|_F \quad \text{if } s_h q_h = 1.$$

This inequality implies [(1-37)(i)].

On the other hand, since $q_h s_h = q_h s_h \hat{q}_h s_h$, we deduce

$$\|q_h s_h f\|_F \leq \|q_h s_h\|_{L(F,F)} \|\hat{q}_h s_h f\|_F,$$

which implies [(1-37)(ii)]. ∎

Now let us define external error functions and external truncation errors of approximations (V_h, p_h, r_h) of W.

Definition 1-6 Let us introduce spaces $\bar{V}$, V, and $\bar{W}$, where

$$(1\text{-}38) \quad \begin{cases} \text{(i)} \ \bar{V} \subset \bar{W} \text{ with a stronger topology,} \\ \text{(ii)} \ V \text{ is a closed subspace of } \bar{V} \text{ and } W \text{ is the closure of } V \text{ in } \bar{W}. \end{cases}$$

We say that

$$(1\text{-}39) \quad t_{\bar{V}}{}^{W}(p_h r_h) = \sup_{u \in \bar{V}} \ \inf_{v \in W} \frac{\|p_h r_h u - v\|_{\bar{W}}}{\|u\|_{\bar{V}}} = \sup_{u \in \bar{V}} \frac{\|(1 - \tau)p_h r_h u\|_{\bar{W}}}{\|u\|_{\bar{V}}}$$

(where τ is the orthogonal projector from $\bar{W}$ onto W) is the *external truncation error* of the approximation. If we choose $r_h = \hat{r}_h$, we say that

$$(1\text{-}40) \qquad\qquad t_{\bar{V}}{}^{W}(p_h) = t_{\bar{V}}{}^{W}(p_h \hat{r}_h)$$

is the *external error function* of the approximation. ▲

The external truncation error and the external error function are related by properties analogous to Lemma 1-1.

LEMMA 1-2 The following inequalities hold:

$$(1\text{-}41) \quad \begin{cases} \text{(i)} \ t_{\bar{V}}{}^{W}(p_h) = \sup_{u_h \in V_h} \ \inf_{v \in W} \dfrac{\|p_h u_h - v\|_{\bar{W}}}{\|p_h u_h\|_{\bar{V}}} = \sup_{u_h \in V_h} \dfrac{\|(1 - \tau)p_h u_h\|_{\bar{W}}}{\|p_h u_h\|_{\bar{V}}} \\[2mm] \text{(ii)} \ t_{\bar{V}}{}^{W}(p_h) \leq t_{\bar{V}}{}^{W}(p_h r_h) \quad \text{if} \ \ r_h p_h = 1, \\[2mm] \text{(iii)} \ t_{\bar{V}}{}^{W}(p_h r_h) \leq \|p_h r_h\|_{L(\bar{V}, \bar{V})} t_{\bar{V}}{}^{W}(p_h). \end{cases}$$

▲

Proof Let us denote by $t_{\bar{V}}{}^{W}(h)$ the expression defined by the right-hand side of [(1-41)(i)]. Then if $r_h p_h = 1$, we deduce that

$$(1\text{-}42) \qquad \frac{\|(1 - \tau)p_h u_h\|_{\bar{W}}}{\|p_h u_h\|_{\bar{V}}} = \frac{\|(1 - \tau)p_h r_h p_h u_h\|_{\bar{W}}}{\|p_h u_h\|_{\bar{V}}} \leq t_{\bar{V}}{}^{W}(p_h r_h).$$

Taking the supremum on V_h in (1-42), we deduce that $t_{\bar{V}}{}^{W}(h) \leq t_{\bar{V}}{}^{W}(p_h r_h)$. On the other hand, since $\|p_h r_h u\| \leq \|p_h r_h\| \|u\|$, we obtain

$$(1\text{-}43) \quad \frac{\|(1 - \tau)p_h r_h u\|_{\bar{W}}}{\|u\|_{\bar{V}}} \leq \|p_h r_h\| \frac{\|(1 - \tau)p_h r_h u\|_{\bar{W}}}{\|p_h r_h u\|_{\bar{V}}} \leq t_{\bar{V}}{}^{W}(h) \, \|p_h r_h\|.$$

Taking the supremum on $\bar{V}$, we deduce that $t_{\bar{V}}{}^{W}(p_h r_h) \leq \|p_h r_h\| \, t_{\bar{V}}{}^{W}(h)$. Then if we take $r_h = \hat{r}_h$, we obtain $t_{\bar{V}}{}^{W}(p_h) = t_{\bar{V}}{}^{W}(h)$, since $\hat{r}_h p_h = 1$ and $\|p_h \hat{r}_h\| = 1$. ■

Naturally, there are duality relations between the external truncation errors of an approximation and its dual approximation. Let us assume that:

$$(1\text{-}44) \qquad\qquad \tilde{V} \subset \overline{W},$$

the injection is continuous and dense, and let us set

$$(1\text{-}45) \quad \tilde{F} = \tilde{V}', \; \tilde{E} = \overline{W}', \; E_0 = W^\perp, \; F_0 = V^\perp, \; E = W'., \text{ and } F = V'.$$

LEMMA 1-3 Assuming (1-44), let (V_h, p_h, r_h) be approximations of $\overline{W}$ and (V_h', q_h, s_h) the dual approximations (where $s_h = p_h'$ and $q_h = r_h'$). Let V be a closed subspace of $\tilde{V}$ and $V' = \tilde{V}'/V^\perp$ its dual. Then the external truncation errors coincide:

$$(1\text{-}46) \quad t_{\tilde{V}}{}^W(p_h r_h) = t_{E_0}{}^F(q_h s_h) \qquad \text{and} \qquad t_{\tilde{V}}{}^W(p_h) = t_{E_0}{}^F(s_h). \quad \blacktriangle$$

Proof Indeed, $t_{E_0}{}^F(q_h s_h) = \|q_h s_h \pi\|_{L(E_0, F)}$, where π denotes the canonical injection from E_0 into $\tilde{E}$. Then, by transposition,

$$(1\text{-}47) \qquad\qquad t_{E_0}{}^F(q_h s_h) = \|\pi' p_h r_h\|_{L(F', E_0')}$$

But since $F' = \tilde{V}$, $E_0' = \overline{W}/W$, and π' is the canonical injection from $\overline{W}$ onto $\overline{W}/W$, we deduce from (1-47) that

$$(1\text{-}48) \qquad t_{E_0}{}^F(q_h s_h) = \sup_{u \in \tilde{V}} \inf_{v \in W} \frac{\|p_h r_h u - v\|_{\overline{W}}}{\|u\|_{\tilde{V}}} = t_{\tilde{V}}{}^W(p_h r_h).$$

Since $\dot{q}_h = \dot{r}_h'$, we deduce from (1-40) that $t_{E_0}{}^F(s_h) = t_{\tilde{V}}{}^W(p_h)$. $\quad\blacksquare$

LEMMA 1-4 Let us assume (1-38) and

$$(1\text{-}49) \qquad \text{The injection from } \tilde{V} \text{ into } \overline{W} \text{ is dense and compact.}$$

Let us also assume that the approximations (V_h, p_h, r_h) of $\overline{W}$ are external approximations of W. Then

$$(1\text{-}50) \qquad \text{the external error functions } t_{\tilde{V}}{}^W(p_h) \text{ converge to } 0. \quad \blacktriangle$$

Proof If we use π' to denote the canonical surjection from $\overline{W}$ onto $\overline{W}/W$, then $\pi' p_h \hat{r}_h$ converges pointwise to 0 in $L(\overline{W}, \overline{W}/W)$: indeed, by Corollary 1-1, $\pi' p_h \hat{r}_h u$ converges to $\pi' \tau u = 0$. Then $\pi' p_h \hat{r}_h$ converges to 0 uniformly over every compact subset of $\overline{W}$, and thus over every bounded subset of $\tilde{V}$ (see Theorem 2.1-9). In other words, $\|\pi' p_h \hat{r}_h\|_{L(\tilde{V}, \overline{W}/W)}$ converges to 0. This implies Lemma 1-4 because $t_{\tilde{V}}{}^W(p_h) \leq t_{\tilde{V}}{}^W(p_h \hat{r}_h) = \|\pi' p_h r_h\|_{L(\tilde{V}, \overline{W}/W)}$ by (1-48). $\quad\blacksquare$

2. EXTERNAL AND PARTIAL APPROXIMATIONS OF VARIATIONAL EQUATIONS

2-1. Partial Approximation of a Split Variational Equation

The examples of variational boundary-value problems we studied in Chapter 7 (see, e.g., Section 7.1-1) are associated with continuous bilinear forms that are "split" in a sum of "partial" bilinear forms.

We use the particular situation in which a space V is the intersection $V = \bigcap V^i$ of a finite number of Hilbert spaces V^i such that

$$(2\text{-}1) \qquad V = \bigcap V^i \subset V^i \subset V^0 = H,$$

the injections being continuous and dense.

We introduce a split continuous bilinear form

$$(2\text{-}2) \qquad a(u, v) = \sum_{i,j} a_{ij}(u, v),$$

where

$$(2\text{-}3) \qquad a_{ij}(u, v) \text{ is a continuous bilinear form of } V^i \times V^j.$$

Let $f \in H = V^0$ be given, with $u \in V$ the solution of the split variational equation

$$(2\text{-}4) \qquad a(u, v) = \sum_{i,j} a_{ij}(u, v) = (f, v) \qquad \text{for any} \quad v \in V.$$

Let (V_h, p_h^i, r_h^i) be approximations of the spaces V^i which are partial approximations of the space V (see Section 1-2).

Definition 2-1 The discrete variational equation

$$(2\text{-}5) \qquad \sum_{i,j} a_{ij}(p_h^i u_h, p_h^j v_h) = (f, p_h^0 v_h) \qquad \text{for any} \quad v_h \in V_h$$

is the partial approximation of the split variational equation (2-4). ▲

Remark 2-1 If V_h is a finite-dimensional space, we can write the prolongations p_h^i in the form

$$(2\text{-}6) \qquad p_h^j u_h = \sum_k u_h^k \mu_{jh}^k \qquad \text{where} \quad (\mu_{jh}^k)_k \text{ is a basis of } V^j.$$

Therefore, the entries of the matrix A_h associated with the bilinear form $\sum_{i,j} a_{ij}(p_h^i u_h, p_h^j v_h)$ are equal to $\sum_{i,j} a_{ij}(\mu_{ih}^k, \mu_{jh}^l)$. ■

Remark 2-2 The definition of partial approximation of a split variational equation is a particular case of an external approximation. Indeed, we can identify V with the diagonal of $\bar{V} = \Pi V^i$ and extend $a(u, v)$ to a continuous bilinear form $\bar{a}(\bar{u}, \bar{v})$ on $\bar{V} \times \bar{V}$ by setting

(2-7)
$$\bar{a}(\bar{u}, \bar{v}) = \sum_{i,j} a_{ij}(u^i, v^j).$$

The form $\bar{a}(\bar{u}, \bar{v})$ defines an operator $\bar{A} \in L(\bar{V}, \bar{V}')$ equal to

(2-8)
$$\begin{cases} \text{(i)} \ (\bar{A}\bar{u})^j = \sum_i A_{ij}u^i \quad \text{where} \\ \text{(ii)} \ (A_{ij}u^i, v^j) = a_{ij}(u^i, v^j) \quad \text{for any} \ \ u^i \in V^i, v^j \in V^j. \end{cases}$$

Let us use π to denote the canonical injection from V into $\bar{V}$ and $\mu = \pi'$ for its transpose. Then if

(2-9)
$$\bar{f} = (f^i)_i \in \bar{V}' = \Pi V^{i'}, \mu(\bar{f}) = \sum_i f^i \in V'.$$

We identify V' with the factor space $\bar{V}'/V^\perp$ and μ with the canonical surjection from $\bar{V}'$ onto $\bar{V}'/V^\perp$.

Therefore, it is clear that the operator $A \in L(V, V')$ defined by the continuous bilinear form $a(u, v)$ on $V \times V$ can be written in the form

(2-10)
$$Au = \mu\bar{A}\pi u = \sum_{j,i} A_{ij}u.$$

In order to construct external approximation of A, we need approximations of $\bar{V}$ and of $F = \bar{V}'$. We will choose for our approximation of $F = \bar{V}'$ the dual approximation (V'_h, q_h, s_h) of an approximation (V_h, p_h, r_h) of the space $\bar{V}$. In this case, the prolongations p_h and the restrictions $s_h = p'_h$ can be written in the form

(2-11)
$$\begin{cases} \text{(i)} \ p_h u_h = (p_h{}^i u_h)_i \\ \text{(ii)} \ s_h f = p'_h f = \sum_i p_h{}^{i\prime} f^i. \end{cases}$$

Then the external approximation of A is the operator A_h defined by

(2-12)
$$A_h u_h = s_h \bar{A} p_h u_h = \sum_{i,j} p_h{}^{j\prime} A_{ij} p_h{}^i u_h$$

and it is the operator associated with the continuous bilinear form

(2-13)
$$(p_h u_h, p_h v_h) = \sum_{i,j} a_{ij}(p_h{}^i u_h, p_h{}^j v_h).$$

Finally, we have defined partial approximations of V as approximations of $\bar{V}$ which are external approximations of V. By Theorem 1-1, this amounts to saying that the dual approximations of $\bar{V}'$ are external approximations of V'. ∎

Next we study convergence properties and estimates of error in the more general framework of external approximations of variational equations.

2-2. External Approximation of Variational Equations

In this section we investigate the convergence and estimates of error of external approximations of $\bar{V}$-elliptic operators $\bar{A}$. We call $\bar{a}(\bar{u}, \bar{v})$ a continuous bilinear form of $\bar{V} \times \bar{V}$, and $a(u, v)$ its restriction to $V \times V$, where V is a closed subspace of $\bar{V}$.

Let $\bar{f} \in \bar{V}'$ be a given element and u a solution of the variational equation

$$(2\text{-}14) \qquad \begin{cases} \text{(i)} \ u \in V, \\ \text{(ii)} \ a(u, v) = (\bar{f}, v) \quad \text{for any} \ v \in V. \end{cases}$$

Let (V_h, p_h, r_h) be approximations of $\bar{V}$ satisfying

$$(2\text{-}15) \qquad \begin{cases} \text{(i)} \ (V_h, p_h, r_h) \text{ are convergent in } V, \\ \text{(ii)} \ (V_h, p_h, r_h) \text{ are external approximations of } V. \end{cases}$$

Let us approximate (2-14) by the discrete variational equation

$$(2\text{-}16) \qquad \begin{cases} \text{(i)} \ u_h \in V_h, \\ \text{(ii)} \ \bar{a}(p_h u_h, p_h v_h) = (\bar{f}, p_h v_h) \quad \text{for any} \ v_h \in V_h, \end{cases}$$

which is its external approximation.

THEOREM 2-1 Let us assume (2-15) and the $\bar{V}$-ellipticity of $\bar{a}(\bar{u}, \bar{v})$:

$$(2\text{-}17) \qquad \bar{a}(\bar{v}, \bar{v}) \geq c \, \|\bar{v}\|_{\bar{V}}^2 \quad \text{for any} \ \bar{v} \in \bar{V}, c > 0.$$

Then (2-14) and (2-16) have unique solutions u and u_h, and

$$(2\text{-}18) \qquad \lim_{h \to 0} \|u - p_h u_h\|_{\bar{V}} = 0. \qquad \blacktriangle$$

Proof The existence and uniqueness of solutions u and u_h of (2-14) and (2-16) follow from (2-17) and the Lax-Milgram theorem (see Theorem 1.1-4). However, the operators A_h defined by $\bar{a}(p_h u_h, p_h v_h)$ are equal to $p'_h \bar{A} p_h$ and thus are external approximations of $A = \pi' \bar{A} \pi$, where π is the canonical injection from V into $\bar{V}$. The $\bar{V}$-ellipticity of $\bar{A}$ implies the stability of the operators A_h: indeed, the inequality

$$c \, \|p_h u_h\|_{\bar{V}}^2 \leq \bar{a}(p_h u_h, p_h u_h) = (p'_h \bar{A} p_h u_h, u_h)_h \leq \|A_h u_h\|_{V_h'} \, \|p_h u_h\|$$

implies the stability inequality

$$(2\text{-}19) \qquad \|p_h u_h\|_{\bar{V}} \leq c^{-1} \|A_h u_h\|_{V_h'} = c^{-1} \|\hat{q}_h s_h A p_h u_h\|_{\bar{V}'},$$

where $s_h = p'_h$ and $\hat{q}_h = \hat{r}'_h$.

Therefore, Theorem 2-1 follows from Theorem 1-2 since the operators A_h are stable, the approximations of $\bar{V}$ are convergent in V in accordance with [(2-15)(i)], and the dual approximations of $\bar{V}'$ are external approximations of V'—by [(2-15)(ii)] and Theorem 1-1. ∎

In the same way, Theorem 1-4 and Lemma 1-3 imply the following estimate of the error, where U and $\bar{W}$ are Hilbert spaces such that

$$(2\text{-}20) \qquad U \subset V, \bar{V} \subset \bar{W},$$

the injections being continuous and dense.

Let us consider the error and external error functions $e_U{}^V(p_h)$ and $t_{\bar{V}}{}^W(p_h)$ (see Section 1-4 and 1-5).

THEOREM 2-2 Let us assume (2-17), (2-20), and

$$(2\text{-}21) \qquad f \in \bar{W}', u \in U, \text{ and } \bar{A}u \in \bar{W}'.$$

Then the error $u - p_h u_h$ is estimated by

$$(2\text{-}22) \quad \|u - p_h u_h\|_{\bar{V}} \leq M(e_U{}^V(p_h) \|u\|_U + t_{\bar{V}}{}^{\bar{W}}(p_h) \|\bar{f} - \bar{A}u\|_{\bar{W}'}). \quad \blacktriangle$$

Now to give estimates of error in larger spaces, let X and $\bar{X}$ be Hilbert spaces such that

$$(2\text{-}23) \quad \begin{cases} \text{(i) } X \subset V, \bar{X} \subset \bar{V}; \text{ the injections being continuous and dense,} \\ \text{(ii) } X \text{ is closed in } \bar{X} \text{ and } V \text{ is closed in } \bar{V}. \end{cases}$$

Let us notice that if π is the canonical injection from V into $\bar{V}$, then $\pi'\bar{A}$ maps $\bar{V}$ into $V' \subset X'$.

THEOREM 2-3 Let us assume (2-21). Then the following estimate holds:

$$(2\text{-}24) \quad \begin{aligned} \|\pi'\bar{A}(u - p_h u_h)\|_{X'} &\leq M(e_X{}^V(p_h r_h) \|u - p_h u_h\|_{\bar{V}} \\ &\quad + t_X{}^{\bar{V}}(p_h r_h) \|\bar{f} - \bar{A}u\|_{\bar{V}'}) \end{aligned}$$

for any restriction r_h. ▲

Proof For any $v \in X \subset V$, if τ denotes the orthogonal projector from $\bar{V}$ onto V, we can write

$$(2\text{-}25) \quad \bar{a}(u - p_h u_h, v) = \bar{a}(u - p_h u_h, v - p_h r_h v) + (\bar{f} - \bar{A}u, p_h r_h v - \tau p_h r_h v),$$

since the following identity holds:

$$\bar{a}(u - p_h u_h, p_h r_h v) = (\bar{A}u, (1 - \tau)p_h r_h v) + (f, (\tau - 1)p_h r_h v).$$

Now we use the estimates

$$(2\text{-}26) \begin{cases} \text{(i)} \ |\bar{a}(u - p_h u_h, v - p_h r_h v)| \le M \, \|u - p_h u_h\| \, e_X^V(p_h r_h) \, \|v\|_X, \\ \text{(ii)} \ |(\bar{f} - \bar{A}u, (1 - \tau)p_h r_h v)| \le \|\bar{f} - \bar{A}u\|_{\bar{V}'} \, t_X^{\bar{V}}(p_h r_h) \, \|v\|_X. \end{cases}$$

Since $\|\pi' \bar{A}(u - p_h u_h)\|_{X'} = \sup\limits_{v \in X} |\bar{a}(u - p_h u_h, v)| / \|v\|_X$, inequality (2-24)

follows from (2-25) and (2-26). ∎

We can mix the results given by Theorems 2-2 and 2-3 in order to obtain Corollary 2-1.

COROLLARY 2-1 Let us suppose the assumptions of Theorems 2-2 and 2-3. Then the error $u - p_h u_h$ is estimated by

$$(2\text{-}27) \begin{cases} \|\pi' \bar{A}(u - p_h u_h)\|_{X'} \le M(e_X^V(p_h r_h) e_U^V(p_h) \, \|u\|_U \\ \qquad\qquad + [e_X^V(p_h r_h) t_{\bar{V}}^W(p_h) + t_X^W(p_h r_h)] \, \|f - \bar{A}u\|_{\bar{W}'}) \end{cases} \ \blacktriangle$$

Proof If $\bar{f}$ and $\bar{A}u \in \bar{W}'$, we can replace (2-24) by

$$\|\pi' \bar{A}(u - p_h u_h)\|_{X'} \le M(e_X^V(p_h r_h) \, \|u - p_h u_h\|_V + t_X^W(p_h r_h) \, \|\bar{f} - \bar{A}u\|_{\bar{W}'})$$

since we can replace $\bar{V}$ by $\bar{W}$ in [(2-26)(ii)].

Then we estimate $\|u - p_h u_h\|_V$ by inequality (2-22) in order to obtain (2-27). ∎

2-3. Partial Approximation of Neumann Problems

Let $a(u, v)$ be a continuous bilinear form on $V \times V$ and let γ, T, and $H = V^0$ satisfying

$$(2\text{-}28) \begin{cases} \text{(i)} \ \gamma \text{ maps } V \text{ onto } T, \\ \text{(ii)} \ V_0 = \ker \gamma \text{ is dense in } H = V^0, \\ \text{(iii)} \ V^0 = H \text{ is identified with its dual.} \end{cases}$$

Let Λ and δ be the formal operator and the Neumann operator associated with $a(u, v)$ (see Section 6.2-1), and let us assume, moreover, that

$$2\text{-}29) \begin{cases} \text{(i)} \ V = \cap V^i \subset V^i \subset V^0 = H, \text{ the injections being continuous,} \\ \qquad\qquad\qquad\qquad\qquad\qquad\qquad \text{and dense,} \\ \text{(ii)} \ a(u, v) = \sum a_{ij}(u, v) \quad \text{where} \quad a_{ij}(u, v) \text{ is continuous } on \\ \qquad\qquad\qquad\qquad\qquad\qquad\qquad\qquad\qquad V^i \times V^j. \end{cases}$$

Therefore, we identify V to the diagonal of $\bar{V} = \Pi V^i$ and $a(u, v)$ to the restriction to V of $\bar{a}(\bar{u}, \bar{v}) = \sum_{ij} a_{i,j}(u^i, v^j)$, continuous on $V \times V$. If f is given in H, the solution u of

$$(2\text{-}30) \qquad \begin{cases} \text{(i)} \ u \in V, \\ \text{(ii)} \ a(u, v) = (f, v) \qquad \text{for any} \ \ v \in V \end{cases}$$

is the solution u of the homogeneous Neumann problem

$$(2\text{-}31) \qquad\qquad u \in V(\Lambda), \ \Lambda u = f, \text{ and } \delta u = 0.$$

We can approximate (2-31) by using partial approximations $(V_h, p_h{}^i, r_h{}^i)$ of the spaces V which are convergent on V (see Section 2-1). We approximate u by the solution u_h of

$$(2\text{-}32) \qquad \begin{cases} \text{(i)} \ u_h \in V_h, \\ \text{(ii)} \ (A_h u_h, v_h)_h = \bar{a}(p_h u_h, p_h v_h) = \sum a_{ij}(p_h{}^i u_h, p_h{}^j v_h) = (f, p_h{}^0 v_h) \\ \qquad\qquad\qquad\qquad\qquad\qquad\qquad\qquad\qquad \text{for any} \ \ v_h \in V_h. \end{cases}$$

Then we can apply theorems of Section 2-2 to this approximate problem. ∎

Now to study the "partial" approximation of nonhomogeneous Neumann problems, let us introduce spaces U, R, S satisfying

$$(2\text{-}33) \qquad \begin{cases} \text{(i)} \ U \subset V(\Lambda) \cap V(\Lambda^*), \ S \subset T \subset R, \text{ the injections being} \\ \qquad\qquad\qquad\qquad\qquad\qquad\qquad\qquad\qquad \text{continuous and dense,} \\ \text{(ii)} \ D, D^*, D_0, \text{ and } D_0^* \text{ are contained in } U, \\ \text{(iii)} \ (\gamma, \delta) \text{ and } (\gamma, \delta^*) \text{ map } U \text{ onto } S \times R', \\ \text{(iv)} \ U_0 = \ker(\gamma, \delta) = \ker(\gamma, \delta^*) \text{ is dense in } H. \end{cases}$$

(where Λ^* and δ^* are the formal operator and Neumann operator associated with $a_*(u, v) = a(v, u)$; see Sections 6.2-7 and 6.2-8).

Let us identify H with the diagonal of $\bar{H} = \prod_i H_i$ and assume that

$$(2\text{-}34) \qquad\qquad \bar{A} \text{ maps } U \text{ into } \bar{H} = \prod_i H.$$

If $f \in H$ and $t \in S'$ are given, we are in a position to approximate the solution u of the nonhomogeneous Neumann problem

$$(2\text{-}35) \qquad\qquad u \in H(\Lambda), \ \Lambda u = f, \text{ and } \delta u = t.$$

For that purpose, we introduce partial approximations $(V_h, p_h{}^i, r_h{}^i)$ of the space V and a prolongation $p_h{}^1$ mapping V_h into U.

Let us consider the following discrete equation:

$$(2\text{-}36) \quad \begin{cases} \text{(i) } u_h \in V_h, \\ \text{(ii) } (p_h u_h, p_h v_h) = (f, p_h^1 v_h) + \langle t, \gamma p_h^1 v_h \rangle \quad \text{for any} \quad v_h \in V_h. \end{cases}$$

Finally, let us assume that

$$(2\text{-}37) \quad \begin{cases} \text{(i) } \|u - p_h^1 r_h u\|_V \text{ converges to } 0 \quad \text{for any} \quad v \in V, \\ \text{(ii) } \sup_h s_U^V(p_h^1, p_h)(e_U^V(p_h r_h) + t_V^H(p_h)) \leq M, \\ \text{(iii) } \sup s_V^V(p_h^1, p_h) \leq M \quad \text{and} \quad \sup e_U^U(p_h^1 r_h) \leq M, \end{cases}$$

where $p_h u_h = (p_h^i u_h)_i \in \bar{V} = \Pi V^i)$.

THEOREM 2-4 Let us assume (2-17), (2-28), (2-29), (2-33), (2-34), and (2-37). Let $f \in H$, $t \in S'$ be given with, $u \in H$ the solution of (2-35) and $u_h \in V_h$ the solution of (2-36). Then

$$(2\text{-}38) \qquad \|u - p_h^0 u_h\|_H \text{ converges to } 0.$$

Furthermore, if $t \in T'$ (which is contained in S'), then

$$(2\text{-}39) \quad \begin{cases} \|u - p_h^0 u_h\|_H \leq M(e_U^V(p_h^1 r_h) \\ \qquad + [e_U^V(p_h r_h) + t_P^H(p_h)](\|f\|_H^2 + \|t\|_{T'}^2)^{\frac{1}{2}}. \quad \blacktriangle \end{cases}$$

Proof Let u and u_h be the solutions of (2-30) and (2-32), where $a(u, v)$ is replaced by $a_*(u, v) = a(v, u)$. Let $D^* = A^{*-1}(H) = \{u \in U,$ such that $\delta^* u = 0\}$, with ω the injection from D^* into U.

Let us consider the operator $B_h \in L(H, U)$ defined by

$$(2\text{-}40) \qquad B_h f = \omega u - p_h^1 u_h = \omega A^{*-1} f - p_h^1 A_h^{*-1} p_h^{0'} f.$$

To begin with, we prove that

$$(2\text{-}41) \qquad \|B_h\|_{L(H,U)} = \|B_h'\|_{L(U',H)} \qquad \text{are bounded.}$$

Indeed, by Theorem 2-2, we have the following inequality:

$$(2\text{-}42) \quad \begin{cases} \|B_h f\|_U \leq e_U^U(p_h^1 r_h) \|u\|_U + M s_U^V(p_h^1, p_h) \\ \qquad \cdot (e_U^V(p_h r_h) \|u\|_U + t_V^H(p_h) \|f - \bar{A}^* u\|_H). \end{cases}$$

But $\|u\|_U = \|A^{*-1} f\|_U \leq M \|f\|_H$ and $\|f - \bar{A}^* u\|_H = \|f - \bar{A}^* A^{*-1} f\|_H \leq M \|f\|$. Then (2-41) follows from (2-42) and [(2-37)(ii) and (iii)].

On the other hand, the following inequality holds:

$$(2\text{-}43) \quad \begin{cases} \|B_h\|_{L(H,V)} = \|B_h'\|_{L(V',H)} \\ \qquad \leq M(e_U^V(p_h^1 r_h) + e_U^V(p_h r_h) + t_V^H(p_h)). \end{cases}$$

This inequality can be proved as (2-42) by replacing U by V. Then (2-43) follows in accordance with [(2-37)(iii)].

Thus inequalities (2-40) and (2-43) imply that

$$(2\text{-}44) \qquad B_h' l \text{ converges to } 0 \text{ in } H \qquad \text{for any } l \in U'.$$

Indeed, the operators B_h' are bounded in $L(U', H)$, and by (2-43), $B_h' l$ converges to 0 in H for any $l \in V'$ which is dense in U'. Then Theorem 2.1-9 implies (2-44). Let us compute $B_h' l = A^{-1}\omega' l - p_h{}^0 A_h^{-1} p_h{}^1{}' l$ when $l = f + \gamma' t \in U'$ (where $f \in H$ and $t \in S'$). It is clear that $p_h{}^0 A_h^{-1} p_h{}^{2'} l = p_h{}^0 u_h$, where u_h is the solution of (2-36). The Green formula shows that $A^{-1}\omega' l = u$ is the solution of the Neumann problem (2-35) (see Sections 6.2-7, 6.2-8, and 8.1-2). Since $B_h' l = u - p_h{}^0 u_h$, then (2-44) implies (2-38) and (2-43) implies (2-39) (because when $t \in T'$, $l = f + \gamma' t$ belongs to V'). ∎

Remark 2-3 Since the operators B_h are bounded in $L(H, U)$, we deduce from assumptions of Theorem 2-4 that

$$(2\text{-}45) \qquad u - p_h{}^1 u_h \text{ converges weakly to } 0 \text{ in } U \text{ when } t \in R'.$$

We obtain strong convergence in U if we replace assumptions (2-37) by assumptions (1-36) of Theorem 1-5 with $\bar{F} = \bar{V}'$, $\bar{E} = \bar{H} = \bar{H}'$, $E_0 = H^\perp$, $s_h = p_h'$, and $\hat{q}_h = \hat{r}_h'$. ∎

2-4. Perturbed Partial Approximation of Boundary-Value Problems

This section deals with other boundary-value problems associated with a split bilinear form $a(u, v)$.

Let us assume (2-28) and (2-29), with σ_1 a projector of T and $\sigma_2 = 1 - \sigma_1$,

$$(2\text{-}46) \qquad \gamma_i = \sigma_i \gamma, \ \delta_i = \sigma_i' \delta, \ T_i = \sigma_i T, \qquad \text{and} \qquad T_i' = \sigma_i' T'.$$

Let us introduce the following data:

$$(2\text{-}47) \qquad f \in H, \ t_1 \in T_1 \qquad \text{and} \qquad t_2 \in T_2'.$$

We approximate the solution u of the boundary-value problem

$$(2\text{-}48) \qquad \begin{cases} \text{(i)} \ u \in V(\Lambda) \subset V, \\ \text{(ii)} \ \Lambda u = f, \\ \text{(iii)} \ \gamma_1 u = t_1 \qquad \text{and} \qquad \delta_2 u = t_2 \end{cases}$$

(see Sections 6.2-5 and 9.1-2).

Let us assume also that there exist spaces W and Q such that

(2-49) $\begin{cases} \text{(i) } V \subset W, T \subset Q, Q = Q'; \text{ the injections being continuous} \\ \qquad \text{and dense,} \\ \text{(ii) } \gamma \text{ maps } W \text{ into } Q. \end{cases}$

Let us consider partial approximations $(V_h, p_h{}^i, r_h{}^i)$ of the space V, approximations $(V_h, p_h{}^1, r_h)$ of the space V and a function $\varepsilon(h)$ converging to 0 satisfying

(2-50) $\begin{cases} \text{(i) } \|u - p_h r_h u\|_{\bar{V}} \text{ converges to } 0 \qquad \text{for any } u \in V, \\ \text{(ii) } \|p_h{}^1 u_h\|_V \leq M \|p_h u_h\|_{\bar{V}} \qquad \text{where } M \text{ is independent of } h, \\ \text{(iii) if } p_h u_h \text{ converges weakly to } u, \text{ then } p_h{}^1 u \text{ converges weakly to } u, \\ \text{(iv) } \|u - p_h{}^1 r_h u\|_V \text{ converges to } 0 \qquad \text{for any } u \in V, \\ \text{(v) } \lim_{h \to 0} \varepsilon(h)^{-\frac{1}{2}} e_V{}^W(p_h{}^1 r_h) = 0 \end{cases}$

Remark 2-4 Assumptions [(2-50)i–iv)] amount to saying that if we identify V with the diagonal of $\bar{V} = \Pi V^i$, the approximations $(V_h, p_h{}^1, r_h)$, $(V_h, p_h{}^i, r_h{}^i)$ are convergent partial approximations of the space V. ∎

Let A_h be the partial approximation of A defined by

(2-51) $(A_h u_h, v_h)_h = \bar{a}(p_h u_h, p_h v_h) = \sum a_{ij}(p_h{}^i u_h, p_h{}^j v_h)$ for any $v_h \in V_h$.

We approximate the boundary-value problem (2-48) by the following discrete variational equation:

(2-52) $\begin{cases} \text{(i) } u_h \in V_h, \\ \text{(ii) } \bar{a}(p_h u_h, p_h v_h) + \varepsilon(h)^{-1} \langle \gamma_1 p_h{}^1 u_h, \gamma_1 p_h{}^1 v_h \rangle = (f, p_h{}^0 v_h) \\ \qquad + \langle t_2, \gamma_2 p_h{}^1 v_h \rangle + \varepsilon(h)^{-1} \langle t_1, \gamma_1 p_h{}^1 v_h \rangle \qquad \text{for any } v_h \in V_h. \end{cases}$

THEOREM **2-5** Let us assume (2-17) through (2-29), (2-49), and (2-50), with $f \in H$, $t_1 \in T_1$, $t_2 \in T_2'$, $u \in V$ the solution of (2-48) and $u_h \in V_h$ the solution of (2-52). Then

(2-53) $\begin{cases} \text{(i) } \|u - p_h{}^1 u_h\|_V \text{ converges to } 0, \\ \text{(ii) } \|\gamma_1(u - p_h{}^1 u_h)\|_{T_1'} \leq M\varepsilon(h), \\ \text{(iii) } \varepsilon(h)^{-1}\gamma_1(u - p_h{}^1 u_h) \text{ converges weakly to } \delta_1 u \text{ in } T_1'. \end{cases}$ ▲

Proof The proof of Theorem 2-5 is analogous to the proof of Theorem 9.1-1. Let us denote by τ the orthogonal projector from $\bar{V}$ onto V. Then the following identity holds for any $v_h \in V_h$:

$$(2\text{-}54) \quad \begin{cases} \bar{a}(p_h u_h, p_h v_h) + \varepsilon(h)^{-1} \langle \gamma_1 p_h^1(u_h - r_h u), \gamma_1 p_h^1 v_h \rangle = \\ (f - \bar{A}u, (1 - \tau)p_h v_h) + \langle t_2, \gamma_2 p_h^1 v_h - \gamma_2 \tau p_h v_h \rangle + \bar{a}(u - p_h r_h u, p_h v_h) \\ \qquad + \varepsilon(h)^{-1} \langle \gamma_1(u - p_h^1 r_h u), \gamma_1 p_h^1 v_h \rangle - \langle \delta_1 u, \gamma_1 \tau p_h v_h \rangle. \end{cases}$$

This identity follows from (2-48) and (2-52) by

1. Adding and subtracting $\bar{a}(p_h r_h u, p_h v_h) + \varepsilon(h)^{-1} \langle \gamma_1 p_h^1 r_h u, \gamma_1 p_h^1 v_h \rangle$.

2. Adding and subtracting $a(u, p_h v_h) - (f, \tau p_h v_h)$.

3. Using the Green formula $a(u, \tau p_h v_h) = (f, \tau p_h v_h) + \langle \delta_1 u, \gamma_1 \tau p_h v_h \rangle + \langle t_2, \gamma_2 \tau p_h v_h \rangle$.

Let us set in (2-54) $v_h = \psi_h = u_h - r_h u$. We prove that

$$(2\text{-}55) \qquad \|p_h \psi_h\|_V^2 + \varepsilon(h)^{-1} \|\gamma_1 p_h^1 \psi_h\|_Q^2 \leq M$$

by estimating $X_h = a(p_h \psi_h, p_h \psi_h) + \varepsilon(h)^{-1} \langle \gamma_1 p_h^1 \psi_h, \gamma_1 p_h^1 \psi_h \rangle$. We can estimate the right-hand side of (2-54) in the following way:

$$(2\text{-}56) \quad \begin{cases} X_h \leq \|f - \bar{A}u\|_{V'} \|p_h \psi_h\|_V + M \|t_2\|_{T'} \|p_h \psi_h\|_V \\ \qquad + M \|u - p_h r_h u\|_V \|p_h \psi_h\| \\ \qquad + \varepsilon(h)^{-1/2} e_V^W(p_h^1 r_h) \|u\|_V \, \varepsilon(h)^{-1/2} \|\gamma_1 p_h^1 \psi_h\|_Q \\ \qquad + \|\delta_1 u\|_{T'} \|p_h \psi_h\|_V \end{cases}$$

[we have used the assumption (2-49)]. Now we apply the inequality $ab \leq (4\varepsilon)^{-1} a^2 + \varepsilon b^2$ to each terms of the right-hand side of (2-56), and by using the $\bar{V}$-ellipticity of $\bar{a}(\bar{u}, \bar{v})$, we deduce from (2-56) the inequality

$$(2\text{-}57) \quad \begin{cases} \|p_h \psi_h\|_V^2 + \varepsilon(h)^{-1} \|\gamma_1 p_h^1 \psi_h\|_Q^2 \leq M(\|f - \bar{A}u\|_{V'}^2 + \|t_2\|_{T'}^2 \\ \qquad + \|u - p_h r_h u\|_V^2 \\ \qquad + \varepsilon(h)^{-1} e_V^W(p_h^1 r_h)^2 \|u\|_V^2 \\ \qquad + \|\gamma_1 u\|_{T'}^2). \end{cases}$$

Therefore, (2-55) follows from (2-57), [(2-50)(i)], and [(2-50)(v)].

Since $p_h \psi_h$ is in a bounded subset of $\bar{V}$, we deduce from Theorem 2.1-11 that

$$(2\text{-}58) \qquad p_h \psi_h \text{ converges weakly to } \psi \text{ in } \bar{V}.$$

Furthermore,

$$(2\text{-}59) \quad \begin{cases} \text{(i) } \psi \text{ belongs to } V \text{ (and thus, } (1 - \tau)\psi = 0), \\ \text{(ii) } p_h^1 \psi_h \text{ converges weakly to } \psi \text{ in } V, \\ \text{(iii) } \gamma_1 \psi = 0. \end{cases}$$

Indeed, the first statement follows because the approximations of the spaces V^i are partial approximations of V [see (1-15), section 1-2] and the second statement from [(2-50)(ii)]. On the other hand, (2-55) implies that $\|\gamma_1 p_h^1 v_h\|_Q \leq M\sqrt{\varepsilon(h)}$ converges to 0, and [(2-59)(iii)] holds.

Actually, we prove that

(2-60) $\qquad\qquad \psi = 0$ and $p_h \psi_h$ converges strongly to 0.

Indeed, let us reconsider (2-54) with $v_h = \psi_h$. The right-hand side converges to 0, since

1. $(f - \bar{A}u, (1 - \tau)p_h\psi_h) + \langle t_2, \gamma_2 p_h^1 \psi_h - \gamma_2(\tau p_h\psi_h)\rangle$ converges to 0 by (2-58) and [(2-59)(i)].

2. $\bar{a}(u - p_h r_h u, p_h\psi_h)$ converges to 0 by (2-58) and [(3-50)(i)].

3. $\varepsilon(h)^{-1}|\langle \gamma_1(u - p_h^1 r_h u), \gamma_1 p_h^1 \psi_h\rangle| \leq M\varepsilon(h)^{-1/2}\|\gamma_1 p_h^1 \psi_h\|_Q \, \varepsilon(h)^{-1/2} e_V{}^W(p_h^1 r_h)$ converges to 0 by (2-55) and [(2-50)(v)].

4. $\langle \delta_1 u, \gamma_1 \tau p_h\psi_h\rangle$ converges to 0, since $\gamma_1 \tau p_h\psi_h$ converges weakly to $\gamma_1 \tau \psi = \gamma_1 \psi = 0$ by [(2-59)(i) and (iii)].

We thus have proved the first statement of the theorem, since

(2-61) $\begin{cases} \text{(i) } \|u - p_h u_h\|_V \leq \|u - p_h r_h u\|_V + \|p_h\psi_h\|_V \text{ converges to 0,} \\ \text{(ii) } \|u - p_h^1 u_h\|_V \leq \|u - p_h^1 r_h u\|_V + M\,\|p_h\psi_h\|_V \text{ converges to 0.} \end{cases}$

Furthermore, we obtain the estimate

(2-62) $\begin{cases} \|\gamma_1(u - p_h^1 u_h)\|_Q \leq \|\gamma_1(u - p_h^1 r_h u)\|_Q + \|\gamma_1 p_h^1 \psi_h\|_Q \\ \qquad\qquad \leq M(e_V{}^W(p_h^1 r_h)\,\|u\|_V + \sqrt{\varepsilon(h)}) \leq M\sqrt{\varepsilon(h)}. \end{cases}$

Finally, let us prove the two last statements of the theorem. Using (2-52), we can write

(2-63) $\begin{cases} \varepsilon(h)^{-1}\langle \gamma_1(p_h^1 u_h - u), \gamma_1 v\rangle = \varepsilon(h)^{-1}\langle \gamma_1(p_h^1 u_h - u), \gamma_1(v - p_h^1 r_h v)\rangle \\ + (f, p_h^0 r_h v) + \langle t_2, \gamma_2 p_h^1 r_h v\rangle - \bar{a}(p_h u_h, p_h r_h v) \quad \text{for any } v \in V. \end{cases}$

The first term of the right-hand side converges to 0 in accordance with (2-62) and [(2-50)(v)] and then the right-hand side converges to $(f, v) + \langle t_2, \gamma_2 v\rangle - a(u, v) = -\langle \delta_1 u, \gamma_1 v\rangle$ (by the Green formula). This implies that $\varepsilon(h)^{-1}\gamma_1(u - p_h^1 u_h)$ converges weakly to $\delta_1 u$ in T_1' (since γ_1 maps V onto T_1). Then $\varepsilon(h)^{-1}\gamma_1(u - p_h^1 u_h)$ is bounded in T_1', and this implies [(2-53)(ii)]. ∎

Now we estimate the error in "larger spaces" (as in Theorem 2-3). Let $\bar{\Lambda} \in L(\bar{V}, V_1')$ be the operator defined by

(2-64) $\qquad\qquad (\bar{\Lambda}\bar{u}, v) = \bar{a}(\bar{u}, v) \qquad \text{for any } v \in V_0.$

Let $U_0 \subset V_0$ be a Hilbert space dense in V_0, and set

$$(2\text{-}65) \qquad t_{U_0}{}^V(p_h{}^1 r_h, p_h r_h) = \sup_{u \in U_0} \frac{\|p_h{}^1 r_h u - p_h r_h u\|_V}{\|u\|_{U_0}}.$$

THEOREM 2-6 Let us suppose the assumptions of Theorem 2-5. If U_0 is a dense subset of V_0, then, if u and u_h are the solutions of (2-48) and (2-52),

$$(2\text{-}66) \quad \|\overline{\Lambda}(u - p_h u_h)\|_{U_0'} \leq M(e_{U_0}{}^V(p_h r_h) + e_{U_0}{}^V(p_h{}^1 r_h) + t_{U_0}{}^V(p_h{}^1 r_h, p_h r_h)). \quad \blacktriangle$$

Proof We can write

$$(2\text{-}67) \quad \begin{cases} \bar{a}(u - p_h u_h, v) = \bar{a}(u - p_h u_h, v - p_h r_h v) + (\bar{f} - \bar{A}u, p_h r_h v - p_h{}^1 r_h v) \\ \quad + \langle \delta_1 u + \varepsilon(h)^{-1} \gamma_1(p_h{}^1 u_h - u), \gamma_1(p_h{}^1 r_h v - v) \rangle \quad \text{when } v \in U_0, \end{cases}$$

since

$$(2\text{-}68) \quad \begin{aligned} \bar{a}(u - p_h u_h, p_h r_h v) = {}& \bar{a}(u, p_h r_h v) - (f, p_h{}^0 r_h v) - \langle t_2, \gamma_2 p_h{}^1 r_h v \rangle \\ & + \varepsilon(h)^{-1} \langle \gamma_1(p_h{}^1 u_h - u), \gamma_1 p_h{}^1 r_h v \rangle. \end{aligned}$$

Indeed, (2-67) follows from (2-68) by the Green formula

$$(f, p_h{}^1 r_h v) - a(u, p_h{}^1 r_h v) + \langle t_2, \gamma_2 p_h{}^1 r_h v \rangle + \langle \delta_1 u, \gamma_1 p_h{}^1 r_h v \rangle = 0$$

and because $\gamma_1 v = 0$ when $v \in U_0$.

Then (2-66) follows from (2-64) and (2-67), and also because

$$\|u - p_h u_h\|_V \leq M \qquad \text{and} \qquad \|\delta_1 u + \varepsilon(h)^{-1} \gamma_1(p_h{}^1 u_h - u)\|_{T_1'} \leq M. \quad \blacksquare$$

3. PARTIAL APPROXIMATIONS OF SOBOLEV SPACES

Let Ω be a smooth bounded open subset of R^n, and consider the spaces $H(\Omega, D_i)$ defined by

$$(3\text{-}1) \qquad H(\Omega, D_i) = \{u \in L^2(\Omega) \quad \text{such that} \quad D_i u \in L^2(\Omega)\},$$

where D_i is the derivative at the sense of distributions. It is easy to check that $H(\Omega, D_i)$ is a Hilbert space for the norm

$$(3\text{-}2) \quad \|u\|_{H(\Omega, D_i)} = (|u|^2 + |D_i u|^2)^{1/2}; \ |u| = \sqrt{(u, u)} = \left(\int |u(x)|^2 \, dx \right)^{1/2}.$$

(See Section 1.1-6.)

Then, by the very definition of the Sobolev space $H^1(\Omega)$, it is clear that $V = H^1(\Omega)$ is the intersection of the space $V^0 = H = L^2(\Omega)$ and the spaces $V^i = H(\Omega, D_i)$ for $i = 1, \dots, n$.

Then we identify $V = H^1(\Omega)$ with the diagonal of the space $\mathbf{V}$ defined by

(3-3)
$$\mathbf{V} = H \times \Pi V^i = L^2(\Omega) \times \prod_{i=1,n} H(\Omega, D_i)$$

and equipped with the norm

(3-4)
$$\|u\|_{\mathbf{V}} = \left(\sum_{i=0}^{n} \|u^i\|^2_{H(\Omega,D_i)} \right)^{\frac{1}{2}}.$$

We also supply V with the seminorm

(3-5)
$$\|u\|_{\mathbf{V}} = \left(|u^0|^2 + \sum_{i=1}^{n} |D_i u^i|^2 \right)^{\frac{1}{2}}.$$

It is clear that the restriction of $\|u\|_{\mathbf{V}}$ to V is a norm (which is equal to the norm of $V = H^1(\Omega)$) equivalent to the norm $\|\|u\|\|_{\mathbf{V}}$.

Since Ω is smooth, there exists a continuous right inverse ω of the operator $\rho \in L(H^k(R^n), H^k(\Omega))$ which associates with u its restriction to Ω for $k = 1, 2$ (see Section 6.3-5).

We assume also that Ω satisfies the following property: there exist a subsequence of $h = (h_1, \ldots, h_n)$ (again denoted by h) and a constant c such that

(3-6)
$$\int_{(kh,(k+1)h)} |\phi(x)|^2 \, dx \leq c \int_{\Omega \cap (kh,(k+1)h)} |\phi(x)|^2 \, dx$$

for any polynomial ϕ of multidegree ≤ 2 and for any multi-integer $k = (k_1, \ldots, k_n) \in Z^n$ such that $(kh, (k+1)h) = \prod_{i=1}^{n} (k_i h_i, (k_i + 1)h_i)$ intersects Ω.

3-2. Partial Approximations of the Sobolev Space $H^1(\Omega)$

As usual (see Chapter 5), we denote by $\theta_{kh}(x) = \theta_{k_1 h_1}(x_1), \ldots, \theta k_n h_n(x_n)$ the characteristic function of $(kh, (k+1)h)$ and by $\pi_{i,h} = \pi_{\varepsilon_i,h}$ the measure defined by

(3-7)
$$(\pi_{i,h}, \phi) = h_i^{-1} \int_0^{h_i} \phi(0, \ldots, x_i, \ldots, 0) \, dx_i.$$

We construct the simplest partial approximations of $H^1(\Omega)$.

1. *Discrete spaces*
Let us define $\mathscr{R}_h^{\,0}(\Omega)$, $\mathscr{R}_h^{\,i}(\Omega) = \mathscr{R}_h^{\,\varepsilon_i}(\Omega)$, and $\mathscr{R}_h^{\,1}(\Omega)$ by

(3-8)
$$\begin{cases} \text{(i)} \ \ \mathscr{R}_h^{\,0}(\Omega) = \{k \in Z^n \text{ such that } (kh, (k+1)h) \cap \Omega \neq \varnothing \} \\ \text{(ii)} \ \ \mathscr{R}_h^{\,i}(\Omega) = \{k \in Z^n \text{ such that support } (\pi_{i,h} * \theta_{kh}) \cap \Omega \neq \varnothing \} \\ \text{(iii)} \ \ \mathscr{R}_h^{\,1}(\Omega) = \underset{0 \leq i \leq n}{U} \mathscr{R}_h^{\,i}(\Omega). \end{cases}$$

The discrete space $V_h = \mathbf{H}_h^1(\Omega)$ we choose is the subspace of sequences

$$u_h = (u_h^k)_{k \in \mathcal{R}_h^{(1)}(\Omega)} \quad \text{defined on} \quad \mathcal{R}_h^1(\Omega).$$

We supply it with the duality pairing

$$(u_h, v_h)_h = h_1 \cdots h_n \sum_{k \in \mathcal{R}_h^1(\Omega)} u_h^k v_h^k$$

and with the norms

$$(3\text{-}9) \quad |u_h|_h = \sqrt{(u_h, u_h)_h}, \quad \|u_h\|_{1,h} = \left(|u_h|_h^2 + \sum_{1 \le i \le n} |\nabla_{h_i} u_h|^2 \right)^{\!\!1/2}.$$

2. Prolongations p_h^i

Let us set

$$(3\text{-}10) \quad \begin{cases} \text{(i)} \quad p_h^0 u_h = \displaystyle\sum_{k \in Z^n} u_h^k \theta_{kh}(x), \\[2mm] \text{(ii)} \quad p_h^i u_h = \pi_{i,h} * p_h^0 u_h. \end{cases}$$

Since $D_i p_h^i u_h = \nabla_{h_i} p_h^0 u_h$ and since p_h^0 maps V_h into $L^2(\Omega)$, we deduce that the operators p_h^i map V_h into the spaces $V^i = H(\Omega, D_i)$ (see Section 5.2-2).

We will denote by $\mathbf{p}_h^1$ the operator mapping V_h into $\mathbf{V} = L^2(\Omega) \times \Pi H(\Omega, D_i)$ defined by

$$(3\text{-}11) \quad \mathbf{p}_h^1 u_h = (p_h^0 u_h, \ldots, p_h^i u_h, \ldots) \in \mathbf{V}.$$

In order to claim that $\mathbf{p}_h^1$ is a prolongation, we must prove Lemma 3-1.

LEMMA 3-1 The operator $\mathbf{p}_h^1$ defined by (3-11) is an injective operator from V_h into $\mathbf{V}$. ▲

Proof Let us assume that $\mathbf{p}_h^1 u_h = 0$. Then

$$(3\text{-}12) \quad \begin{cases} \text{(i)} \quad p_h^0 u_h = 0 \quad \text{on} \quad \Omega, \\[2mm] \text{(ii)} \quad D_i p_h^i u_h = p_h^0 \nabla_{h_i} u_h = 0 \quad \text{on} \quad \Omega \quad \text{for} \quad i = 1, \ldots, n. \end{cases}$$

Let us show that $u_h^k = 0$ for any $k \in \mathcal{R}_h^1(\Omega)$. If $k \in \mathcal{R}_h^0(\Omega)$, $u_h^k = 0$ by [(3-12)(i)]. If $k \notin \mathcal{R}_h^0(\Omega)$, then it belongs to $\mathcal{R}_h^i(\Omega)$ for at least an index i. Since the support of $\pi_{i,h} * \theta_{kh}$ is the union of $(kh, (k+1)h)$ and of $((k + \varepsilon_i)h, (k + \varepsilon_i + 1)h)$, we deduce that the multi-integer $k + \varepsilon_i$ belongs to $\mathcal{R}_h^0(\Omega)$. Thus [(3-12)(i)] implies that

$$(3\text{-}13) \quad (\nabla_{h_i} u_h)^{k+\varepsilon_i} = h_i^{-1}(u_h^{k+\varepsilon_i} - u_h^k) = 0.$$

Therefore, $u_h^k = u^{k+\varepsilon_i} = 0$, since $k + \varepsilon_i$ belongs to $\mathcal{R}_h^0(\Omega)$. ∎

Now let us prove the following properties of stability.

LEMMA 3-2   Let us assume that Ω satisfies (3-6). There exist constants M independent of h such that

$$(3-14) \qquad \|\mathbf{p}_h^{\,1} u_h\|_{\mathbf{V}} \leq \||\mathbf{p}_h^{\,1} u_h\||_{\mathbf{V}} \leq M \, \|u_h\|_{1,h} \leq M \, \|\mathbf{p}^1 u_h\|_{\mathbf{V}}. \qquad \blacktriangle$$

Proof   The two first inequalities of (3-14) are obvious. The last one follows from (3-6) and from Theorem 5.2-5 (see, e.g., the last part of the proof of Theorem 5.3-1).   ∎

REMARK 3-1   We deduce from Lemma 1-2 that the restriction of the seminorm $\|u\|_{\mathbf{V}}$ to the subspaces $\mathbf{V} + \mathbf{p}_h^{\,1} V_h$ is a norm equivalent to the initial norm $\|u\|_{\mathbf{V}}$ of the space $\mathbf{V}$.   ∎

3. *The restrictions $r_{h,\Omega}^i$*
Let us associate with a function λ satisfying

$$(3-15) \qquad \lambda \in L^\infty(R^n), \ \lambda \text{ has a compact support and } \int \lambda(x)\,dx = 1$$

the restriction r_h defined by

$$(3-16) \quad (r_h u)^k = (\lambda_h^{\,k}, u) = h^{-1} \int u(x) \lambda\!\left(\frac{x}{h} - k\right) dx \qquad \text{for any} \quad k \in Z^n.$$

Let ω be a right inverse of ρ, defining the restrictions $r_{h,\Omega}^i$ and the restriction $\mathbf{r}_{h,\Omega}$ by

$$(3-17) \qquad \begin{cases} \text{(i)} \ \ r_{h,\Omega}^i u = (n+1)^{-1} r_h(\omega u), \\[2mm] \text{(ii)} \ \ \mathbf{r}_{h,\Omega} \mathbf{u} = \displaystyle\sum_{0 \leq i \leq n} r_{h,\Omega}^i u^i = r_h \omega\!\left((n+1)^{-1} \sum_{0 \leq i \leq n} u^i\right). \end{cases}$$

Finally, we must prove that approximations $(\mathbf{H}_h^1(\Omega), p_h^{\,i}, r_{h,\Omega}^i)$ we have constructed are convergent partial approximations of $H^1(\Omega)$.

4. *Convergent partial approximations of $H^1(\Omega)$*

THEOREM 3-1   The approximations $(\mathbf{H}_h^1(\Omega), p_h^{\,i}, r_{h,\Omega}^i)$ are convergent partial approximations of the Sobolev space $H^1(\Omega)$.   ▲

Proof   We have already verified that $\mathbf{p}_h^{\,1}$ is an isomorphism from V_h into $\mathbf{V}$. Now we must verify that the approximations $(V_h, \mathbf{p}_h^{\,1}, \mathbf{r}_{h,\Omega})$ are external approximations of $\mathbf{V}$ (see Section 1-2). Let us assume that $\mathbf{p}_h^{\,1} u_h$ converges weakly to $\mathbf{u}$ in $\mathbf{V}$. Proving that $\mathbf{u} \in \mathbf{V}$ amounts to showing that

$$(3-18) \qquad \begin{cases} \text{(i)} \ \ p_h^0 u_h \text{ converges weakly to } u^0 \text{ in } L^2(\Omega), \\[2mm] \text{(ii)} \ \ D_i p_h^{\,i} u_h \text{ converges weakly to } v^i = D_i u^0 \text{ in } L^2(\Omega). \end{cases}$$

Let $\varphi(x)$ be an infinitely differentiable function with compact support in Ω. Then, for h small enough, the supports of the functions $\nabla_{h_i}\phi$ are contained in Ω. Therefore

$$(D_i p_h^i u_h, \phi) = \int_\Omega \nabla_{h_i} p_h^0 u_h \cdot \phi(x)\text{-}dx = -\int_\Omega p_h^0 u_h \cdot \overline{\nabla}_{h_i}\phi(x)\,dx$$

converges to $\int_\Omega v^i(x)\phi(x)\,dx$ and to $-\int_\Omega u^0(x)D_i\phi(x)\,dx = \int_\Omega D_i u^0(x)\phi(x)\,dx$.

Finally, let us prove that the approximations are convergent on V. If $\mathbf{u} = (u, \ldots, u, \ldots, u)$ belongs to $H^1(\Omega)$, then $\mathbf{r}_{h,\Omega}\,\mathbf{u} = r_h\omega u$, and thus

$$u - p_h^i\,\mathbf{r}_{h,\Omega}\,\mathbf{u} = \rho(1 - p_h^i r_h)\omega u \text{ converges to } \rho\omega u = u \text{ in } H(\Omega, D_i)$$

since $\phi - p_h^i r_h \phi$ and $D_i(\phi - p_h^i r_h \phi)$ converges to 0 in $L^2(R^n)$ for any function $\phi \in H(R^n, D_i)$ (see Theorem 5.2-4). Thus $\mathbf{u} - \mathbf{p}_h^1\mathbf{r}_{h,\Omega}\mathbf{u}$ converges to 0 for any $\mathbf{u} \in H^1(\Omega)$.

3-3. Estimates of Truncation Errors and External Error Functions

In order to estimate truncation errors and external error functions of the partial approximations of $H^1(\Omega)$ we have constructed, let us introduce the following spaces:

(3-19)
$$\begin{cases} \text{(i) } V = H^1(\Omega), \mathbf{V} = L^2(\Omega) \times \prod_{i=1}^n H(\Omega, D_i), \\[2mm] \text{(ii) } H = L^2(\Omega), \mathbf{H} = (L^2(\Omega))^{n+1} \\[2mm] \text{(iii) } U = H^2(\Omega). \end{cases}$$

THEOREM 3-2   Let $(\mathbf{H}_h^1(\Omega), p_h^i, r_{h,\Omega}^i)$ be the partial approximations of $H^1(\Omega)$ constructed in Section 3-2. Then there exist constants M such that

(3-20)
$$\begin{cases} \text{(i) } e_U^\mathbf{V}(\mathbf{p}_h^1\mathbf{r}_{h,\Omega}) \le M\,|h|, \\[2mm] \text{(ii) } t_\mathbf{V}^H(\mathbf{p}_h^1) \le M\,|h|. \end{cases} \qquad \blacktriangle$$

Proof   Beginning with an estimate of the truncation errors, we deduce from Theorem 5.2-4

(3-21)
$$\begin{cases} |D_i(u - p_h^i r_h \omega u)| \le |(1 - p_h^0 s_h)D_i\omega u|_{L^2(R^n)} \le M\,|h|\,\|D_i\omega u\|_{H^1(R^n)} \\[2mm] \qquad \le M\,|h|\,\|D_i u\|_{H^1(\Omega)} \le M\,|h|\,\|u\|_{H^2(\Omega)}, \end{cases}$$

where s_h is the restriction defined by $s_h u = r_h(\pi_{i,h} * u)$ and thus associated with the function $\hat{\pi}_{i,h} * \lambda$. Then

$$\|u - \mathbf{p}_h^1\mathbf{r}_{h,\Omega}u\|_\mathbf{V}^2 = \sum_{i=0}^n |D_i(u - p_h^i r_{h,\Omega}u)|^2 \le M\,|h|^2\,\|u\|_{H^2(\Omega)}^2$$

for any $u \in H_1(\Omega)$, and this implies [(3-20)(i)].

Now let us estimate the external error function. Since $p_h{}^0 u_h \in H = L^2(\Omega)$

$$(3\text{-}22) \quad t_{\mathbf{V}}{}^H(p_h{}^1) = \sup_{u_h} \inf_{v \in H} \frac{(\sum |p_h{}^i u_h - v|^2)^{1/2}}{\|\mathbf{p}_h{}^1 u_h\|_{\mathbf{V}}} \le \sup_{u_h} \frac{\sqrt{(\sum |p_h{}^i u_h - p_h{}^0 u_h|^2)}}{\|\mathbf{p}_h{}^1 u_h\|_{\mathbf{V}}}.$$

A simple computation shows that we can write

$$(3\text{-}23) \qquad p_h{}^i u_h - p_h{}^0 u_h = h_i \sum_k (\nabla_{h_i} u_h)^k \left(\frac{x_i}{h_i} - k_i - 1\right) \theta_{kh}(x).$$

On the other hand, if we set $\alpha(x) = x_i - 1$, we obtain the following estimate:

$$(3\text{-}24) \quad \begin{cases} \left| \sum_k v_h{}^k \alpha\left(\frac{x}{h} - k\right) \theta_{kh}(x) \right|^2 \le \sum_k |v_h{}^k|^2 \theta_{kh}(x) \cdot \sum_k \left|\alpha\left(\frac{x}{h} - k\right)\right|^2 \theta_{kh}(x) \\ \qquad\qquad \le \sum_k |v_h{}^k|^2 \theta_{kh}(x), \end{cases}$$

since the function $\sum_k |\alpha(x/h - k)|^2 \theta_{kh}(x)$ is bounded by 1.

Then, by integrating inequality (3-24) on Ω and by replacing v_h by $h_i \nabla_{h_i} u_h$, we deduce from (3-23) that

$$(3\text{-}25) \qquad |p_h{}^i u_h - p_h{}^0 u_h|_{L^2(\Omega)} \le M h_i \, |\nabla_{h_i} u_h|_h \le M h_i \, \|p_h{}^1 u_h\|_{\mathbf{V}}$$

(the last inequality follows from Lemma 3-2). This implies [(3-20)(ii)]. ∎

Let us end this section by stating the two following results.

COROLLARY 3-1 We assume that Ω satisfies (3-6). Let $(V_h, p_h^{(2)}, r_{h,\Omega})$ be the piecewise-quadratic approximations of the space $U = H^2(\Omega)$ (see Section 5.2-4). Then

$$(3\text{-}26) \quad \begin{cases} \text{(i)} \, s_U{}^V(p_h^{(2)}, \mathbf{p}_h{}^1) \le M \, |h|^{-1}, \\ \text{(ii)} \, \|u - p_h^{(2)} r_{h,\Omega} u\|_U \text{ converges to } 0 \quad \text{for any} \quad u \in U = H^2(\Omega). \end{cases}$$
▲

Proof The first statement follows from Lemma 3-2, since the norm $\|\mathbf{p}_h{}^1 u_h\|_V$ is equivalent to the norm $\|u_h\|_{1,h}$ and the norm $\|p_h^{(2)} u_h\|_U$ is equivalent to the norm $\|u_h\|_{2,h}$. The second statement follows from Theorem 5.2-4). ∎

COROLLARY 3-2 Let $(V_h, p_h^{(1)}, r_{h,\Omega})$ be the piecewise-linear approximations of the Sobolev space $H^1(\Omega)$. Then

$$(3\text{-}27) \quad \begin{cases} \text{(i)} \, s_V{}^V(p_h^{(1)}, \mathbf{p}_h{}^1) \le M, \\ \text{(ii) if } \mathbf{p}_h{}^1 u_h \text{ converges weakly to } \mathbf{u} \text{ in } \mathbf{V}, \text{ then } \mathbf{u} \in V \text{ and } p_h^{(1)} u_h \\ \qquad\qquad\qquad\qquad\qquad\qquad\qquad \text{converges weakly to } \mathbf{u} \text{ in } \mathbf{V}, \\ \text{(iii)} \, \|u - p_h^{(1)} u_h\|_V \text{ converges to } 0 \quad \text{for any} \quad u \in V = H^1(\Omega). \end{cases}$$
▲

Proof The first statement follows from Lemma 3-2, the third from Theorem 5.2-4. The proof of the second statement is analogous to the proof of Theorem 3-1.   ∎

3-4.   Partial Approximations of the Sobolev Spaces $H^m(\Omega)$ and $H_0^m(\Omega)$

We can generalize the construction of the foregoing partial approximations of $H^1(\Omega)$, beginning with a brief description of partial approximations of $H^m(\Omega)$. The space $V = H^m(\Omega)$ is the intersection of the spaces $H(\Omega, D^q)$ for $|q| \leq m$, where $H(\Omega, D^q)$ is the space of functions $u \in L^2(\Omega)$ such that $D^q u \in L^2(\Omega)$. Thus we identify $H^m(\Omega)$ with the diagonal of $V = \prod_{|q| \leq m} H(\Omega, D^q)$. Let $\mathbf{m} = (m_q)_{|q| \leq m}$ be a sequence of multi-integers $m_q \geq q$. We define

$$(3\text{-}28) \quad \begin{cases} \mathscr{R}_h^{\mathbf{m}}(\Omega) = \bigcup_{|q| \leq m} \mathscr{R}_h^{m_q}(\Omega); \\ \mathscr{R}_h^{m_q}(\Omega) = \{k \in Z^n \quad \text{such that support } \pi_{m_q+(1)h} * \theta_{kh} \cap \Omega \neq \varnothing\}. \end{cases}$$

Then we choose the following discrete space $V_h = \mathbf{H}_h^{\mathbf{m}}(\Omega)$ of sequences $u_h = (u_h^k)_{k \in Z^n}$ such that $u_h^k = 0$ for $k \notin \mathscr{R}_h^{\mathbf{m}}(\Omega)$. We define the prolongations $p_h^{m_q}$ by

$$(3\text{-}29) \qquad\qquad p_h^{m_q} u_h = \pi_{m_q, h} * p_h^0 u_h$$

(see Section 5.2-2). Then the prolongation

$$(3\text{-}30) \qquad\qquad \mathbf{p}_h^{\mathbf{m}} u_h = (p_h^{m_q} u_h)_{|q| \leq m} \in V = \prod_{|q| \leq m} H(\Omega, D^q)$$

maps V_h into V.

If $\omega \in L(H^m(\Omega), H^m(R^n))$ is a right inverse of ρ, if r_h is defined by (3-16), and if M is the number of multi-integers q such that $|q| \leq m$, we define the following restriction:

$$(3\text{-}31) \quad \begin{cases} \text{(i) } r_{h,\Omega} u = r_h \omega u \quad \text{if} \quad u \in H^m(\Omega), \\ \text{(ii) } r_{h,\Omega}^q u = M^{-1} r_{h,\Omega} u \quad \text{if} \quad u \in H(\Omega, D^q), \\ \text{(iii) } \mathbf{r}_{h,\Omega} \mathbf{u} = \sum_{q \leq m} r_{h,\Omega}^q u^q \quad \text{if} \quad \mathbf{u} = (u^q)_{|q| \leq m} \text{ belongs to } V. \end{cases}$$

The proof of the following Theorem is left as an exercise.

THEOREM 3-3   Let $\mathbf{m} = (m_q)_{|q| \leq m}$ be a sequence of multi-integers $m_q \geq q$. Then the approximations $(\mathbf{H}_h^{\mathbf{m}}(\Omega), p_h^{m_q}, r_{h,\Omega}^q)$ are convergent partial approximations of the Sobolev space $H^m(\Omega)$.   ▲

In other words, the approximations $(\mathbf{H}_h^{\mathbf{m}}, \mathbf{p}_h^{\mathbf{m}}, \mathbf{r}_{h,\Omega})$ of $V = \Pi H(\Omega, D^q)$ are convergent on $H^m(\Omega)$ and are external approximations of $H^m(\Omega)$.

Now let us describe the construction of convergent partial approximations of the Sobolev space $H_0^m(\Omega)$, using $H_0(\Omega, D^q)$ to denote the closure of the space $\Phi_0(\Omega)$ of infinitely differentiable functions with compact support in Ω in the space $H(\Omega, D^q)$. Then we can prove that

$$(3\text{-}32) \qquad H_0^m(\Omega) = \bigcap_{|q| \leq m} H_0(\Omega, D^q).$$

Thus we identify $H_0^m(\Omega)$ with the diagonal of the product space $\mathbf{V}_0 = \prod_{|q| \leq m} H_0(\Omega, D^q)$.

Let $\mathbf{m} = (m_q)_{|q| \leq m}$ be a sequence of multi-integers $m_q \geq q$. We define

$$\mathscr{R}_{0,h}^{\mathbf{m}}(\Omega) = \{k \in Z^n \quad \text{such that} \quad \text{support} \ (\tau_{m_q+(1),h}) \subset \Omega$$
$$(3\text{-}33) \qquad\qquad\qquad\qquad\qquad\qquad\qquad\qquad \text{for any} \ \ |q| \leq m\}$$

and

$(3\text{-}34) \quad \mathbf{H}_{0,h}^{\mathbf{m}}(\Omega) = \{u_h = (u_h^k)_{k \in Z^n} \quad \text{such that} \quad u_h^k = 0 \ \ \text{if} \ \ k \notin \mathscr{R}_{0,h}^{\mathbf{m}}(\Omega).$

Let us introduce the following prolongations:

$$(3\text{-}35) \qquad \begin{cases} \text{(i)} \ \ p_h^{m_p} u_h = \pi_{m_q,h} * p_h^0 u_h & \text{for} \ \ |q| \leq m, \\ \text{(ii)} \ \ \mathbf{p}_h^{\mathbf{m}} u_h = (p_h^{m_q} u_h)_{|q| \leq m} . \end{cases}$$

Then it is clear that $p_h^{m_q}$ maps $\mathbf{H}_{0,h}^{\mathbf{m}}(\Omega)$ into $H_0(\Omega, D^q)$ and that $\mathbf{p}_h^{\mathbf{m}}$ is an isomorphism from $\mathbf{H}_{0,h}^{\mathbf{m}}(\Omega)$ into $\mathbf{V}_0$. If we associate with a restriction r_h defined by (3-16) the restriction $r_{0,h}$ defined by

$$(3\text{-}36) \qquad (r_{0,h} u)^k = \begin{cases} (r_h \tilde{u})^k & \text{if} \ \ k \in \mathscr{R}_{0,h}^{\mathbf{m}}(\Omega), \\ 0 & \text{if} \ \ k \notin \mathscr{R}_{0,h}^{\mathbf{m}}(\Omega), \end{cases}$$

and if M denotes the number of multi-integers q such that $|q| \leq m$, then we introduce the following restrictions:

$$(3\text{-}37) \qquad \begin{cases} \text{(i)} \ \ r_{0,h}^q u = M^{-1} r_{0,h} u & \text{if} \ \ u \in H_0(\Omega, D^q), \\ \text{(ii)} \ \ \mathbf{r}_{0,h} \mathbf{u} = \sum_{|q| \leq m} r_{0,h}^q u^q & \text{if} \ \ \mathbf{u} = (u^q)_{|q| \leq m} \in \mathbf{V}_0. \end{cases}$$

Finally, let us assume that Ω satisfies the property of $\pi_{(m)}$-convergence. The proof of the following theorem is left as an exercise.

THEOREM 3-4 Let us assume that Ω satisfies the property of $\pi_{(m)}$-convergence, and let $\mathbf{m} = (m_q)_{|q| \leq m}$ be a sequence of multi-integers $m_q \geq q$. Then the approximations $(\mathbf{H}_{0,h}^{\mathbf{m}}(\Omega), p_h^{m_q}, r_{0,h})$ are convergent partial approximations of the Sobolev space $H_0^m(\Omega)$. ▲

In other words, the approximations $(\mathbf{H}_{0,h}^{\mathbf{m}}(\Omega), \mathbf{p}_h^{\mathbf{m}}, r_{0,h})$ of $\mathbf{V}_0 = \prod_{|q| \leq m} H_0(\Omega, D^q)$ are external approximations of $H_0^m(\Omega)$, which are convergent in $H_0^m(\Omega)$.

4.   PARTIAL APPROXIMATION OF BOUNDARY-VALUE PROBLEMS

4-1.   Partial Approximation of Second-Order Linear Operators

Let $\Omega \subset R^n$ be a smooth bounded open subset and Γ its boundary. Let $a_0(x)$, $a_{ij}(x)$ be functions of $L^\infty(\Omega)$, so that we can study partial approximations of the differential operator

$$(4\text{-}1) \qquad \Lambda u = - \sum_{i,j} D_j(a_{ij}(x)D_i u) + a_0(x)u$$

which is the formal operator associated with the bilinear form

$$(4\text{-}2) \quad \left\{ \begin{array}{l} a(u, v) = \sum a_{ij}(u, v) + a_0(u, v) \qquad \text{where} \quad a_{ij}(u, v) \\[2mm] \qquad = \displaystyle\int_\Omega a_{ij}(x)D_i u D_j v \, dx \text{ is continuous on } H(\Omega, D_i) \times H(\Omega, D_j), \\[2mm] a_0(u, v) = \displaystyle\int_\Omega a_0(x)uv \, dx \text{ is continuous on } L^2(\Omega) \times L^2(\Omega). \end{array} \right.$$

Then $a(u, v)$ is continuous on $H^1(\Omega) \times H^1(\Omega)$, since $H^1(\Omega) = L^2(\Omega) \cap \bigcap_{i=1}^n H(\Omega, D_i)$.

Let us consider the partial approximations $(\mathbf{H}_h^1(\Omega), p_h^i, r_{h,\Omega}^i)$ of the space $H^1(\Omega)$ defined and studied in Sections 3-2 and 3-3. Then the partial approximation A_h of the operator A defined by $a(u, v)$ is defined by the discrete bilinear form

$$(4\text{-}3) \quad (A_h u_h, v_h)_h = \sum_{i,j} \int_\Omega a_{ij}(x) D_i p_h^i u_h D_j p_h^j v_h \, dx + \int_\Omega a_0(x) p_h^0 u_h p_h^0 v_h \, dx.$$

Let us introduce the following notations:

$$(4\text{-}4) \quad \left\{ \begin{array}{l} \text{(i) if } a(x) \in L^1(\Omega), \ a^h \text{ is the sequence defined by} \\[2mm] \qquad (a^h)^k = (h_1 \dots h_n)^{-1} \displaystyle\int_\Omega \theta_{kh}(x)a(x) \, dx \\[2mm] \qquad\qquad = (h_1 \cdots h_n)^{-1} \displaystyle\int_{\Omega \cap (kh,(k+1)h)} a(x) \, dx, \\[2mm] \text{(ii) } (\nabla_{h_i} u_h)^k = h_i^{-1}(u_h^k - u_h^{k-\varepsilon_i}), \ (\tilde\nabla_{h_i} u_h)^k = h_i^{-1}(u_h^{k+\varepsilon_i} - u_h^k). \end{array} \right.$$

PROPOSITION 4-1   The partial approximation of the operator A defined by $a(u, v)$, where $a(u, v)$ is given by (4-2), is the following finite-difference operator A_h:

$$(4\text{-}5) \qquad A_h u_h = - \sum_{i,j=1}^n \tilde\nabla_{h_i}(a_{ij}^h \nabla_{h_i} u_h) + a_0^h u_h. \qquad\qquad \blacktriangle$$

Proof First, we recall that $D_i p_h^t u_h = p_h^0 \nabla_{h_i} u_h$ on Ω.

Then

$$
\int_\Omega a_{ij}(x) p_h^0 \nabla_{h_i} u_h \cdot p_h^0 \nabla_{h_j} v_h \, dx
$$

$$
= h_1 \cdots h_n \sum_{k \in Z^n} (\nabla_{h_i} u_h)^k (\nabla_{h_j} v_h)^k (h_1 \cdots h_n)^{-1} \int_\Omega a_{ij}(x) \theta_{kh}(x) \, dx
$$

$$
= (a_{ij}^{\ h} \nabla_{h_i} u_h, \nabla_{h_j} v_h)_h.
$$

We end the proof by noticing that $(u_h, \nabla_{h_j} v_h)_h = -(\tilde{\nabla}_{h_j} u_h, v_h)_h.$ ∎

PROPOSITION 4-2   The number of levels of A_h is less or equal to 3^n. If the coefficients $a_{ij}(x)$ are equal to 0 for $i \neq j$, then the number of levels is equal to $2n + 1$.

The sum of the entries of the kth row (equal to the sum of the entries of the kth column) of A_h is equal to $(a_0^h)^k$. If the coefficient of Λ satisfy

(4-6)
$$
\begin{cases}
\text{(i)} \ \sum_{i,j} a_{ij}(x) z^i z^j \geq c \, |z|^2, c > 0, \\
\text{(ii)} \ a_0(x) \geqslant c > 0
\end{cases}
$$

almost everywhere on Ω, then A_h is positive definite and the entries of its principal diagonal are positive. Moreover, if Ω satisfies property (3-6), then the condition number $\chi(A_h)$ of A_h is estimated by

(4-7)
$$
\chi(A_h) \leq M \, |h|^2. \qquad \blacktriangle
$$

Proof The statements of Proposition 4-2 follow obviously from (4-3) and (4-5) and from the properties of the partial approximations (Lemmas 3-1 and 3-2). See also Sections 8.2-1 through 8.2-3. ∎

Remark 4-1 If the coefficients $a_{ij}(x)$ are different from 0 for $i \neq j$, then the $(2^{n+1} - 1)$-level approximations of $a(u, v)$ have a smaller number of non-zero diagonals (see Section 8.2-4). ∎

Remark 4-2 The entries of the partial approximation A_h involve only the following integrals $(a_{ij}^h)^k = \int_\Omega a_{ij}(x) \theta_{kh}(x) \, dx$. They are "simpler" than the entries of the $(2m + 1)^n$-level and $(2(2m)^n - (2m - 1)^n)$-level approximation, which involve the integrals $\int_\Omega a_{ij}(x) x^p \theta_{kh}(x) \, dx$ for $p \leq (2m)$. (See Sections 8.2-3 and 8.2-4.) ∎

4-2. Partial Approximation of the Neumann Problem

We illustrate here the results of Section 2-3. Let us begin by the homogeneous Neumann problem (see Section 7.1-4), assuming that f is given in $L^2(\Omega)$ and $u \in H^1(\Omega, \Lambda)$ is the solution of the homogeneous Neumann problem

$$(4\text{-}8) \qquad \Lambda u = f \quad \text{on} \quad \Omega \quad \text{and} \quad \frac{\partial u}{\partial n_\Lambda} = 0 \quad \text{on} \quad \Gamma.$$

Let us associate with f the sequence f^h defined by [(4-4)(i)] and let $u_h \in H_h^1(\Omega)$ be the solution of the approximate equation

$$(4\text{-}9) \qquad A_h u_h = - \sum_{i,j=1}^n \tilde{\nabla}_{hj}(a_{ij}{}^h \nabla_{hi} u_h) + a_0{}^h u_h = f^h.$$

THEOREM 4-1 Let us assume that Ω is smooth. Let us assume also that the ellipticity assumptions (4-6) are satisfied. If u and u_h are the solutions of (4-8) and (4-9), then

$$(4\text{-}10) \quad \begin{cases} \text{(i)} \ \|u - p_h{}^0 u_h\|_{L^2(\Omega)} \text{ converges to } 0, \\ \text{(ii)} \ \|D_i u - p_h{}^0 \nabla_{hi} u_h\|_{L^2(\Omega)} \text{ converges to } 0 \quad \text{for} \quad i = 1, \dots, n. \end{cases}$$

Furthermore, if the coefficients $a_{ij}(x)$ and $a_0(x)$ are continuously differentiable, then the solution u belongs to $H^2(\Omega)$ and

$$(4\text{-}11) \quad \begin{cases} \text{(i)} \ \|u - p_h{}^0 u_h\|_{L^2(\Omega)} \leq M \, |h| \, \|u\|_{H^2(\Omega)}, \\ \text{(ii)} \ \|D_i u - p_h{}^0 \nabla_{hi} u_h\|_{L^2(\Omega)} \leq M \, |h| \, \|u\|_{H^2(\Omega)}. \end{cases} \quad \blacktriangle$$

Proof The first statement follows from Theorems 2-1 and 3-1. Now, if the coefficients of Λ are continuously differentiable, the solution u belongs to $H^2(\Omega)$ (see Theorem 6.1-1). Then the estimates of error follow from Theorem 2-2 and 3-2. ∎

Let us approximate the nonhomogeneous Neumann problem. The assumptions (2-28), (2-33), and (2-34) are satisfied when the coefficients of Λ are continuously differentiable and we set

$$(4\text{-}12) \quad \begin{cases} \text{(i)} \ U = H^2(\Omega), \ U_0 = H_0^2(\Omega), \ V = H^1(\Omega), \ H = V^0 = L^2(\Omega), \\ \text{(ii)} \ T = H^{1/2}(\Gamma), \ S = H^{3/2}(\Gamma) \quad \text{and} \quad R = H^{-1/2}(\Gamma), \\ \text{(iii)} \ \gamma u = u|_\Gamma \quad \text{and} \quad \delta u = \frac{\partial u}{\partial n_\Lambda}. \end{cases}$$

Let us introduce the piecewise-quadratic approximations $(V_h, p_h^{(2)}, r_{h,\Omega})$ of the space $U = \dot{H}^2(\Omega)$. Then Corollary 3-1 implies that assumptions (2-37) are satisfied.

With $f \in L^2(\Omega)$ and $t \in H^{-3/2}(\Gamma)$ given, we can approximate the solution $u \in H^1(\Omega, \Lambda)$ of

$$(4\text{-}13) \qquad \Lambda u = f \quad \text{on} \quad \Omega \quad \text{and} \quad \frac{\partial u}{\partial n_\Lambda} = t \quad \text{on} \quad \Gamma$$

by the solution $u_h \in \mathbf{H}_h^1(\Omega)$ of

$$(4\text{-}14) \qquad A_h u_h = - \sum_{i,j} \tilde{\nabla}_{h_j}(a_{ij}{}^h \nabla_{h_i} u_h) + a_0{}^h u_h = l_h^{(2)},$$

where the components $l_h^{(2),k}$ of the vector $l_h^{(2)}$ are defined by

$$(4\text{-}15) \quad (h_1 \cdots h_n)^{-1} \left(\int_\Omega f(x) \pi_{(3)} \left(\frac{x}{h} - k \right) dx + \int_\Gamma t(x) \pi_{(3)} \left(\frac{x}{h} - k \right) d\sigma(x) \right).$$

Then Theorem 2-4 implies the following result:

THEOREM 4-2 Let us assume that Ω is smooth and satisfies (3-6). Let us assume also that the coefficients of Λ are continuously differentiable and satisfy the ellipticity assumptions (4-6). Let $f \in L^2(\Omega)$ and $t \in H^{-3/2}(\Gamma)$ be given, with u and u_h the solutions of (4-13) and (4-14). Then

$$(4\text{-}16) \qquad \|u - p_h^0 u_h\|_{L^2(\Omega)} \text{ converges to } 0.$$

Furthermore, if t actually belongs to $H^{-1/2}(\Omega)$, then u belongs to $H^1(\Omega, \Lambda)$ and

$$(4\text{-}17) \qquad \|u - p_h^0 u_h\|_{L^2(\Omega)} \leq M \, |h| \, \|u\|_{H^1(\Omega, \Lambda)}. \qquad \blacktriangle$$

Remark 4-3 By applying Remark 2-3, we obtain Corollary 4-1. ∎

COROLLARY 4-1 Let us suppose the assumptions of Theorem 4-2. Let $u \in H^2(\Omega)$ and u_h be the solution of (4-8) and (4-9). Then

$$(4\text{-}18) \qquad p_h^{(2)} u_h \text{ converges weakly to } u \text{ in } H^2(\Omega). \qquad \blacktriangle$$

Remark 4-4 The convergence properties and the estimates of error of the partial approximations of the Neumann problem for a second-order differential operator are comparable to the ones obtained by taking internal approximations by finite-element approximations associated with a function μ satisfying the criterion of 1-convergence (see Section 8.2-2). ∎

Remark 4-5 The finite-difference scheme described in Section 1.4-2 is the particular case of the Neumann problem (4-8) when $n = 1$, and in that case, the conclusions of Theorem 4-1 and Corollary 4-1 hold. ∎

4-3.   Perturbed Partial Approximation of Mixed Boundary-Value Problems

If Γ_1 is a smooth open subset of the boundary Γ of Ω and Γ_2 the interior of $\Gamma - \Gamma_1$, let us consider the following data:

(4-19) $\qquad f \in L^2(\Omega)$, $t_1 \in H^{\frac{1}{2}}(\Gamma)$, and $t_2 \in H^{-\frac{1}{2}}(\Gamma)$.

We approximate the solution $u \in H^1(\Omega, \Lambda)$ of the mixed problem

(4-20)
$$\begin{cases} \text{(i) } \Lambda u = f, \\[4pt] \text{(ii) } u|_{\Gamma_1} = t_1 \quad \text{on } \Gamma_1, \\[4pt] \text{(iii) } \dfrac{\partial u}{\partial n_\Lambda}\bigg|_{\Gamma_2} = t_2 \quad \text{on } \Gamma_2. \end{cases}$$

This is a particular case of problem (2-48) studied previously (see Section 7.1-5): the assumptions (2-28), (2-29), and (2-46) are satisfied when we choose

(4-21)
$$V = H^1(\Omega), \ H = L^2(\Omega), \ V_0 = H_0^1(\Omega), \ T = H^{\frac{1}{2}}(\Gamma), \ \gamma u = u|_\Gamma$$
$$\text{and} \quad \sigma_1 = \omega\rho,$$

where $\rho \in L(H^{\frac{1}{2}}(\Gamma), H^{\frac{1}{2}}(\Gamma_1))$ is the operator of restriction to Γ_1 and ω is a continuous right inverse of ρ.

Assumptions (2-49) are satisfied with

(4-22) $Q = L^2(\Gamma)$ and $K = H^{\alpha+\frac{1}{2}}(\Gamma)$ for any $\alpha > 0$.

(see Theorem 6.3-9.)

Once we introduce the piecewise-linear approximations $(V_h, p_h^{(1)}, r_{h,\Omega})$ of the space $V = H^1(\Omega)$ (see Section 5.3-2), then Corollary 3-2 implies that the four first assumptions of (2-50) are satisfied. On the other hand, there exists a restriction $r_{h,\Omega}$ such that

(4-23) $e_V^H(p_h^{(1)} r_{h,\Omega}) \leq c\,|h|$ and $e_V^K(p_h^{(1)} r_{h,\Omega}) \leq c\,|h|^{(1-2\alpha)/2}$,

since $K = H^{\alpha+\frac{1}{2}}(\Omega)$ is a space of order $\theta = (1 - 2\alpha)/2$ between $V = H^1(\Omega)$ and $H = L^2(\Omega)$. Then assumption [(2-50)(v)] is satisfied if we choose

(4-24)
$$\varepsilon(h) = \lambda^{-1}\,|h|^{1-4\alpha}$$

where λ is a constant independent of h. Let $M_h^{(1)}$ be the matrix associated with the bilinear form

(4-25)
$$(M_h^{(1)} u_h, v_h)_h = \int_\Gamma p_h^{(1)} u_h \cdot p_h^{(1)} v_h \, d\sigma(x).$$

This is a matrix with at most 3^n-nonzero diagonals whose entries $c_h^{(1)}(k, l)$ vanish whenever either the support of $\pi_{(2)}(x/h - k)$ or the support of $\pi_{(2)}(x/h - l)$ is contained in Ω.

Finally, let us denote by $l_h^{(1)}$ the vector of components $l_h^{(1),k}$ defined by

$$
(4\text{-}26) \quad
\begin{cases}
l_h^{(1),k} = (h_1 \cdots h_n)^{-1} \left(\int_\Omega f(x) \pi_{(2)}\left(\frac{x}{h} - k\right) dx \right. \\[2mm]
\qquad + \lambda |h|^{4\alpha-1} \int_{\Gamma_1} t_1(x) \pi_{(2)}\left(\frac{x}{h} - k\right) d\sigma(x) \\[2mm]
\qquad \left. + \int_{\Gamma_2} t_2(x) \pi_{(2)}\left(\frac{x}{h} - k\right) d\sigma(x) \right).
\end{cases}
$$

Then we shall approximate the solution u of (4-20) by the solution $u_h \in H_h^1(\Omega)$ of

$$
(4\text{-}27) \qquad -\sum_{i,j} \tilde{\nabla}_{h_i}(a_{ij}^{\ h} \nabla_{h_j} u_h) + a_0^{\ h} u_h + \lambda |h|^{4\alpha-1} M_h^{(1)} u_h = l_h^{(1)}.
$$

Then Theorem 2-5 implies the following result:

THEOREM 4-3   Let us assume (4-6) and that Ω is smooth; we suppose further that Γ_1 is a smooth open subset of the boundary Γ of Ω. Let $f \in L^2(\Omega)$, $t_1 \in H^{\frac{1}{2}}(\Gamma)$, and $t_2 \in H^{-\frac{1}{2}}(\Gamma)$ be given, with $u \in H^1(\Omega, \Lambda)$ and $u_h \in H_h^1(\Omega)$ the solutions of (4-20) and (4-27). Then the following conclusions hold:

$$
(4\text{-}28) \quad
\begin{cases}
\text{(i)} \ \|u - p_h^{(1)} u_h\|_{H^1(\Omega)} \text{ converges to } 0, \\[2mm]
\text{(ii)} \ \displaystyle\int_{\Gamma_1} |u(x) - p_h^{(1)} u_h|^2 \, d\sigma(x) \le M \, |h|^{(1-4\alpha)}, \\[2mm]
\text{(iii)} \ \lambda |h|^{4\alpha-1}(u - p_h^{(1)} u_h)|_{\Gamma_1} \text{ converges weakly to } \partial u/\partial n_\Lambda \,|_{\Gamma_1} \text{ in} \\[1mm]
\qquad\qquad\qquad\qquad\qquad\qquad\qquad\qquad\qquad H^{-\frac{1}{2}}(\Gamma).
\end{cases}
$$

▲

Remark 4-6   The Dirichlet problem is the particular case of the mixed problem (4-20) when $\Gamma_1 = \Gamma$. Then the conclusions of Theorem 4-3 hold for the Dirichlet problem.   ■

Let us consider the homogeneous Dirichlet problem

$$
(4\text{-}29) \quad \Lambda u = f \quad \text{on} \ \Omega \quad \text{and} \quad \gamma u = u|_\Gamma = 0 \quad \text{on} \ \Gamma,
$$

which is equivalent to the variational problem

$$
(4\text{-}30) \quad
\begin{cases}
\text{(i)} \ u \in H_0^1(\Omega), \\[2mm]
\text{(ii)} \ a(u, v) = (f, v) \quad \text{for any} \ v \in H_0^1(\Omega),
\end{cases}
$$

where $a(u, v)$ is defined by (4-2) and where $f \in L^2(\Omega)$ (see Section 7.1-3).

Then we can approximate (4-29) [or (4-30)] by using the partial approximations $(\mathbf{H}_{0,h}^1(\Omega), p_h{}^i, r_{0,h}^i)$ of $H_0^1(\Omega)$ constructed in Section 3-4 (we take $\mathbf{m} = (0, \varepsilon_i)_{1 \leq i \leq n}$ and set $p_h{}^i = p_h{}^{\varepsilon_i}$). Then Proposition 4-1 implies that the partial approximation of (4-30) is equivalent to the discrete problem in which we look for $u_h \in \mathbf{H}_{0,h}^1(\Omega)$ satisfying

$$(4\text{-}31) \qquad\qquad -\sum_{i,j=1}^n \tilde{\nabla}_{h_j}(a_{ij}{}^h \, \nabla_{h_i} u_h) + a_0{}^h u_h = f^h.$$

Then Theorems 2-1 and 3-4 imply the following result:

THEOREM 4-4   Let us assume that Ω is smooth and satisfies the property of $\pi_{(m)}$-convergence. Let us also suppose that the coefficients of Λ satisfy the ellipticity assumptions (4-6). Let $f \in L^2(\Omega)$ be given, with u and $u_h \in \mathbf{H}_{0,h}^1(\Omega)$ the solutions of (4-29) and (4-31). Then

$$(4\text{-}32) \qquad \begin{cases} \text{(i) } \|u - p_h{}^0 u_h\|_{L^2(\Omega)} \text{ converges to } 0, \\[4pt] \text{(ii) } \|D_i u - p_h{}^0 \nabla_{h_i} u_h\|_{L^2(\Omega)} \text{ converges to } 0 \qquad \text{for } 1 \leq i \leq n. \quad \blacktriangle \end{cases}$$

4-4.   Estimates of Error in the Interior

Let us begin by applying Theorem 2-6 to (4-20), which is approximated by (4-27). If we choose $U_0 = H_0^2(\Omega)$, which is dense in $V_0 = H_0^1(\Omega)$, then

$$(4\text{-}33) \quad \Lambda \mathbf{u} = -\sum_{i,j=1}^n D_j(a_{ij}(x)D_i u^i) + a_0 u^0 \qquad \text{when} \quad \mathbf{u} = (u^0, u^i)_{1 \leq i \leq n},$$

since, when $v \in H_0^1(\Omega)$, closure of the space of infinitely differentiable functions with compact support, we have by definition of the derivative a distribution $a(\mathbf{u}, v) = (\Lambda \mathbf{u}, v)$.

We deduce from Theorem 2-6 the following estimate of the error:

THEOREM 4-5   Let us suppose the assumptions of Theorem 4-3. Let $u \in H^1(\Omega, \Lambda)$ and $u_h \in \mathbf{H}_h^1(\Omega)$ be the solutions of (4-20) and (4-27). Then

$$(4\text{-}34) \qquad\qquad \|\Lambda(u - \mathbf{p}_h^1 u_h)\|_{H^{-2}(\Omega)} \leq c\,|h|. \qquad\qquad \blacktriangle$$

Proof   We already know that

$$\begin{cases} e_{U_0}{}^V(\mathbf{p}_h^1 r_{h,\Omega}) \leq e_U{}^V(\mathbf{p}_h^1 r_{h,\Omega}) \leq c\,|h|, \\[4pt] e_{U_0}{}^V(p_h^{(1)} r_{h,\Omega}) \leq e_U{}^V(p_h^{(1)} r_{h,\Omega}) \leq c\,|h| \end{cases}$$

(see Theorem 3-2).   ∎

It remains to estimate $t_{U_0}{}^V(\mathbf{p}_h^1 r_{h,\Omega}, p_h^{(1)} r_{h,\Omega})$.

LEMMA 4-1

If $U = H^2(\Omega)$, $V = H^1(\Omega)$, and $\mathbf{V} = L^2(\Omega) \times \prod_{1 \le i \le n} H(\Omega, D_i)$, then

(4-35)
$$t_U{}^\mathbf{V}(\mathbf{p}_h{}^1 r_{h,\Omega}, p_h^{(1)} r_{h,\Omega}) \le c \,|h| \,. \qquad \blacktriangle$$

Proof We have to estimate in $L^2(\Omega)$ the differences

(4-36)
$$D_i(p_h{}^i u_h - p_h^{(1)} u_h) = p_h{}^0 \nabla_{h_i} u_h - p_h^{(1)-\varepsilon_i} \nabla_{h_i} u_h,$$

where $u_h = r_{h,\Omega} u = r_h \omega u$.

Thus we have to estimate $p_h{}^0 v_h - p_h^{(1)-\varepsilon_i} v_h$ in $L^2(\Omega)$. We will use the fact that the space of sequences of the form $v_h = (v_h{}^k)_k$ where

(4-37)
$$v_h{}^k = v_{h_1}^{k_1} \ldots v_{h_n}^{k_n}; \; v_h = v_{h_1} \ldots v_{h_n} = (v_h{}^k)_{k \in Z^n}$$

is dense in V_h. For such sequences we can write

(4-38)
$$\begin{cases} \text{(i)} \;\; p_h{}^0 v_h = p_{h_1}{}^0 v_{h_1} \ldots p_{h_n}{}^0 v_{h_n}, \\ \text{(ii)} \;\; p_h^{(1)-\varepsilon_1} v_h = p_{h_1}{}^0 v_{h_1} p_{h_2}{}^1 v_{h_2} \ldots p_{h_n}{}^1 v_{h_n}. \end{cases}$$

Therefore we obtain

(4-39)
$$\begin{aligned} p_h^{(1)-\varepsilon_1} v_h - p_h{}^0 v_h &= p_{h_1}{}^0 v_{h_1}(p_{h_2}{}^1 v_{h_2} - p_{h_2}{}^0 v_{h_2}) p_{h_3}{}^1 v_{h_3} \cdots p_{h_n}{}^1 v_{h_n} + \cdots \\ &+ p_{h_1}{}^0 v_{h_1} \cdots p_{h_{n-1}}{}^0 v_{h_{n-1}}(p_{h_n}{}^1 v_{h_n} - p_{h_n}{}^0 v_{h_n}). \end{aligned}$$

Since $|p_{h_i}{}^1 v_{h_i} - p_{h_i}{}^0 v_{h_i}|_{L^2(R)} \le ch_i \,|\nabla_{h_i} v_{h_i}|_{h_i}$ by Theorem 3-2 for $n = 1$, we deduce that

(4-40)
$$|p_h^{(1)-\varepsilon_1} v_h - p_h{}^0 v_h|_{L^2(R^n)} \le c \left(\sum_{2 \le i \le n} h_i \,|\nabla_h v_h|_h \right) \le c \,|h| \, \|v_h\|_{1,h}.$$

Since the space of sequences $v_h = v_{h_1} \ldots v_{h_n}$ is dense, inequality (4-40) holds for any sequence v_h. Now, since $v_h = \nabla_{h_i} r_{h,\Omega} u = r_h \nabla_{h_i} \omega u$, we deduce from (4-40) that

(4-41)
$$\begin{cases} |D_i(p_h^{(1)} r_{h,\Omega} u - p_h{}^0 r_{h,\Omega} u)|_{L^2(\Omega)} \le c \,|h| \, \|r_h \nabla_{h_i} \omega u\|_{1,h} \\ \qquad\qquad\qquad \le c \,|h| \, \|\nabla_{h_i} \omega u\|_{H^1(R^n)} \\ \qquad\qquad\qquad \le c \,|h| \, \|u\|_{H^2(\Omega)} \,. \end{cases}$$

These inequalities imply Lemma 4-1, and thus Theorem 4-5 is proved. ∎

4-5. Partial Approximations of Higher-Order Differential Operators

Let us consider the differential operator

(4-42)
$$\Lambda u = \sum_{|p|,|q| \le m} (-1)^{|q|} D^q(a_{pq}(x) D^p u),$$

which is the formal operator associated with the bilinear form $a(u, v)$ defined by

(4-43) $a(u, v) = \sum\limits_{|p|,|q| \leq m} a_{pq}(u, v)$ where $a_{pq}(u, v) = \int_{\Omega} a_{pq}(x) D^p u D^q v \, dx$

is continuous on $H(\Omega, D^p) \times H(\Omega, D^q)$, if we suppose that the functions $a_{pq}(x)$ belong to $L^{\infty}(\Omega)$. If we use the partial approximations $(H_h^m(\Omega), p_h^{m_q}, r_{h,\Omega}^q)$ of $H^m(\Omega)$ defined in Section 3-4, the partial approximation of the operator A associated with $a(u, v)$, is the operator A_h defined by

(4-44) $(A_h u_h, v_h)_h = \sum\limits_{|p|,|q| \leq m} \int_{\Omega} a_{pq}(x) D^p p_h^{m_p} u_h D^q p_h^{m_q} v_h \, dx.$

In particular, if we choose $m_q = q$, the operator A_h is the following finite-differences operator:

(4-45) $A_h u_h = \sum\limits_{|p|,|q| \leq m} (-1)^{|q|} \tilde{\nabla}_h^q (a_{pq}{}^h \nabla_h^p u_h).$

This is the most interesting case because it gives us the simplest finite-differences scheme. For instance, let us apply Theorem 2-1 and Theorems 3-3 and 3-4 to the homogeneous Neumann and Dirichlet problems for Λ. When $f \in L^2(\Omega)$ is given, we can approximate the solution $u \in H^m(\Omega, \Lambda)$ of the Neumann problem

(4-46) $\Lambda u = f$ in Ω, $\delta_j u = 0$ in Γ for $m \leq j \leq 2m - 1$

(see Section 7.2-3) by the solution $u_h \in H_h^m(\Omega)$ of the finite-differences problem

(4-47) $\sum\limits_{|p|,|q| \leq m} (-1)^{|q|} \tilde{\nabla}_h^q (a_{pq}{}^h \nabla_h^p u_h) = f^h.$

THEOREM 4-6   Let us assume that Ω is smooth, and also that

(4-48) $\sum\limits_{|p|,|q| \leq m} a_{pq}(u^p, u^q) \geq c \sum\limits_{|p| \leq m} \|u^p\|^2_{H(\Omega, D^p)}$ for any $u \in V.$

If u and u_h are the solutions of (4-46) and (4-47), then

(4-49) $\|D^q u - p_h^0 \nabla_h^q u_h\|_{L^2(\Omega)}$ converges to 0 for any q such that $|q| \leq m.$ ▲

Let us introduce the partial approximations $(H_{0,h}^m(\Omega), p_h^q, r_{0,h}^q)$ of the space $H_0^m(\Omega)$, so that we can approximate the solution $u \in H^m(\Omega, \Lambda)$ of the homogeneous Dirichlet problem

(4-50) $\Lambda u = f$ in Ω, $\gamma_j u = 0$ in Γ for $0 \leq j \leq m - 1$

(see Section 7.2-2) by the solution $u_h \in H_{0,h}^m(\Omega)$ of the finite-differences problem

(4-51) $$\sum_{|p|,|q| \le m} (-1)^{|q|} \tilde{\nabla}_h^q (a_{pq}^{h} \nabla_h^p u_h) = f^h.$$

THEOREM 4-7 Let us assume (4-48) and that Ω is smooth. If $u \in H^m(\Omega, \Lambda)$ and $u_h \in H_{0,h}^m(\Omega)$ are the solutions of (4-50) and (4-51), then

(4-52) $\| D^q u - p_h^0 \nabla_h^q u_h \|_{L^2(\Omega)}$ converges to 0 for any q such that $|q| \le m$.

▲

Comments

Chapter 2

In the study of the stability of discrete schemes for parabolic equations and other evolution equations, Raviart [1] introduced the concept of stability function. The role played by these stability functions in the study of the regularity of the convergence was studied by Aubin [7]. A systematic study of stability and error functions in Banach spaces is made in Aubin [8].

The concept of n-width was introduced by Kolmogorov, and its characterization in terms of eigenvalues was investigated by Golomb [1]. Estimates of n-width in Sobolev spaces can be found in Jerome [1]–[3]. Connections between spaces H_θ and "spaces of interpolation" are studied in the book by Lions and Magenes, Chapter 2, in a more general setting.

Dual approximations were implicitly introduced long ago in the study of quadrature formulas. The concept of optimal prolongations was discovered by Schoenberg when he studied properties of spline functions. He also found duality relations between optimal interpolation properties and best quadrature formulas, in the sense of Sard.

Formalization of these concepts were made by Golomb and Weinberger, Golomb [3], Aubin [4], Anselone and, Laurent, Attéia, and others. The definition of spaces of order θ was introduced by Lions and Peetre. It is sufficient for our purpose to use spaces of order θ instead of spaces of interpolation.

Chapter 3

The internal approximation of operators began with the study of the Rayleigh-Ritz-Galerkin method for approximating symmetric operators from V onto V', and many authors contributed to it. See, for instance, the book of Kantorovich and Krylov. Both the introduction of operators p_h and r_h and their application to the construction of variational equation are due to Céa.

A systematic study comparable to that of Chapter 3 can be found in Aubin [1]. For extensions to the nonlinear case, see Aubin [1], Browder, Petryshyn [1]–[4], and so on, to the case of equations of evolution, Raviart [1], [2], Temam [1], and so on.

An investigation of the regularity of convergence in which stability functions were employed can be found in Aubin [7]; studies in which elliptic regularization was used were made by Aubin and Lions. Optimal properties of internal approximations are covered in Aubin [5].

Chapters 4 and 5

The concept of finite-element approximation method has a long history and has appeared in connections of numerous problems and under various names (e.g., spline functions, Hill functions, etc.). We mention several papers dealing with this topic in the bibliography (but not all!).

The main motivation for the introduction of finite-element approximation was the need of replacing approximants defined on the whole domain by approximants with compact support.

The criterion of m-convergence was introduced in Aubin [3] in the case of piecewise-polynomial approximation; it is also discussed in Babuska [2], [7], [8], in Bramble and Hilbert [1], [2], Di Guglielmo [1]–[3], Strang and Fix, [1], [2], and so on.

The main theorem of characterization of the convergent finite-element method is due to Strang and Fix [1].

A systematic study of $(2m + 1)^n$-level piecewise-polynomial approximations appears in Aubin [1]–[3], and an examination of the $2(2m)^n$-$(2m - 1)^n$-level piecewise approximation is furnished in Di Guglielmo [1]–[4].

In Schultz [1] can be found another way of constructing piecewise-polynomial approximations of Sobolev spaces $H_0^m(\Omega)$.

Chapters 6 and 7

A systematic study of Sobolev spaces and boundary-value problems can be found in the books of Lions and Magenes. The abstract Green formula is due to Aubin [6].

Chapter 9

Perturbation methods in the study of differential problems were systematically used by Lions and other authors. (See, e.g., the book by Lions and the books by Lions and Magenes).

In particular, Aubin and Lions used elliptic regularization to obtain the regularity of the convergence, Mignot used perturbation methods for approximation of boundary-value problems by discrete problems on a smoother (or simpler) domain; Aubin [9], [10] and Babuska [1] used penalization methods for approximating boundary-value problems by Neumann-type problems.

Elliptic regularization can also be used successfully in the study of approximation of degenerate problems (see, e.g., Baouendi, Baouendi, and Goulaouic, Kohn and Nirenberg, Mertz and Rivkind, etc.).

The results of Section 3 concerning least-squares approximations of boundary-value problems are due to Bramble and Shatz. Future papers of the same authors will develop these important results.

Chapter 10

The results on conjugate problems are due to Aubin and Burchard. We refer the reader to this paper for further comments.

Chapter 11

External and partial approximations were introduced by Céa in order to obtain the 5-level difference scheme of the Laplacian by means of prolongations and restrictions. Iterative methods for solving sums of operators of the "alternate directions methods" type were devised by several methods (see, e.g., Temam [1]).

References

Anselone, P. M., and P. J. Laurent. A general method for the construction of interpolating or smoothing spline functions. *Mathematical Research Center, Summary Report,* University of Wisconsin, 1969.

Atteia, M. Théorie et application des fonctions splines en analyse numérique. Thesis, University of Grenoble, 1966.

Aubin, J. P. Approximation des espaces de distributions et des opérateurs différentiels. *Bull. Soc. Math. France Mém.* 12 (1967), 1–139. [1]

Aubin, J. P. Behavior of the error of the approximate solutions of boundary-value problems for linear elliptic operators by Galerkin's and finite-difference methods. *Ann. Sci. Norm. Pisa.* 21 (1967), 599–637. [2]

Aubin, J. P. Evaluation des erreurs de troncature des approximations des espaces des Sobolev. *J. Math. Anal. Appl.* 21 (1968), 356–368. [3]

Aubin, J. P. Approximation et interpolation optimales et spline functions. *J. Math. Anal. Appl.* 24 (1968), 1–24. [4]

Aubin, J. P. Best approximation of linear operators in Hilbert spaces. *SIAM J. Numer. Anal.* 5 (1968), 518–521. [5]

Aubin, J. P. Abstract boundary-value operators and their adjoint. *Rend. Seminario. Mat. Padova.* 43 (1970), 1–33. [6]

Aubin, J. P. Approximation of nonhomogeneous Neumann problems, regularity of the convergence and estimates of error in terms of n-width. *Mathematical Research Center, Summary Report* 924, 1968, 1–43. [7]

Aubin, J. P. Optimal approximation and characterization of the error and stability functions. *J. Approx. Theory* 3 (1970). [8]

Aubin, J. P. Approximation des problèmes aux limites nonhomogènes et régularité de la convergence. *Calcolo* 6 (1969), 117–140. [9]

Aubin, J. P. Approximation des problèmes aux limites non homogènes pour des opérateurs non linéaires. *J. Math. Anal. Appl.* 30 (1970). [10]

Aubin, J. P., and H. Burchard. Some aspects of the method of the hypercircle applied to elliptic variational problems. *Proceedings of SYNSPADE.* Academic Press, 1971.

Aubin, J. P., and J. L. Lions. Remarques sur l'approximation régularisée des problèmes variationnels elliptiques. *Cours CIME,* 1967.

Babuska, I. Numerical solutions of boundary-value problems by the perturbed variational principle. *Technical Note* BN 624, IFDAM. University of Maryland Press, 1969. [1]

349

Babuska, I. Error bounds for finite element method. *Technical Note* BN 630, IFDAM. University of Maryland Press, 1969. [2]

Babuska, I. The finite element method for elliptic equations with discontinuous coefficients. *Technical Note* BN 631, IFDAM. University of Maryland Press, 1969. [3]

Babuska, I. Finite element method for domains with corners. *Technical Note* BN 636, IFDAM. University of Maryland Press, 1970. [4]

Babuska, I. Computation of derivatives in the finite element method. *Technical Note* BN 650, IFDAM. University of Maryland Press, 1970. [5]

Babuska, I. The rate of convergence for the finite element method. *Technical Note* BN 646, IFDAM. University of Maryland Press, 1970. [6]

Babuska, I. Approximation by Hill functions. *Technical Note* BN 648, IFDAM. University of Maryland Press, 1970. [7]

Babuska, I. The finite element method for elliptic differential equations. *Proceedings of SYNSPADE.* Academic Press, 1971. [8]

Babuska, I., and S. L. Sobolev. The optimization of numerical processes. *Appl. Mat.*, 1965, 96–129.

Baouendi, M. S. Sur une classe d'opérateurs elliptiques dégénérés. *Bull. Soc. Math. France.* **95** 1967, 45–87.

Baouendi, M. S., and C. Goulaouic. Régularité et théorie spectrale pour une classe d'opérateurs elliptiques dégénérés. *Arch. Ration. Mecha. Anal.* (on press.)

Birkhoff, G., and C. De Boor. Error bounds for spline interpolation. *J. Math. Mech.* **13** (1964), 827–836.

Birkhoff, G., M. H. Schultz, and R. S. Varga. Piecewise Hermite interpolation in one and two variables with applications to partial differential equations. *Numer. Math.* **11** (1968), 232–256.

De Boor, C. Bicubic spline interpolation. *J. Math. Phys.* **41** (1962), 212–218. [1]

De Boor, C. The method of projections as applied to the numerical solutions of two-point boundary-value problems using cubic splines. Ph.D. thesis, University of Michigan, 1966. [2]

De Boor, C. On uniform approximation by splines. *J. Approx. Theory* **1** (1968), 219–235. [3]

Bramble, J. H. Error estimates for elliptic boundary-value problems. *Cours CIME*, 1967.

Bramble, J. H., and S. Hilbert. Estimation of linear functionals on Sobolev spaces with application to Fourier transforms and spline interpolation. *SIAM J. Numer. Anal.* (to appear). [1]

Bramble, J. H., and S. Hilbert. Bounds for a class of linear functionals with applications to Hermite interpolation (to appear). [2]

Bramble, J. H., and A. H. Schatz. Rayleigh-Ritz-Galerkin methods for Dirichlet's problems using subspaces without boundary conditions. *Commun. Pure Appl.* (to appear).

Bramble, J. H., and M. Zlamal. Triangular elements in the finite element method. *Math. Comp.*

Bréziz, H., and M. Sibony. Méthodes d'approximation et d'itération pour les opérateurs monotones. *Arch. Ration. Mech. Anal.* **28** (1968), 59–82.

Browder, F. E. Approximation-solvability of nonlinear functional equations in normed linear spaces. *Arch. Ration. Mech. Anal.* **26** (1967), 33–42.

Céa, J. Approximation variationnelle des problèmes aux limites. *Ann. Inst. Fourier.* **14** (1964), 345–444.

Ciarlet, P. Discrete variational Green's functions. *Aeg. Math.* **4** (1970), 74–82.

Ciarlet, P., M. H. Schultz, and R. S. Varga. Numerical methods of higher order accuracy for nonlinear boundary-value problems. Part I, One-dimensional problems. *Numer. Math.* **9** (1967), 394–430; Part II, Nonlinear boundary-conditions. *Numer. Math.* **11** (1968), 331–345; Part III, Eigenvalue problems. *Numer. Math.* **12** (1968), 120–133; Part IV, Periodic boundary conditions. *Numer. Math.* **12** (1968), 266–279; Part V, Monotone operator theory. *Numer. Math.* **13** (1969), 51–77.

Demjanovitch, Iu. K. Net method in certain problems of mathematical physics. *Dokl. Akad. Nauk SSSR* **159** (1964), 250–264. [1]

Demjanovitch, Iu. K. Approximation and convergence of the net method in elliptic problems. *Dokl. Acad. Nauk SSSR* **170** (1966); *Soviet Math. "Dokl."* **7** (1966), 1129–1133. [2]

Demjanovitch, Iu. K. Estimates of the rate of convergence of certain projection methods for the solution of elliptic equations. *Dokl. Acad. Nauk SSSR* **174** (1967); *Soviet Math. "Dokl."* **8** (1967), 658–661. [3]

Di Guglielmo, F. Construction d'approximations des espaces de Sobolev $H^m(R^n)$, m entier positif, sur des réseaux en simplexes. *C.R. Acad. Sci.* **268** (1969), 314–317. [1]

Di Guglielmo, F. Construction d'approximations des espaces de Sobolev sur des réseaux en simplexes. *Calcolo* **6** (1969), 279–331. [2]

Di Guglielmo, F. Méthode des éléments finis: une famille d'approximation des espaces de Sobolev par les translatés de p fonctions. *Calcolo* **7** (1970), 185–234. [3]

Di Guglielmo, F. Approximation des espaces de Sobolev par les translatés de plusieurs fonctions et applications (to appear). [4]

Dunford, N., and J. Schwartz. Linear operators. Wiley-Interscience. Part 1, 1958; Part 2, 1963; Part 3, 1971.

Forsythe, G. E., and W. Wasow. Finite difference methods for partial differential equations. Wiley-Interscience, 1960.

Fix, G. Higher order Rayleigh approximations. *J. Math. Mech.* **18** (1969), 645–658. [1]

Fix, G. Fourier analysis of the finite element method in Ritz-Galerkin theory. *Studies Appl. Math.* **48** (1969), 265–273. [2]

Friedrichs, K. O., and H. Keller. A finite difference scheme for generalized Neumann problems. In Numerical solutions of partial differential equations. Academic Press, 1966.

Godunov, S. K., and V. S. Rabienkii. The theory of difference schemes. Wiley-Interscience, 1964.

Golomb, M. Optimal approximating manifolds in L^2 spaces. *J. Math. Anal. Appl.* **12** (1965), 505–512. [1]

Golomb, M. Optimal and nearly optimal linear approximation. In Approximation of functions. American Elsevier Publishing Company, 1965. [2]

Golomb, M. Splines, n-width and optimal approximation. Mathematical Research Center, Summary Report 784, 1967. [3]

Golomb, M. Approximation by periodic spline interpolants on uniform meshes. *J. Approx. Theory* **1** (1968), 26–65. [4]

Golomb, M., and F. Weinberger. Optimal approximation and error bounds. In Numerical approximation. University of Wisconsin Press, 1959.

Jerome, J. On the L^2 n-width of certain classes of functions of several variables. *J. Math. Anal. Appl.* **20** (1967), 110–123. [1]

Jerome, J. Asymptotic estimates on the L^2 n-width. *J. Math. Anal. Appl.* **22** (1968), 449–464 [2]

Jerome, J. On n-width in Sobolev spaces and applications to elliptic boundary value problems. Mathematical Research Center, Summary Report 917, 1968. [3]

Kantorovich, L. V., and V. I. Krylov. Approximate methods of higher analysis. Moorhoff Publishing Company, 1958.

Keller, H. B. Numerical methods for two point boundary-value problems. Blaisdell Publishing Company, 1969.

Kellog, R. B. Difference equations on a mesh arising from a general triangulation. *Math. Comp.* **18** (1964), 203–210.

Kohn, J. J., and L. Nirenberg. Degenerate elliptic-parabolic equations of second order. *Commun. Pure Appl. Math.* **20** (1967), 797–802.

Kolmogorov, A. N. Über die beste Annäherung von Funktionen einer gegebener Funktionklasse. *Ann. Math.* **37** (1936), 107–111.

Lions, J. L. Quelques méthodes de résolution de problèmes aux limites nonlinéaires. Dunod-Gauthier-Villars, 1969.

Lions, J. L., and E. Magenes. Problèmes aux limites nonhomogènes et applications. Volumes I, II, and III. Dunod 1968.

Lions, J. L., and J. Peetre. Sur une classe d'espaces d'interpolation. *Inst. Hautes Etudes Sci. Publ. Math.* **19** (1964), 5–68.

Merts, R., and V. Rivkind. Finite difference method for degenerate elliptic and parabolic equations. *Dokl. Akad. Nauk SSSR* **172** (1967), 783–786.

Mignot, A. Méthodes d'approximation des solutions de certains problèmes aux limites linéaires. *Rend. Seminario. Mat. Padova.* **15** (1968), 1–38.

Mikhlin, S. G. The stability of the Ritz method. *Soviet Math. "Dokl."* **1** (1960), 1230–1233.

Nitshe, J. Zur Konvergenz des Ritzschen Verfahrens und der Fehlerquadratmethod. *ISNM* **9** (1968), 97–103. [1]

Nitshe, J. Bermerkungen zur Approximationsgute bei projektiven Verfahren. *Z. Mat.* **106** (1968), 327–331. [2]

Nitshe, J. Ein Kriterium fur die quasioptimilital des Ritzschen Verfahrens. *Numer, Math* **11** (1968), 346–348. [3]

Nitshe, J. Verfahren von Ritz und Spline-Interpolation bei Sturm-Liouville-Randwertprobleme. *Numer. Math* **13** (1969), 260–265. [4]

Nitshe, J. Interpolation in Soboleschen Funktionenraumen. *Numer. Math* **13** (1969), 339–343. [5]

Oganesjan, L. A. Convergence of difference schemes in case of improved approximation of the boundary Zh. *Vycisl. Mat. Fiz.* **6** (1966), 1029–1042. [1]

Oganesjan, L. A. Convergence of variational difference schemes under improved approximation to the boundary. *Soviet Math. "Dokl."* **7** (1966), 1146–1150. [2]

Oganesjan, L. A., and L. A. Ruchovec. Variational difference schemes for second-order linear elliptic equations in a two dimensional region with a piecewise-smooth boundary. *Zh. Vycisl. Mat. Fiz.* **8** (1968), 97–114. [1]

Oganesjan, L. A., and L. A. Ruchovec. A study of rates of convergence of some variational difference schemes for elliptic equations of second order in a two-dimensional domain with smooth boundary. *Zh. Vycisl. Mat. Fiz.* 9 (1969), 1102–1119. [2]

Perrin, F. M., H. S. Price, and R. S. Varga. On higher order numerical methods for nonlinear two point boundary value problems. *Numer. Math.* 13 (1968), 180–198.

Petryshyn, W. V. On a class of Kpd and non-Kpd operators and operator equations. *J. Math. Anal. Appl.* 19 (1965), 1–24. [1]

Petryshyn, W. V. Projection methods in nonlinear numerical functional analysis. *J. Math. Mech.* 17 (1967), 353–372. [2]

Petryshyn, W. V. Remarks on the approximation-solvability of nonlinear functional equations. *Arch. Ration. Mech. Anal.* 26 (1967), 43–49. [3]

Petryshyn, W. V. On the approximation-solvability of nonlinear equations. *Mat. Ann.* 177 (1968), 156–164. [4]

Price, H. S., and R. S. Varga. Numerical analysis of simplified mathematical models of fluid flow in porous media. *Proceedings of the XX Symposium on Applied Mathematics,* 1969.

Raviart, P. A. Sur l'approximation de certaines equations d'évolution linéaires et non-linéaires. *J. Math. Pures Appl.* 46 (1967), 11–183. [1]

Raviart, P. A. Sur la résolution et l'approximation de certaines equations paraboliques non linéaires dégénérées. *Arch. Ration. Mech. Anal.* 25 (1967), 64–80. [2]

Sard, A. Linear approximation. *Math. Surv. AMS* 9 (1963).

Schaeffer, D. Approximation of elliptic boundary-value problems by difference equations, *I*—Factorization of the symbol. *J. Funct. Anal.* (to appear).

Schoenberg, I. J. On monosplines of least deviation and best quadrature formulae. Part I, *SIAM J. Numer. Anal.* 2 (1965), 144–170; Part II, *SIAM J. Numer. Anal.* 3 (1966). 321–328.

Schultz, M. H. Rayleigh-Ritz-Galerkin method for multidimensional problems. *SIAM J, Numer. Anal.* 6 (1969), 523–538. [1]

Schultz, M. H. Error bounds for Rayleigh-Ritz-Galerkin method. *J. Math. Anal. Appl.* 27 (1969), 524–533. [2]

Schultz, M. H. Elliptic spline functions and the Rayleigh-Ritz-Galerkin method. *Math. Comp.* 24 (1970), 65–80. [3]

Strang, G. The finite element method and approximation theory. *Proceedings of SYNSPADE.* Academic Press, 1971.

Strang, G., and G. Fix. A Fourier analysis of the finite element variational method (to appear). [1]

Strang, G., and G. Fix. The finite element variational method (to appear). [2]

Temam, R. Analyse Numérique. Presses Universitaires de France, 1970. [1]

Temam, R. Sur la stabilité et la convergence de la méthode des pas fractionnaires. [2]

Thomée, V. Elliptic difference operators and Dirichlet's problem. *Contrib. Differential. Equat.* 3 (1964), 301–324. [1]

Thomée, V. Discrete interior Schauder estimates for elliptic difference operators. *SIAM J. Numer. Anal.* 5 (1968), 626–645. [2]

Treves, F. Topological vector spaces, distributions, and kernels. Academic Press, 1967.

Varga, R. S. Accurate numerical methods for nonlinear boundary value problems. *Proceedings of the XX. Symposium on Applied Mathematics,* 1969.

Yosida, K. Functional analysis. Springer Verlag, 1968.

Index

A CATALOG OF SELECTED

DOVER BOOKS
IN SCIENCE AND MATHEMATICS

Astronomy

BURNHAM'S CELESTIAL HANDBOOK, Robert Burnham, Jr. Thorough guide to the stars beyond our solar system. Exhaustive treatment. Alphabetical by constellation: Andromeda to Cetus in Vol. 1; Chamaeleon to Orion in Vol. 2; and Pavo to Vulpecula in Vol. 3. Hundreds of illustrations. Index in Vol. 3. 2,000pp. 6⅛ x 9¼.
Vol. I: 0-486-23567-X
Vol. II: 0-486-23568-8
Vol. III: 0-486-23673-0

EXPLORING THE MOON THROUGH BINOCULARS AND SMALL TELESCOPES, Ernest H. Cherrington, Jr. Informative, profusely illustrated guide to locating and identifying craters, rills, seas, mountains, other lunar features. Newly revised and updated with special section of new photos. Over 100 photos and diagrams. 240pp. 8¼ x 11. 0-486-24491-1

THE EXTRATERRESTRIAL LIFE DEBATE, 1750–1900, Michael J. Crowe. First detailed, scholarly study in English of the many ideas that developed from 1750 to 1900 regarding the existence of intelligent extraterrestrial life. Examines ideas of Kant, Herschel, Voltaire, Percival Lowell, many other scientists and thinkers. 16 illustrations. 704pp. 5⅜ x 8½. 0-486-40675-X

THEORIES OF THE WORLD FROM ANTIQUITY TO THE COPERNICAN REVOLUTION, Michael J. Crowe. Newly revised edition of an accessible, enlightening book recreates the change from an earth-centered to a sun-centered conception of the solar system. 242pp. 5⅜ x 8½. 0-486-41444-2

A HISTORY OF ASTRONOMY, A. Pannekoek. Well-balanced, carefully reasoned study covers such topics as Ptolemaic theory, work of Copernicus, Kepler, Newton, Eddington's work on stars, much more. Illustrated. References. 521pp. 5⅜ x 8½.
0-486-65994-1

A COMPLETE MANUAL OF AMATEUR ASTRONOMY: TOOLS AND TECHNIQUES FOR ASTRONOMICAL OBSERVATIONS, P. Clay Sherrod with Thomas L. Koed. Concise, highly readable book discusses: selecting, setting up and maintaining a telescope; amateur studies of the sun; lunar topography and occultations; observations of Mars, Jupiter, Saturn, the minor planets and the stars; an introduction to photoelectric photometry; more. 1981 ed. 124 figures. 25 halftones. 37 tables. 335pp. 6½ x 9¼. 0-486-40675-X

AMATEUR ASTRONOMER'S HANDBOOK, J. B. Sidgwick. Timeless, comprehensive coverage of telescopes, mirrors, lenses, mountings, telescope drives, micrometers, spectroscopes, more. 189 illustrations. 576pp. 5⅝ x 8¼. (Available in U.S. only.)
0-486-24034-7

STARS AND RELATIVITY, Ya. B. Zel'dovich and I. D. Novikov. Vol. 1 of *Relativistic Astrophysics* by famed Russian scientists. General relativity, properties of matter under astrophysical conditions, stars, and stellar systems. Deep physical insights, clear presentation. 1971 edition. References. 544pp. 5⅝ x 8¼. 0-486-69424-0

Mathematics

FUNCTIONAL ANALYSIS (Second Corrected Edition), George Bachman and Lawrence Narici. Excellent treatment of subject geared toward students with background in linear algebra, advanced calculus, physics and engineering. Text covers introduction to inner-product spaces, normed, metric spaces, and topological spaces; complete orthonormal sets, the Hahn-Banach Theorem and its consequences, and many other related subjects. 1966 ed. 544pp. 6⅛ x 9¼. 0-486-40251-7

ASYMPTOTIC EXPANSIONS OF INTEGRALS, Norman Bleistein & Richard A. Handelsman. Best introduction to important field with applications in a variety of scientific disciplines. New preface. Problems. Diagrams. Tables. Bibliography. Index. 448pp. 5⅜ x 8½. 0-486-65082-0

VECTOR AND TENSOR ANALYSIS WITH APPLICATIONS, A. I. Borisenko and I. E. Tarapov. Concise introduction. Worked-out problems, solutions, exercises. 257pp. 5⅜ x 8¼. 0-486-63833-2

AN INTRODUCTION TO ORDINARY DIFFERENTIAL EQUATIONS, Earl A. Coddington. A thorough and systematic first course in elementary differential equations for undergraduates in mathematics and science, with many exercises and problems (with answers). Index. 304pp. 5⅜ x 8½. 0-486-65942-9

FOURIER SERIES AND ORTHOGONAL FUNCTIONS, Harry F. Davis. An incisive text combining theory and practical example to introduce Fourier series, orthogonal functions and applications of the Fourier method to boundary-value problems. 570 exercises. Answers and notes. 416pp. 5⅜ x 8½. 0-486-65973-9

COMPUTABILITY AND UNSOLVABILITY, Martin Davis. Classic graduate-level introduction to theory of computability, usually referred to as theory of recurrent functions. New preface and appendix. 288pp. 5⅜ x 8½. 0-486-61471-9

ASYMPTOTIC METHODS IN ANALYSIS, N. G. de Bruijn. An inexpensive, comprehensive guide to asymptotic methods—the pioneering work that teaches by explaining worked examples in detail. Index. 224pp. 5⅜ x 8½ 0-486-64221-6

APPLIED COMPLEX VARIABLES, John W. Dettman. Step-by-step coverage of fundamentals of analytic function theory—plus lucid exposition of five important applications: Potential Theory; Ordinary Differential Equations; Fourier Transforms; Laplace Transforms; Asymptotic Expansions. 66 figures. Exercises at chapter ends. 512pp. 5⅜ x 8½. 0-486-64670-X

INTRODUCTION TO LINEAR ALGEBRA AND DIFFERENTIAL EQUATIONS, John W. Dettman. Excellent text covers complex numbers, determinants, orthonormal bases, Laplace transforms, much more. Exercises with solutions. Undergraduate level. 416pp. 5⅜ x 8½. 0-486-65191-6

RIEMANN'S ZETA FUNCTION, H. M. Edwards. Superb, high-level study of landmark 1859 publication entitled "On the Number of Primes Less Than a Given Magnitude" traces developments in mathematical theory that it inspired. xiv+315pp. 5⅜ x 8½. 0-486-41740-9

CALCULUS OF VARIATIONS WITH APPLICATIONS, George M. Ewing. Applications-oriented introduction to variational theory develops insight and promotes understanding of specialized books, research papers. Suitable for advanced undergraduate/graduate students as primary, supplementary text. 352pp. 5⅜ x 8½.
0-486-64856-7

COMPLEX VARIABLES, Francis J. Flanigan. Unusual approach, delaying complex algebra till harmonic functions have been analyzed from real variable viewpoint. Includes problems with answers. 364pp. 5⅜ x 8½.
0-486-61388-7

AN INTRODUCTION TO THE CALCULUS OF VARIATIONS, Charles Fox. Graduate-level text covers variations of an integral, isoperimetrical problems, least action, special relativity, approximations, more. References. 279pp. 5⅜ x 8½.
0-486-65499-0

COUNTEREXAMPLES IN ANALYSIS, Bernard R. Gelbaum and John M. H. Olmsted. These counterexamples deal mostly with the part of analysis known as "real variables." The first half covers the real number system, and the second half encompasses higher dimensions. 1962 edition. xxiv+198pp. 5⅜ x 8½. 0-486-42875-3

CATASTROPHE THEORY FOR SCIENTISTS AND ENGINEERS, Robert Gilmore. Advanced-level treatment describes mathematics of theory grounded in the work of Poincaré, R. Thom, other mathematicians. Also important applications to problems in mathematics, physics, chemistry and engineering. 1981 edition. References. 28 tables. 397 black-and-white illustrations. xvii + 666pp. 6⅛ x 9¼.
0-486-67539-4

INTRODUCTION TO DIFFERENCE EQUATIONS, Samuel Goldberg. Exceptionally clear exposition of important discipline with applications to sociology, psychology, economics. Many illustrative examples; over 250 problems. 260pp. 5⅜ x 8½.
0-486-65084-7

NUMERICAL METHODS FOR SCIENTISTS AND ENGINEERS, Richard Hamming. Classic text stresses frequency approach in coverage of algorithms, polynomial approximation, Fourier approximation, exponential approximation, other topics. Revised and enlarged 2nd edition. 721pp. 5⅜ x 8½.
0-486-65241-6

INTRODUCTION TO NUMERICAL ANALYSIS (2nd Edition), F. B. Hildebrand. Classic, fundamental treatment covers computation, approximation, interpolation, numerical differentiation and integration, other topics. 150 new problems. 669pp. 5⅜ x 8½.
0-486-65363-3

THREE PEARLS OF NUMBER THEORY, A. Y. Khinchin. Three compelling puzzles require proof of a basic law governing the world of numbers. Challenges concern van der Waerden's theorem, the Landau-Schnirelmann hypothesis and Mann's theorem, and a solution to Waring's problem. Solutions included. 64pp. 5⅜ x 8½.
0-486-40026-3

THE PHILOSOPHY OF MATHEMATICS: AN INTRODUCTORY ESSAY, Stephan Körner. Surveys the views of Plato, Aristotle, Leibniz & Kant concerning propositions and theories of applied and pure mathematics. Introduction. Two appendices. Index. 198pp. 5⅜ x 8½.
0-486-25048-2

TENSOR CALCULUS, J.L. Synge and A. Schild. Widely used introductory text covers spaces and tensors, basic operations in Riemannian space, non-Riemannian spaces, etc. 324pp. 5⅜ x 8¼. 0-486-63612-7

ORDINARY DIFFERENTIAL EQUATIONS, Morris Tenenbaum and Harry Pollard. Exhaustive survey of ordinary differential equations for undergraduates in mathematics, engineering, science. Thorough analysis of theorems. Diagrams. Bibliography. Index. 818pp. 5⅜ x 8½. 0-486-64940-7

INTEGRAL EQUATIONS, F. G. Tricomi. Authoritative, well-written treatment of extremely useful mathematical tool with wide applications. Volterra Equations, Fredholm Equations, much more. Advanced undergraduate to graduate level. Exercises. Bibliography. 238pp. 5⅜ x 8½. 0-486-64828-1

FOURIER SERIES, Georgi P. Tolstov. Translated by Richard A. Silverman. A valuable addition to the literature on the subject, moving clearly from subject to subject and theorem to theorem. 107 problems, answers. 336pp. 5⅜ x 8½. 0-486-63317-9

INTRODUCTION TO MATHEMATICAL THINKING, Friedrich Waismann. Examinations of arithmetic, geometry, and theory of integers; rational and natural numbers; complete induction; limit and point of accumulation; remarkable curves; complex and hypercomplex numbers, more. 1959 ed. 27 figures. xii+260pp. 5⅜ x 8½. 0-486-63317-9

POPULAR LECTURES ON MATHEMATICAL LOGIC, Hao Wang. Noted logician's lucid treatment of historical developments, set theory, model theory, recursion theory and constructivism, proof theory, more. 3 appendixes. Bibliography. 1981 edition. ix + 283pp. 5⅜ x 8½. 0-486-67632-3

CALCULUS OF VARIATIONS, Robert Weinstock. Basic introduction covering isoperimetric problems, theory of elasticity, quantum mechanics, electrostatics, etc. Exercises throughout. 326pp. 5⅜ x 8½. 0-486-63069-2

THE CONTINUUM: A CRITICAL EXAMINATION OF THE FOUNDATION OF ANALYSIS, Hermann Weyl. Classic of 20th-century foundational research deals with the conceptual problem posed by the continuum. 156pp. 5⅜ x 8½. 0-486-67982-9

CHALLENGING MATHEMATICAL PROBLEMS WITH ELEMENTARY SOLUTIONS, A. M. Yaglom and I. M. Yaglom. Over 170 challenging problems on probability theory, combinatorial analysis, points and lines, topology, convex polygons, many other topics. Solutions. Total of 445pp. 5⅜ x 8½. Two-vol. set. Vol. I: 0-486-65536-9   Vol. II: 0-486-65537-7

Paperbound unless otherwise indicated. Available at your book dealer, online at **www.doverpublications.com**, or by writing to Dept. GI, Dover Publications, Inc., 31 East 2nd Street, Mineola, NY 11501. For current price information or for free catalogues (please indicate field of interest), write to Dover Publications or log on to **www.doverpublications.com** and see every Dover book in print. Dover publishes more than 500 books each year on science, elementary and advanced mathematics, biology, music, art, literary history, social sciences, and other areas.